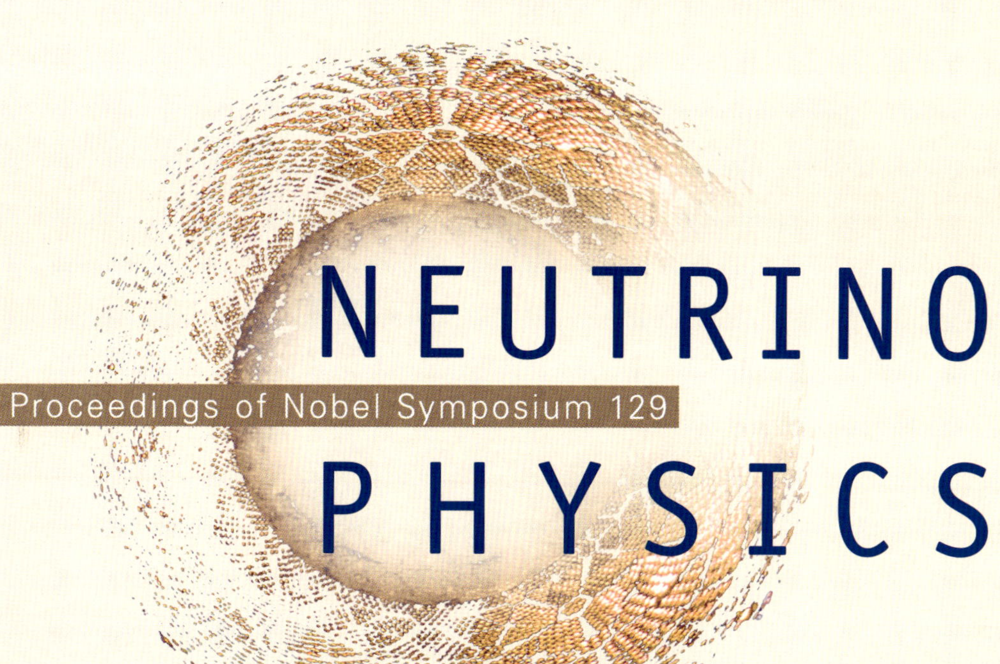

NEUTRINO

PHYSICS

Haga Slott, Enköping, Sweden
19 – 24 August 2004

edited by

L Bergström

O Botner

P Carlson

P O Hulth

T Ohlsson

Physica Scripta

Vol. **T121** 2005

Recognized by the European Physical Society 

World Scientific

NEW JERSEY · LONDON · SINGAPORE · BEIJING · SHANGHAI · HONG KONG · TAIPEI · CHENNAI

Physica Scripta

Published jointly by

The Royal Danish Academy of Sciences and Letters in cooperation with the Danish Physical Society
The Delegation of the Finnish Academies of Science and Letters in cooperation with the Finnish Physical Societies
The Icelandic Scientific Society in cooperation with the Icelandic Physical Society
The Norwegian Academy of Science and Letters in cooperation with the Norwegian Physical Society
The Royal Swedish Academy of Sciences in cooperation with the Swedish Physical Society

Editorial Office
Physica Scripta
The Royal Swedish Academy of Sciences
Box 50005, SE-104 05 Stockholm, Sweden
email: physica@kva.se
web: www.physica.org

Production Editor
Katarina Lundin
Telephone: +46 8673 9528
email: kat@physica.kva.se

Scientific Editors
Managing Editor:
Prof. Roger Wäppling
Department of Physics, University of Uppsala
Box 530, SE-751 21 Uppsala, Sweden
Facsimile: +46 1851 2227
email: roger.wappling@fysik.uu.se

Technical Editor:
Dr Agneta Seidel
Department of Physics, University of Uppsala
Box 530, SE-751 21 Uppsala, Sweden
email: agnetas@physica.kva.se

Subscriptions
The Royal Swedish Academy of Sciences
PO Box 168, Didcot D.O., Oxford, OX11 7WD, United Kingdom

Subscription enquiries
May be made to the above address or by email to
subscriptions@physica.org

Back numbers and single issues
Can be ordered separately. Price information on request to the above
address or by email to subscriptions@physica.org

Manuscripts
Shall be submitted electronically to submissions@physica.org
Please see Instructions to Authors or consult
http://www.physica.org. Enquiries as to the status of submitted
manuscripts should be made to physica@kva.se

Offprints
The corresponding author will receive a .pdf file upon completion of
publication

Editors
General and cross-disciplinary physics
Prof. K. Mork, Department of Physics,
Norwegian University of Science and Technology,
N-7491 Trondheim, Norway
email: Kjell.Mork@phys.ntnu.no

Prof. J. J. Rasmussen, Optics and Plasma Research
Department, OPL 128,
Risø National Laboratory,
Box 49, DK-4000 Roskilde, Denmark
email: jens.juul.rasmussen@risoe.dk

High Energy Physics
Prof. G. Ingelman, Department of Radiation Sciences,
Uppsala University,
Box 535, SE-75121 Uppsala, Sweden
email: Gunnar.Ingelman@tsl.uu.se

Atomic, molecular and optical physics
Prof. L. J. Curtis, Department of Physics & Astronomy,
University of Toledo, Toledo, OH 43606, USA
email: ljc@physics.utoledo.edu

Prof. J. Javanainen, Department of Physics,
University of Connecticut, Storrs, CT 06269-3046, USA
email: jj@phys.uconn.edu

Prof. S. Mannervik, Department of Physics,
Stockholm University, AlbaNova University Center
SE-106 91 Stockholm, Sweden
email: Mannervik@physto.se

Plasma physics
Prof. H. L. Pécseli, Institute of Physics,
University of Oslo, POB 1048, Blindern,
N-3016 Oslo, Norway
email: hans.pecseli@fys.uio.no

Prof. L. Stenflo, Department of Plasma Physics,
Umeå University, SE-90187 Umeå, Sweden
email: lennart.stenflo@physics.umu.se

Dr M. Y. Yu, Theoretische Physik I,
Ruhr-Universität Bochum, D-44780 Bochum, Germany
email: yu@tpl.ruhr-uni-bochum.de

Condensed matter physics and material sciences
Prof. L. Dobrzynski, University of Bialystok,
Institute of Physics, Lipowa 41, PL-15-424 Bialystok, Poland
email: ludwik@ipj.gov.pl or ludwik@alpha.uwb.edu.pl

Prof. S. W. Lovesey, Rutherford Appleton Laboratory,
Chilton, Didcot, Oxfordshire, OX11 0QX, UK
email: S.W.Lovesey@rl.ac.uk

Prof. R. Niemienen, Laboratory of Physics,
Helsinki University of Technology, FIN-02150 Espoo, Finland
email: risto.nieminen@hut.fi

© 2005 Royal Swedish Academy of Sciences

Typeset by J. W. Arrowsmith Ltd and Printed in Great Britain by Marston Book Services Ltd, Didcot, Oxon.

Contents

Published jointly by

Physica Scripta
The Royal Swedish Academy of Sciences
Box 50005, S-104 05, Stockholm, Sweden

and

World Scientific Publishing Co. Pte. Ltd.
5 Toh Tuck Link, Singapore 596224
USA office: 27 Warren Street, Suite 401-402, Hackensack, NJ 07601
UK office: 57 Shelton Street, Covent Garden, London WC2H 9HE

British Library Cataloguing-in-Publication Data
A catalogue record for this book is available from the British Library.

NEUTRINO PHYSICS
Proceedings of Nobel Symposium 129
Hoga Slott, Enköping, Sweden

ISSN   Royal Swedish Academy of Sciences   0031-8949 (0281-1847)
ISBN   Royal Swedish Academy of Sciences   91-89621-27-1
ISBN   World Scientific   981-256-737-2

Printed by Mainland Press Pte Ltd

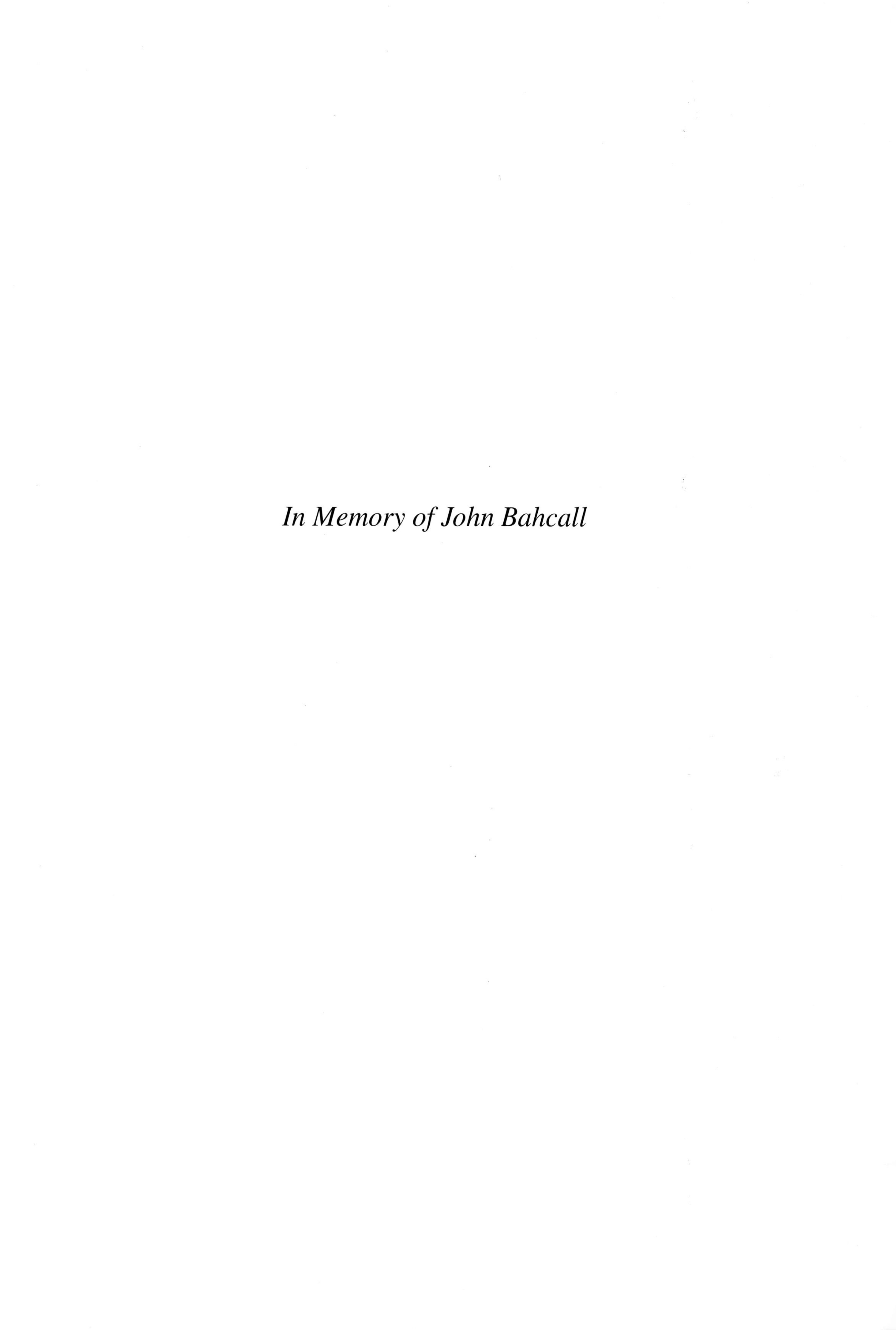

In Memory of John Bahcall

John N. Bahcall (1934–2005)

Courtesy of the institute for Advanced Study, Princeton, USA.

John Norris Bahcall, passed away on August 17, 2005, in New York City, USA. He was born on December 30, 1934, in Shreveport, Louisiana, USA. He was Richard Black Professor of Astrophysics in the School of Natural Sciences at the Institute for Advanced Study (IAS) in Princeton, New Jersey, USA and a recipient of the National Medal of Science. In addition, he was President of the American Astronomical Society, President-Elect of the American Physical Society, and a prominent leader of the astrophysics community.

John had a long and prolific career in astronomy and astrophysics, spanning five decades and the publication of more than five hundred technical articles, books, and popular papers.

John's most recognized scientific contribution was the novel proposal in 1964, together with Raymond Davis Jr., that scientific mysteries of our Sun "how it shines, how old it is, how hot it is" could be examined by measuring the number of neutrinos arriving on Earth from the Sun. Measuring the properties of these neutrinos tests both our understanding of how stars shine and our understanding of fundamental particle physics.

However, in the 1960s and 1970s, the observations by Raymond Davis Jr. showed a clear discrepancy between John's theoretical predictions, based on standard solar and particle physics models, and what was experimentally measured. This discrepancy, known as the "Solar Neutrino Problem," was examined by hundreds of physicists, chemists, and astronomers over the subsequent three decades. In the late 1990s through 2002, new large-scale neutrino experiments in Japan, Canada, Italy, and Russia culminated in the conclusion that the discrepancy between John's theoretical predictions and the experimental results required a modification of our understanding of particle physics: neutrinos must have a mass and "oscillate" among different particle states.

In addition of neutrino astrophysics, John contributed to many areas of astrophysics including the study of dark matter in the Universe, properties of quasar, structure of the galaxies, the evolution of stars, and the identification of the first neutron star companion.

John was an active member of the International Advisory Committee of the Nobel Symposium 129 on Neutrino Physics in Enköping, Sweden between August 19 and August 24, 2004, but he was unfortunately not able to attend the Symposium himself due to his illness. He will be hugely missed in the scientific community and especially among neutrino physicists. We, the members of the Local Organizing Committee of the Symposium, will always remember his large enthusiasm and creativity, warm friendship, and sharp intellect.

Lars Bergström, Olga Botner, Per Carlson, Per Olof Hulth, and Tommy Ohlsson

Some of the text has been adopted from a press release from the Institute for Advanced Study, Princeton, USA.

Preface

Nobel Symposium 129 on Neutrino Physics was held at Haga Slott in Enköping, Sweden during August 19–24, 2004. Invited to the symposium were around 40 globally leading researchers in the field of neutrino physics, both experimental and theoretical. In addition to these participants, some 30 local researchers and graduate students participated in the symposium.

The dominant theme of the lectures was neutrino oscillations, which after several years were recently verified by results from the Super-Kamiokande detector in Kamioka, Japan and the SNO detector in Sudbury, Canada. Discussion focused especially on effects of neutrino oscillations derived from the presence of matter and the fact that three different neutrinos exist. Since neutrino oscillations imply that neutrinos have mass, this is the first experimental observation that fundamentally deviates from the standard model of particle physics. This is a challenge to both theoretical and experimental physics. The various oscillation parameters will be determined with increased precision in new, specially designed experiments. Theoretical physics is working intensively to insert the knowledge that neutrinos have mass into the theoretical models that describe particle physics. It will probably turn out that the discovery of neutrino oscillations signifies a breakthrough in the description of the very smallest constituents of matter. The lectures provided a very good description of the intensive situation in the field right now. The topics discussed also included mass models for neutrinos, neutrinos in extra dimensions as well as the "seesaw mechanism," which provides a good description of why neutrino masses are so small.

Also discussed, besides neutrino oscillations, was the new field of neutrino astronomy. Among the questions that neutrino astronomy hopes to answer are what the dark matter in the Universe consists of and where cosmic radiation at extremely high energies comes from. For this purpose, large neutrino telescopes are built deep in the Antarctic ice, in the Baikal Lake, and in the Mediterranean Sea.

Among prominent unanswered questions, highlighted as one of the most important, was whether neutrinos are Dirac or Majorana particles. By studying neutrino double beta decay, researchers hope to answer this question, but it will put very large demands on detectors.

The programme also included ample time for lively and valuable discussions, which cannot normally be held at ordinary conferences.

The symposium concluded with a round-table discussion, where participants discussed the future of neutrino physics. Without a doubt, neutrino physics today is moving toward a very exciting and interesting period.

An important contribution to the success of the symposium was the wonderful setting that the Haga Slott manor house hotel and conference center offered to the participants.

Lars Bergström, Olga Botner, Per Carlson, Per Olof Hulth, Tommy Ohlsson

Editors

Top 1st row: Petter Hofverberg, Alan Watson, Don Perkins, Håkan Snellman, Manfred Lindner, Christian Spiering, Carlos de los Heros, Max Tegmark, Mark Pearce, Per Olof Hulth, Christian Bohm

2nd row: Jean-Jacques Aubert, Atsuto Suzuki, Alain Blondel, Janet Conrad, Per Hansson, Christian Walck, Per Carlson, Gary Steigman, Jukka Maalampi, Boris Kayser, David Saltzberg, Christian Weinheimer, Yoichiro Suzuki

3rd row: Lars Brink, Steve King, Evgeny Akhmedov, Tom Gaisser, Mike Shaevitz, Georg Raffelt, Concha Gonzalez-Garcia, Steve Barwick, Stephan Hundertmark, Pierre Ramond, Thomas Burgess

4th row: Samoil Bilenky, Øystein Elgarøy, Tommy Ohlsson, Art McDonald, Cecilia Jarlskog, Serguey Petcov, Graciela Gelmini, Olga Botner, Allan Hallgren, Christin Burgess-Wiedemann

Bottom 5th row: Rabi Mohapatra, Mattias Blennow, Tomas Hällgren, Alexei Smirnov, Anna Davour, Joakim Edsjö, Carlos Peña-Garay, Tsutomu Yanagida, Francis Halzen, Eli Waxman

List of Participants

E. Akhmedov
Universitat de València
Valencia, Spain

J. Aubert
Centre de Physique des Particules
de Marseille
Marseille, France

J. Bahcall
IAS
Princeton, USA

S. Barwick
University of California Irvine
Irvine, USA

S. Bergenius
Royal Institute of Technology
Stockholm, Sweden

L. Bergström
Stockholm University
Stockholm, Sweden

S. Bilenky
SISSA
Trieste, Italy

M. Blennow
Royal Institute of Technology
Stockholm, Sweden

M. Blom
Royal Institute of Technology
Stockholm, Sweden

A. Blondel
Geneva University
Geneva, Switzerland

C. Bohm
Stockholm University
Stockholm, Sweden

O. Botner
Uppsala University
Uppsala, Sweden

A. Bouchta
Uppsala University
Uppsala, Sweden

L. Brink
Chalmers University of Technology
Gothenburg, Sweden

T. Burgess
Stockholm University
Stockholm, Sweden

C. Burgess-Wiedemann
Stockholm University
Stockholm, Sweden

P. Carlson
Royal Institute of Technology
Stockholm, Sweden

J. Conrad
Columbia University
New York, USA

A. Davour
Uppsala University
Uppsala, Sweden

C. de los Heros
Uppsala University
Uppsala, Sweden

J. Edsjö
Stockholm University
Stockholm, Sweden

Ø. Elgarøy
University of Oslo
Oslo, Norway

E. Fiorini
Università di Milano – Bicocca
Milan, Italy

C. Fuglesang
European Space Agency (ESA)
Houston, USA

T. Gaisser
Bartol Research Institute, Uni. of Delaware
Newark, USA

G. Gelmini
UCLA
Los Angeles, USA

M. C. Gonzalez-Garcia
Stony Brook
New York, USA

A. Hallgren
Uppsala University
Uppsala, Sweden

F. Halzen
University of Wisconsin
Madison, USA

P. Hansson
Royal Institute of Technology
Stockholm, Sweden

P. Hofverberg
Royal Institute of Technology
Stockholm, Sweden

P. O. Hulth, chairman
Stockholm University
Stockholm, Sweden

K. Hultqvist
Stockholm University
Stockholm, Sweden

S. Hundertmark
Stockholm University
Stockholm, Sweden

T. Hällgren
Royal Institute of Technology
Stockholm, Sweden

C. Jarlskog
Lund University
Lund, Sweden

R. Johansson
Royal Institute of Technology
Stockholm, Sweden

B. Kayser
Fermilab
Batavia, USA

S. King
University of Southampton
Southampton, UK

M. Lindner
Technische Uni. München
Munich, Germany

J. Maalampi
University of Jyväskylä
Jyväskylä, Finland

A. McDonald
Queen's University
Kingston, Canada

Y. Minaeva
Stockholm University
Stockholm, Sweden

R. Mohapatra
University of Maryland
College Park, USA

T. Ohlsson
Royal Institute of Technology
Stockholm, Sweden

M. Pearce
Royal Institute of Technology
Stockholm, Sweden

C. Peña-Garay
IAS
Princeton, USA

D. Perkins
University of Oxford
Oxford, UK

S. Petcov
SISSA
Trieste, Italy

G. Raffelt
Max-Planck-Institut für Physik
Munich, Germany

P. Ramond
University of Florida
Gainesville, USA

H. Rubinstein
Stockholm University
Stockholm, Sweden

D. Saltzberg
UCLA
Los Angeles, USA

M. Shaevitz
Columbia University
New York, USA

A. Smirnov
ICTP
Trieste, Italy

H. Snellman
Royal Institute of Technology
Stockholm, Sweden

C. Spiering
DESY-Zeuthen
Zeuthen, Germany

G. Steigman
The Ohio State University
Columbus, Ohio

A. Suzuki
Tohoku University
Sendai, Japan

Y. Suzuki
Kamioka Observatory, ICRR,
University of Tokyo
Kamioka, Japan

M. Tegmark
University of Pennsylvania
Philadelphia, USA

Y. Totsuka
KEK
Tsukuba, Japan

A. Watson
University of Leeds
Leeds, UK

C. Walck
Stockholm University
Stockholm, Sweden

E. Waxman
Weizmann Institute
Rehovot, Israel

C. Weinheimer
Rheinische Friedrich-Willhelms-Universität
Bonn, Germany

T. Yanagida
University of Tokyo
Tokyo, Japan

Committees

Local Organizing Committee

L. Bergström
O. Botner
P. Carlson
P. O. Hulth, chairman
T. Ohlsson

International Advisory Committee

J. Bahcall
J. Conrad
E. Fiorini
M. C. Gonzalez-Garcia
F. Halzen

B. Kayser
M. Lindner
A. McDonald
G. Raffelt
Y. Totsuka

Nobel Symposium on Neutrino Physics Program

Thursday, August 19, 2004

Time	Event/Speaker	Session/Title
12:30–14:00	*Lunch*	
14:00–14:10	*Welcome address*	
Chairman: D. Perkins		Introduction
14:10–15:00	S. Bilenky	History of neutrino oscillations
15:00–15:30	*Discussion*	
15:30–16:00	*Break*	
Chairman: P. Carlson		Neutrino Oscillations – Experiments
16:00–16:30	Y. Suzuki	Super-Kamiokande results
16:30–17:00	A. McDonald	SNO results
17:00–17:30	*Discussion*	
17:30–18:00	A. Suzuki	KamLAND results
18:00–18:30	J. Conrad	Accelerator results
18:30–19:00	*Discussion*	
19:00–	*Welcome reception*	

Friday, August 20, 2004

Time	Event/Speaker	Session/Title
Chairman: H. Snellman		Neutrino Oscillations – Theory I
09:00–09:30	J. Bahcall (presented by C. Peña-Garay)	Solar models and neutrinos
09:30–10:00	T. Gaisser	Overview of atmospheric neutrino fluxes
10:00–10:30	*Discussion*	
10:30–11:00	*Break*	
Chairman: J. Maalampi		Neutrino Oscillations – Theory II
11:00–11:30	A. Smirnov	The MSW effect and matter effects in neutrino oscillations
11:30–12:00	E. Akhmedov	Three flavor effects and CP and T violation in neutrino oscillations
12:00–12:30	*Discussion*	
12:30–14:00	*Lunch*	
Chairman: T. Ohlsson		Neutrino Oscillations – Phenomenology
14:00–14:30	M. C. Gonzalez-Garcia	Three flavor global fits of neutrino data
14:30–15:00	A. Blondel	Future oscillation experiments
15:00–15:30	M. Lindner	Theoretical implications of future precision oscillation experiments
15:30–16:00	*Discussion*	
16:00–16:30	*Break*	
Chairman: H. Rubinstein		Neutrino Double Beta Decay
16:30–17:00	E. Fiorini	Experimental prospects of neutrino double beta decays
17:00–17:30	S. Petcov	Theoretical prospects of neutrino double beta decays
17:30–18:00	*Discussion*	
19:00–	*Dinner*	

Saturday, August 21, 2004

Time	Event/Speaker	Session/Title
Chairman: C. Peña-Garay		Neutrino Astronomy, Astrophysics, and Cosmology I
09:00–09:30	G. Raffelt	Supernova neutrinos
09:30–10:00	F. Halzen	Neutrino astronomy
10:00–10:30	*Discussion*	
10:30–11:00	*Break*	
Chairman: S. Barwick		High-Energy Neutrino Experiments I
11:00–11:30	C. Spiering	AMANDA/Baikal/IceCube
11:30–12:00	J. Aubert	Antares/Nemo/Nestor
12:00–12:30	*Discussion*	
12:30–14:00	*Lunch*	
Chairman: P. O. Hulth		High-Energy Neutrino Experiments II
14:00–14:30	D. Saltzberg	Radio and acoustic detection of high-energy neutrinos
14:30–15:00	A. Watson	Detection of neutrino induced air-showers

15:00–15:30	*Discussion*	
15:30–16:00	*Break*	
19:00–	*Special dinner – Crayfish Party*	

Sunday, August 22, 2004

Time	Event/Speaker	Session/Title
<u>Chairman: L. Bergström</u>		<u>Neutrino Astronomy, Astrophysics, and Cosmology II</u>
10:30–11:00	G. Gelmini	Prospects of relic neutrino searches
11:00–11:30	T. Yanagida	Leptogenesis in the early Universe
11:30–12:00	G. Steigman	Neutrinos and Big Bang nucleosynthesis
12:00–12:30	*Discussion*	
12:30–14:00	*Lunch*	
14:00–19:00	*Excursion*	
19:00–	*Dinner VASA museum*	

Monday, August 23, 2004

Time	Event/Speaker	Session/Title
<u>Chairman: Ø. Elgarøy</u>		<u>Neutrino Astronomy, Astrophysics, and Cosmology III</u>
09:00–09:30	E. Waxman	High-energy neutrino sources
09:30–10:00	M. Tegmark	Neutrino masses and other parameters from cosmological observations
10:00–10:30	*Discussion*	
10:30–11:00	*Break*	
<u>Chairman: O. Botner</u>		<u>Absolute Masses and Intrinsic Properties of Neutrinos I</u>
11:00–11:30	B. Kayser	Neutrino intrinsic properties
11:30–12:00	M. Shaevitz	Implication of the NuTeV result on intrinsic properties of neutrinos
12:00–12:30	*Discussion*	
12:30–12:45	*Symposium photo*	
12:45–14:00	*Lunch*	
<u>Chairman: L. Brink</u>		<u>Absolute Masses and Intrinsic Properties of Neutrinos II/Neutrino Mass Models I</u>
14:00–14:30	C. Weinheimer	Experimental results and future possibilities
14:30–15:00	P. Ramond	The see-saw mechanism
15:00–15:30	*Discussion*	
15:30–16:00	*Break*	
<u>Chairman: C. Jarlskog</u>		<u>Neutrino Mass Models II</u>
16:00–16:30	S. King	Neutrino mass models
16:30–17:00	R. Mohapatra	Neutrinos in extra dimensions and GUT
17:00–17:30	*Discussion*	
19:00–	*Conference banquet*	

Tuesday, August 24, 2004

Time	Event/Speaker	Session/Title
10:00–12:00	*Round table discussion*	The future of neutrino physics
	A. Blondel	
	F. Halzen	
	B. Kayser	
	A. Smirnov	
	Y. Suzuki	
12:00–13:00	*Lunch*	

Physica Scripta. Vol. T121, 17–22, 2005

The History of Neutrino Oscillations

S. M. Bilenky

Joint Institute for Nuclear Research, Dubna, R-141980, Russia, SISSA, via Beirut 2-4, I-31014, Trieste, Italy

Received October 6, 2004; accepted in revised form February 11, 2005

PACS number: 13.15.+g

Abstract

The early history of neutrino mixing and oscillations is briefly reviewed.

1. Introduction

After many years of heroic efforts of many physicists we have now model independent evidence for neutrino oscillations. The evidence of neutrino oscillations was obtained in the atmospheric Super Kamiokande experiment [1], in the solar SNO experiment [2], in the reactor KamLAND experiment [3], and also in solar neutrino experiments [4, 5, 6, 7], atmospheric neutrino experiments [8, 9], and in the first long baseline accelerator K2K experiment [10].

Neutrino oscillations are the signature of small neutrino masses and neutrino mixings. It took more than 40 years to discover this phenomenon.

The first idea of neutrino oscillations was put forward by B. Pontecorvo in 1957–58 [11, 12]. I worked with B. Pontecorvo for more than 15 years, starting from the time when the majority of physicists believed that neutrinos are massless two-component particles. I will consider mainly the evolution of the *original ideas* of neutrino masses, mixings and oscillations.

When Pauli introduced the neutrino in 1930 he assumed that the neutrino ("neutron") is a neutral weakly interacting particle with spin 1/2 and a mass smaller than electron mass. The first method to measure the neutrino mass was proposed in 1933 by Fermi [13] and Perrin [14]. They proposed to search for effects of a neutrino mass via detailed investigation of the high-energy part of β-spectra which corresponds to the emission of neutrinos with a small energy.

Usually an effect of the neutrino mass is searched for through the investigation of the β-spectrum of the decay

$$^3\text{H} \rightarrow {}^3\text{He} + e^- + \bar{\nu}_e. \tag{1}$$

Up to now no effects of a neutrino mass were found in these experiments. In the first experiments for an upper bound of the neutrino mass one obtained [15]

$$m_\nu \lesssim 500 \, \text{eV}.$$

With further experiments this bound was decreasing and at the end of the fifties one found for the upper bound of the neutrino mass the value

$$m_\nu \lesssim (100\text{–}200) \, \text{eV}. \tag{2}$$

The two-component neutrino theory, proposed by Landau [16], Lee and Yang [17] and Salam [18] in 1957, after the violation of the parity in the β-decay was discovered [19], *was the first theoretical idea about a neutrino mass*.

In order to demonstrate the idea of a two-component neutrino let us consider the Dirac equation for the field of a neutrino with mass m_ν

$$i\gamma^\alpha \partial_\alpha \nu(x) - m_\nu \nu(x) = 0. \tag{3}$$

From Eq. (3) we have for the left-handed and right-handed components $\nu_L(x)$ and $\nu_R(x)$ two coupled equations

$$i\gamma^\alpha \partial_\alpha \nu_L(x) - m_\nu \nu_R(x) = 0, \tag{4}$$

and

$$i\gamma^\alpha \partial_\alpha \nu_R(x) - m_\nu \nu_L(x) = 0. \tag{5}$$

Taking into account the bound (2) it looked natural in the fifties that the neutrino mass is equal to zero. This assumption was made by Landau, Lee and Yang and Salam.

For $m_\nu = 0$ we obtain from (4) and (5) two decoupled Weyl equations

$$i\gamma^\alpha \partial_\alpha \nu_{L,R}(x) = 0, \tag{6}$$

and in this case the neutrino field can be

$$\nu_L(x) \qquad \text{or} \qquad \nu_R(x).$$

The authors of the two-component neutrino theory made this choice.

If the neutrino field is $\nu_L(x)$ $(\nu_R(x))$,

1. The general Hamiltonian of the β-decay has the form

$$\mathcal{H}_I^\beta = \sum_i{}' G_i (\bar{p} O_i n)(\bar{e} O^i \tfrac{1}{2}(1 \mp \gamma_5)\nu) + h.c., \tag{7}$$

   where the index i runs over S, V, T, A, P (scalar, vector etc).

   Thus, the two-component neutrino theory ensures large violation of parity, as observed in the β-decay.

2. The neutrino helicity is equal to -1 $(+1)$ and the antineutrino helicity is equal to $+1$ (-1) in the case of $\nu_L(x)$ $(\nu_R(x)$).

The neutrino helicity was measured in 1958 in a spectacular experiment by M. Goldhaber *et al.* [20]. In this experiment the circular polarization of γ-quanta from the chain of reactions

$$e^- + \text{Gd} \rightarrow \nu_e + \text{Sm}^*$$
$$\downarrow$$
$$\text{Sm} + \gamma$$

was measured. The measurement of the polarization of the γ-quanta allowed them to determine the longitudinal polarization of the neutrino. It was found that the neutrino is a left-handed particle. Thus, the neutrino field is $\nu_L(x)$.

It is interesting to note that equations (6) for a massless particle were discussed by Pauli in his Handbuch der Physik article "General Principles of Quantum Mechanics" (1933). Pauli wrote that because the equation for v_L (v_R) is not invariant under space reflection it is "not applicable to the physical reality".

From the point of view of the two-component theory, large violation of parity in the β-decay and other leptonic processes is ultimately connected with the neutrino mass being equal to zero. This point of view changed after Feynman and Gell-Mann [21], Marshak and Sudarshan [22] in 1958 proposed the $V - A$ theory.

This theory was based on the assumption that *in the Hamiltonian of the weak interaction only left-handed components of all fields enter*. This means that the violation of parity in the weak interaction is not connected with exceptional properties of the neutrinos. There exist other reasons for left-handed fields in the Hamiltonian. Moreover, after the V-A theory it was natural to turn up arguments and consider the neutrino as a particle with mass different from zero (see later).

Nevertheless, the two-component neutrino theory was a nice and also the simplest theoretical possibility. It was in a perfect agreement with numerous experiments on the investigation of weak processes.

From my point of view this was the main reason why during many years there was a common opinion that neutrinos are massless particles. The Glashow-Weinberg-Salam Standard Model was build under the assumption of massless two-component neutrinos.

2. B. Pontecorvo

The first idea of neutrino masses, mixings and oscillations was suggested by B. Pontecorvo in 1957 [11]. He thought that there is an analogy between leptons and hadrons and he believed that in the lepton world there would exist a phenomenon analogous to the famous $K^0 \rightleftarrows \bar{K}^0$ oscillations.

The only possible candidate were neutrino oscillations. At that time only one neutrino type was known. Possible oscillations in this case are

$$v_L \rightleftarrows \bar{v}_L \quad \text{and} \quad \bar{v}_R \rightleftarrows v_R.$$

According to the two-component neutrino theory the states $\bar{v}_L$ and v_R do not exist. Such states were a problem for B. Pontecorvo. We will see how he solved it.

In 1957–58 R. Davis [23] was doing an experiment searching for the production of ^{37}Ar in the process

$$\bar{v}_e + {}^{37}\text{Cl} \rightarrow e^- + {}^{37}\text{Ar}, \tag{8}$$

with antineutrinos from a reactor. A rumor reached B. Pontecorvo that Davis observed production of ^{37}Ar. B. Pontecorvo, who was thinking about neutrino oscillations at that time, decided that production of ^{37}Ar could be due to antineutrino $\rightleftarrows$ neutrino transitions in vacuum. He published a paper on neutrino oscillations [12]. In this paper he wrote:

> "Recently the question was discussed whether there exist other *mixed* neutral particles beside the K^0 mesons, i.e. particles that differ from the corresponding antiparticles, with the transitions between particle and antiparticle states not being strictly forbidden. It was noted that the neutrino might be such a mixed particle, and consequently, there exists the possibility of real neutrino $\rightleftarrows$ antineutrino transitions in vacuum, provided that lepton (neutrino) charge is not conserved. This means that the neutrino and antineutrino are *mixed* particles, i.e., a symmetric and antisymmetric combination of two truly neutral Majorana particles v_1 and v_2."

B. Pontecorvo came to the conclusion that

> "The flux of neutral leptons consisting mainly of antineutrino when emitted from a reactor will consist at some distance R from the reactor of half neutrinos and half antineutrinos".

In the fifties Reines and Cowan [24] were doing their famous experiment in which $\bar{v}_e$ was discovered via the observation of e^+ and neutrons produced in the reaction

$$\bar{v}_e + \text{p} \rightarrow e^+ + \text{n}. \tag{9}$$

In order to see effects of neutrino oscillations B. Pontecorvo proposed that

> "It will be extremely interesting to perform Cowan and Reines experiment at different distances from a reactor".

In the paper [12], which was written at the time when the two-component theory had just appeared and the Davis experiment was not finished, B. Pontecorvo wrote

> "... it is not possible to state apriori that some part of the flux can initiate the Davis reaction".

Thus, he admitted at that time that the two-component theory could be violated. Later, after the Davis experiment was finished and no production of ^{37}Ar was observed, B. Pontecorvo understood that due to oscillations neutrino (antineutrino) could transfer into $\bar{v}_L$ (v_R), particles which do not participate in the standard weak interaction. B. Pontecorvo was thus the first who introduced the notion of *sterile neutrinos* so popular nowadays.

After the second neutrino v_μ was discovered in the Brookhaven experiment [25] it was very natural (and not difficult) for B. Pontecorvo to generalize his idea of neutrino oscillations to the case of two neutrinos [26]. He considered in the paper [26] oscillations into active and sterile states $v_\mu \rightleftarrows v_e$, $v_\mu \rightleftarrows \bar{v}_{\mu L}$ etc.

In the paper [26] B. Pontecorvo discussed oscillations of the solar neutrinos. In 1967 the Davis solar neutrino experiment only started. Three years before the first results of Davis experiment were published. B. Pontecorvo pointed out that due to neutrino oscillations the observed flux of solar neutrinos could be two times smaller than the expected flux.

> "From an observational point of view the ideal object is the sun. If the oscillation length is smaller than the radius of the sun region effectively producing neutrinos, (let us say one tenth of the sun radius $R_\odot$ or 0.1 million km for ^8B neutrinos, which will give the main contribution in the experiments being planned now), direct oscillations will be smeared out and unobservable. The only effect on the earth's surface would be that the flux of observable sun neutrinos must be two times smaller than the total neutrino flux".

In the Davis experiment only very rare high energy solar neutrinos mainly from the decay $^8\text{B} \rightarrow {}^8\text{Be} + v_e$ can be observed. In [26] B. Pontecorvo wrote

> "Unfortunately, the relative weight of different thermonuclear reactions in the sun and its central temperature are not known well enough to permit a comparison of the expected and observed solar neutrino intensities".

In 1967 it was impossible to envisage the NC result of the SNO experiment [2] in which the total flux of the ^{8}B neutrinos was measured. At that time not even neutral current interactions were known. It is interesting to notice that in 1988 when the SNO experiment was in its initial stage B. Pontecorvo enthusiastically supported the experiment. This is a letter which B. Pontecorvo wrote at that time.

Dr. Walter F. Davidson
High Energy Physics Section,
National Research Council, Canada

Dear Dr. Davidson,

Thank you very much for sending me the SNO proposal. Below I am writing a short comment on SNO in the hope that opinion of a person who already in 1946 worked in Canada on neutrinos may be of some value. The SNO proposal (1000 tons of D_2O immersed in H_2O in a mine 2 km deep) in my opinion is a wonderful proposal for several reasons.

First it is new in the sense that with the help of a large D_2O detector immersed in H_2O there becomes possible the investigation of reactions 1. $\nu_e d \to e^- pp$ 2. $\nu_x e \to \nu_x e$ 3. $\nu_x d \to \nu_x np$ 4. $\bar{\nu}_e d \to e^+ nn$ 5. $\bar{\nu}_e p \to e^+ n$ with main application to solar and star collapse neutrinos (1,2,3) and star collapse antineutrinos (4,5).

Second, the proposal is realistic, in a sense that at least one large Cerenkov counter filled with H_2O is known to work properly (Kamiokande).

Third, the proposal can be realized only in Canada, where for historical reasons large quantities of D_2O are available during a period of several years.

Finally, in my opinion the neutral current reaction 3. yielding the total number of neutrinos of all flavors, can be investigated in spite of serious difficulties of registration of neutrons.

In conclusion the SNO proposal is progressive and should be supported by all means.

Yours sincerely,
Bruno Pontecorvo,
Dubna August 18, 1988.

3. Z. Maki, M. Nakagawa and S. Sakata

Two-neutrino mixing was proposed in 1962 by Z. Maki, M. Nakagawa and S. Sakata in a paper [27] which was practically unknown for many years. The approach adopted in this paper was based on the Nagoya model. According to this model, in the hadronic current there enter fields of three fundamental baryons, p, n and Λ. These particles were considered as bound states of leptons and a boson B^+ ("a new sort of matter"):

$$p = \langle \nu B^+ \rangle, \qquad n = \langle e^- B^+ \rangle, \qquad \Lambda = \langle \mu^- B^+ \rangle.$$

A natural consequence of the model was baryon-lepton symmetry, a symmetry of the weak current under the exchange

$$\nu \leftrightarrow p, \qquad e^- \leftrightarrow n, \qquad \mu^- \leftrightarrow \Lambda.$$

The paper [27] was writen at the time when there was an indication [28], obtained from the analysis of the data of experiments on the search for $\mu \to e\gamma$, that ν_e and ν_μ are different particles. The Brookhaven neutrino experiment [25] which proved that ν_e and ν_μ are different particles was in preparation at that time.

The possible existence of two different neutrinos was a problem for the Nagoya model (four leptons and three fundamental hadrons). MNS proposed the following solution of the problem. The leptonic weak current

$$j_\alpha = 2(\bar{\nu}_{eL}\gamma_\alpha e_L + \bar{\nu}_{\mu L}\gamma_\alpha\mu_L) \tag{10}$$

determines the weak neutrinos ν_e and ν_μ. They wrote

"...the definition of the particle state of neutrino is quite arbitrary; we can speak of "neutrinos" which are different of weak neutrinos but expressed by the linear combinations of the latter. We assume that there exists a representation which defines *the true neutrinos*

$[\nu_1$ and $\nu_2]$ through some orthogonal transformation... :

$$\nu_1 = +\nu_e \cos\delta + \nu_\mu \sin\delta,$$
$$\nu_2 = -\nu_e \sin\delta + \nu_\mu \cos\delta. \tag{11}$$

"...*the true neutrinos should be so defined that B^+ can be bound to ν_1 to form a proton but can not be bound to ν_2*".

Thus, MNS proposed a modified Nagoya model:

$$p = \langle \nu_1 B^+ \rangle, \qquad n = \langle e^- B^+ \rangle, \qquad \Lambda = \langle \mu^- B^+ \rangle.$$

To the lepton current, written in terms of "true neutrinos",

$$j_\alpha = 2(\bar{\nu}_{1L}\gamma_\alpha e_L \cos\delta + \bar{\nu}_{1L}\gamma_\alpha\mu_L \sin\delta) + \cdots. \tag{12}$$

corresponds the hadronic current

$$j_\alpha = 2(\bar{p}_L\gamma_\alpha n_L \cos\delta + \bar{p}_L\gamma_\alpha\Lambda_L \sin\delta) + \cdots. \tag{13}$$

which was identical to the Gell-Mann-Levy current [29]. In fact this was a pre-Cabibbo discussion of the hadron mixing.

Further, MNS assumed that there exists an additional interaction of ν_2 with a field X of heavy bosons

$$\mathcal{L} = g\bar{\nu}_2\nu_2 X^+ X \tag{14}$$

which provides the difference of the masses of ν_2 and ν_1.

In connection with the Brookhaven neutrino experiment MNS wrote

"Weak neutrinos[1]

$$\begin{bmatrix} \nu_e = \nu_1 \cos\delta - \nu_2 \sin\delta \\ \nu_\mu = \nu_1 \sin\delta + \nu_2 \cos\delta \end{bmatrix} \tag{15}$$

are *not stable* due to occurrence of virtual transmutation $\nu_e \leftrightarrow \nu_\mu$ induced by the interaction (14)....... a chain of reactions such as

$$\pi^+ \to \mu^+ + \nu_\mu$$
$$\nu_\mu + Z(\text{nucleus}) \to Z' + (\mu^- \text{ and/or } e^-)$$

is useful to check the two-neutrino hypothesis only when $|m_{\nu_2} - m_{\nu_1}| < 10^{-6}$ MeV under a conventional geometry of experiments. Conversely, the absence of e^- in the reactions [above] will be able not only to verify two-neutrino hypothesis but also to provide an upper limit of the mass of the second neutrino ν_2 if the present scheme should be accepted".

4. B. Pontecorvo and collaborators

The first phenomenological theory of two-neutrino mixing was proposed by V. Gribov and B. Pontecorvo in 1969 [30]. They assumed that the left-handed fields ν_{eL} and $\nu_{\mu L}$ enter not only into the weak interaction but also into a mass term (which they called an additional neutrino interaction).

Let us notice that there was a wide-spread prejudice at that time that in the case of left-handed neutrino fields neutrino masses must be equal to zero. This is correct if the total lepton number is conserved. In [30] it was shown that if the total lepton number is changed by 2 neutrino masses can be introduced even in the case of left-handed fields in the Lagrangian.

Gribov and Ponecorvo assumed that the following neutrino mass term enters into the Lagrangian

$$\mathcal{L}^M = -\tfrac{1}{2}(m_{e\bar{e}}\overline{(\nu_{eL})^c}\nu_{eL} + m_{\mu\bar{\mu}}\overline{(\nu_{\mu L})^c}\nu_{\mu L}$$
$$+ m_{e\bar{\mu}}(\overline{(\nu_{eL})^c}\nu_{\mu L} + \overline{(\nu_{\mu L})^c}\nu_{eL})) + \text{h.c.} \tag{16}$$

---

[1]Let us stress that according to MNS δ is the Cabibbo angle.

Here $m_{\mu\bar\mu}$, $m_{e\bar e}$ and $m_{e\bar\mu}$ are real parameters and $(v_{lL})^c = C(\bar v_{lL})^T$ is the conjugated field.

After the diagonalization of the mass term (16) the following mixing relations were obtained

$$v_{eL} = \cos\theta\, v_{1L} + \sin\theta\, v_{2L},$$
$$v_{\mu L} = -\sin\theta\, v_{1L} + \cos\theta\, v_{2L}, \tag{17}$$

where $v_{1,2}$ are fields of *Majorana neutrinos* with masses $m_{1,2}$. The mass term (16) is called the Majorana mass term.

Possible oscillations in the case of the mixing (17) are $v_e \rightleftarrows v_\mu$ and $\bar v_e \rightleftarrows \bar v_\mu$. There are no transitions into sterile states in the case of the Majorana mass term. This was one of the main ideas of Gribov and Pontecorvo.

It was shown in [30] that the observables (the mixing angle θ and Majorana neutrino masses $m_{1,2}$) are connected with the parameters of the theory by the relations

$$\tan 2\theta = \frac{2m_{e\bar\mu}}{m_{\mu\bar\mu} - m_{e\bar e}}, \tag{18}$$

$$m_{1,2} = \tfrac{1}{2}\left|m_{\mu\bar\mu} + m_{e\bar e} \mp \sqrt{(m_{\mu\bar\mu} - m_{e\bar e})^2 + 4m_{e\bar\mu}^2}\right|. \tag{19}$$

From (19) it follows that if $\mu - e$ symmetry holds, so that

$$m_{\mu\bar\mu} = m_{e\bar e}; \qquad m_{e\bar\mu} \neq 0, \tag{20}$$

then $\theta = \pi/4$ (maximal mixing). Vacuum oscillations of solar neutrinos were discussed in [30] for this case.

The full phenomenological theory of neutrino mixing and the theory of neutrino oscillations in vacuum were developed in the seventies.

In the papers [31] neutrino mixing was introduced in analogy with Cabibbo-GIM mixing of quarks (lepton-quark analogy).

The main idea of the paper [31] was the following: neutrinos like all other fundamental fermions (leptons and quarks) are massive particles. A mixing of massive fermions is a general feature of gauge theories with spontaneous symmetry breaking. Thus, it is natural to assume that *the phenomenon of mixing is common for quarks and massive neutrinos*. For the case of two neutrinos we have

$$v_{eL} = \cos\theta\, v_{1L} + \sin\theta\, v_{2L},$$
$$v_{\mu L} = -\sin\theta\, v_{1L} + \cos\theta\, v_{2L}. \tag{21}$$

Here $v_{1,2}$ are 4-component fields of *Dirac neutrinos* with masses $m_{1,2}$.

Possible values of the neutrino mixing angle were discussed in [31]. We have concluded

> "... it seems to us that the special values of the mixing angle $\theta = 0$ (the usual scheme in which muonic charge is strictly conserved) and $\theta = \pi/4$ (maximal mixing) are of the greatest interest."

The Dirac mass term has the form

$$\mathcal{L}^D = -\sum_{ll'} \bar v_{l'R} M^D_{l'l} v_{lL} + \text{h.c.} \tag{22}$$

where $M^D_{l'l}$ is a complex matrix. If the mass term (22) enters into the Lagrangian, the total lepton number L is conserved. Due to the conservation of L in the case of the Dirac mass term there are no transitions into sterile states. For two neutrinos v_e and v_μ the possible oscillations are $v_e \rightleftarrows v_\mu$ and $\bar v_e \rightleftarrows \bar v_\mu$, the same as in the Majorana case.

In [32] we have introduced the most general mass term. If we assume that v_{lL} and v_{lR} both enter into the mass term and the total lepton number L is not conserved for the mass term we obtain

$$\mathcal{L}^{D+M} = -\tfrac{1}{2}\sum_{l,l'}\overline{(v_{l'L})^c} M^L_{l'l} v_{lL} - \sum_{l'l'}\bar v_{l'R} M^D_{l'l} v_{lL}$$
$$-\tfrac{1}{2}\sum_{l,l'}\bar v_{l'R} M^R_{l'l}(v_{lR})^c + \text{h.c.} \tag{23}$$

Here M^L and M^R are complex symmetric matrices and M^D is a complex matrix. Eq. (23) includes a left-handed Majorana mass term, a Dirac mass term and a right-handed Majorana mass term. It is called the Dirac and Majorana mass term.

For the mixing in the general case of n flavors we have from (23)

$$v_{lL} = \sum_{i=1}^{2n} U_{li} v_{iL}, \qquad (v_{lR})^c = \sum_{i=1}^{2n} U_{\bar l i} v_{iL}, \tag{24}$$

where U is a unitary $2n \times 2n$ mixing matrix, and v_i is the field of the Majorana neutrino with mass m_i. Now after the LEP experiments we know that $n = 3$.

From (24) it follows that in the case of small masses m_i, it is possible to have transitions $v_l \to v_{l'}$ (flavor-flavor) and $v_l \to \bar v_{l'L}$ (flavor-sterile). Vacuum oscillations of the solar neutrinos in the general case of the Dirac and Majorana mass term were considered in [32]. As is well known, the Dirac and Majorana mass terms constitute the framework for the see-saw mechanism of neutrino mass generation.

In the seventies also the theory of neutrino oscillations in vacuum was developed, which is widely used today for analysis of the data of neutrino oscillation experiments.

In the case of neutrino mixing the lepton numbers L_e, L_μ and L_τ are not conserved. What are flavor neutrinos v_e, v_μ, v_τ and corresponding antineutrinos in this case?

From the very beginning we determine flavor neutrinos and antineutrinos as particles which take part in the standard CC weak processes with corresponding leptons. For example, the neutrino which is produced together with μ^+ in the decay $\pi^+ \to \mu^+ + v_\mu$ is the muon neutrino v_μ, whereas an electron antineutrino $\bar v_e$ produces e^+ in the process $\bar v_e + p \to e^+ + n$, etc.

The states of flavor neutrinos are given by

$$|v_l\rangle = \sum_i U^*_{li}|v_i\rangle, \tag{25}$$

where $|v_i\rangle$ is the state of a neutrino with mass m_i, momentum $\vec p$ and energy $E_i = \sqrt{p^2 + m_i^2} \simeq p + \frac{m_i^2}{2p}$. Thus, flavor neutrinos are described by *mixed coherent states*.

The relation (25) is based on the assumption that neutrino mass-squared differences are so small that due to the uncertainty relation it is impossible to distinguish production (detection) of neutrinos with different masses. This condition can be presented in the form

$$L_{osc} \gg d, \tag{26}$$

where d is the quantum mechanical dimension of a neutrino source and

$$L_{osc} = 4\pi\frac{E}{\Delta m^2} \tag{27}$$

is the oscillation length. If this condition is satisfied, neutrino cross sections and decay probabilities are given by the Standard Model.

If we now apply the evolution equation of the field theory to the flavor states

$$i\frac{\partial|\Psi(t)\rangle}{\partial t} = H|\Psi(t)\rangle \tag{28}$$

we come to the standard expression for the transition probability

$$P(\nu_l \to \nu_{l'}) = \left| \delta_{ll'} + \sum_{i \geq 2} U_{l'i} U_{li}^* (e^{-i\Delta m_{i1}^2 \frac{L}{2E}} - 1) \right|^2. \tag{29}$$

Here L is the source-detector distance, E is neutrino energy and $\Delta m_{i1}^2 = m_i^2 - m_1^2$.

A necessary condition for the observation of neutrino oscillations has the form

$$\Delta m_{i1}^2 \frac{L}{E} \geq 1. \tag{30}$$

From this condition we see that experiments searching for neutrino oscillations have enormous sensitivity to neutrino mass squared differences (for example, reactor experiments of the KamLAND type with $L \simeq 100\,\mathrm{km}$ and $E \simeq 1\,\mathrm{MeV}$ are sensitive to $\Delta m^2 \simeq 10^{-5}\,\mathrm{eV}^2$ etc).

For us this was the main reason and motivation for neutrino oscillation experiments: due to the interference nature of the phenomenon of neutrino oscillations and the possibility to perform neutrino experiments with large values of the parameter $\frac{L}{E}$ the investigation of neutrino oscillations is an extremely sensitive method to search for small Δm^2. This strategy brought success. We summarized it in the first review on neutrino oscillations published in 1977 [33]. Except for the papers of B. Pontecorvo, MNS and B. Pontecorvo and collaborators at that time only a few papers on neutrino oscillations ([34, 35, 36, 37]) were published.

At the end of the seventies a common interest for the problem of the neutrino masses and mixings started. It was connected with the appearance of the GUT models and the invention of the see-saw mechanism for the neutrino mass generation. Neutrino masses started to be considered as a signature of a new physics beyond the Standard Model. At that time special reactor and accelerator neutrino experiments on search of neutrino oscillations started. The investigation of matter effects in the case of neutrino mixing in the seventies [38] and disclosure of the MSW effect in the eighties [39] had very important impact on the field.

5. Conclusion

Today all existing neutrino oscillation data with the exception of the data of the LSND experiment [40], which need confirmation, are described by the three-neutrino mixing scheme.

For neutrino oscillation parameters the following values were obtained [1, 3, 41]:

$$\Delta m_{21}^2 = (8.2^{+0.6}_{-0.5}) \cdot 10^{-5}\mathrm{eV}^2; \qquad \tan^2 \theta_{12} = (0.40^{+0.09}_{-0.07}),$$

$$1.9 \cdot 10^{-3} \leq \Delta m_{32}^2 \leq 3.0 \cdot 10^{-3}\mathrm{eV}^2; \qquad \sin^2 2\theta_{23} \geq 0.90,$$

$$\sin^2 \theta_{13} \leq 5 \cdot 10^{-2}. \tag{31}$$

Thus, the neutrino oscillation parameters satisfy the inequalities

$$\Delta m_{21}^2 \ll \Delta m_{32}^2; \qquad \sin^2 \theta_{13} \ll 1. \tag{32}$$

It follows from (32) (see [42]) that the dominant transitions, governed by Δm_{32}^2, are $\nu_\mu \to \nu_\tau$ and $\bar{\nu}_\mu \to \bar{\nu}_\tau$. The dominant transitions governed by Δm_{21}^2 are $\nu_e \to \nu_{\mu,\tau}$ and $\bar{\nu}_e \to \bar{\nu}_{\mu,\tau}$. This is a present-day picture of neutrino oscillations.

The main question to be answered is: What physics was discovered? What are the implications of the discovered phenomenon? From my point of view, to answer these fundamental questions we need to know

1. Are massive neutrinos Majorana or Dirac particles?
2. What is the neutrino mass spectrum (hierarchical, inverted, degenerate etc)?
3. What is the mass of the lightest neutrino?
4. How many massive neutrinos exist in nature? Do sterile neutrinos exist?
5. What is the value of the parameter $\sin^2 \theta_{13}$?
6. What is the value of the CP phase?
7. What are precise values of oscillation parameters?
8. ...

I think that the first period of the investigation of neutrino properties is basically finished. In spite of the fact that it took many years to discover neutrino oscillations, in a sense up to now we have been lucky. It seems that the next decisive step will be difficult and will require a lot of efforts. Existing data allow us to conclude that to probe the nature of neutrinos via the investigation of neutrinoless double β-decay is a challenge. It will be extremely difficult to observe this process if the Majorana neutrino mass spectrum is hierarchical, which is a plausible possibility. The solution of other problems also seem a difficult and challenging task. After the discovery of neutrino masses and mixings we know, however, what to look for. This is a great advantage of the present stage.

The history of neutrino oscillations is an illustration of a complicated and thorny way of science.

- Correct pioneer ideas sometimes can have a wrong basis or can be accompanied by a wrong one.
- Courageous general ideas have good chances to be correct.
- Analogy is an important guiding principle in physics.

I will finish with a quotation from the report by S. L. Glashow [43] at the Venice "Neutrino Telescope Workshop" (March 2003).

"...If only Bruno Pontecorvo could have seen how far we have come towards understanding the pattern of neutrino masses and mixing! Way back in 1963 he was among the first to have envisaged the possibility of neutrino flavor oscillations. For that reason the analog to the Cabibbo-Kobayashi-Maskawa matrix pertinent to neutrino oscillations should be known as the PMNS matrix to honor four neutrino visionaries: Pontecorvo, Maki, Nakagawa, and Sakata".

References

1. Super-Kamiokande Collaboration, Fukuda, S. *et al.*, Phys. Rev. Lett. **81**, 1562 (1998); Fukuda, S. *et al.*, Phys. Rev. Lett. **82**, 2644 (1999); Fukuda, S. *et al.*, Phys. Rev. Lett. **85**, 3999 (2000); Kearns, E., Proc. 21th International Conference on Neutrino Physics and Astrophysics (Neutrino 2004), 13–19 June (2004), Paris, France.
2. SNO collaboration, Ahmad, Q. R. *et al.*, Phys. Rev. Lett. **87**, 071301 (2001); Phys. Rev. Lett. **89**, 011301 (2002); nucl-ex/0204008; Phys. Rev. Lett. **89**, 011302 (2002); nucl-ex/0204009, nucl-ex/0309004.
3. KamLAND Collaboration, Araki, T. *et al.*, hep-ex/0406035, submitted to Phys. Rev. Lett.
4. Cleveland, B. T. *et al.*, Astrophys. J. **496**, 505 (1998).
5. GALLEX Collaboration, Hampel, W. *et al.*, Phys. Lett. **B 447**, 127 (1999); GNO Collaboration, Altmann, M. *et al.*, Phys. Lett. **B 490**, 16 (2000); Nucl. Phys. Proc. Suppl. **91**, 44 (2001).
6. SAGE Collaboration, Abdurashitov, J. N. *et al.*, Phys. Rev. **C 60**, 055801 (1999); Nucl. Phys. Proc. Suppl. **110**, 315 (2002).
7. Super-Kamiokande Collaboration, Fukuda, S. *et al.*, Phys. Rev. Lett. **86**, 5651 (2001).

8. Soudan 2 Collaboration, Allison, W. W. M. *et al.*, Phys. Lett. **B 449**, 137 (1999).
9. MACRO Collaboration, Ambrosio, M. *et al.*, hep-ex/0106049; Phys. Lett. **B517**, 59 (2001); Ambrosio, M. *et al.* NATO Advanced Research Workshop on Cosmic Radiations, Oujda (Morocco), 21–23 March (2001).
10. K2K Collaboration, Nakaya, T., Proc. 21th International Conference on Neutrino Physics and Astrophysics (Neutrino 2004), 13–19 June (2004), Paris, France.
11. Pontecorvo, B., J. Exp. Theor. Phys. **33**, 549 (1957) [Sov. Phys. JETP **6**, 429 (1958)].
12. Pontecorvo, B., J. Exp. Theor. Phys. **34**, 247 (1958) [Sov. Phys. JETP **7**, 172 (1958)].
13. Fermi, E., Ricerca Scientifica **2**, N12 (1933), Z. Physik **88**, 161 (1934).
14. Perrin, F., Comptes Rerdues **197**, 1625 (1933).
15. Hanna, G. and Pontecorvo, B., Phys. Rev. **75**, 983 (1949); Curran, S. *et al.*, Phil. Mag. **40**, 53 (1949).
16. Landau, L. D., Nucl. Phys. **3**, 127 (1957).
17. Lee, T. D. and Yang, C. N., Phys. Rev. **105**, 1671 (1957).
18. Salam, A., Nuovo Cimento **5**, 299 (1957).
19. Wu, C. S. *et al.*, Phys. Rev. **105**, 1413 (1957).
20. Goldhaber, M., Grodzins, L. and Sunyar, A. W., Phys. Rev. **109**, 1015 (1958).
21. Feynman, R. P. and Gell-Mann, M., Phys. Rev. **109**, 193 (1958).
22. Sudarshun, E. C. G. and Marshak, R., Phys. Rev. **109**, 1860 (1958).
23. Davis, R., Bull. Am. Phys. Soc. (Washington, meeting, 1959).
24. Reines, F. and Cowan, C., Phys. Rev. **113**, 273 (1959).
25. Danby, G. *et al.*, Phys. Rev. Lett. **9**, 36 (1962).
26. Pontecorvo, B., J. Exp. Theor. Phys. **53**, 1717 (1967) [Sov. Phys. JETP **26**, 984 (1968)].
27. Maki, Z., Nakagava, M. and Sakata, S., Prog. Theor. Phys. **28**, 870 (1962).
28. Feinberg, G., Phys. Rev., **110**, 1482 (1958).
29. Gell-Mann, M. and M. Levi, Nuovo Cimento **10**, 705 (1960).
30. Gribov, V. and Pontecorvo, B., Phys. Lett. **B28**, 493 (1969).
31. Bilenky, S. M. and Pontecorvo, B., Phys. Lett. **B61**, 248 (1976); Yad. Fiz. **3**, 603 (1976).
32. Bilenky, S. M. and Pontecorvo, B., Lett. Nuovo Cim. **17**, 569 (1976).
33. Bilenky, S. M. and Pontecorvo, B., Phys. Rep. **41**, 225 (1978).
34. Bahcall, J. and Frautschi, S., Phys. Lett. **29**, 623 (1969).
35. Fritzsch, H. and Minkowski, P., Phys. Lett. **B62**, 72 (1976).
36. Eliezer, S. and Swift, A., Nucl. Phys. **B105**, 45 (1976).
37. Bilenky, S. M. and Pontecorvo, B., JINR Preprint E2-10032, Dubna, (1976), Proc. Intern. Conf. High Energy Physics, Tbilisi, July (1976).
38. Wolfenstein, L., Phys. Rev. D **17**, 2369 (1978); Phys. Rev. D **20**, 2634 (1979).
39. Mikheyev, S. P. and Smirnov, A.Yu.,Yad. Fiz. **42**, 1441 (1985) [Sov. J. Nucl. Phys. **42**, 913 (1985)]; Il Nuovo Cim. C **9**, 17 (1986); Zh. Eksp. Teor. Fiz. **91**, 7 (1986) [Sov. Phys. JETP **64**, 4 (1986)].
40. LSND Collaboration, Aguilar, A. *et al.*, Phys. Rev. **D64**, 112007 (2001); hep-ex/0104.
41. CHOOZ Collaboration, Apollonio, M. *et al.*, Phys. Lett. B **466**, 415 (1999).
42. Bilenky, S. M., Giunti, C. and Grimus, W., Prog. Part. Nucl. Phys. **43**, 1 (1999); hep-ph/9812360.
43. Glashow, S. L., Proc. the 10th International Workshop Neutrino Telescopes, Venice, Italy, 11–14 Mar. (2003), vol. 2, p. 611; hep-ph/0306100.

Physica Scripta. Vol. T121, 23–28, 2005

Super-Kamiokande Results on Neutrino Oscillations

Y. Suzuki[1]

Kamioka Observatory, Institute for Cosmic Ray Research, University of Tokyo, Higashi-Mozumi, Kamioka-cho, Hida-city, Gifu 506-1205, Japan

Received January 3, 2005; accepted January 12, 2005

PACS number: 1460Pq, 1215Ff

Abstract

Super-Kamiokande (SK) is able to measure neutrino interactions over the 5 decades of the energy from 5 MeV to a few hundreds GeV. In 1998, SK has obtained strong evidence of neutrino oscillations by observing a deficit of the atmospheric v_μ flux coming upward to the detector. This discovery has revealed that neutrinos have finite masses. The mass difference is $\sim 0.002\,\mathrm{eV}^2$ and a mixing angle is nearly maximal. The experimental evidence on the atmospheric neutrino oscillation has been strengthened by the observation of the oscillatory pattern as a function of L/E.

The non-observation of the day-night flux differences and the non-observation of the spectrum distortions of solar neutrinos have placed a strong constraint on the oscillation parameters. In 2000, the so called small mixing angle solutions were rejected and it was shown that the solar neutrino oscillation should be large mixing.

Conclusive evidence of the solar neutrino oscillation was obtained by comparing the precisely measured flux by the neutrino-electron scattering in SK and the flux obtained by the charged current interaction of electron neutrinos in SNO in 2001. The result of the fit by using all the solar neutrino experiments has selected a MSW Large Mixing Angle solution.

1. Super-Kamiokande

1.1. *Detector*

Super-Kamiokande (SK) [1][2] is a 50,000 ton imaging water Cherenkov detector located at 1000 m underground at Kamioka Observatory, Institute for Cosmic Ray Research, University of Tokyo, sited in the Kamioka Mine, Japan. It is a cylindrical detector with 39.3 m in diameter and 42 m in height. The neutrino interactions in the water tank were viewed by 11,146 50 cm-diameter photomultiplier tubes (PMTs) arranged with 2 PMTs per m^2 on the entire inner surface of the 32,000 ton inner detector (ID) volume which is used for the physics analysis. Surrounding the inner volume is an outer detector (OD) with an average thickness of 2.5 m and 1,885 20 cm PMTs are arranged facing outwards.

1.2. *Brief history*

The experiment started to take data on 1st of April in 1996 after five years of construction. Two years later, the first announcement of the discovery of neutrino oscillation was made based on the observation of strong deficits of upward coming atmospheric muon neutrinos [3]. In 2000, there was a strong indication that the solar neutrino oscillation is a large mixing [4] and a year later, together with the SNO charged current data, the definitive evidence of the solar neutrino oscillation was obtained [5][6][7].

By a tragic accident on 12th of November, 2002, we lost 6777 PMTs in two seconds. About a year later, the detector was rebuilt with about 5200 PMTs, nearly half of the original number of PMTs. SK then resumed to take data (SK-II) and the K2K experiment (the long baseline accelerator neutrino experiment) [8] had started again.

In 2004, we observed an oscillatory behavior in the atmospheric neutrino data [9]. The evidence of the oscillation observed by the K2K experiment became stronger in 2004 to the level of 99.99% C.L. [10].

The full restoration of SK will be scheduled to start at the end of the year 2005. About 6000 PMTs will additionally be installed during the 5 months of the construction time. The SK-III with full number of PMTs will start to take data before the summer of 2006, well before the commission of the J-PARC neutrino experiment (the high intensity long baseline man-made neutrino experiment).

1.3. *Characteristics of the neutrino events*

SK can detect neutrino interactions from 5 MeV to O($\sim$TeV). Its fiducial volume is 22,500 tons, and the typical effective area is $\sim 1,200\,\mathrm{m}^2$.

In the low energy region, less than ~ 20 MeV, the neutrinos are detected through neutrino electron scattering and the detection of the electron anti-neutrinos are enhanced through $\bar{v}_e \mathrm{p} \to \mathrm{e}^+ \mathrm{n}$. The energy response in this region was very well calibrated with 0.64% accuracy by the electron LINAC [11] placed near the detector and the $^{16}\mathrm{N}$ source [12]. The direction of the electrons from $v + \mathrm{e} \to v + \mathrm{e}$ scattering is very forward, $\theta^2 < 2m_\mathrm{e}/E_\mathrm{e}$ (300 mrad at 10 MeV), thus keeps the information of the direction of neutrinos, while the positrons from the electron anti-neutrino interactions scatter uniformly. The angular resolutions for the neutrino electron scattering are mostly determined by multiple Coulomb scattering of electrons in the water, typically 26 degrees at 10 MeV.

In the energy region from a few hundred MeV to O(GeV), the accuracy of the energy determination is a few % and the direction of the incoming neutrinos is determined within 30 degrees on the average. In this energy region, the muons and electrons are efficiently separated by the information of the diffuseness of the edge of the Cherenkov ring patterns. The miss-identification probabilities are $0.6\pm0.1\%$ and 2%, for events with visible energy, $E_\mathrm{vis} < 1.33$ GeV and $E_\mathrm{vis} > 1.33$ GeV, respectively. Those probabilities were confirmed by a test experiment by using the well calibrated beams at the KEK (High Energy Accelerator Research Organization) PS.

Very high energy neutrinos, above a few hundred GeV, are detected as upward through going muons which are produced by neutrino interaction in the rocks beneath the detector. The energy of such events cannot be determined, but the directions of the in-coming neutrinos are determined within 4 degrees.

2. Atmospheric neutrinos

2.1. *Event classification*

The atmospheric neutrinos have good characteristics for studying neutrino oscillations. The incoming direction reflects the flight length of neutrinos from the production source to the detector, which ranges from 10 km to 13,000 km and the neutrino energy

---

[1] suzuki@suketto.icrr.u-tokyo.ac.jp

ranges over 5 orders of magnitude. The wide range of the pathlength and energy provides good sensitivity for L/E.

The atmospheric neutrino events having an event vertex in the target water are classified as fully contained events (FC) if all the produced particles stop within the inner detector (ID) and partially contained events (PC) if some of the tracks escape to the outer detector. The averaged parent neutrino energy is $\sim 1\,\mathrm{GeV}$ for FC and $\sim 10\,\mathrm{GeV}$ for PC, respectively. The FC events are further categorized based on their visible energy into sub-GeV ($E_{\mathrm{vis}} < 1.33\,\mathrm{GeV}$) and multi-GeV ($E_{\mathrm{vis}} > 1.33\,\mathrm{GeV}$).

The upward going muons are divided into upward through going muons and upward stopping muons. The energies of the parent neutrinos are $\sim 100\,\mathrm{GeV}$ and $\sim 10\,\mathrm{GeV}$ on an average, respectively.

2.2. Atmospheric neutrino flux and uncertainty

The primary cosmic rays entering the earth's atmosphere collide and produce secondary particles, which subsequently decay into neutrinos. We have used the calcuration where the three dimensional effects for the secondary particles are incooperated [13].

The flux of primary cosmic rays below $\sim 10\,\mathrm{GeV}$ which produce neutrinos below $\sim 1\,\mathrm{GeV}$ is well measured and the uncertainty is small. But it suffers from the solar activity modulation and the geo-magnetic effect through the rigidity cut off. In the high energy region, on the contrary, though the effects from the solar activity and the geo-magnetic effect is negligible, the primary cosmic ray flux is not known very well. We have prepared three MC data sets corresponding to the solar minimum – 3 years from 1996 to the summer 1999, to solar maximum – 1 year from the summer 2000 to 2001, and to the transition time – 1 year from the summer 1999 to the summer 2000. In general we have used the atmospheric neutrino calculation of reference [13] and occasionally we have compared other flux model calculations [14][15].

The uncertainty of the primary cosmic ray flux and their interactions are the major contributions to the uncertainty of the absolute neutrino flux of 15% for $E_\nu < 10\,\mathrm{GeV}$. The uncertainty of the high energy neutrino flux is given as a spectrum index error of 0.05. The systematic uncertainty is largely reduced in the ratio of $(\nu_\mu + \bar{\nu}_\mu)/(\nu_e + \bar{\nu}_e)$. For neutrino energy smaller than 5 GeV, the ratio is mostly determined by the $\pi \to \mu \to e$ decay chain and the uncertainty is estimated to be 3%. In the higher energy, the K production and the subsequent decay become important and we have assigned 15% uncertainty at 100 GeV by comparing the flux values from the three different calculations.

The zenith angle distribution is basically up-down symmetric. But in the low energy region less than a few GeV, the up-going neutrinos have a slight excess over the down going neutrinos due to the directional dependence of the effect of the rigidity cut off. The horizontal enhancement seen in the zenith angle distribution is due to the longer path length and the three-dimensional effect. But this enhancement cannot be seen in charged leptons in the low energy region below 1 GeV due to the poor angular correlation between the charged leptons produced and the incoming neutrinos. The uncertainty on the up-down ratio is estimated to be small, $1\sim 2\%$ for sub-GeV and $<2\%$ for multi-GeV. The uncertainty of the horizontal and the vertical ratio for the up-going muons is estimated to be 3% mainly due to the uncertainty of the K production.

2.3. Atmospheric neutrino oscillation

The deviation from unity of the double ratio, $R = (\mu/e)_{\mathrm{data}}/(\mu/e)_{\mathrm{MC(no\text{-}osc)}}$, and the asymmetry of the zenith angle distribution

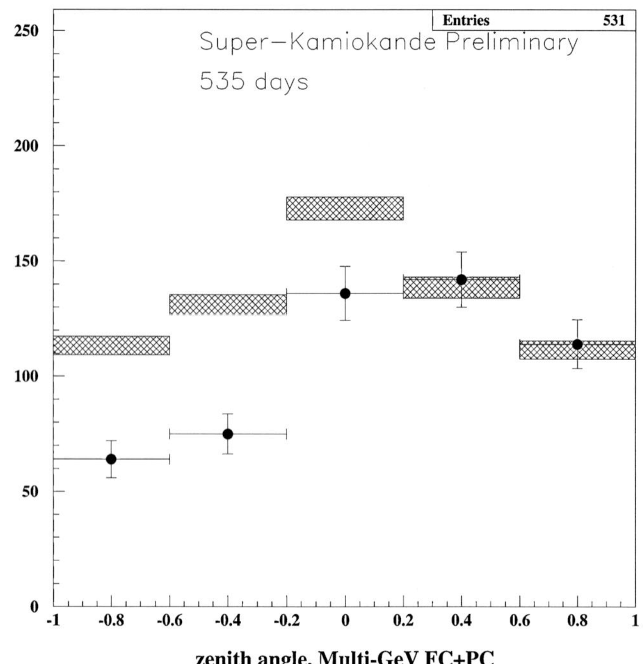

Fig. 1. The zenith angle distribution from 535 days of SK data in 1998. The shaded boxes are expected symmetrical distribution of atmospheric neutrinos for no oscillation case. The deficits of the up-coming neutrinos are clearly seen.

are evidence of neutrino oscillations. Since the zenith angle distribution is of less dependence on the absolute flux calculations, it is taken as the most stringent evidence for the neutrino oscillation.

Super-Kamiokande has announced the discovery of the neutrino oscillation in 1998 based on the observation of the strong deficit of the up-going muon neutrinos in the zenith angle distribution from the 535 days of data [3] as shown in Fig. 1.

2.4. Atmospheric neutrino results from SK-I

The data taken from May, 1996 to July, 2001 were used for the analysis for SK-I. We here present final results on the two-flavor analysis and preliminary results on the three-flavor analysis. A new result from the L/E analysis is shown which demonstrates that the flavor conversion follows a sinusoidal function, predicted solely by the neutrino oscillation.

The effective 1489 days of data are used for the FC and PC samples, while a slightly longer 1646 days of data are used for the upward going muons due to less susceptibility to the detector effects. The numbers of events observed are listed in Table I. The event rate for fully contained events, partially contained events, upward through going muons and upward stopping muons are, 8.2, 0.61, 1.1 and 0.28 events per day, respectively. A detailed description of the data analysis will be presented elsewhere [16].

Table I. *Summary of the events observed in SK-I.*

		e-like events	μ-like events
sub-GeV	1 Ring	3353	3227
Multi-GeV	1 Ring	746	651
Partially Contained		–	647
Stopping up μ		–	417.7
Through going up μ		–	1841.6

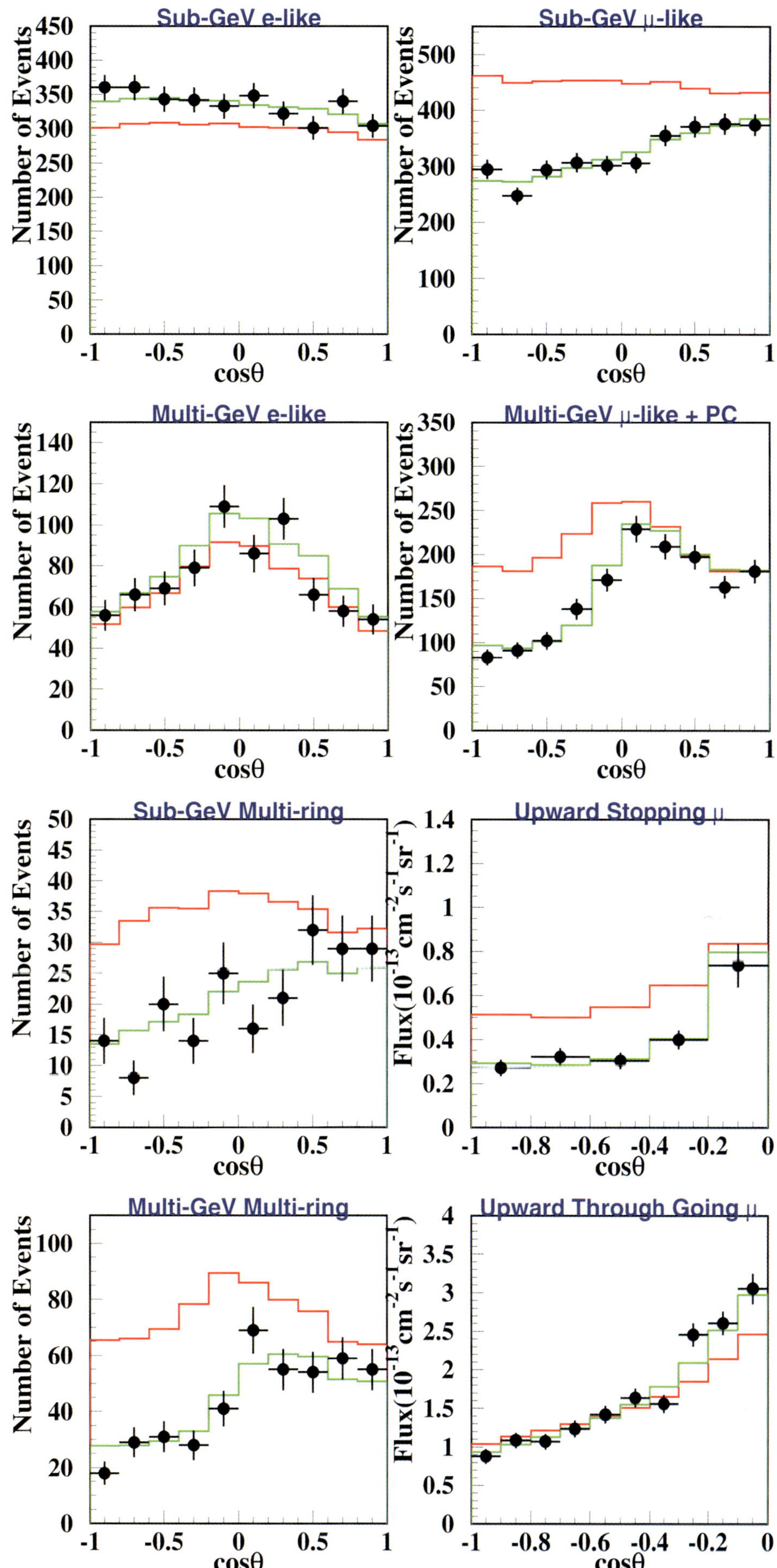

Fig. 2. Zenith angle distributions of the atmospheric neutrinos from 1489 days of data. The lines following the data points are predictions for the best fit oscillation parameters of $\Delta m^2 = 2.1 \times 10^{-3}\,\text{eV}^2$ and $\sin^2 2\theta = 1.02$. The best fit lines are also affected by the shift due to the pull within the systematic errors, for example, the overall flux normalization, electron and muon neutrino ratio, spectrum shape and so on.

2.5. Two flavor analysis

We have produced Monte Carlo events equivalent to 100 years of exposure which are then processed in the same procedure as real data. The measured double ratios, $R = (\mu/e)_{\text{data}}/(\mu/e)_{\text{MC(no-osc)}}$ for sub-GeV and multi-GeV are $R_{\text{sub}} = 0.658 \pm 0.016(\text{stat.}) \pm 0.032(\text{syst.})$ and $R_{\text{multi}} = 0.702 + 0.032/ - 0.030(\text{stat}) \pm 0.099(\text{syst.})$. The systematic errors are estimated to be 4.8% and 14.1% for sub-GeV and multi-GeV, respectively. The amount of deficits in R observed correspond to the oscillation with Δm^2 ranging from 10^{-3} to 10^{-2} eV^2.

The zenith angle distributions of the μ-like and electron-like events for different event categories are shown in Fig. 2. The total 180 bins, divided by the zenith angles and momenta, are used for the $\nu_\mu \to \nu_\tau$ oscillation fits. The results of the fits are shown in Fig. 3. The minimum $\chi^2 = 174.9$ (177 d.o.f.) is located at $\Delta m^2 = 2.1 \times 10^{-3}$eV^2 and $\sin^2 2\theta = 1.02$ while the $\chi^2 = 465$ is obtained for no oscillations. The significance of the oscillation, $\Delta\chi^2$ is 290. The allowed parameter ranges at 90% C.L. are 1.5×10^{-3}eV$^2 < \Delta m^2 < 3.4 \times 10^{-3}$ eV^2 and $\sin^2 2\theta > 0.92$. We also point out that the χ^2 distribution is rather flat between 2 and 2.5×10^{-3} eV^2, therefore the minimum value does not have a strong meaning.

2.6. Three flavor analysis

Although θ_{13} is known to be small ($\sin^2 \theta_{13} < 0.05$) and it is neglected in the two-flavor analysis, there will be non-negligible effects if experiments get more statistics and become more accurate. The ingredients for the 3 flavor analysis are the matter effects which enhance the oscillation probability from ν_μ to ν_e around several GeV regions and the mass hierarchy which determines which neutrinos, either neutrinos or anti-neutrinos, are affected in the matter. Another important factor to consider is the effect of Δm_{12} and its interference effects though this is known to be small. The results of the MC calculation indicates that Δm_{12} and its interference effects are less than 2% in the low energy region and less than 1% in the multi-GeV region. Therefore we have neglected the effect of Δm_{12} and treated Δm_{23}, θ_{23} and θ_{13} as independent parameters (we are currently working on the analysis including the effect on Δm_{12}). The best fit paramet ers for Δm_{23} and θ_{23} are consistent with the results from 2 flavor analysis and we set the limit, $\sin^2 \theta_{13} < 0.16$ at $\Delta m_{23} = 2 \times 10^{-3}$ eV^2.

2.7. L/E analysis [17]

After the discovery of the neutrino oscillation in 1998, one of the remaining tasks was the direct observation of the oscillatory behavior in the L/E plot that strengthens the evidence of neutrino oscillation. Also the observation of the sinusoidal pattern completely distinguishes the neutrino oscillation hypothesis from other hypotheses like neutrino decay or de-coherence which have marginally fit to SK zenith angle distributions.

In the analysis, we need to obtain L, the flight length and the neutrino energy, E_ν. A Monte Carlo calculation was used to make an expected visible energy distribution for a given neutrino energy. This MC information was then inversely used to determine the neutrino energy, E_ν, from the observed energy, E_{obs}. The flight length, L, is determined by a much more straightforward way where the direction of the vector sum of the charged particles was extended back to the atmosphere.

It is difficult to observe the dip for all the atmospheric neutrino data without any selections because of the relatively poor determination of L and E. It is essential to choose events with good L/E resolution.

Before we make event selections, we have improved the energy assignments of PC events which play an important role in this analysis. The data was categorized as OD stopping and OD through-going depending on the charge observed in the Outer Detector (OD). A different energy assignment for those events was made. This process has improved the energy resolution of the PC events. The fiducial volume was also expanded from 22.5 kt to 26.4 kt in order to include more high energy muon events. We then apply cuts on the resolution, $\Delta(L/E) < 70\%$. The resultant data sample consist of 2726 events, about 1/5 of the full data set. The resolution cuts have basically removed horizontally going events because of the large uncertainty in determining L, and low energy events because of the large scattering angle. The resultant L/E plot is shown in Fig. 4. The first dip has clearly been observed. This observation has strengthened the evidence of neutrino oscillation and the observed dip cannot be explained by other hypotheses. The data have excluded the other hypotheses with 3.4 σ for decay and 3.8 σ for de-coherence.

Fits are also made to obtain the allowed parameter regions. The results are shown in Fig. 3 together with the results of the standard two flavor analysis. The allowed region at 90% C.L. is $1.9 \times 10^{-3} < \Delta m^2 < 3.0 \times 10^{-3}$ eV^2, $\sin^2 2\theta > 0.90$ which is consistent with the results from the standard zenith angle analysis, but gives a much restricted region for Δm^2 even with 1/5 of the entire data. This demonstrates the usefulness of this analysis.

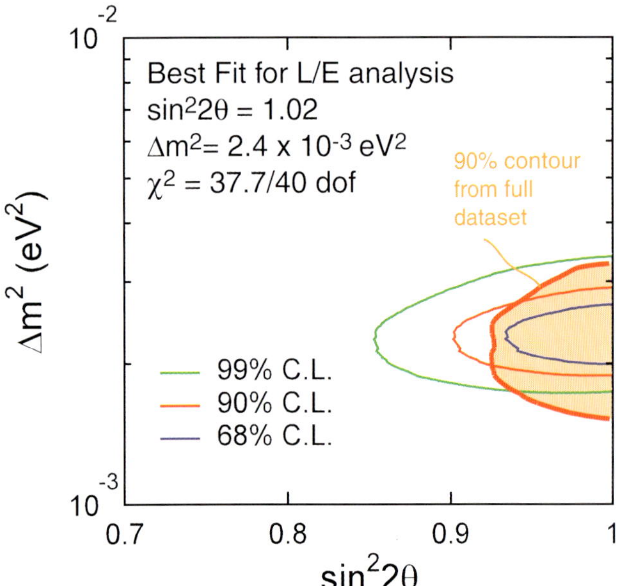

Fig. 3. The painted contour shows the allowed parameter regions for the standard two flavor analysis at 90% C.L. The narrower allowed regions shown by lines are the results for the L/E analysis at 68, 90 and 99% C.L. from inside to outside. Tighter range for Δm^2 comparing to the standard zenith angle analysis is obtained.

3. Solar neutrinos

The original aim of the SK solar neutrino measurement is to look for the flux independent evidence of the neutrino oscillations, namely the day-night flux difference, spectrum distortion and seasonal variations. SK could not observe such evidence significantly for the last five years of observation.

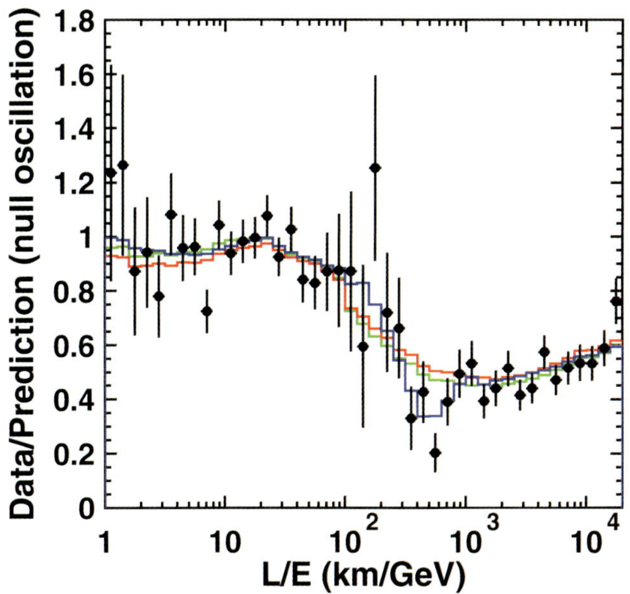

Fig. 4. L/E distribution after the event selection. The first dip is clearly seen. The line is for the best fit oscillation parameters. The lines for decay and de-coherence are also shown, which decrease monotonically as L/E increases.

3.1. Solar neutrino oscillation

Non-observation of the day-night flux difference and the energy spectrum distortion, however, has placed a strong constraint on the neutrino oscillation parameters. By using the data accumulated for 1026 days in 2000, SK has first excluded the Small Mixing Angle solutions at 95% C.L. and indicated that the mixing angle must be large [4].

The results from SK based on the 1258 days of data published in June, 2001 [5][6] and the first charged current results reported by SNO on the same day [7], together, have provided convincing evidence for solar neutrino oscillation. SK measures the solar neutrino flux through the $v + e \rightarrow v + e$ scattering which is sensitive not only to v_e, but also v_μ and v_τ although the sensitivity to v_μ and v_τ is reduced to 15% of that of v_e. Comparing the measured flux value of SK, $2.32 \pm 0.03(\text{stat.}) +0.08/-0.07(\text{syst.}) \times 10^6 \, \text{cm}^{-2}\,\text{s}^{-1}$, and SNO CC flux value, $1.75 \pm 0.07(\text{stat.}) +0.12/-0.11(\text{syst.}) \pm 0.05(\text{theor.}) \times 10^6 \, \text{cm}^{-2}\,\text{s}^{-1}$, it is evident that there are non electron neutrino components in the SK data.

3.2. SK-I results on solar neutrinos

The 1496 live-days of solar neutrino data taken between May-31st, 1996 and July-15th, 2001 are used for the final SK-I analysis. A total of 22,400 solar neutrino events corresponding to 14.5 events per day are observed. The measured flux is $2.35 \pm 0.02(\text{stat.}) \pm 0.08(\text{syst.}) \times 10^6 \, \text{cm}^{-2}\,\text{s}^{-1}$ [18]. Since the predicted ^{8}B solar neutrino flux from the standard solar model (SSM) of BP2004 [19] is $5.82(1 \pm 0.23) \times 10^6 \, \text{cm}^{-2}\,\text{s}^{-1}$, the flux ratio of the data to the prediction is $0.406 \pm 0.005(\text{stat.}) +0.014/-0.013(\text{syst.})$.

The enhancement of the solar neutrino flux at night time is expected to be large for oscillation parameters around $\Delta m^2 \sim 10^{-6} \sim 10^{-5} \, \text{eV}^2$ due to the earth's regeneration effect. The measured difference is $(\text{night–day})/2(\text{day} + \text{night}) = -0.021 \pm 0.020(\text{stat.}) +0.013/-0.012(\text{syst.})$, which is consistent to be zero and thus prefers large mass difference above $\Delta m^2 >$ a few $\times 10^{-5} \, \text{eV}^2$. The most systematic error comes from the up-down asymmetry of the energy scale error. More statistics and more precise energy calibration is needed for future improvements.

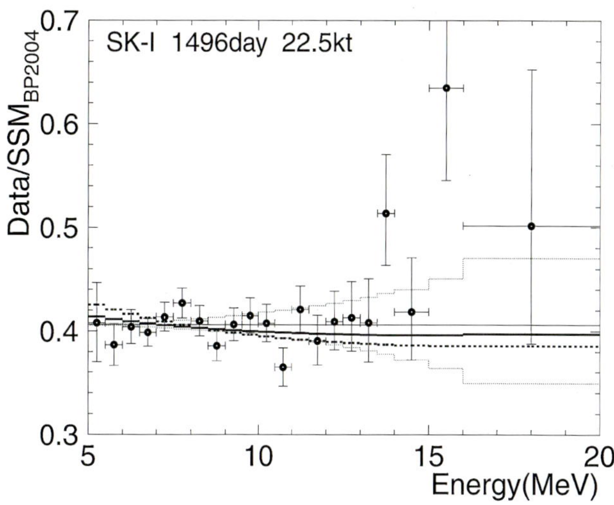

Fig. 5. The ratio of the measured recoil electron spectrum to the prediction. No significant distortions are seen though there is a small excess in high energy data. The solid line and dotted line show the case for $\Delta m^2 = 6.3 \times 10^{-5} \, \text{eV}^2$ and $\tan^2\theta = 0.55$, and $\Delta m^2 = 7.2 \times 10^{-5} \, \text{eV}^2$ and $\tan^2\theta = 0.38$, respectively.

The energy spectrum is also useful for constraining the oscillation parameters. The measured recoil electron spectrum is consistent with the convolution of the predicted ^{8}B spectrum, the interaction of solar neutrinos on electrons and the detector responses. The χ^2 value for non-distortion is 20.2 for 20 degrees of freedom (44.3% C.L.). But there are some excesses in the higher energy region as shown in Fig. 5. The excesses may come from a possible contribution from hep neutrinos. A 90% C.L. hep neutrino flux limit of $73 \times 10^3 \, \text{cm}^{-2}$ is obtained by using the data between 18 and 21 MeV. The upper bound is 7.9 times the SSM prediction.

Although SK did not observe large day-night effects and spectrum distortions, we are able to reject model independently those regions of the oscillation parameters which reveal large effect. We make full use of the time information [20]. The excluded region only by using the zenith angle spectrum is shown in Fig. 6 as a blank region. So-called small mixing angle solutions and the lower part of the LMA and most of the LOW solutions are

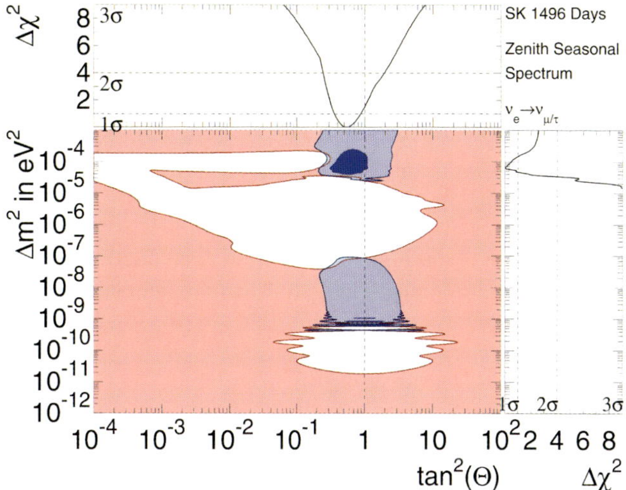

Fig. 6. Allowed oscillation parameter region obtained by the Super-Kamiokande data. The white region is excluded by the non-observation of day-night flux difference and spectrum distortions. The thick parts are the allowed regions obtained by constraining the absolute flux to the prediction of BP2004.

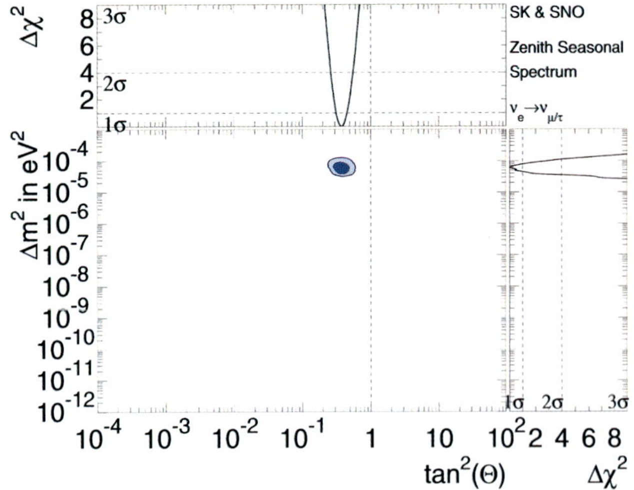

Fig. 7. Allowed parameter region obtained by using only Super-K and SNO. The flux constraint from the solar model calculation is not used.

excluded at more than 99% C.L. By constraining the SK data with the absolute flux predicted from the standard solar model, large mixing angle regions are selected. The allowed regions around LMA in Fig. 6 show 68% and 95% C.L.

The uncertainty of BP2004 on the ^{8}B solar neutrino flux is 23% and the SNO neutral current measurement gives the flux with an accuracy of 7.7%. There is no reason not to use SNO Neutral Current measurement as an absolute flux. The combined analysis of SK and SNO is, therefore, the most accurate method which is independent on solar model calculations. Fig. 7 shows the allowed region obtained by the SK and SNO data.

4. Super-Kamiokande II and III

One year after the tragic accident in November, 2001, SK was reconstructed (SK-II) with 47% of the PMTs ($\sim$5,200) uniformly distributed in the detector. All the inner PMTs now have plastic shells which prevent against future similar accidents. The outer detector has been fully reconstructed. The detector calibrations have been performed extensively as before. For the initial 311.5 days of the atmospheric neutrino data, the event rate of the FC atmospheric neutrino events was 8.22±0.16 events/day and that of the PC events was 0.51±0.04 events/day. Those rates are consistent with SK-I. Based on the first 325 days of solar neutrino data taken with a minimum energy threshold of 8 MeV, a flux of 2.38±0.09(stat) was obtained. Although the energy threshold is higher than SK-I, the measured flux is consistent with the SK-I result.

We are preparing about 6,000 PMTs to be used for the full restoration of SK. The restoration will begin in November, 2005 and will last for about 5 months. Water will be filled from April, 2006, and by the summer, 2006, SK-III will re-start and be taking data.

5. Future of Super-Kamiokande

5.1. *Atmospheric neutrinos*

A most important subject in future is to look for an indication of the electron appearance in a few GeV region where the matter effect enhances the appearance through θ_{13}, which eventually enables us to determine θ_{13}. SK will probably run 20 years. The MC simulation predicts that if $\sin^2\theta_{13}$ is larger than 0.01 and if $\sin^2\theta_{23}$ around 0.6, then the positive observation with a confidence level larger than 3σ is possible.

5.2. *Solar neutrinos*

A crucial issue is to look for the up-tern of the energy spectrum in low energy, which is a characteristic of the LMA solution. If the energy threshold reaches down to 4 MeV, the effect becomes as large as $\sim$10%. Although the current energy threshold is 5 MeV and we are already able to identify solar neutrino signals above 4.5 MeV. It, however, requires further reduction of the backgrounds to achieve 4.0 MeV. It is a challenge, but not impossible. One of the sources of the backgrounds in the low energy region is Radon emanated from the PMT glass. Fortunately, SK-III will have acrylic covers over all the PMT glass, then we expect that the Radon from the water remains within the cover and never flows into the detector fiducial volume and thus the reduction of the background will be possible. After 5 years of measurement, we are able to see the spectrum up-turn at a confidence level larger than 3σ, if the solution of the LMA is either $\Delta m^2 > 7.2 \times 10^{-5}$ eV^2 or $\tan^2\theta < 0.38$ and if we are able to reduce the current background level to 1/3 and make the systematic error on the energy be 1/2.

References

1. Fukuda, S. *et al.*, Nucl. Instr. Meth. **A501**, 418 (2003).
2. SK Collaboration consists of 138 physicists from 17 Japanese, 15 US, 1 Polish and 3 Korean Institutions.
3. Fukuda, Y. *et al.*, Phys. Rev. Lett. **81**, 1562 (1998).
4. Suzuki, Y., in Proc. Neutrino 2000, Sudbury, Canada, June, (2002).
5. Fukuda, S. *et al.*, Phys. Rev. Lett. **86**, 5651 (2001).
6. Fukuda, S. *et al.*, Phys. Rev. lett. **86**, 5656 (2001).
7. Ahmad, Q. R. *et al.*, Phys. Rev. Lett. **87**, 071301 (2001).
8. Ahn, S. H. *et al.*, Phys. Lett. **B511**, 178 (2001).
9. Ashie, Y. *et al.*, Phys. Rev. Lett. **93**, 101801 (2004).
10. Ahn, M. H. *et al.*, Phys. Rev. Lett. **90**, 041801 (2003).
11. Nakahata, M. *et al.*, Nucl. Instr. Meth. **A421**, 113 (1999).
12. Blaufuss, E. *et al.*, Nucl. Instr. Meth. **A458**, 638 (2001).
13. Honda, M., Kajita, T., Kasahara, K. and Midorikawa, S., (2004); astro-ph/0404457.
14. Barr, G. D., Gaisser, T. K., Lipari, P., Robbins, S. and Stanev, T., (2004); astro-ph/0305208.
15. Battistone, G., Ferrari, A., Montauli, T. and Sala, P. R., (2003); hep-ph/0305208Barr.
16. The Super-Kamiokande Collaboration, to be published.
17. Ashie, Y. *et al.*, Phys. Rev. Lett. **93**, 101801 (2004).
18. Fukuda, S. *et al.*, Phys. Lett. **B539**, 179 (2002).
19. Bahcall, J. N. and Pinsonneault, M. H., Phys. Rev. Lett. **92**, 121301 (2004).
20. Smy, M. B. *et al.*, Phys. Rev. **D69**, 01104 (2004).

Physica Scripta. Vol. T121, 29–32, 2005

Sudbury Neutrino Observatory Results

A. B. McDonald[*]

Physics Department, Queen's University, Kingston, Ontario, Canada K7L 3N6, For the SNO Collaboration[1]

Received December 2, 2004; accepted in revised form December 24, 2004

PACS numbers: 14.60.Pq, 13.15.+g, 96.60.Vg, 26.65.+t

Abstract

The Sudbury Neutrino Observatory uses 1000 tonnes of heavy water in an ultra-clean Cherenkov detector situated 2 km underground in Ontario, Canada to study neutrinos from the Sun and other astrophysical sources. The Charged Current (CC) reaction on deuterium is sensitive only to electron neutrinos whereas the Neutral Current (NC) is equally sensitive to all active neutrino types. By measuring the flux of neutrinos from ^{8}B decay in the Sun with the CC and NC reactions it has been possible to establish clearly, through an appearance measurement, that electron neutrinos change to other active neutrino types, properties that are beyond the Standard Model of elementary particles. The observed total flux of active neutrinos agrees well with solar model flux calculations for ^{8}B. This provides a clear answer to the "Solar Neutrino Problem". When these results are combined with other measurements, the oscillation of massive neutrinos is strongly defined as the primary mechanism for flavor change and oscillation parameters are well constrained.

1. Introduction

The Sudbury Neutrino Observatory (SNO) [1] was designed to make separate measurements of the flux of electron neutrinos from the Sun and the flux of all active neutrino types. By a direct comparison of these two fluxes it is possible to be definitive about whether solar neutrinos are changing their types before reaching the Earth. By situating the detector 2 km underground in an active nickel mine near Sudbury, Ontario, Canada and taking extreme care with radioactivity in the construction and operation of the experiment, it has been possible to make these neutrino flux measurements with a very small contribution from background events. The results of these measurements [2, 3, 4, 5] show clearly that about 2/3 of the electron neutrinos from ^{8}B decay in the Sun change into other active neutrino types in transit to the Earth. The total ^{8}B neutrino flux observed is in very good agreement with calculated [6] fluxes. Including the results of other measurements of solar [7–9] and reactor [10] neutrinos defines the oscillation of massive neutrinos as the dominant mechanism for flavor change, with mixing parameters for active neutrinos restricted to a very narrow range and oscillation to sterile neutrinos strongly disfavored. The present paper will describe results to date from the Sudbury Neutrino Observatory and indicate the physics potential for measurements planned in future.

2. Resolving the solar neutrino "Problem"

Since the pioneering measurements of Ray Davis and collaborators in the 1960's, measurements [7–9] of solar neutrinos using reactions that were solely or predominantly sensitive to electron neutrinos had observed fluxes that were factors of two or three smaller than predicted by solar model predictions [6]. This discrepancy was widely studied and came to be known as the "Solar Neutrino Problem". The experiments included radiochemical measurements with Chlorine [7] and Gallium [8] and measurements of the elastic scattering from electrons in light water Cherenkov detectors [9]. These experiments were sensitive to fluxes from different neutrino source reactions in the Sun. Solar models provided very good agreement with other properties of the Sun. However, it was not known whether the neutrino flux discrepancy was due to limitations in the solar models or unknown neutrino properties. Measurements that could distinguish the two possibilities were required.

The leading explanation arising from new neutrino physics was flavor change through the oscillation of massive neutrinos. The formalism for this process was put forward by Maki, Nakagawa, Sakata and Pontecorvo (MNSP) [11] and involved physics beyond the Standard Model of elementary particles, namely neutrino flavor change and non-zero rest mass. This theory was augmented by the possibility of enhanced oscillation through the matter-enhancement of neutrino oscillation in the Sun or Earth [12]. This process, proposed by Mikheyev and Smirnov, building on work by Wolfenstein, provided the possibility of strong enhancement, distortion of the energy spectra of oscillated neutrinos and differences in daytime and nighttime fluxes of neutrinos arising from the matter-enhancement of neutrinos passing through the Earth. These effects were sought by the experiments discussed above with no finite effects observed, even with the high sensitivity provided by the Super-Kamiokande experiment.

The Sudbury Neutrino Observatory is designed to make direct observations of solar neutrino flavor change through the following reactions in a heavy water Cherenkov detector:

$$v_e + d \rightarrow p + p + e^-(CC),$$
$$v_x + d \rightarrow p + n + v_x(NC),$$
$$v_x + e^- \rightarrow v_x + e^-(ES).$$

The charged current (CC) reaction is sensitive only to electron neutrinos, while the neutral current (NC) reaction is sensitive to all active neutrino flavors ($x = $ e, μ, τ) above the energy threshold of 2.2 MeV. The elastic scattering (ES) reaction is sensitive to all flavors as well, but with reduced sensitivity to v_μ and v_τ. The Cherenkov light produced by the electrons in the final state is used to observe the CC and ES reactions. The NC reaction is observed through the detection of the neutron in the final state of the reaction.

By comparing the flux observed with the CC reaction (electron neutrinos) with the flux observed with the NC reaction (all active neutrino types), it is possible to make a direct measurement of neutrino flavor change. A less sensitive measurement can be made through a comparison of the CC reaction with the ES reaction. Because all reactions are sensitive to the flux of neutrinos from ^{8}B decay in the Sun, the results are not sensitive to details of solar models. In fact, the flux of all neutrino types tests solar model calculations for ^{8}B neutrino flux, independent of neutrino flavor change to active neutrinos. The SNO experiment

[*]e-mail: art@snolab.ca
[1]SNO Collaboration members are listed on references [1–5].

has provided direct evidence of neutrino flavor change via these flux comparisons, and through the NC reaction, provided the first "appearance" observation of oscillated neutrino flavors.

3. The Sudbury Neutrino Observatory

The SNO detector [1] uses 1000 tonnes of ultra pure heavy water to observe Cherenkov light from the CC and ES neutrino reactions and to observe the production of neutrons by the NC reaction through three subsidiary techniques. The detector consists of about 9500 20-cm-diameter photomultipliers supported on an 18 meter diameter geodesic sphere viewing heavy water contained in a 12-meter diameter, 5-cm thick acrylic sphere. These are suspended in ultra-pure light water filling a 22-meter diameter barrel-shaped cavity coated with a polyurethane liner impermeable to water and to radon from the surrounding rock. The entire laboratory was constructed under Class 3000 clean-room conditions. All workers have showered and worn lint-free clothing during construction and operation of the experiment. The detector materials were carefully chosen for low radioactivity and extensive water purification and assay systems have been used to measure very low levels of U and Th. These radioactive backgrounds are of particular interest because the 2.4 to 2.6 MeV gamma rays from the decay chains of these naturally-occurring isotopes can photo-disintegrate deuterium. Therefore these isotopes produce a neutron background for the NC reaction, as well as low energy Cherenkov backgrounds for the CC and ES reactions. With the extreme care applied to cleanliness and material selection, it was possible to restrict the backgrounds from radioactivity to less than about 15% of the neutrino signals in all cases and to measure those backgrounds with good accuracy [1, 13]. Details of instrumental background reduction, calibration, signal extraction and systematic uncertainties are contained in references [2 to 5].

The SNO experimental plan involves three phases with different techniques for the detection of neutrons from the NC reaction. In the first phase, using pure heavy water, the Cherenkov light from the 6.25 MeV gammas produced when neutrons are captured on deuterium was used to detect the neutrons. For the second phase, about 2 tons of salt was added to the heavy water. Neutron detection efficiency was enhanced through the higher capture cross section on Cl, and the fact that the 8.6 MeV cascade of gammas produces more Cherenkov light. This light also produces a more isotropic distribution of hit PMT's than the CC events. The fluxes from the CC and NC reactions are extracted on a statistical basis with no constraint on the shape of the CC energy spectrum. For the third phase, the salt was removed and an array of ^{3}He-filled proportional counters has been installed to provide neutron detection independent of the PMT array.

4. SNO measurements to date

In 2001, the SNO collaboration reported a measurement [2] of the flux of electron neutrinos from ^{8}B decay in the Sun of $\Phi_{CC} = 1.59 \pm 0.07(\text{stat})^{+0.12}_{-0.11}(\text{syst}) \pm 0.05(\text{theor})$. This result is significantly lower than the measurement by the SuperKamiokande experiment [9] using the elastic scattering of neutrinos from electrons $\Phi_{ES} = 2.32 \pm 0.03(\text{stat})^{+0.08}_{-0.07}(\text{syst})$. These fluxes and those following in this section are quoted in units of $10^6\,\text{cm}^{-2}\,\text{s}^{-1}$. The difference in these fluxes is 3.3 σ from zero, providing evidence for a non-electron active component in the solar neutrino flux. In 2002, more accurate measurements were

reported by SNO for the pure heavy water phase [3, 4], including data from the CC, ES and NC reactions. The non-v_e active component was found to be $\Phi_{\mu\tau} = 3.41 \pm 0.45(\text{stat})^{+0.48}_{-0.45}(\text{syst})$, 5.3 sigma greater than zero, providing strong evidence for solar v_e flavor transformation to active neutrinos.

For CC events, assuming an undistorted ^{8}B spectrum, the night minus day rate was found to be $14.0\% \pm 6.3\%(\text{stat})^{+1.5}_{-1.4}$ (syst) of the average rate. If the total flux of active neutrinos was additionally constrained to have no asymmetry, the v_e asymmetry was found to be $7.0\% \pm 4.9\%(\text{stat})^{+1.4}_{-1.2}(\text{syst})$.

With the addition of salt it was possible to break the strong correlation between the CC and NC event types because the pattern of light on the PMT's is different for the two reactions. An event angular distribution parameter β_{14} (defined in reference [5]) was used to separate NC and CC events on a statistical basis. The data from the salt phase [5] are shown in Figure 1, together with the best fit to the data of the NC (8.6 MeV gamma) shape, the CC and ES reactions and a small background component. The fluxes determined from this fit are:

$$\Phi_{CC} = 1.59^{+0.08}_{-0.07}(\text{statistical})^{+0.06}_{-0.08}(\text{systematic}),$$

$$\Phi_{NC} = 5.21 \pm 0.27(\text{stat}) \pm 0.38(\text{syst}),$$

$$\Phi_{ES} = 2.21^{+0.31}_{-0.26}(\text{stat}) \pm 0.10(\text{syst}).$$

The fluxes show clearly that about two-thirds of the electron neutrinos have changed their flavor to other active neutrino types, violating a hypothesis test for no flavor change at greater than 7 σ. The observed total flux of active neutrinos (NC) is in excellent agreement with the flux of ^{8}B neutrinos obtained [6] from solar models: $\Phi_{SSM} = 5.82 \pm 1.3;\ 5.31 \pm 0.6$. Oscillation purely to sterile neutrinos is strongly disfavored. By comparison of the active total flux of ^{8}B solar neutrinos observed by the NC reaction with the ^{8}B flux calculated with solar models, restrictions are placed on sub-dominant oscillations to sterile neutrinos.

The data for solar neutrinos have been analyzed by many authors in terms of the MNSP [11] mixing matrix wherein flavor eigenstates for the three active neutrino types ($l = $ e, μ, τ) are related to mass eigenstates (i) via the matrix $\mathbf{U}_{li}$:

$$|v_l\rangle = \sum \mathbf{U}_{li} |v_i\rangle$$

The MNSP matrix $\mathbf{U}_{li}$ can be written for three active neutrino types as:

$$\mathbf{U}_{li} = \begin{pmatrix} c_{12} & s_{12} & 0 \\ -s_{12} & c_{12} & 0 \\ 0 & 0 & 1 \end{pmatrix} \cdot \begin{pmatrix} 1 & 0 & 0 \\ 0 & c_{23} & s_{23} \\ 0 & -s_{23} & c_{23} \end{pmatrix} \cdot \begin{pmatrix} 1 & 0 & 0 \\ 0 & 1 & 0 \\ 0 & 0 & e^{-i\delta} \end{pmatrix} \cdot \begin{pmatrix} c_{13} & 0 & s_{13} \\ 0 & 1 & 0 \\ -s_{13} & 0 & c_{13} \end{pmatrix}$$

where $c_{ij} = \cos\theta_{ij}$, and $s_{ij} = \sin\theta_{ij}$.

For non-degenerate mass eigenstates, and for small θ_{13}, oscillations of solar neutrinos are dominated by the first sub-matrix involving θ_{12} and this approximation is often used for analyses. The second sub-matrix dominates the oscillation of atmospheric neutrinos, the third sub-matrix involves the CP violating angle δ and the fourth sub-matrix is tested by reactor and accelerator neutrino measurements. For oscillations in vacuum in the two-neutrino mixing approximation, the survival probability for solar neutrinos with energy E, that have traveled a distance L is:

$$P(v_e \to v_e) = 1 - \sin^2 2\theta_{12} \sin^2 \left(1.27 \frac{\Delta m_{12}^2 L}{E} \right).$$

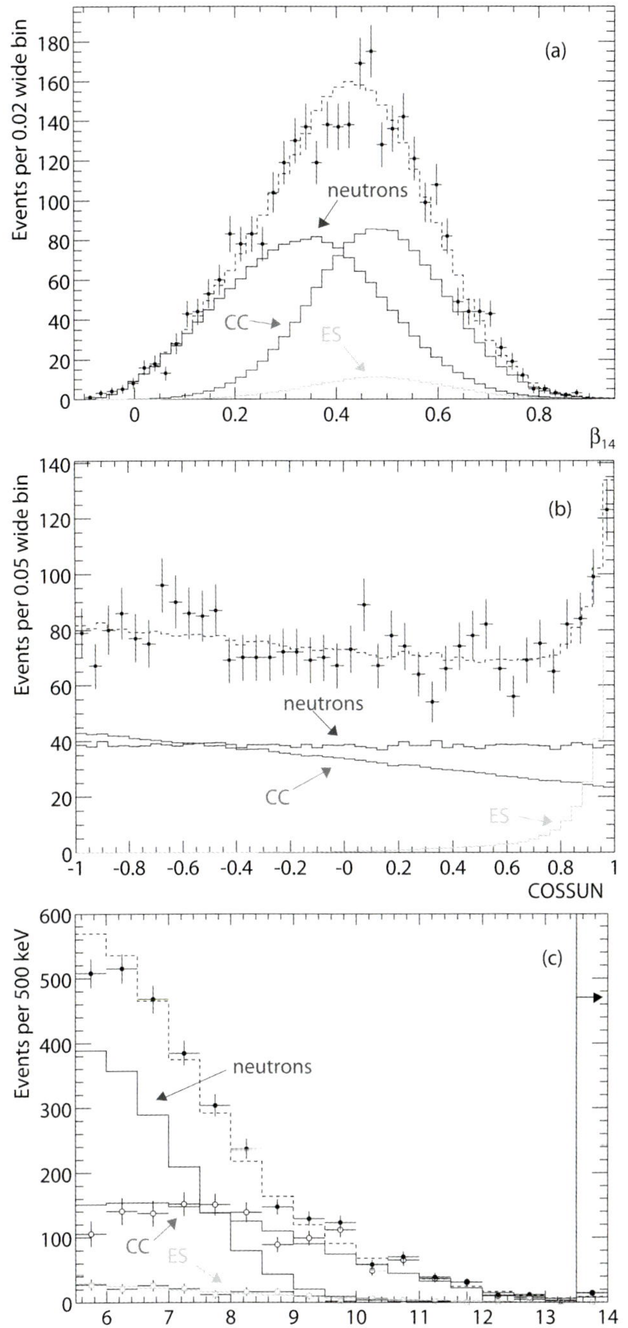

Fig. 1. Distribution of (a) isotropy parameter β_{14}, (b) $\cos\theta_{sun}$ and (c) Recoil electron kinetic energy, for the selected events in the fiducial volume for the salt phase of SNO. The CC and ES spectra are extracted from the data using β_{14} and $\cos\theta_{sun}$ distributions in each energy bin. Also shown are the Monte Carlo predictions for CC, ES, NC+ internal and external-source neutron events, all scaled to the fit results. The dashed lines represent the summed components. All distributions are for events with effective energy >5.5 MeV and Radius >550 cm. Differential systematics are not shown.

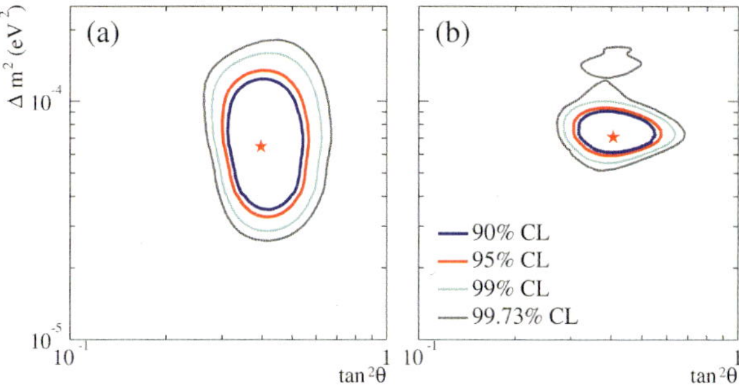

Fig. 2. Global neutrino oscillation contours for two parameter mixing. The four contour lines correspond to 90, 95, 99, 99.73% confidence limits. (a) Solar global: D_2O day and night spectra, salt CC, NC, ES fluxes, Super-Kamiokande spectral and day-night data, Cl, Ga fluxes. The best-fit point is $\Delta m^2 = 6.5 \times 10^{-5}$ eV^2, $\tan^2\theta_{12} = 0.40$. (b) Solar global + KamLAND. The best-fit point is $\Delta m^2 = 6.7 \times 10^{-5}$ eV^2, $\tan^2\theta_{12} = 0.41$.

The results for an analysis of all solar neutrino data for ν_1, ν_2 mixing [5], following the SNO salt data is shown in Figure 2 (a). The mixing is found to be non-maximal (mixing angle less than $\pi/4$) with a confidence level corresponding to 5.5 standard deviations. The LMA region, involving matter enhancement of the oscillation via the MSW effect is the preferred solution and that provides a determination of the sign of Δm_{12}. It is found that m_2 is greater than m_1.

Following the SNO evidence for neutrino flavor change, the KAMLAND collaboration reported their first results [10] showing the oscillation of reactor neutrinos with parameters that overlap with the solar neutrino results for part of their allowed region. Figure 2 (b) shows the allowed region when the initial Kamland data is included along with the solar data. The agreement between the solar neutrino oscillation parameters and the reactor anti-neutrino oscillation parameters confirms MNSP mass oscillations as the dominant process for flavor change. The agreement provides a confirmation of CPT for neutrinos. If full CPT invariance is assumed, the agreement confirms matter enhancement for the solar neutrinos and restricts the possibility of flavor change arising from Resonant Spin Flavor Precession [14] in the Sun. Sub-dominant transitions to sterile neutrinos and other flavor-changing mechanisms are also restricted significantly [15]. Possible processes such as neutrino decay, decoherence, violation of the equivalence principle, violation of Lorentz Invariance and flavor-changing neutral currents are strongly restricted by the observations of flavor change in solar, reactor and atmospheric neutrino experiments and by limits set in other measurements.

The principal restriction on the θ_{12} axis in Figure 2 (b) comes from the solar (mainly SNO) measurements and for the Δm^2 axis, from the KamLAND results. This arises because the ^{8}B solar neutrinos in the LMA region are undergoing a resonant MSW transition and emerge from the Sun in essentially a pure ν_2 state. Then, for small θ_{13}, the electron neutrino survival probability (CC/NC fluxes for SNO) is almost exactly $\sin^2\theta_{12}$. On the other hand, the vacuum oscillation probability for the reactor anti-neutrinos $P(\nu_e \to \nu_e)$ is very sensitive to Δm^2.

In addition to the measurements of solar neutrinos, the SNO collaboration has reported limits on the fluxes of electron anti-neutrinos from the Sun [16] and limits on "invisible" proton decays and neutron decays that are respectively 10 and 400 times lower than previous measurements [17].

If the neutrinos travel in a region of high electron density, such as in regions within the Sun or Earth, the interaction of neutrinos with electrons can produce matter enhancement of the oscillation process, via the MSW effect [12]. The MSW process produces adjustments to the effective values for θ_{12} and Δm_{12}^2, depending on the electron density. The sign of Δm_{12}^2 can be determined if matter interactions are substantial and distortions of the energy spectrum can occur. Interactions in the Earth can produce differences in the measured fluxes at detectors for day and night time periods.

5. Future SNO and SNOLAB measurements

The SNO experiment has entered its third phase with an array of ^{3}He-filled proportional detectors installed in the heavy water. In this phase, the NC and CC reactions will be detected independently, removing the correlations between them and providing a more accurate measure of the NC flux. As the SNO data provide an almost direct determination of the θ_{12} parameter, the new measurements will improve the accuracy for this mixing parameter. In addition, the combination of new SNO measurements with further KamLAND reactor anti-neutrino measurements will further restrict the range of θ_{13} for low values of Δm_{23}^2.

By 2007, the sensitivity of the SNO experiment will be limited primarily by systematic uncertainties and it is planned to terminate the heavy water measurements. Discussions are underway concerning the possible replacement of the heavy water with liquid scintillator for future measurements of lower energy solar neutrinos, geo-neutrinos, reactor neutrinos and possibly double beta decay. The laboratory 2 km underground is being expanded to create a long term international facility known as SNOLAB, expected to be finished in 2007. This will provide the deepest available site in future for a number of additional underground measurements. Letters of Interest have been received from over 15 international experiments proposing a very interesting future research program at SNOLAB, including measurements of solar and supernova neutrinos, dark matter and double beta decay.

Acknowledgements

This research was supported by: Canada: NSERC, Industry Canada, National Research Council, Northern Ontario Heritage Fund, Inco, AECL, Ontario Power Generation, HPCVL, CFI; US: Dept. of Energy; UK: PPARC. We thank the SNO technical staff for their strong contributions.

References

1. SNO Collaboration, Boger, J. *et al.*, Nucl. Instrum. Meth. **A449**, 172 (2000).
2. SNO Collaboration, Ahmad, Q. R. *et al.*, Phys. Rev. Lett. **87**, 071301 (2001).
3. SNO Collaboration, Ahmad, Q. R. *et al.*, Phys. Rev. Lett. **89**, 011301 (2002).
4. SNO Collaboration, Ahmad, Q. R. *et al.*, Phys. Rev. Lett. **89**, 011302 (2002).
5. SNO Collaboration, Ahmad, S. N. *et al.*, Phys. Rev. Lett. **92**, 181301 (2004).
6. Bahcall, J. N. and Pinsonneault, M. H., Phys. Rev. Lett. (2004), in press, astro-ph/0402114; J. Turck. Chieze *et al.*, Phys. Rev. Lett. **93**, 21102 (2004).
7. Davis, R. *et al.*, Phys. Rev. Lett. **14**, 20 (1968); Cleveland, B. T. *et al.*, Astrophys. J. **496**, 505 (1998).
8. Gavrin, V. N. *et al.* Nucl. Phys. B (Proc. Suppl.) **118**, 39 (2003); Hampel, W. *et al.*, Phys. Lett. **B 447**, 127 (1999); Kirsten, T. Proc. Neutrino 2002 Conference, Munich, to be published (2002).
9. Fukuda, Y. *et al.*, Phys. Rev. **D 44**, 1683 (1996); Fukuda, S. *et al.*, Phys. Rev. Lett. **86**, 5651 (2001); Fukuda, S. *et al.*, Phys. Lett. **B 539**, 179 (2002); Smy, M. for the Super-Kamiokande collaboration, Nucl. Phys. B (Proc. Suppl.) **118**, 25 (2003).
10. Eguchi, E. *et al.*, Phys. Rev. Lett. **90**, 021802 (2003).
11. Maki, Z., Nakagawa, N. and Sakata, S., Prog. Theor. Phys. **28**, 870 (1962); Pontecorvo, B., J. Expt. Theor. Phys. **33**, 549 (1957); J. Expt. Theor. Phys. **34**, 247 (1958); Gribov, V. and Pontecorvo, B., Phys. Lett. **B28**, 493 (1969).
12. Mikheyev, S. P. and Smirnov, A. Y. in "Massive Neutrinos in Astrophysics and in Particle Physics", Proc. Moriond Workshop, (edited by O. Fackler and J. Tran Thanh Van), (Editions Frontieres, Gif-sur-Yvette (1986), 335; Mikheyev, S. P. and Smirnov, A. Yu., Sov. J. Nucl. Phys. **42**, 1441 (1985); Wolfenstein, L., Phys. Rev. **D 17**, 2369 (1978).
13. Blevis, I. *et al.*, Nucl. Instrum. and Meth. **A517**, 139 (2003); Anderson, T. C. *et al.*, Nucl. Instrum. Meth. **A501**, 399 (2003).
14. Miranda, O. G. *et al.*, Nucl. Phys. **B 595**, 360 (2001); Chauhan, B. C. and Pulido, J. Phys. Lett. **66**, 053006 (2002); Akhmedov, E. Kh. and Pulido, J., Phys. Lett. **B 553**, 7 (2003).
15. Bahcall, J. N., Gonzales-Garcia, C. and Pena-Garay, C., J. High. Energy Phys. **030**, 009 (2003).
16. SNO Collaboration: Aharmim, B. *et al.*, submitted to Phys. Rev. **D70**, 093014 (2004).
17. SNO Collaboration: Ahmed, S. N. *et al.*, Phys. Rev. Lett. **92**, 102004 (2004).

Physica Scripta. Vol. T121, 33–38, 2005

Results from KamLAND Reactor Neutrino Detection

Atsuto Suzuki* for the KamLAND Collaboration

Research Center for Neutrino Science, Tohoku University, Aramaki, Aoba, Sendai, 980-8578, Japan

Received February 14, 2005; accepted in revised form March 9, 2005

PACS numbers: 14.60.Pq, 26.65.+t, 28.50.Hw, 91.65.Dt

Abstract

Data corresponding to a KamLAND detector exposure of 766.3 ton-years to reactor neutrinos was used to search for neutrino oscillations. Based on this data-sample, KamLAND demonstrates the reactor $\bar{\nu}_e$ disappearance at the 99.998% significance and finds the reactor $\bar{\nu}_e$ spectrum distortion at the 99.6% significance. The reactor neutrino anomaly defined as the flux disappearance and spectrum shape distortion is confirmed at the 99.999995% significance. A two-flavor neutrino oscillation analysis including the rate and energy spectrum yields a best fit at $\Delta m^2 = 7.9 \times 10^{-5}\,\mathrm{eV}^2$ and $\sin^2 2\theta = 0.86$. A global analysis of data from KamLAND and solar neutrino experiments yields $\Delta m^2 = 7.9^{+0.6}_{-0.5} \times 10^{-5}\,\mathrm{eV}^2$ and $\tan^2 \theta = 0.40^{+0.10}_{-0.07}$. The present result gives the most precise determination of Δm^2 to date. To test the goodness of the oscillation hypothesis, the ratio of observed anti-neutrino spectrum to the expected for no-oscillation as a function of L_0/E is compared with the predictions which give neutrino-flavor changes. The neutrino oscillation is much favored with more than 99% C.L., while the decay and decoherence are excluded at the 95% and 94% C.L.

1. Introduction

The main objectives of KamLAND (<u>Kam</u>ioka <u>L</u>iquid scintillator <u>A</u>nti-<u>N</u>eutrino <u>D</u>etector) [1] are to aim at studying reactor electron anti-neutrino ($\bar{\nu}_e$) oscillations with more than 100 km baseline, and simultaneously at searching for neutrinos from terrestrial sources: the so-called geoneutrinos which are generated from ^{238}U and ^{232}Th decays inside the Earth. In the future stage of KamLAND, it is planned to detect ^{7}Be solar neutrinos, accompanying with patient efforts on reducing background events with energies below 1 MeV.

The KamLAND project was proposed in 1994. In 1997 the full budget was funded by the Japanese Ministry of Education. In 1999 the United States Department of Energy approved the US-KamLAND proposal. Following a five year period for constructing the detector and the underground facility, KamLAND launched into data-taking in January 22, 2002. At present the KamLAND project is being carried out in collaboration with China, France, Japan and U.S.

Data taking has been quite stable since March 2002. The first reactor neutrino result demonstrated the reactor anti-neutrino disappearance at the 99.95% C.L. with a data sample corresponding to 162 ton-year exposure [2]. In the context of the $\bar{\nu}_e \rightarrow \bar{\nu}_x$ oscillation with CPT invariance, the combined results of KamLAND and solar neutrino experiments excluded all regions except for the large mixing angle (LMA) solution to the solar neutrino problem. In this symposium the reactor neutrino results are presented, based on analyzing the 766.3 ton-year data-sample which was taken through March 2002 to January 2004.

2. KamLAND detector

KamLAND is a large neutrino detector built in the Kamioka mine. The detector is physically located at the site of the old

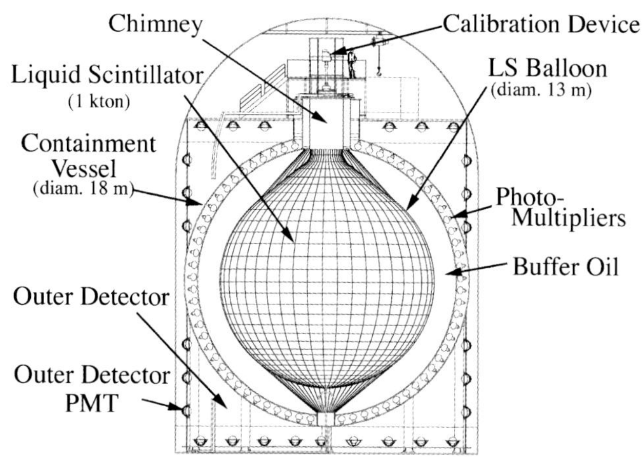

Fig. 1. Schematic view of the KamLAND detector.

Kamiokande: the 3000 m^3 water Cerenkov detector which played a leading role in the study of neutrinos produced via cosmic rays and also helped to pioneer the subject of neutrino astronomy. After dismantling the Kamiokande detector, the rock cavity was enlarged to be 20 m in diameter and 20 m in height. The KamLAND detector consists of a series of concentric spherical shells. Fig. 1 shows a conceptual drawing of the detector. The neutrino detector/target is 1000 ton of ultra pure liquid scintillator (LS) located at the center of the detector. The KamLAND LS is a chemical cocktail of 80% dodecane, 20% 1,2,4-trimethylbenzene, and 1.52 g/liter 2,5-diphenyloxazole as a fluor. The scintillator is housed in a 13 m diameter spherical balloon made of 3 layers of nylon with a total thickness of 135 μm, and supported by a cargo net structure at the top of a stainless steel vessel. This balloon system hangs inside an 18 m-diameter stainless-steel spherical vessel. A buffer mixture of dodecane and isoparaffin oils fills the volume between the stainless steel vessel and the balloon. Its density is kept at 0.04% lower than that of LS to reduce the weight-load on the balloon. The entire inner surface of the vessel (Inner Detector: ID) is covered by an array of a total of 1879 photomultiplier tubes (PMTs), 1325 of which are specially developed of 17 inches diameter and 554 of which are old Kamiokande 20 inch diameter devices. The total photocathode coverage is 34%. The light yield at the detector center is 500 p.e./MeV. A 3 mm thick acrylic barrier at 16.6 m diameter prevents radon emanating from PMT glasses from invading into LS. This central detector stands in a cylindrical rock cavity. The volume between the sphere vessel and the cavity is filled with $\sim$3200 m^3 of pure water where 225 Kamiokande PMTs are placed to detect cosmic-ray muons by their Cerenkov light. This outer detector (OD) absorbs γ-rays and neutrons from the surrounding rock and provides a tag for cosmic-ray μs.

Each PMT signal is recorded with the analog-transient-waveform-digitizer (ATWD). The ATWDs are self launching with a threshold $\sim$1/3 p.e. and operated with 3 different gains allowing

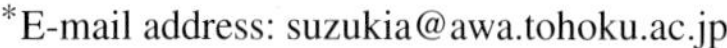

*E-mail address: suzukia@awa.tohoku.ac.jp

a dynamic range of ($\sim$1–1000) mV. There are 128 samples per waveform with a sampling time of 1.5 ns. 2 ATWD sets for each PMT are equipped to reduce detector dead time. The primary ID trigger is set at 200 PMT hits, corresponding to about 0.7 MeV with a 50% detection efficiency and 0.89 MeV with a 99% efficiency. This threshold is lowered to 120 hits ($\sim$0.46 MeV with a 99% efficiency) for 1 ms after the primary trigger to detect lower energy delayed signals.

3. Detector performance

The KamLAND detector performance is investigated, using laser and LED light-sources, radioactive sources of ^{203}Hg with 1 γ-ray of 0.279 MeV, ^{68}Ge with 2 annihilation γ-rays of 0.511 MeV, ^{65}Zn with 1 γ-ray of 1.116 MeV, ^{60}Co with 2 γ-rays of 1.173 and 1.333 MeV, Am-Be with 2 γ-rays of 4.438 MeV and $\sim$8 MeV from a thermal neutron capture on Ni or Fe, and cosmic-ray μ-associated events of spallation-products (^{12}B and ^{12}N), neutrons and γ-rays with 2.20 MeV and 4.95 MeV from thermal neutron captures on protons and ^{12}C. Cosmic-ray μ-induced events provide a monitor to examine the position dependence and the time variation of detector performance, since these events distribute uniformly in space and time.

Determining the position reconstruction uncertainty is carried out by deploying γ-ray sources along the vertical axis. Deviations of reconstructed positions from the sources are plotted as a function of the vertical position in Fig. 2. $\pm$5 cm uncertainty is obtained inside the fiducial volume of the present analysis (-5.5 m $< Z < 5.5$ m). The position resolution for 2.506 MeV γ from the ^{60}Co source is 19 cm. The energy dependence of the position resolution is estimated to $\sim$30 cm/$\sqrt{E(\mathrm{MeV})}$ for energies up to $\sim$8 MeV.

The uncertainties for determining the energy scale come mainly from the non-uniformity in position to position, the time variation due to the detector-operation condition, the non-linearity of the 20 inch PMT response, the additional light yield of Cerenkov light, and the non-linearity of scintillation photons, so-called quenching effect. Combining all these effects, the systematic uncertainty in the energy scale at the 2.6 MeV analysis threshold is 2.0%. The energy resolution is 6.2%/$\sqrt{E(\mathrm{MeV})}$. Fig. 3 (a) and (b) show the energy spectra of calibration sources and $\Delta E/E$ in the energy range between 0.3 and 9 MeV.

Radioactive contaminants inside the liquid scintillator are possible background sources for reactor anti-neutrino events. A Monte Carlo study shows that the required contamination levels of ^{238}U, ^{232}Th and ^{40}K in the liquid scintillator should be below 10^{-13} g/g, 10^{-13} g/g, and 10^{-14} g/g respectively. Detecting the

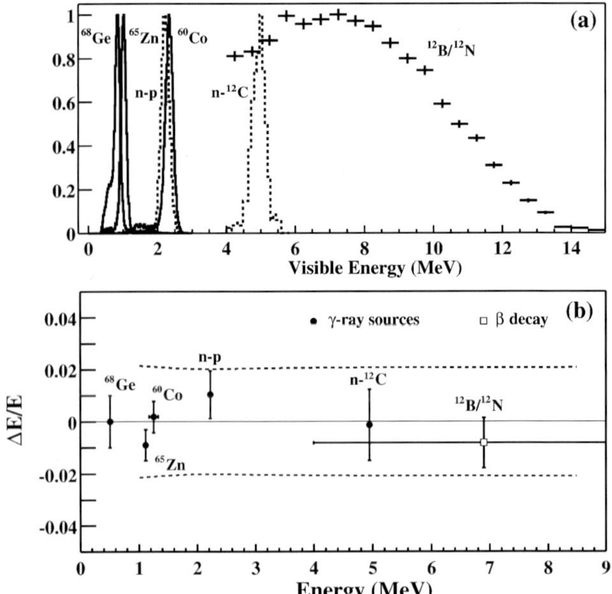

Fig. 3. (a) Energy spectra used in calibrations. (b) The fractional difference of the reconstructed average energies and known energies of the source γ-rays and the β-rays from ^{12}B/^{12}N, where the plotted energies at ^{68}Ge and ^{60}Co are the average of 2 γ-rays.

sequential chain-decays of ^{214}Bi $\rightarrow$ ^{214}Po $\rightarrow$ ^{210}Pb and ^{212}Bi $\rightarrow$ ^{212}Po $\rightarrow$ ^{208}Pb are used to estimate the ^{238}U and ^{232}Th concentrations. The results are $(3.4 \pm 0.4) \times 10^{-18}$ g/g for ^{238}U and $(5.2 \pm 0.8) \times 10^{-17}$ g/g for ^{232}Th. The ^{40}K concentration is extracted from the visible energy distribution of single events, subtracting the contributions from ^{238}U, ^{232}Th and μ-induced products. It gives the upper limit of 2.7×10^{-16} g/g. These results tell us that the contaminations of ^{238}U, ^{232}Th and ^{40}K inside the liquid scintillator are well below the requirements.

4. Reactor anti-neutrino detection

Nuclear reactors are very intense sources of $\bar{\nu}_e$ produced through β-decays of neutron rich fission fragments. Studying neutrino oscillations is done to look for a deficit of the neutrino flux and observe a deformation in the energy spectrum characterized by oscillation parameter, $(\Delta m^2, \sin^2 2\theta)$. The KamLAND detector is exposed to a large flux of $\bar{\nu}_e$'s produced by 52 local reactors in 18 Japanese commercial power-stations. The 26 reactors located at distances of 140–210 kilometers away from Kamioka generate $\sim$70 GW which is $\sim$7% of the world nuclear power generation. With such desirable conditions as a huge reactor power, an extremely long and definite base-line ($\sim$180 km) and the relatively low neutrino energy, KamLAND improves the Δm^2 detection sensitivity by more than 2 orders of magnitudes for the previous reactor experiments. In addition, KamLAND is accessible to the LMA oscillation parameter region ($3 \times 10^{-5} < \Delta m^2 < 2 \times 10^{-4}$ eV^2, $\sin^2 2\theta$) to solve the solar neutrino problem.

To determine the reactor $\bar{\nu}_e$ flux, the knowledge of the instantaneous thermal power, fuel burn-up, exchange and enrichment records for all Japanese power reactors is required. The fission rate for each fissile element is calculated from the data. The thermal power generation is checked with independent records of electric power generation. Fig. 4 shows one example of thermal power data and the corresponding fission-rate calculations for fissile elements of ^{235,238}U and ^{239,241}Pu elements which

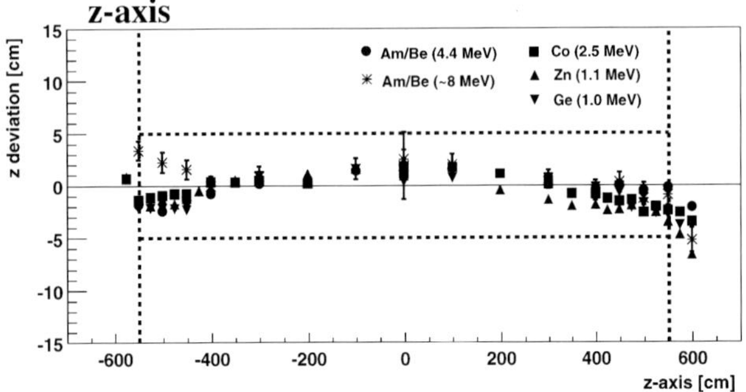

Fig. 2. Deviation of reconstructed vertexes from radioactive source positions.

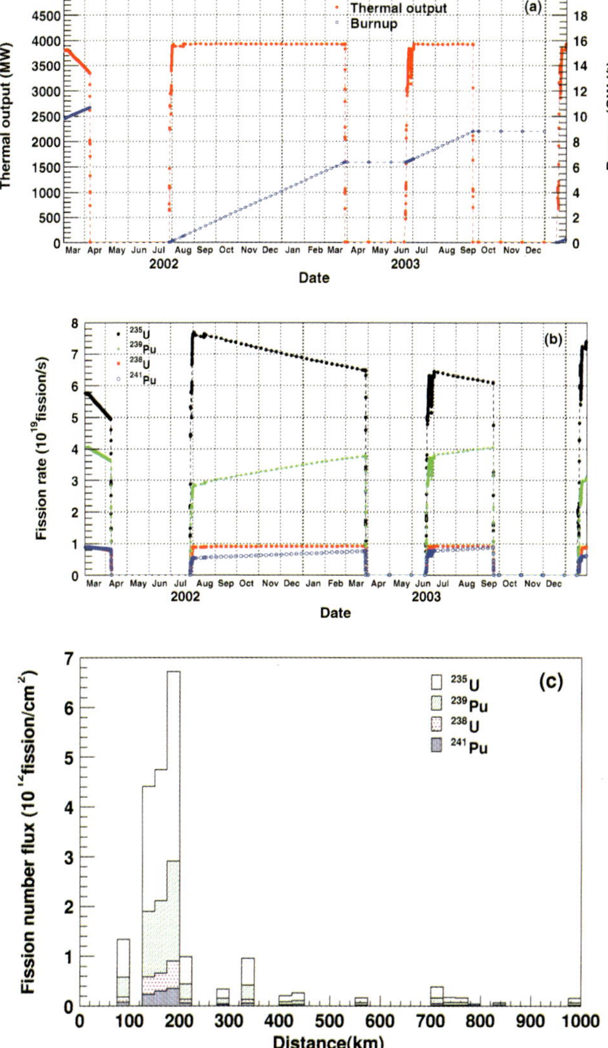

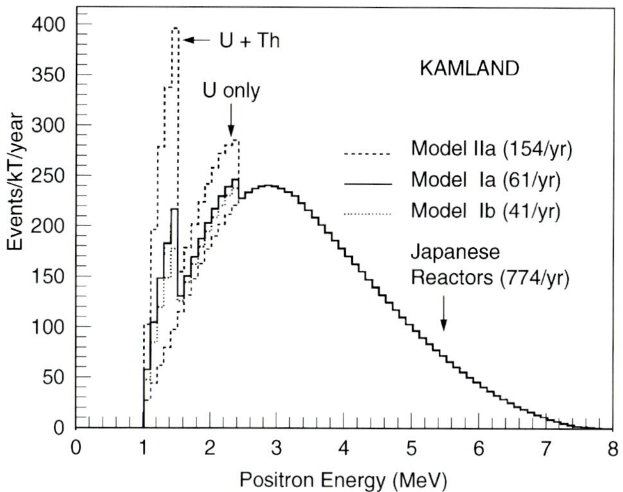

Fig. 5. Expected energy spectra of positrons produced by reactor neutrinos and geoneutrinos [4].

Fig. 4. Example of instantaneous thermal power (a) and fuel burn-up (b) records for one of the Japanese commercial reactors. (c) is the fission yields at Kamioka from 4 fissile nuclei. The accumulation period is the same as the data-taking interval of March 9, 2002 to January 11, 2004. These data are provided according to the special agreement between Tohoku Univ. and the Japanese nuclear power-reactor organization.

contribute to 99.9% of the $\bar{\nu}_e$ flux generation. The time-integrated fission flux at Kamioka given by these fuel elements in units of fission number/cm^2 is also plotted in Fig. 4 as a function of the distance. Here the accumulation time of these data corresponds to the detector live time of the present reactor analysis. More than 79% of the total fission flux arises from 26 reactors within a distance of 138–214 km from Kamioka. The relatively narrow band of distances allows KamLAND to be sensitive to spectral distortions for certain oscillation parameters. The contribution to the $\bar{\nu}_e$ flux from Korean reactors is estimated to be $(2.46 \pm 0.25)\%$ based on the reported electric power generation rates. That from other reactors around the world is $(0.70 \pm 0.35)\%$ as an average. Once the fission rates of fissile isotopes are obtained. The expected $\bar{\nu}_e$ flux is calculated using the measurement of $\bar{\nu}_e$ energy spectrum per fission [3].

Upon entering the detector, $\bar{\nu}_e$ is captured by a free proton and an inverse β-decay reaction occurs, $\bar{\nu}_e + p \rightarrow e^+ + n$. The positron deposits its energy and then annihilates, yielding two γ-rays (each 511 keV). The neutron thermalizes and is captured by a proton in $(211.2 \pm 2.6)\,\mu$sec. releasing a 2.22 MeV γ from

the $n + p \rightarrow d + \gamma$ reaction. Thus the inverse β-decay reaction provides a clear sequential signature of the prompt e^+ and delayed γ with the definite time- and close space-correlations. Although the need to reduce all potential background that can imitate a $\bar{\nu}_e$ signal is imperative, these signal correlations give a high rejection-power for background events. Nevertheless, with the expectation of the energy spectrum, the background from geoneutrinon is irreducible. KamLAND has the first chance to search for geoneutrinos which are $\bar{\nu}_e$'s originated from U/Th decays inside the Earth. The radiogenic heat by U/Th decays plays a dominant role in the energy generation of the Earth. We evaluated the detection rate of geoneutrinos in KamLAND, using various geophysical and geochemical models [4]. In Fig. 5 a smooth broad histogram is the expected visible energy spectrum of positrons produced by reactor neutrinos, and two sharp peaks in the energy below 2.5 MeV are expected from geoneutrinos. In the reactor neutrino oscillation analysis, positrons with energies above 2.6 MeV are used to avoid geoneutrino pollution.

5. Data analysis

The data sample which we used was taken in March 9, 2002 to January 11, 2004 [5]. The total exposure time is 776.3 ton-yr after eliminating various noise events. In the present analysis we applied more elaborate $\bar{\nu}_e$ event selections than those of the first reactor data analysis [2]. The selection of the inverse β-decay events uses $2.6 < E_{\text{prompt}} < 8.5$ MeV, no outer detector hits, $0.5 < \Delta T(|\text{prompt} - \text{delayed}|) < 1000\,\mu$sec, $\Delta R(|\text{prompt} - \text{delayed}|) < 2$ m and $1.8 < E_{\text{delayed}} < 2.6$ MeV. The overall tagging efficiency is $(89.8 \pm 1.5)\%$ which is reconfirmed by the Am-Be radioactive source and LED calibrations. The μ-induced spallation events are eliminated, using $2\,\text{s} > \Delta T\,(|\mu - \text{prompt}|)$ with an energy deposit of $E_\mu > 3$ GeV or $\Delta L\,(|\mu \cdot \text{path} - \text{prompt}|) < 3$ m. Finally, events produced inside the fiducial volume defined as the volume of $R < 5.5$ m are selected. The corresponding fiducial mass is 543.7 ton which contains 4.61×10^{31} free target protons. The systematic uncertainty of the fiducial mass is estimated by comparing the number of μ-induced ^{12}B/^{12}N events reconstructed in the fiducial volume to the total number in the entire LS. Here the total volume of LS is $(1171 \pm 25)\,\text{m}^3$, as measured by the flow meters during detector filling.

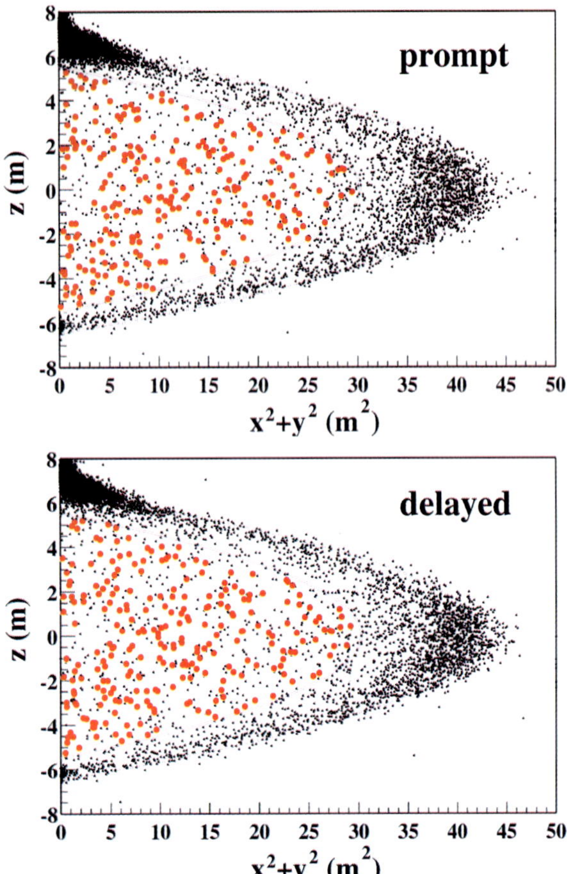

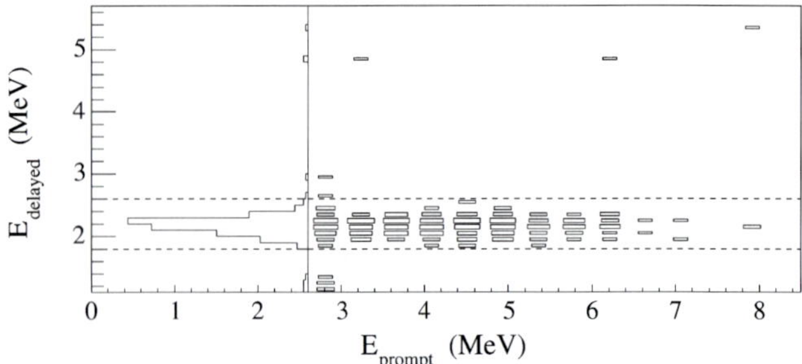

Fig. 7. Scatter plot of E_{prompt} and $E_{delayed}$ for the $\bar{\nu}_e$ candidate events.

Table I. *Summary of systematic errors (%).*

Fiducial Volume	4.7	Reactor power	2.1
Energy threshold	2.3	Fuel composition	1.0
Efficiency of cuts	1.6	$\bar{\nu}_e$ spectra [3]	2.5
Livetime	0.06	Cross section [6]	0.2
Total systematic error		6.5	

Fig. 6. Vertex distributions of the prompt and delayed events before applying each energy cut. The solid curve stands for the fiducial limit ($R = 5.5$ m) and the dotted curve is the balloon position ($R = 6.5$ m).

Fig. 6 shows the vertex distributions of the prompt and delayed events before applying their energy cuts. The red-dotted events in this figure are the inverse β-decay candidates.

The number of expected background events passing through the above event-selection criteria is estimated to be (17.8 ± 7.3) during the present exposure time. The dominant contribution of (10.3 ± 7.1) comes from the reactions of $^{13}C\,(\alpha, n)^{16}O^*$ and ^{12}C $(n, n')\,^{12}C^*$ induced by α particles of the radon daughter ^{210}Po in the LS. The observation of ^{210}Po concentration establishes that there are $(1.47 \pm 0.20) \times 10^9$ α-decays during the present lifetime of data taking. The μ-induced spallation products of $^9Li/^8He$ also give a sizable contribution of 4.8 ± 0.9. These nuclei emit a β and a neutron and thus mimic the inverse β-decay event. Contributions from the accidental backgrounds and fast neutrons are (2.69 ± 0.02) and (< 0.89).

The correlation of prompt and delayed energies for the $\bar{\nu}_e$ candidates before applying $E_{delayed}$ cut is plotted in Fig. 7. A clear event-isolation in the delayed energy window can be seen. Events concentrated in $E_{delayed} < 1$ MeV are expected to be accidental backgrounds. The event rate of $E_{delayed} \sim 5$ MeV is consistent with the expected neutron radiative capture rate on ^{12}C. These events are not used in the present analysis.

For the 766.3 ton-year exposure we observe $258\bar{\nu}_e$ events with energies > 3.4 MeV which should be compared to 365.2 ± 23.7(syst) events expected in the absence of neutrino oscillation. The systematic uncertainty is summarized in Table I. The overall systematic error is 6.5% for the events with $E_{prompt} > 2.6$ MeV. The significance of the disappearance is evaluated by taking the ratio of observed $\bar{\nu}_e$ events minus background events to expected. The result is 0.658 ± 0.044(stat) ± 0.047(syst), confirming $\bar{\nu}_e$ disappearance at the 99.998% significance level. Applying the present event selection procedure to the previously reported result data gives 0.601 ± 0.069(stat) ± 0.042(syst), in agreement with 0.589 ± 0.085(stat) ± 0.042(syst), after correction for the (α, n) background.

The time variation of observed and estimated reactor $\bar{\nu}_e$ event rates is plotted at 6 time-periods in Fig. 8 (a). The time-behaviors

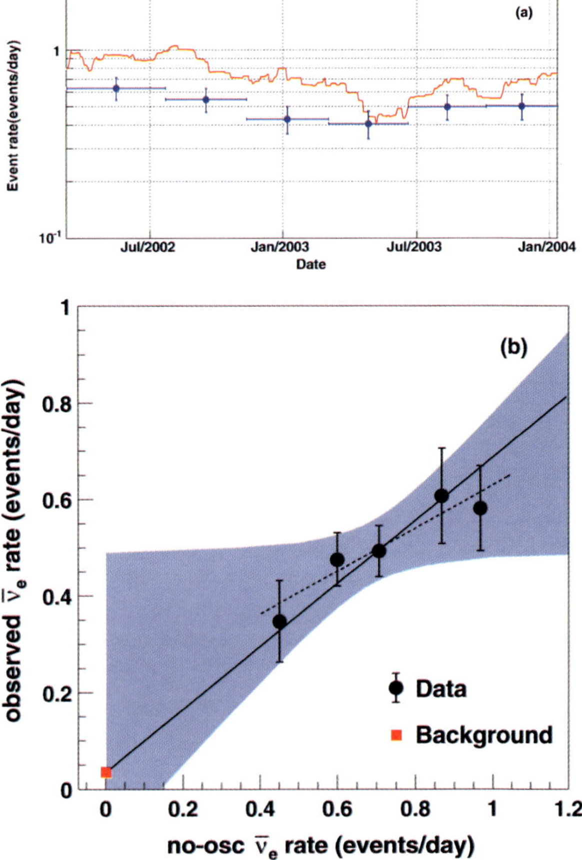

Fig. 8. Time dependence (a) and correlation (b) of the observed and expected event rates. The dashed line is a fit, the 90% C.L. is shown in gray.

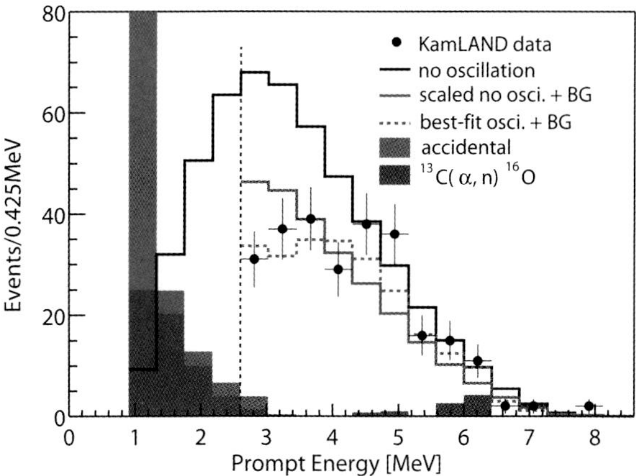

Fig. 9. Energy spectrum of prompt events. The no-oscillation, best-fit scaled no-oscillation and best-fit oscillation spectra are compared to the data.

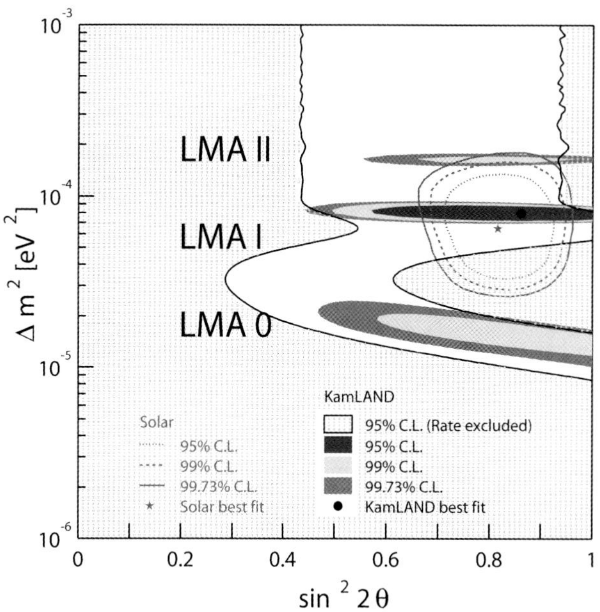

Fig. 10. Excluded regions for the rate analysis and allowed regions for the combined rate and shape analysis at several confidence levels.

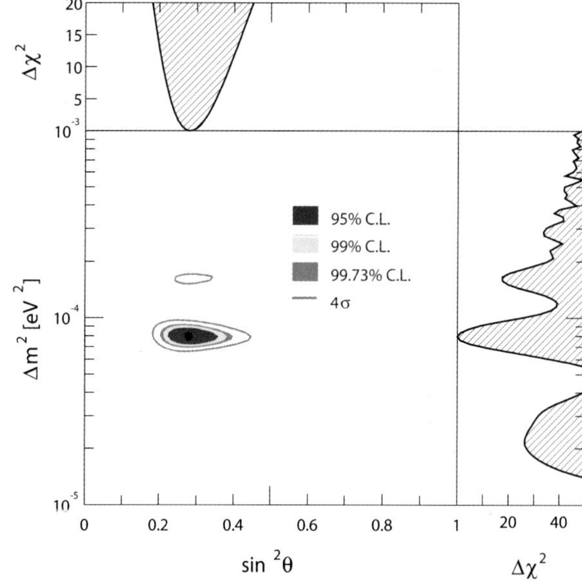

Fig. 11. Allowed region obtained by the global analysis combined the KamLAND data and solar neutrino fluxes. $\Delta\chi^2$ profiles on $\sin^2 2\theta$ and Δm^2 are also shown.

are almost consistent with each other. The correlation between both rates is clearly seen in Fig. 8 (b) and well approximated by a straight line. The intercept is compatible with known background, but cannot exclude larger backgrounds with the present data statistics.

The prompt energy spectrum above 2.6 MeV is shown in Fig. 9, comparing with the spectra estimated from the reactor $\bar{\nu}_e$ events and the accidental and $^{13}C(\alpha, n)^{16}O$ backgrounds. The data clearly deviate from the expected spectrum in the absence of neutrino oscillations, particularly in the region below 4 MeV. It is shown that by fitting the data to the reactor neutrino spectrum without oscillations with a free scale factor, the data excludes the expected spectral shape in the absence of neutrino oscillation with a 99.6% significance. Here the spectral distortion due to systematic uncertainties for the backgrounds is considered in the following: for the $^{13}C(\alpha, n)^{16}O$ background, a free-scale uncertainty around 6 MeV and a 32% scale uncertainty of the estimated rate around 2.6 and 4.4 MeV are applied to fitting. KamLAND gives the first evidence of the spectrum distortion in neutrino experiments with a confidence level of $\sim 3\sigma$. Thus the prompt energy spectrum prefers the distortion expected from $\bar{\nu}_e$ oscillation effects. The reactor neutrino anomaly defined as the combined effect of the rate disappearance and spectrum distortion is found at the high confidence level of 99.999995%.

6. Neutrino oscillation analysis

A two-flavor oscillation analysis of the KamLAND data and the observed solar neutrino fluxes [7] is carried out with CPT invariance. The best-fit parameter of Δm^2 and $\sin^2 2\theta$ are given by $7.9 \times 10^{-5}\,\text{eV}^2$ and 0.86. The corresponding best-fit spectrum together with the backgrounds is drawn also in Fig. 9. The allowed and excluded regions of the oscillation parameter are shown in Fig. 10. The allowed region of the large mixing angle (LMA) solution of solar neutrino experiment [8] is also shown in this figure. At the 3σ confidence KamLAND squeezes the allowed Δm^2 into 3 narrow bands named LMA-0, LMA-I and LMA-II. However LMA-0 is disfavored at the 97.5% C.L. and LMA-II at 98.0% C.L. The best fit stands at LMA-I, which is consistent with the solar neutrino result assuming CPT invariance. A global analysis of data from KamLAND and solar neutrino experiments gives an stringent constraint to the oscillation parameter as shown in Fig. 11, where the best fit

parameter is $\Delta m^2 = 7.9^{+0.6}_{-0.5} \times 10^{-5}\,\text{eV}^2$ and $\tan^2 \theta = 0.40^{+0.10}_{-0.07}$. The $\Delta\chi^2$ distributions for Δm^2 and $\sin^2 2\theta$ show that the sensitivity in Δm^2 is dominated by the observed distortion in the KamLAND spectrum, while solar neutrino data provide the best constraint on θ. This result gives the most precise determination of Δm^2 to date.

To test the goodness of the oscillation hypothesis, the ratio of observed $\bar{\nu}_e$ spectrum to the expected for no-oscillation as a function of L_0/E is compared with the prediction from the neutrino oscillation with the best-fit values of Δm^2 and $\sin^2 2\theta$. The constant L_0 chooses 180 km, as if a single reactor at this distance contributes to KamLAND. Two alternative hypotheses for neutrino disappearance, neutrino decay [9] and decoherence [10] give different L_0/E. Fig. 12 shows the data and best-fits of oscillation, decay and decoherence. The neutrino oscillation is much favored with more than 99% C.L., while the decay and decoherence are excluded at the 95% and 94% C.L.

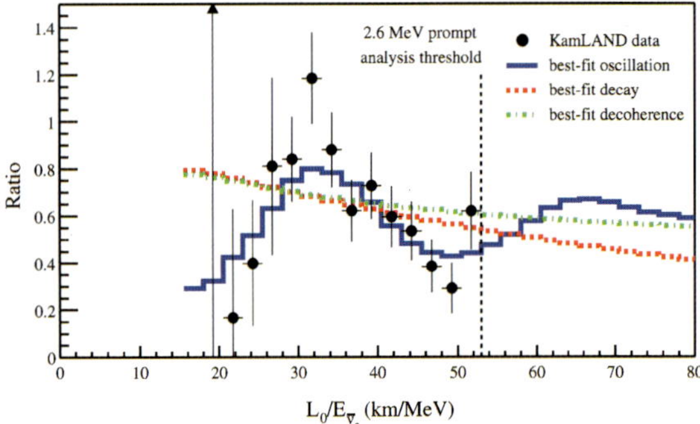

Fig. 12. Ratio of the observed $\bar{\nu}_e$ spectrum to the expectation for no-oscillation versus L_0/E with $L_0 = 180\,$km. The histogram is the expectation for the best-fit oscillation, and the curves show the energy dependence for the best-fit decay and best-fit decoherence.

7. Conclusions

KamLAND has reconfirmed the reactor neutrino disappearance with much higher significance. The spectrum distortion and L_0/E behavior observed in the present data strongly support neutrino oscillations to solve the disappearance. Statistical uncertainties are now close to the systematic uncertainties. In the near future KamLAND will provide more precise measurements of the reactor $\bar{\nu}_e$'s, with the aid of current efforts on full-volume source calibrations for improving systematic uncertainties and on purifying LS with distillation and degasification for reducing α-decays from ^{210}Po.

References

1. Suzuki, A., Talk at the XVIII International Conference on Neutrino Physics and Astrophysics, Takayama, Japan, (1998); Nucl. Phys. B (Proc. Suppl.) **77**, 171 (1999).
2. Eguchi, K. *et al.*, (KamLAND Collaboration), Phys. Rev. Lett. **90**, 021802 (2003).
3. Schreckenbach, K. *et al.*, Phys. Lett. B**160**, 325 (1985); Hahn, A. A. *et al.*, Phys. Lett. B**218**, 365 (1989); Vogel, P. *et al.*, Phys. Rev. C**24**, 1543 (1989).
4. Raghavan, R. S. *et al.*, Phys. Rev. Lett. **80**, 635 (1998).
5. Araki, T. *et al.*, (KamLAND Collaboration), to be published in Phys. Rev. Lett. (2005).
6. Vogel, P. and Beacom, J. F., Phys. Rev. D**60**, 053003 (1999); Kurylov, A. *et al.*, Phys. Rev. C**67**, 035502 (2003).
7. Bahcall, J. N. and Pena-Garay, C., New J. Phys. **6**, 63 (2004); Maltoni, M. *et al.*, New J. Phys. **6**, 122 (2004).
8. Ahmed, S. N. *et al.*, (SNO Collaboration), Phys. Rev. Lett. **92**, 181301 (2004).
9. Barger, V. D. *et al.*, Phys. Rev. Lett. **82**, 2640 (1999).
10. Lisi, E. *et al.*, Phys. Rev. Lett. **85**, 1166 (2000).

Physica Scripta. Vol. T121, 39–45, 2005

New Opportunities for Surprise

Janet Conrad

[CU] Columbia University, 538 W. 120th St, New York, NY 10027, USA

Received February 6, 2005; accepted in revised form February 9, 2005

Abstract

If history is any guide, then there is a good chance that our present description of neutrinos is wrong. Over the past 30 years, neutrino physics has again and again produced results which took us by surprise, and forced us to change our theoretical direction. In light of this, it is worth examining the experimental clues we now have to see if they lead to ideas outside our "New Standard Model."

Our present paradigm says that there are only three neutrinos participating in the weak interaction. They are light, with masses less than 0.1 eV, and predicted to be Majorana neutrinos, that is, each is its own antiparticle. The light mass is motivated by the see-saw model, which postulates partners which are very massive ($\sim$hundreds of GeV).

By contrast with this "conventional wisdom," this talk discusses five examples of experimental results which, if confirmed, would cause us to rethink our theory. It focuses only on oscillations, and on those questions which can be addressed at accelerators. Other ways neutrinos may be surprising, such as the anomalous measurement of $\sin^2 \theta_W$ from NuTeV [1], are left for other talks. The ideas discussed here are best explored through accelerator-produced neutrino beams. So the reasons why beam-based experiments are becoming the venue-of-choice are then examined. Lastly, the first two of the five ideas are explored in detail.

1. Five clues and ideas

There are many ideas and clues for physics beyond our present neutrino model. Choosing only five is artificial, but serves to illustrate the rich potential for exploring ideas outside of the usual bounds of discussion.

- *What if LSND is correct?* The LSND experiment [2] observed the appearance of anti-electron-neutrinos in an anti-muon-neutrino beam. This oscillation cannot be accommodated within the standard three-neutrino picture [3]. If LSND is confirmed by the MiniBooNE experiment, which expects results in 2005, we will be forced to substantially rethink our three-neutrino picture.

- *What if there are sterile neutrinos?* Additional non-interacting neutrinos represent one way of accommodating the LSND signal. The data accommodate good fits to two or three sterile neutrinos [4]. Grand Unified Theories and models of Extra Dimensions can incorporate these particles. Cosmological measurements can also accommodate sterile neutrinos with modification to the "standard" picture; these were discussed by other speakers at this conference [5].

- *What if neutrinos decay?* Introducing neutrino mass opens the possibility not just for oscillations but for decay as well [6]. The atmospheric data has provided a first search, which gave a null result [8]; to settle the question we need a high statistics long baseline experiment such as Minos [7].

- *Is the oscillation parameter θ_{23} exactly 45 degrees?* This parameter is the largest of all of the mixing angles and may possibly be exactly $45°$. At this point, the angle is only measured to the 20% level [8], but Minos will improve [7] this and T2K will achieve about a 5% measurement [9]. If this parameter is exactly or very near 45 degrees, then we will need to look to a new symmetry to explain this special value [10].

- *What if CP violation is large in the lepton sector?* The present paradigm calls for CP violation in neutral heavy lepton decays in order to explain the matter-antimatter asymmetry present in the Universe today, via leptogenesis. But no theory yet connects this to CP violation among the light neutrinos. Because CP violation in the quark sector is small, prejudice leads one to imagine it will also be a small effect, if it is there at all. Thus it would be a big surprise if CP violation is observed to be large in $\nu_\mu \to \nu_e$ versus $\bar{\nu}_\mu \to \bar{\nu}_e$ studies. If the CP violation angle is close to 90 or 270 degrees (thus, large), and the mixing parameter, θ_{13} is also large ($\theta_{13} > 0.3$), then programs such as T2K and Nova could observe CP violation in the lepton sector at 2 to 3σ [11]. For smaller values of θ_{13}, we will need a Very Long Baseline Neutrino Oscillation experiment [12]. This would have a baseline three times longer than Minos and a detector five times the size of Super K.

2. Why accelerator beams?

Accelerator-based neutrino beams are best for exploring these clues and ideas because they combine high intensity with good control of systematics. These two aspects are crucial for rare searches. All of the questions listed above will be answered through accelerator-based studies.

Several recent technological breakthroughs are leading to very powerful accelerator-based neutrino sources. First, we are entering the era of megawatt beams. For example, the new SNS facility at Oak Ridge, Tennessee will supply a beam of 1.4 MW at 1.3 GeV [13]. The promise of multi-megawatt facilities, based on the new technology of super-conducting RF cavities, is not far away [14]. Second, many engineering issues associated with fast-pulsed, high current horns have been solved [15], which allows substantially improved secondary focusing.

The ultra-high intensities of present and future beams allow for new experimental techniques. For example, the highest intensity beams already make long-baseline experiments possible. In these experiments, beam timing is crucial in order to separate signal from the environmental backgrounds. Hence, the development of GPS has been a crucial technology breakthrough for long-baseline experiments. Soon accelerators will have sufficient rates to allow secondary particle energy selection and still maintain a reasonable level of statistics. An effective method of energy selection is to place the detector at an angle θ off the beam direction axis. The resulting beam energy is given by $E_\nu = (0.43 E_\pi)/(1 + \gamma^2 \theta^2)$.

The final component which has made accelerator-based sources viable for precision studies is the recent investment in the knowledge needed to describe the events in the few GeV energy range. These studies fall into two categories. First, several experiments, BNL-910 [16], HARP [17] and MIPP [18], are studying secondary particle production for neutrino beams so that the neutrino flavor and energy distribution content can be accurately understood. Second, experiments like FINeSSE

[19] and MiniBooNE are making accurate measurements of cross sections in the MeV to GeV range where most proposed accelerator experiments run. This latter information is being combined into a highly detailed simulation called NUANCE [20], which many experiments (K2K [21], MiniBooNE [22], SNO-atmospheric [23], and T2K) now use. The fact that many experiments are using the same open-source code makes makes meaningful cross comparisons possible.

At present, there are only two accelerator-based oscillation searches underway: MiniBooNE (at Fermilab) and K2K (at KEK). However, in the near term, Minos [7] (Fermilab, 2005), Opera [24] (CERN, 2006), T2K [9] (JHF, 2008), and possibly Nova [25] (Fermilab, 2012?) will begin taking data. In the longer term more ambitious facilities such as a Neutrino Factory or a β-Beam Source are under discussion. We are on the threshold of at least a decade of exciting studies from accelerator-based neutrino sources.

3. One example surprise: what if LSND is correct?

The LSND experiment [2] observed the appearance of anti-electron-neutrinos in a anti-muon-neutrino beam. This experiment ran at the LAMPF accelerator at Los Alamos National Laboratory between 1993 and 1998. The decay-at-rest (DAR) beam was produced by impinging 800 MeV protons on a beam dump. These produced π^+'s which stop and decay to produce μ^+'s, which also stop and decay to produce $\bar{\nu}_\mu$ and ν_e, which were studied in the range of 20 to 55 MeV. The π^-'s capture, so the beam has a $<8 \times 10^{-4}$ contamination of $\bar{\nu}_e$. The neutrino events were observed in a detector located 30 m downstream of the beam dump. 1220 phototubes surrounded the periphery of a cylindrical detector filled with 167 tons of mineral oil, lightly doped with scintillator. The signature of a $\bar{\nu}_e$ appearance was $\bar{\nu}_e + p \rightarrow e^+ n$. This resulted in a two-component signature: the initial Cerenkov and scintillation light associated with the e^+, followed later by the scintillation light from the n capture on hydrogen, producing a 2.2 MeV γ. The experiment observed $87.9 \pm 22.4 \pm 6.0$ events, a 4σ excess [2].

LSND is a short baseline experiment, with an $L/E \sim 1$ m/MeV. Thus, from the two-generation oscillation formula

$$P = \sin^2 2\theta \sin^2(1.27\Delta m^2 L/E) , \qquad (1)$$

one can see that this experiment is sensitive to $\Delta m^2 \geq 0.1$ eV^2. Other experiments have searched for oscillations at high Δm^2, and the two most relevant to LSND are Karmen [26] and Bugey [27]. The Karmen experiment, which also used a DAR muon beam and was located 17.7 m from the beam dump, had sensitivity to address only a portion of the LSND, and did not see a signal in that region. Since the design of the experiments are very similar (see Table I), one can think of the Karmen experiment as a "near detector," which measures the flux before oscillation. The results were combined in a joint analysis performed by collaborators from both experiments; the allowed range for oscillations is shown in Figure 1 (top) [28]. The Bugey experiment was a reactor-based $\bar{\nu}_e$ disappearance search which set a limit on oscillations. Because this is disappearance and not explicitly $\bar{\nu}_e \rightarrow \bar{\nu}_\mu$, its limit is applicable to LSND in many, but not all, oscillation models. This limit is shown in Figure 1 (bottom).

The LSND signal cannot be accommodated within the standard three-neutrino picture, given the other evidence for oscillations. There are three signals for oscillations: the "solar" signal ($\Delta m^2 \sim 5 \times 10^{-5}$ eV^2), the "atmospheric" signal ($\Delta m^2 \sim 2.5 \times 10^{-3}$ eV^2), and LSND ($\Delta m^2 \sim 0.5$ eV^2, if one applies the

Table I. *A comparison of the LSND and KARMEN experiments.*

Property	LSND	KARMEN
Proton Energy	798 MeV	800 MeV
Proton Intensity	1000 μA	200 μA
Protons on Target	28,896 C	9425 C
Duty Factor	6×10^{-2}	1×10^{-5}
Total Mass	167 t	56 t
Neutrino Distance	30 m	17.7 m
Particle Identification	yes	no
Energy Res. at 50 MeV	6.6%	1.6%
Events, 100% $\bar{\nu}_\mu \rightarrow \bar{\nu}_e$ Transmutation	33,300	14,000

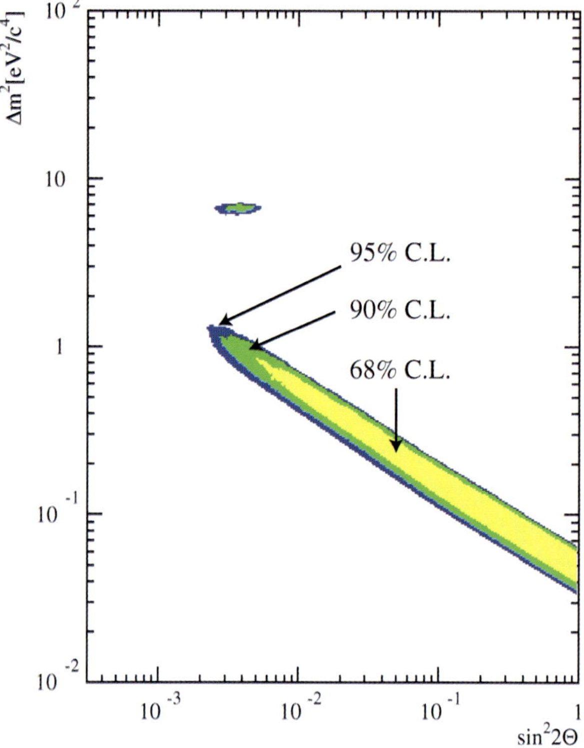

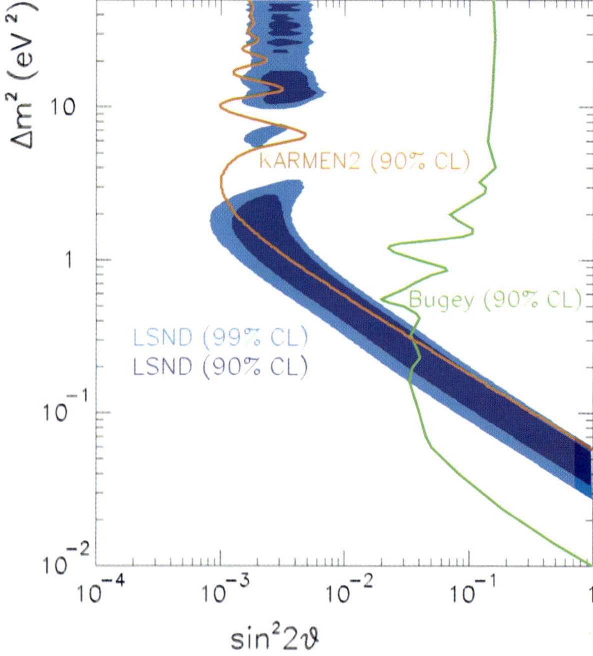

Fig. 1. Top: Allowed range from the Karmen-LSND joint analysis. Bottom: LSND allowed range compared to short baseline experiment limits.

Bugey limit). To see the incompatibility, first consider the case where the oscillation signals de-couple into, effectively, two-generation oscillations (eq. (1)). For three generations

$$\Delta m_{31}^2 = \Delta m_{32}^2 + \Delta m_{21}^2, \tag{2}$$

which is clearly not the the case for these three signals. If we go to the general case, then

$$\begin{pmatrix} \nu_e \\ \nu_\mu \\ \nu_\tau \end{pmatrix} = \begin{pmatrix} U_{e1} & U_{e2} & U_{e3} \\ U_{\mu 1} & U_{\mu 2} & U_{\mu 3} \\ U_{\tau 1} & U_{\tau 2} & U_{\tau 3} \end{pmatrix} \begin{pmatrix} \nu_1 \\ \nu_2 \\ \nu_3 \end{pmatrix}.$$

This gives the oscillation probability (assuming no CP violation):

$$\text{Prob}(\nu_\alpha \to \nu_\beta)$$

$$= \delta_{\alpha\beta} - 4 \sum_{j>i} U_{\alpha i} U_{\beta i} U_{\alpha j} U_{\beta j} \sin^2 \left(\frac{1.27 \, \Delta m_{ij}^2 \, L}{E} \right) \tag{3}$$

where $\Delta m_{ij}^2 = |m_i^2 - m_j^2|$. Equation(2) still holds; in this case, however, the atmospheric signal could be a combination of a high Δm^2 oscillation (LSND-range) and a low Δm^2 oscillation (solar range). In order for this model to succeed, the atmospheric Δm^2 from a shape analysis must shift up from its present value of $\sim 2 \times 10^{-3} \, \text{eV}^2$ and the chlorine experiment must have overestimated the deficit of ^{7}Be solar neutrinos. However, the largest clash between data and this model arises from the Super-K ν_e events. This model requires that Super-K has missed a ν_e appearance signal of approximately the same size and shape as the ν_μ deficit before detector smearing and cuts. Neutrino measurements are experimentally difficult and parameters do sometimes shift with time as systematics are better understood, but it seems unlikely that all of the above results could change sufficiently to accommodate LSND. Thus, if LSND is correct, then this must imply Beyond-the-Standard-Model physics.

4. An example explanation: sterile neutrinos

Additional neutrinos which do not interact via exchange of W or Z are called "sterile"; they may mix with active neutrinos, and thereby can be produced in neutrino oscillations. Experimental evidence of this would be the disappearance of the active flavor from the beam. Many Beyond-the-Standard-Model theories introduce sterile neutrinos, but the sterile neutrinos observable in present experiments must be light (in the eV range), and this narrows the class of acceptable theories. Nevertheless, a number of possible explanations remain, listed in reference [29]. Solar neutrino data may show some slight evidence for a sterile neutrino [30]. At present, however, the main experimental motivation is to accommodate the LSND signal, and this will be the main focus of this section.

Sterile neutrinos solve the LSND problem by adding extra mass splittings. The additional mass states must be mostly sterile, with only a small admixture of the active flavors in order to accommodate the limits on sterile neutrinos from the atmospheric and solar experiments, as shown in Figure 2 (top). The method for fitting the data is described in reference [4]. In the fit, the three mostly active neutrinos are approximated as degenerate. The data that drive the fits are the "short baseline" experiments that provide information on high Δm^2 oscillations. The combination of $\bar{\nu}_\mu \to \bar{\nu}_e$ (LSND [2], Karmen II [26]), $\nu_\mu \to \nu_e$ (Nomad [31]), ν_μ disappearance (CDHS [32], CCFR84 [33]), and ν_e disappearance (Bugey [27], CHOOZ [34]) must all be accommodated within the model. None of the short baseline experiments except for LSND provide evidence for oscillations beyond 3σ. However, it should be noted that CDHS has a 2σ (statistical and systematic, combined) effect consistent with a high Δm^2 sterile neutrino when the data are fit for a shape dependence, and Bugey has a 1σ pull at $\Delta m^2 \sim 1\text{eV}^2$. This is shown in Figure 3. As a result, these two experiments define the best fit combination of high and low Δm^2 for the $3+2$ model. However, there are acceptable solutions with a combination of low Δm^2 values.

Allowed combinations of Δm^2 for the two mostly sterile states, labeled mass states 4 and 5, are shown in Figure 2 (middle). The best fit has $\Delta m_{41}^2 = 0.92\,\text{eV}^2$, $\Delta m_{51}^2 = 22\,\text{eV}^2$, although there are combinations which work with two relatively low Δm^2 values. A wide range of mixing angles can be accommodated, and the best fit has $U_{e4} = 0.121$, $U_{\mu 4} = 0.204$, $U_{e5} = 0.036$ and $U_{\mu 4} = 0.224$. The other mixing angles involving the sterile states are not probed by the ν_μ disappearance, ν_e disappearance and $\nu_\mu \to \nu_e$ appearance experiments listed above.

In order to show that the data are consistent within this model, Figure 2 (bottom) shows the oscillation probability for LSND compared the the oscillation probability for all of the other experiments, which we call "Null Short Baseline," since they all have less than 3σ evidence for a signal. Despite the small preference for oscillations in CDHS and Bugey, the global NSBL oscillation probability peaks close to zero probability. But there is 30% compatibility for these experiments with probabilities allowed by LSND.

Introducing extra neutrinos, including sterile ones, would have cosmological implications, compounded if the extra neutrinos have significant mass ($>1\,\text{eV}$). This is covered in other talks at this conference [5]. However, there are several ways around the problem. The first is to note that while the best fit requires a high mass sterile neutrino, there are low-mass fits which work within the $3+2$ model. The second is to observe that there are a variety of classes of theories where the neutrinos do not thermalize in the early universe [35]. In this case, there is no conflict with the cosmological data, since the cosmological neutrino abundance is substantially reduced. Lastly, some analyses have called into question the strictness of the cosmological bounds [36], thus sterile neutrinos would still be allowed even within the present standard cosmology.

If more than one Δm^2 contributes to an oscillation appearance signal, then the data can be sensitive to a CP-violating phase in the mixing matrix. Experimentally, for this to occur, the Δm^2 values must be within less than about two orders of magnitude of one another. This is the motivation for CP violation searches at T2K [9] and Nova [25], when only three active neutrinos are considered (see question 5 of section 1, above).

In $3+2$ CP-violating models [37]:

$$P(\overset{(-)}{\nu_\mu} \to \overset{(-)}{\nu_e}) = 4|U_{e4}|^2|U_{\mu 4}|^2 \sin^2 x_{41} + 4|U_{e5}|^2|U_{\mu 5}|^2 \sin^2 x_{51}$$

$$+ 8|U_{e4}||U_{\mu 4}||U_{e5}||U_{\mu 5}| \sin x_{41} \sin x_{51} \cos(x_{54} \mp \phi_{54}) \tag{4}$$

where we have defined:

$$x_{ji} \equiv 1.27 \Delta m_{ji}^2 L/E, \qquad \phi_{54} \equiv arg(U_{e4}^* U_{\mu 4} U_{e5} U_{\mu 5}^*)$$

Thus the oscillation probability is affected by CP violation through the term ϕ_{54}. The CP conserving cases are $\phi_{54} = 0$ and 180 degrees.

If evidence for sterile neutrinos is observed in the near future, most particle physicists and cosmologists will be very surprised. If they are observed to be CP-violating, this will be even more surprising. This model simply does not fit neatly into the new

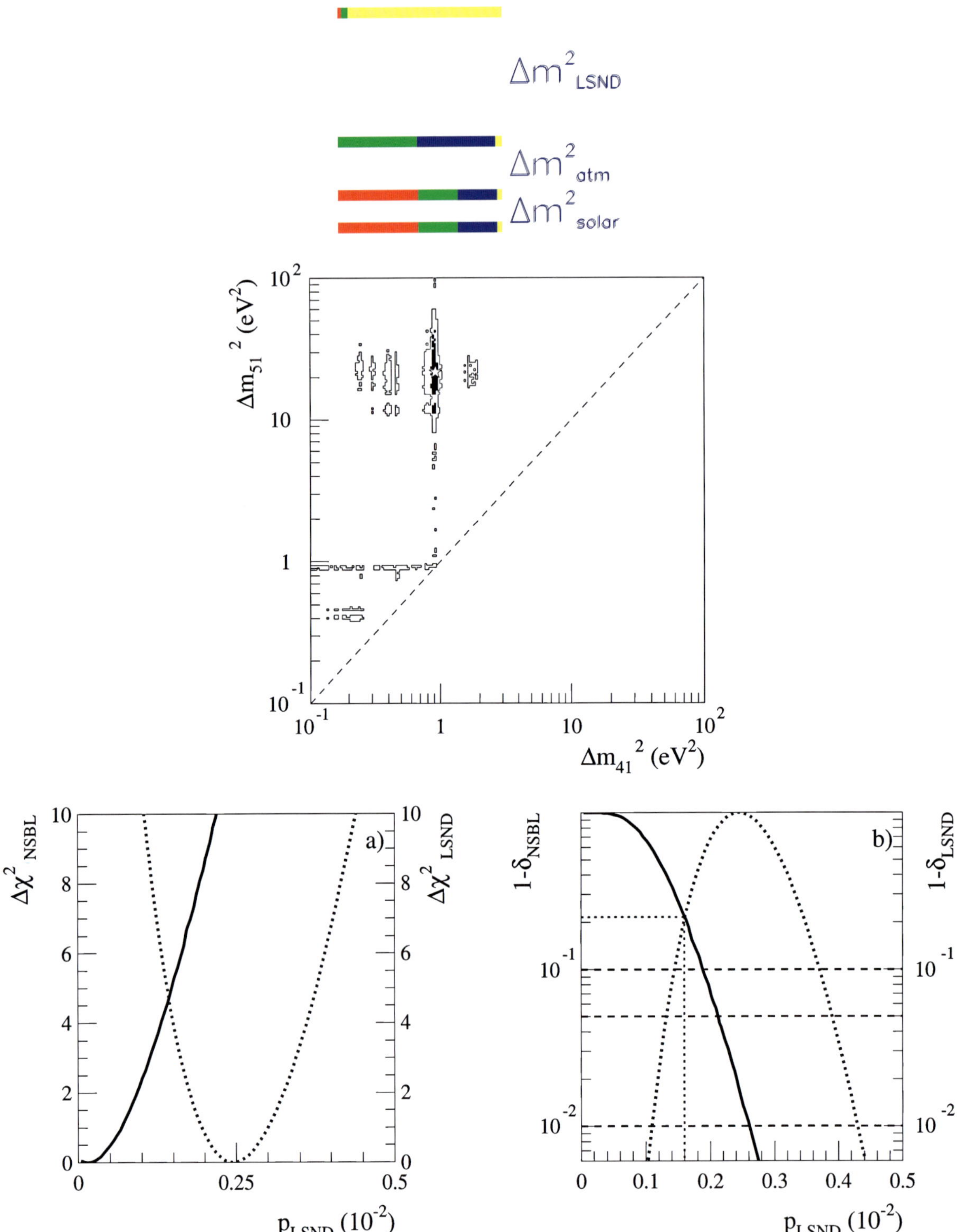

Fig. 2. Top: Cartoon of mass spectra for a 3+2, normal hierarchy. The flavor mixtures are illustrative. Middle: Combinations of pairs of Δm^2's for the sterile states in 3+2 models. Bottom: Overlap in oscillation probability between the "Null Short Baseline" experiments and LSND.

paradigm, adopted to explain neutrino mass. But this may be one more surprise from the neutrino sector, and so, given the hints, it is a question worth pursuing.

5. Pursuing these questions: MiniBooNE and beyond

The main purpose of the MiniBooNE experiment, now running at Fermilab is to resolve the question of the LSND signal. This experiment can answer the generic question: is LSND correct? It can also begin to explore possible explanations, including the

$3 + 2$ model described above. If it is run in antineutrino mode, MiniBooNE can also address CP violation.

In order to test the LSND result, the MiniBooNE design maintains $L/E \sim 1\,\mathrm{m/MeV}$ while substantially changing the systematic errors associated with the experiment. This is accomplished by increasing both L and E by an order of magnitude from the LSND design. The MiniBooNE detector is located $L = 541\,\mathrm{m}$ from the primary target, and the neutrino flux has average energy of $\sim 0.75\,\mathrm{GeV}$. This changes the source of the neutrinos (ν_μ from energetic pions rather than $\bar\nu_\mu$ from

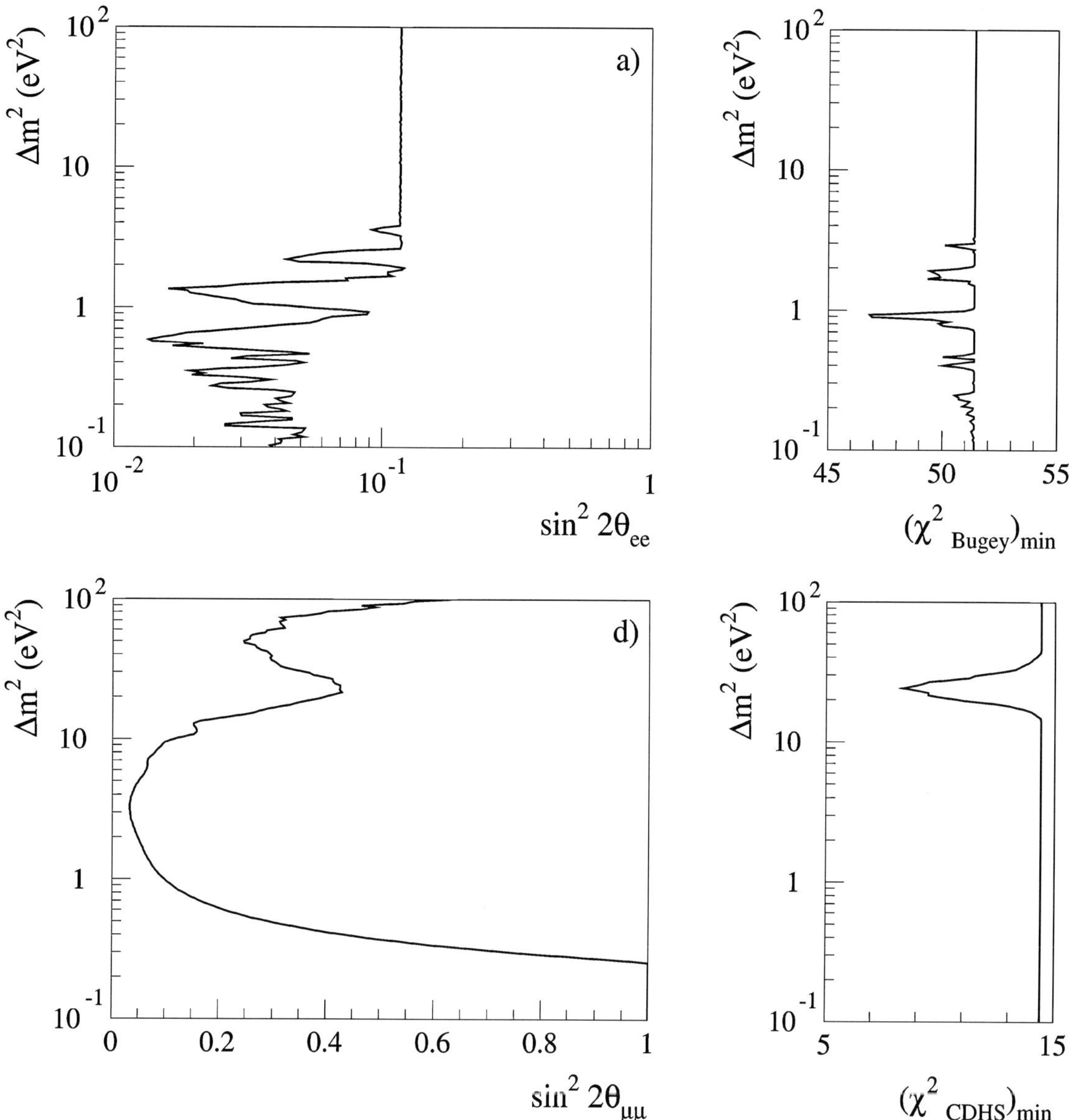

Fig. 3. For Bugey (top) and CDHS (bottom), 90% CL upper limits on oscillations derived in the 3+2 analysis (left) and the the $(\chi^2)_{\min}$ values as a function of Δm^2 (right). The best fit is pulled to the favored Δm^2 values of these two experiments.

stopped muons), the signature for the signal, and the major backgrounds in the detector. A 450 t fiducial volume pure-oil detector is used. The tank is a 12 m diameter sphere with an outer veto region and an inner detector containing 1280 phototubes. Neutrino events in the detector produce both Cerenkov and scintillation light.

At present, MiniBooNE has received about 4.2×10^{20} protons on target. The first phase run, using a beam of neutrinos, will continue through 2005. The result from the first phase of MiniBooNE will be statistics limited and will be bracketed between 5×10^{20} protons on target and 1×10^{21} protons on target. Figure 4 (top) shows the expected sensitivity for the MiniBooNE signal for this range of protons on target compared to the LSND allowed region. If a signal is observed, Figure 4 (bottom) gives examples of the measurement capability for the upper and lower statistical range.

MiniBooNE has been approved for data-taking through 2006 and has been invited to present a proposal for further running. This will represent the beginning of the second phase of the experiment. This phase is expected to continue for three or more years, with running in antineutrino mode.

If a signal is observed in MiniBooNE running, then there will be three immediate follow-up questions to be addressed:

1. Does more than one Δm^2 contribute to the signal?
2. What, if any, is the sterile neutrino content?
3. Is the oscillation probability the same for neutrinos and antineutrinos (*i.e.,* is there evidence for CP violation)?

There are several possible accelerator-based follow-ups to address questions 1 and 2, should a signal be observed. These include further development of MiniBooNE, as well as exploiting other facilities world-wide. Fermilab can pursue a MiniBooNE signal in the near term through continued running of MiniBooNE; this can evolve into BooNE, a set of detectors placed strategically along the MiniBooNE line to establish the Δm^2(s) of the signal. Also at Fermilab, the Minos [7] long baseline experiment can address the sterile neutrino content through comparison of the NC/CC oscillation signal to the near-to-far oscillation signal. In Japan, the T2K near detector complex [9] is proposed to consist of one small detector at 200 m and another large detector at 2 km. The small detector can be used to measure the flux, and the large detector is well placed for observing an LSND signal. This

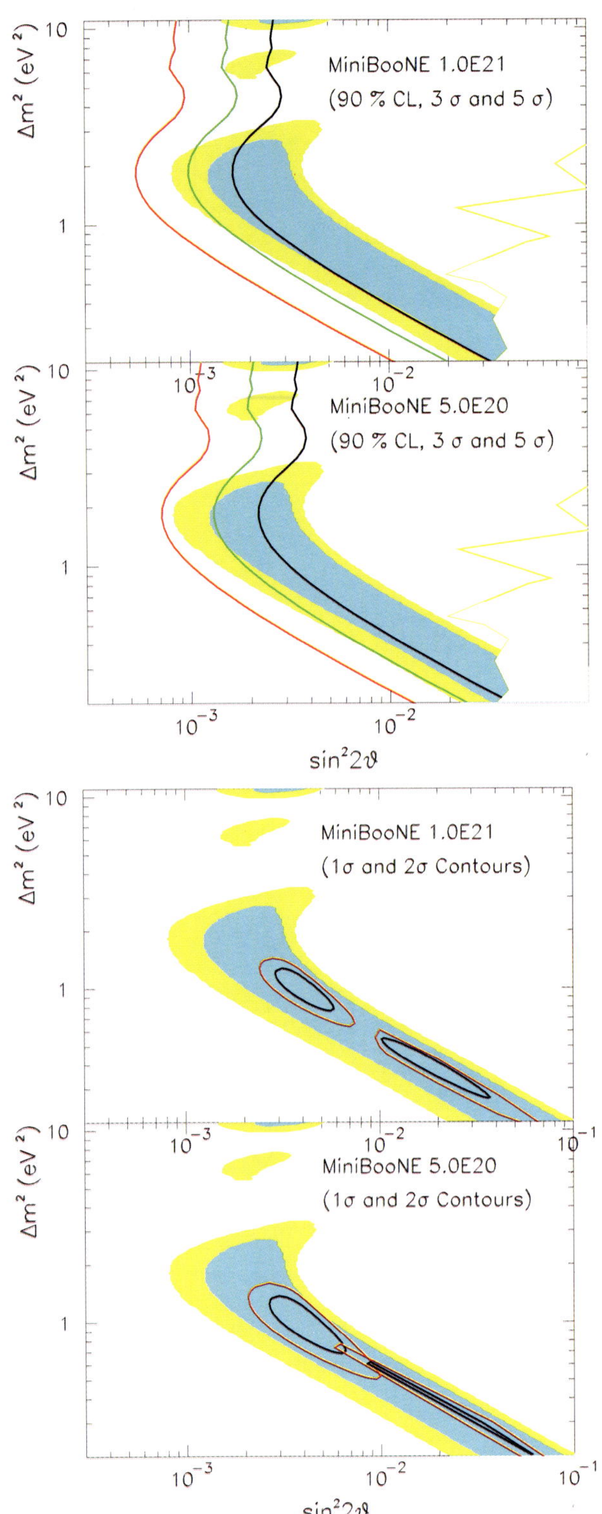

Fig. 4. Top Pair: Sensitivity of MiniBooNE for 1×10^{21} and 5×10^{20} protons on target. Bottom Pair: Examples of the measurement-capability, should a signal be observed.

configuration may be able to accurately address the number of contributing Δm^2's. Lastly, an elegant method of addressing the LSND question is to return to a decay-at-rest beam using the SNS beam dump [13]. This has a fast pulse of ν_μ beam from pion DAR, followed by $\bar{\nu}_\mu$ and ν_e from muon DAR. Sensitivity would be to the same range in Δm^2 as LSND and the well defined flux shape allows the number of contributing mass states to be addressed.

The most immediate way to address question 3 is to run MiniBooNE with an antineutrino beam. This is straightforward

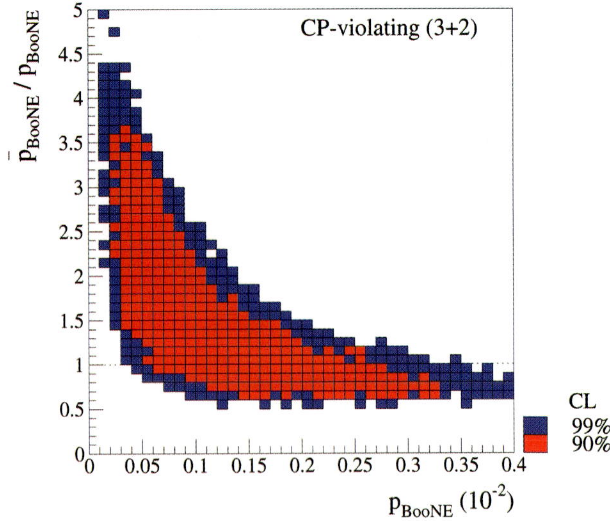

Fig. 5. Each point represents an allowed oscillation probability in neutrino vs. antineutrino mode for MiniBooNE, based on 3+2 models with CP violation.

to do given the experimental layout, and involves changing only one charging supply in the beamline. For MiniBooNE oscillation probability for ν versus $\bar{\nu}$ running can be significantly different in $3+2$ models. Figure 5 provides an example of the expectations for CP violating models (preliminary) [37]. Each point represents a choice for masses, mixings and the CP-violating phase which is consistent with the NSBL and LSND experiment. The calculation uses the MiniBooNE neutrino and antineutrino-mode fluxes, thereby taking the energy dependence and wrong-sign contamination correctly into account. The x-axis is the MiniBooNE oscillation probability in neutrino mode. The y-axis is the ratio of the oscillation probability in antineutrino to neutrino mode for MiniBooNE. Once can see that CP violation can potentially enhance or decrease the oscillation probability in antineutrino mode, depending on the mixing parameters.

6. Conclusions

History more than suggests – it almost insists that unexpected results will appear in the next round of neutrino experiments. These may or may not be one of the topics discussed above, but they will surely appear. Accelerator-based experiments are ideal places for these beyond-the-standard-paradigm observations, because of the high intensity and control of systematics. And so, onward . . . to the next neutrino theory!

References

1. See talk by M. Shaevitz.
2. Aguilar, A. *et al.*, Phys. Rev. D **64**, 112007 (2001), hep-ex/0104049.
3. Maltoni, M. *et al.*, hep-ph/0305312; Maltoni, M. *et al.*, Nucl. Phys. Proc. Suppl. **114**, 203 (2003).
4. Sorel, M., Conrad, J. and Shaevitz, M., Phys. Rev. D **70**, 073004 (2004), hep-ph/0305255.
5. See talks by Gelmini, G. and Steigman, G., this volume.
6. Barger, V. *et al.*, Phys. Lett. B **462**, 109 (1999), hep-ph/9907421.
7. Shanahan, P., Eur. Phys. J. C **33**, S834 (2004).
8. Ashie, Y. *et al.*, Phys. Rev. Lett. **93**, 101801 (2004).
9. T2K Letter of Intent, http://neutrino.kek.jp/jhfnu
10. Grimus, W. and Lavoura, L., hep-ph/0405261.
11. McConnel, K. and Shaevitz, M., hep-ex/0409028.
12. Diwan, M. *et al.*, Phys. Rev. D **68**, 012002 (2003); hep-ph/0303081.
13. Garvey, G. T. *et al.*, hep-ph/0501013.

14. See, for example, Brice, S. J. *et al.*, hep-ex/0408135.
15. Stancu, I. *et al.*, Horn Technical Design Report, available from http://www-boone.fnal.gov/publicpages/.
16. http://www.phy.bnl.gov/~e910/html/home.html.
17. http://harp.web.cern.ch/harp.
18. http://ppd.fnal.gov/experiments/e907/e907.htm.
19. home.fnal.gov/~bfleming/finese.html.
20. Casper, D., Nucl. Phys. Proc. Suppl. **112**, 161 (2002).
21. http://neutrino.kek.jp.
22. http://www-boone.fnal.gov.
23. http://www.sno.phy.queensu.ca.
24. http://www.cern.ch/opera.
25. http://www-nova.fnal.gov. (2001) 112007.
26. Armbruster, B. *et al.*, Phys. Rev. D **65**, 112001 (2002).
27. Achkar, B. *et al.*, Nucl. Phys. B **434**, 503 (1995).
28. Church, E. D., Eitel, K., Mills, G. B. and Steidl, M., Phys. Rev. D **66**, 013001 (2002), hep-ex/0203023.
29. McKellar, B. H. J. *et al.*, hep-ph/0106121; Mohapatra, R. N., Phys. Rev. D **64**, 091301 (2001); hep-ph/0107264; Ioannisian, A. and Valle, J. W. F., Phys. Rev. D **63**, 073002 (2001); Ma, E., Rajasekaran, G. and Sarkar, U., Phys. Lett. B **495**, 363 (2000); hep-ph/0006340; Berezhiani, Z. and Mohapatra, R., Phys. Rev. D **52**, 6607 (1995); Fardon, R., Nelson, A. E. and Weiner, N., arXiv:astro-ph/0309800.
30. Smirnov, A. Yu., Nucl. Phys. Proc. Suppl. **118**, 87 (2003) [arXiv:hep-ph/0209131].
31. Astier, P. *et al.*, Phys. Lett. B **570**, 19 (2003); hep-ex/0306037.
32. Dydak, F. *et al.*, Phys. Lett. B **134** 281 (1984).
33. Stockdale, I. E. *et al.*, Phys. Rev. Lett. **52**, 1384 (1984); Z. Phys. C **27**, 53 (1985).
34. Apollonio, M. *et al.*, hep-ex/0301017.
35. See talks by Gelmini, G. and Steigman. G., Also see: Beacom, J. F., Bell, N. F., Dodelson, S., astro-ph/0404585; Chacko, Z., Hall, L. J., Oliver, S. J. and Perelstein, M., hep-ph/0405067; Abazajian, K., Bell, N. F., Fuller, G. M. and Wong, Y. Y. Y., astro-ph/0410175.
36. Cyburt, R. *et al.*, astro-ph/0408033; Mohapatra, R. N. and Nasri, S., hep-ph/0407194; Hannestad, S., astro-ph/0303076.
37. Aguilar-Arevalo, A., Barger, V., Conrad, J., Sorel, M. and Whisnant, K., paper in preparation.

Physica Scripta. Vol. T121, 46–50, 2005

Solar Models and Solar Neutrinos

J. N. Bahcall[*,1]

[1] Institute for Advanced Study, School of Natural Sciences, Princeton, NJ 08540, USA

Received November 27, 2004; accepted in revised form March 1, 2005

PACS numbers: 96.60.Jw, 26.65.+t

Abstract

I provide a summary of the current theoretical knowledge of solar neutrino fluxes as derived from precise solar models.

1. Introduction

I summarize in this talk the present state of solar model research as the subject relates to solar neutrino investigations. This is not a review talk. I will focus on the latest developments. [In a few places, I will add comments within square brackets on results that have been obtained since the symposium took place.]

To establish the appropriate context for our discussion, I note that measurements of the sound speed inside the Sun (via a technique called helioseismology, which is similar to terrestrial seismology) agree with solar model predictions to a root-mean-square accuracy of better than 0.1%. This excellent agreement between the measured and predicted sound speeds established as early as 1996 that the solution of the "solar neutrino problem" lay in new particle physics, not new astrophysics [1].

There are two specific reasons for continuing to study the solar neutrino fluxes that are predicted by precise solar models. First, the model predictions must be refined (uncertainties reduced) in order to use optimally the neutrino observations to provide information about the interior of the Sun and about some of the parameters used in solar and stellar models. Second, the theoretical knowledge of the neutrino fluxes can be used, in combination with solar neutrino experiments, to refine neutrino parameters such as Δm_{21}^2 and $\tan^2 \theta_{12}$ (see, e.g., [2]).

More generally, we want to use solar models and solar observations as a laboratory for testing and improving the theory of stellar evolution, the theory of how stars shine and evolve. The Sun is the nearest star to Earth and we have much more information about the Sun than about any other star. We can use the comparison between the predictions of precise solar models and the observations of solar properties to explore and refine the theory of stellar evolution. Almost every branch of astronomy and astrophysics uses stellar evolution theory, in one way or another, to interpret observations of distant stars and the elements they produce.

Recent measurements of the surface chemical composition of the Sun have posed a new 'solar problem', a problem that does not have significant implications for neutrinos but may be of importance for the theory of stellar evolution and perhaps for other branches of astronomy. My own work has been focused recently on this aspect of solar models and I will comment briefly on the subject toward the end of my talk.

I begin in Section 2 by summarizing the standard solar model predictions of solar neutrino fluxes. I also summarize the uncertainties in the predicted fluxes. For the first two and a half decades of the solar neutrino problem, the most critical theoretical issue was to establish that the uncertainties in the predictions were less than the discrepancies between the solar model predictions and the solar neutrino measurements. This goal was achieved somewhere between 1982 and 1966, depending upon the degree of skepticism toward astronomical calculations and measurements of the person making the judgment. Actually, for decades most physicists and a surprisingly large number of astronomers believed that inaccurate solar model calculations rather than neutrino properties were responsible for the solar neutrino problem. This situation changed dramatically with the publication of the first SNO experimental results in June, 2001, when it became clear to all that the solution of the solar neutrino problem was new physics rather than inadequate astronomy.

In Section 3, I comment briefly on the implications of recent measurements of the surface chemical composition of the Sun. I conclude in Section 4 with a discussion of the role of neutrinos as dark matter, a subject treated at this symposium by Max Tegmark more extensively and from a different perspective.

2. Solar model fluxes

I base the discussion in this section on the results reported in the recent paper [3]. Full numerical details of the solar models, BP04 and BP00 that are discussed below are presented, together with earlier solar models in this series, at the Web site: http://www.sns.ias.edu/~jnb. In particular, the primordial hydrogen and helium mass fractions for the BP04 model discussed below are, respectively, $X_0 = 0.7078$ and $Y_0 = 0.2734$, while the current surface abundances are $X_S = 0.7397$ and $Y_S = 0.2434$.

2.1. *Fluxes from different solar models*

Table I, taken from Ref. [3], gives the calculated solar neutrino fluxes for a series of solar models calculated with different plausible assumptions about the input parameters. The range of fluxes shown for these models illustrates the systematic uncertainties in calculating solar neutrino fluxes. The second (third) column, labeled BP04 (BP04+), of Table I presents the current best solar model calculations for the neutrino fluxes. The uncertainties in the calculated neutrino fluxes are given in column 2. The model BP04 utilizes the older heavy element abundances on the surface of the Sun as summarized in the paper by Grevesse and Sauval (1998) [6]. The BP04+ model uses recently determined heavy element abundances [7], which are discussed in more detail in Section 3.

Figure 1 presents the neutrino energy spectrum predicted by the BP04 solar model for the most important solar neutrino sources.

The model BP04+ was calculated with the use of new input data for the equation of state, nuclear physics, and solar

*Email address: jnb@sns.ias.edu

Table I. *Predicted solar neutrino fluxes from solar models. The table presents the predicted fluxes, in units of $10^{10}(pp)$, $10^9(^7Be)$, $10^8(pep, ^{13}N, ^{15}O)$, $10^6(^8B, ^{17}F)$, and $10^3(hep)$ $cm^{-2}s^{-1}$. Columns 2–4 show BP04, BP04+, and the previous best model BP00 [4]. Columns 5–7 present the calculated fluxes for solar models that differ from BP00 by an improvement in one set of input data: nuclear fusion cross sections (column 5), equation of state for the solar interior (column 6), and surface chemical composition for the Sun (column 7). Column 8 uses the same input data as for BP04 except for a recent report of the $^{14}N+p$ fusion cross section. References to the improved input data are given in the text. The last two rows ignore neutrino oscillations and present for the chlorine and gallium solar neutrino experiments the capture rates in SNU (1 SNU equals 10^{-36}events per target atom per sec). Due to oscillations, the measured rates are smaller: 2.6 ± 0.2 and 69 ± 4, respectively. The neutrino absorption cross sections and their uncertainties are given in Ref. [5].*

Source	BP04	BP04+	BP00	Nucl	EOS	Comp	^{14}N
pp	$5.94\,(1 \pm 0.01)$	5.99	5.95	5.94	5.95	6.00	5.98
pep	$1.40\,(1 \pm 0.02)$	1.42	1.40	1.40	1.40	1.42	1.42
hep	$7.88\,(1 \pm 0.16)$	8.04	9.24	7.88	9.23	9.44	7.93
^{7}Be	$4.86\,(1 \pm 0.12)$	4.65	4.77	4.84	4.79	4.56	4.86
^{8}B	$5.82\,(1 \pm 0.23)$	5.28	5.05	5.79	5.08	4.62	5.77
^{13}N	$5.71\,(1\,^{+0.37}_{-0.35})$	4.06	5.48	5.69	5.51	3.88	3.23
^{15}O	$5.03\,(1\,^{+0.43}_{-0.39})$	3.54	4.80	5.01	4.82	3.36	2.54
^{17}F	$5.91\,(1\,^{+0.44}_{-0.44})$	3.97	5.63	5.88	5.66	3.77	5.85
Cl	$8.5^{+1.8}_{-1.8}$	7.7	7.6	8.5	7.6	6.9	8.2
Ga	131^{+12}_{-10}	126	128	130	129	123	127

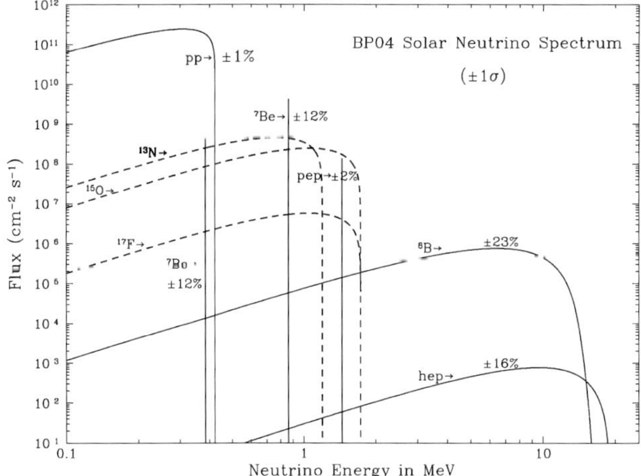

Fig. 1. The predicted solar neutrino energy spectrum. The figure shows the energy spectrum of solar neutrinos predicted by the BP04 solar model [3]. For continuum sources, the neutrino fluxes are given in number per $cm^{-2}s^{-1}MeV^{-1}$ at the Earth's surface. For line sources, the units are number per $cm^{-2}s^{-1}$. The total theoretical uncertainties for the neutrinos in the p-p chain are taken from column 2 of Table I and are shown for each source. In order not to complicate the Fig., I have omitted the very large uncertainties for the difficult-to-detect CNO neutrino fluxes (see Table I).

composition. The model BP04, the currently preferred model, is the same as BP04+ except that BP04 does not include the most recent analyses of the solar surface composition [7]. My collaborators and I prefer the model BP04 over the model BP04+ because the lower heavy element abundance used in calculating BP04+ causes the calculated sound speeds and the depth of the solar convective zone to conflict with helioseismological measurements [3, 8].

The error estimates, which are the same for the three models labeled BP04, BP04+, and ^{14}N in Table I, include the recent composition analyses. The differences between the neutrino fluxes calculated for the solar models BP04 and BP04+ are all well within the estimated uncertainties in the calculated fluxes. This is one way of seeing that the 'new solar problem' associated with the lower heavy element abundances that have been reported recently does not affect solar neutrino questions in a significant way.

Column four of Table I presents the fluxes calculated using the preferred solar model, BP00 [4], that was posted on the archives in October 2000 (prior to the first reported SNO measurement). The BP04 best-estimate neutrino fluxes and their uncertainties have not changed markedly from their BP00 values despite refinements in input parameters. The only exception is the CNO flux uncertainties which have almost doubled due to the larger systematic uncertainty in the surface chemical composition estimated in this paper.

The columns 5–7 of Table I describe improvements in the input data relative to BP00, i.e., relative to our best standard solar model constructed in 2000. Quantities that are not discussed here are the same as for BP00. Each class of improvement is represented by a separate column, columns 5–7. The magnitude of the changes between the fluxes listed in the different columns of Table I are one measure of the sensitivity of the calculated fluxes to the input data.

Column 5 contains the fluxes computed for a solar model that is identical to BP00 except that improved values for direct measurements of the ^{7}Be$(p,\gamma)^8$B cross section [9, 10], and the calculated p-p and hep cross sections [10]. The reactions that produce the ^{8}B and hep neutrinos are rare; changes in their production cross sections only affect, respectively, the ^{8}B and hep fluxes. The 15% increase in the calculated ^{8}B neutrino flux, which is primarily due to a more accurate cross section for ^{7}Be$(p,\gamma)^8$B, is the only significant change in the best-estimate fluxes.

The fluxes in Column 6 were calculated using a refined equation of state, which includes relativistic corrections and a more accurate treatment of molecules [11]. The equation of state improvements between 1996 and 2001, while significant in some regions of parameter space, change all the solar neutrino fluxes by less than 1%. Solar neutrino calculations are insensitive to the present level of uncertainties in the equation of state.

The most important changes in the astronomical data since BP00 result from new analyses of the surface chemical composition of the Sun. The input chemical composition affects the radiative opacity and hence the physical characteristics of the solar model, and to a lesser extent the nuclear reaction rates. New values for C, N, O, Ne, and Ar have been derived [7] using three-dimensional rather than one-dimensional atmospheric models, including hydrodynamical effects, and paying particular attention to uncertainties in atomic data and observational spectra. The new abundance estimates, together with the previous best-estimates for other solar surface abundances [6], imply a ratio of heavy elements to hydrogen by mass of $Z/X = 0.0176$, much less than the previous value of $Z/X = 0.0229$ [6]. Column 7 gives the fluxes calculated for this new composition mixture. The largest change in the neutrino fluxes for the p-p chain is the 9% decrease in the predicted ^{8}B neutrino flux. The N and O fluxes are decreased by much more, $\sim 35\%$, because they reflect directly the inferred C and O abundances. [Subsequent to the completion of the calculations described here, additional changes in the recently-determined heavy element abundances have been made and are reported in Ref. [12]. These additional changes are sufficiently small that they do not affect any of the conclusions expressed in the present review.]

The CNO nuclear reaction rates are less well determined than the rates for the more important (in the Sun) p-p reactions [13]. The rate for ^{14}N(p, γ)^{15}O is poorly known, but is important for calculating CNO neutrino fluxes. Extrapolating to the low energies relevant for solar fusion introduces a large uncertainty. Column 8 gives the neutrino fluxes calculated with input data identical to BP04 except for the cross section factor $S_0(^{14}$N$ + p) = 1.77 \pm 0.2$ keV b that is about half the current best-estimate; this value assumes a particular R-matrix fit to the experimental data [14]. The recent LUNA measurement [15] of the cross section factor $S_0(^{14}$N$ + p)$ provides a firm experimental basis for the assumed smaller value of the cross section factor. The p-p cycle fluxes are changed by only $\sim$1%, but the ^{13}N and ^{15}O neutrino fluxes are reduced by 40%–50% relative to the BP04 predictions. CNO nuclear reactions contribute 1.6% of the solar luminosity in the BP04 model and only 0.8% in the model with a reduced $S_0(^{14}$N$ + p)$.

2.2. *Flux uncertainties*

Table II, also taken from Ref. [3], shows the individual contributions to the flux uncertainties. These uncertainties are useful in deciding how accurately we need to determine a given input parameter in order to improve the overall accuracy of the calculated neutrino fluxes. Moreover, the theoretical flux uncertainties continue to play a significant role in some determinations of neutrino parameters from solar neutrino experiments (see, e.g., Ref. [16]).

Columns 2–5 present the fractional uncertainties from the nuclear reactions whose measurement errors are most important for calculating neutrino fluxes. Unless stated otherwise, the uncertainties in the nuclear fusion cross sections are taken from Ref. [13].

The measured rate of the ^{3}He$-^{3}$He reaction, which changed by a factor of 4 after the first solar model calculation of the solar neutrino flux [17], and the measured rate of the ^{7}Be $+$ p reaction, which for most of this series has been the dominant uncertainty in predicting the ^{8}B neutrino flux, are by now very

well determined. If the current published systematic uncertainties for the ^{3}He$-^{3}$He and ^{7}Be $+$ p reactions are correct, then the uncertainties in these reactions no longer contribute in a crucial way to the calculated theoretical uncertainties (see column 2 and column 4 of Table II). This felicitous situation is the result of an enormous effort extending over four decades, and represents a great collective triumph, for the nuclear physics community.

At the present time, the most important nuclear physics uncertainty in calculating solar neutrino fluxes is the rate of the ^{3}He$-^{4}$He reaction (see column 3 of Table II). The systematic uncertainty in the rate of ^{3}He(^{4}He, γ)^{7}Be reaction (see Ref. [13]) causes an 8% uncertainty in the prediction of both the ^{7}Be and the ^{8}B solar neutrino fluxes. It is scandalous that there has not been any progress in the past 15 years in measuring this rate more accurately. [A very recent precise measurement has been reported for this reaction at center-of-mass energies above 400 keV [18], but so far there have not been any other measurements that could test the accuracy of this result.]

For ^{14}N(p, γ)^{15}O, we have continued to use in Table II the uncertainty given in Ref. [13], although the recent reevaluation in Ref. [14] suggests that the uncertainty could be somewhat larger (see column 7 of Table I).

The dominant uncertainty, 15.1% (1σ), for the hep neutrino flux results from the difficult theoretical calculation of the low energy cross section factor for this reaction [19].

The uncertainties due to the calculated radiative opacity and element diffusion, as well as the measured solar luminosity (columns 6–8 of Table II), are all moderate, non-negligible but not dominant. For the ^{8}B and CNO neutrino fluxes, the uncertainties that are due to the radiative opacity, diffusion coefficient, and solar luminosity are all in the range 2% to 6%.

The surface composition of the Sun is the most problematic and important source of uncertainties. Systematic errors dominate: the effects of line blending, departures from local thermodynamic equilibrium, and details of the model of the solar atmosphere. In the absence of detailed information to the contrary, it is assumed that the uncertainty in all important element abundances is approximately the same. The 3σ range of Z/X is defined as the spread over all modern determinations (see Refs. [4, 17, 20]), which implies that at present $\Delta(Z/X)/(Z/X) = 0.15$ (1σ), 2.5 times larger than the uncertainty adopted in discussing the predictions of the model BP00 [4]. The most recent uncertainty quoted for oxygen, the most abundant heavy element in the Sun, is similar: 12% [7].

Heavier elements like Fe affect the radiative opacity and hence the neutrino fluxes more strongly than the relatively light elements [4]. This is the reason why the difference between the fluxes calculated with BP04 and BP04+ (or between BP00 and Comp, see Table I) is less than would be expected for the 26% decrease in Z/X. The abundances that have changed significantly since BP00 (C, N, O, Ne, Ar) are all for lighter volatile elements for which meteoritic data are not available.

The dominant uncertainty listed in Table II for the ^{8}B and CNO neutrinos is the chemical composition, represented by Z/X (see column 9). The uncertainty ranges from 20% for the ^{8}B neutrino flux to $\sim$35% for the CNO neutrino fluxes. Since the publication of BP00, the best published estimate for Z/X decreased by 4.3σ (BP00 uncertainty) and the estimated uncertainty due to Z/X increased for ^{8}B (^{15}O) neutrinos by a factor of 2.5 (2.8). Over the past three decades, the changes have almost always been toward a smaller Z/X. The monotonicity is surprising since different sources of improvements have

Table II. *Principal sources of uncertainties in calculating solar neutrino fluxes. Columns 2–5 present the fractional uncertainties in the neutrino fluxes from laboratory measurements of, respectively, the ^{3}He$-^{3}$He, ^{3}He$-^{4}$He, p$-^{7}$Be, and p$-^{14}$N nuclear fusion reactions. The last four columns, 6–9, give, respectively, the fractional uncertainties due to the calculated radiative opacity, the calculated rate of element diffusion, measured solar luminosity, and the measured heavy element to hydrogen ratio.*

Source	3–3	3–4	1–7	1–14	Opac	Diff	$L\odot$	Z/X
pp	0.002	0.005	0.000	0.002	0.003	0.003	0.003	0.010
pep	0.003	0.007	0.000	0.002	0.005	0.004	0.003	0.020
hep	0.024	0.007	0.000	0.001	0.011	0.007	0.000	0.026
^{7}Be	0.023	0.080	0.000	0.000	0.028	0.018	0.014	0.080
^{8}B	0.021	0.075	0.038	0.001	0.052	0.040	0.028	0.200
^{13}N	0.001	0.004	0.000	0.118	0.033	0.051	0.021	0.332
^{15}O	0.001	0.004	0.000	0.143	0.041	0.055	0.024	0.375
^{17}F	0.001	0.004	0.000	0.001	0.043	0.057	0.026	0.391

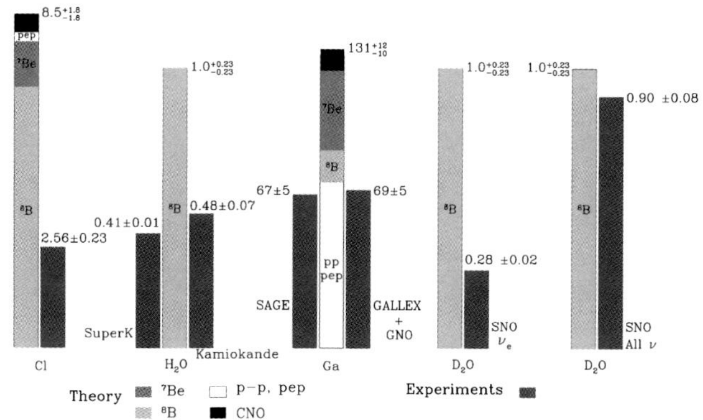

Fig. 2. Theory versus experiment. The figure compares the predictions of the standard solar model plus the standard model of electroweak interactions with the measured rates in all solar neutrino experiments. The solar neutrino experimental results are discussed in talks by Y. Suzuki and A. McDonald in this proceeding.

caused successive changes. Nevertheless, since the changes are monotonic, the uncertainty estimated from the historical record is large.

The total ^{8}B neutrino flux measured by the neutral current mode of the SNO experiment [21] is

$$\phi(^8\text{B}, \text{SNO}) = 0.90\phi(^8\text{B}, \text{BP04 solar model})$$
$$\times [1.0 \pm 0.09 \pm 0.23], \qquad (1)$$

where the first uncertainty listed in equation (1) is the 1σ measurement error and the second (larger) uncertainty is the estimated 1σ uncertainty in the solar model calculation (taken from Table II). If all the data from solar neutrino and reactor experiments are combined together (see figure 2 below), the above relation becomes [16]

$$\phi(^8\text{B}, \text{SNO}) = 0.87\phi(^8\text{B}, \text{BP04 solar model})$$
$$\times \left[1.0 \pm 0.05 \pm 0.23\right]. \qquad (2)$$

The calculated ^{8}B neutrino flux [3] agrees with the measured flux to better than 1σ. The theoretical uncertainty is much larger than the uncertainty in the measurements. The dominant theoretical uncertainty for the ^{8}B neutrino flux is due to uncertainties in the measured surface heavy element abundances of the Sun.

[For decades, the composition uncertainty has been calculated by considering the uncertainty in the total heavy element abundance by mass, Z, or the uncertainty in the total heavy element to hydrogen ratio, Z/X. In an article now in preparation, Bahcall and Serenelli 2005, we have made a more refined calculation of the abundance uncertainties by evaluating separately the abundance uncertainty of each heavy element. The final result for the total uncertainty to be used in equation (1) and equation (2) is ± 0.16 instead of ± 0.23.]

The p–p solar neutrino flux has been measured by combining the results from all the relevant solar neutrino and reactor experiments, together with the imposition of the luminosity constraint [22]. The result is [23]

$$\phi(\text{p} - \text{p}, \text{all neutrino experiments})$$
$$= 1.01\phi(\text{p} - \text{p}, \text{BP04 solar model})[1.0 \pm 0.02 \pm 0.01], \qquad (3)$$

where the first uncertainty listed in equation (3) is the 1σ measurement error and the second (smaller) uncertainty is the

estimated 1σ uncertainty in the solar model calculation (taken from Table II).

3. Recent developments regarding the solar surface composition

In the last several years, determinations of the solar surface abundances of heavy elements have become more refined and detailed [7, 12, 24]. These recent determinations yield significantly lower values than were previously adopted (e.g., by Grevesse and Sauval 1998 [6]) for the abundances of the volatile heavy elements: C, N, O, Ne, and Ar. However, these recent abundance determinations also lead to solar models that disagree with helioseismological measurements ([3, 8]).

Standard solar models constructed with the newly-determined heavy element abundances predict incorrectly the solar sound speeds, the depth of the solar convective zone, and the surface helium abundance (see especially Ref. [8]). The discrepancies occur in the temperature region below the solar convective zone, 2×10^6 K to 4.5×10^6 K [8]. In this temperature domain, the volatile heavy elements, C, N, O, Ne, and Ar, are partially ionized and their abundances significantly affect the radiative opacities. Fortunately, neutrinos are produced in the deep solar interior not in this outer region, $0.45 R_\odot$ to $0.72 R_\odot$, where the discrepancies with helioseismology exist. This is the reason why the new abundances do not affect significantly the calculated neutrino fluxes, i.e., the reason why the models BP04 and BP04+ in Table I yield similar neutrino fluxes.

What is wrong? I don't know. One possibility we have considered is that the standard radiative opacity needs to be increased (by about 11%) near the base of the convective zone. However, detailed and refined recalculations of the radiative opacity by the Opacity Project collaboration disfavor [25, 26] the suggestion [8, 27] that the origin of the discrepancy might be the adopted opacities rather than the adopted heavy element abundances. Another possibility that astronomers have begun to discuss in the literature is that something is systematically wrong with the new abundance determinations. After all, the old standard abundances lead to solar models that are in excellent agreement everywhere with the helioseismological measurements. However, no one has identified a serious problem with the new abundance measurements. But, it would certainly be healthy for the field if more than one group were to undertake a thorough reevaluation of the surface heavy element abundances of the Sun.

As we continue to measure more precisely the properties of the Sun and to compare the results with theoretical solar models, we continue to learn new things. Nature seems to surprise us each time. The present conflict engendered by the new heavy element abundance determinations may have significant implications for other parts of stellar physics. After all, if we can't get the abundances or the opacity or something else significant correct for the Sun, about which we know much more than any other star, how can we have confidence in our quantitative inferences for other less well studied stars?

4. Neutrinos as dark matter

Neutrinos are the first cosmological dark matter to be discovered. I cannot resist making a few remarks on this subject, because of the importance of dark matter to both physics and astronomy.

Solar and atmospheric neutrino experiments show that neutrinos have mass but these oscillations experiments only determine the differences between masses, not the absolute values.

If we make the plausible but unproven assumption that the lowest neutrino mass, m_1, is much less than the square root of $\Delta m^2_{\text{solar}}$, then we can conclude that the mass of cosmological neutrino background is dominated by the mass of the heaviest neutrino. This heaviest neutrino mass is then determined by $\Delta m^2_{\text{atmospheric}}$. With this assumption the cosmological mass density in neutrinos is only [28, 29, 30]

$$\Omega_\nu = (0.0009 \pm 0.0001), \qquad m_1 \ll \sqrt{(\Delta m^2_{\text{solar}})}. \tag{4}$$

Although the mass density given in Equation (4) is small, it is of the same order of magnitude as the observed mass density in stars and gas.

The major uncertainty in determining by neutrino experiments the value of Ω_ν is the unknown value of the lowest neutrino mass. It is possible that neutrino masses are nearly degenerate and cluster around the highest mass scale allowed by direct beta-decay experiments. If, for example, all neutrino masses are close to 1 eV, then $\Omega_\nu(1\text{ ev}) \sim 0.03$, which would be cosmologically significant, but would not explain the bulk of the dark matter.

More sensitive neutrino beta-decay experiments and neutrino-less double beta-decay experiments offer the best opportunities for determining the mass of the lowest mass neutrino and hence establishing the value of Ω_ν from purely laboratory measurements.

For most of the period of 1968–2001, between the first suggestion of a solar neutrino problem and its final resolution as indicating neutrino oscillations, the great majority of physicists believed that solar neutrino experiments indicated the need for improvements in the solar model, not in the standard model of particle physics. In part, this was a matter of aesthetics; the standard model of particle physics was beautiful and experimentally successful. Why mess up this beautiful theory with an ugly neutrino mass? By contrast with the elegant laboratory experiments that confirmed the standard electroweak model, the interior of the appeared Sun remote and complicated. Also, the belief in inadequate astrophysics rather than incomplete physics was, I believe, partially a result of a lack of familiarity with the accuracy of the solar model calculations. For some of the physicists that believed in solar neutrino oscillations for particle physics reasons (see the discussion by Pierre Ramond in this volume), the most attractive possibility was that neutrino masses accounted for the missing dark matter. One could argue persuasively, using Ockham's razor, that this was the only 'natural' value. Unfortunately, as has happened so often in the history of weak interaction physics, our aesthetic principles were violated. The neutrino has a mass, but the neutrino mass does not account for nearly all of the dark matter if we adopt the most plausible mass hierarchy (see equation (4)).

References

1. Bahcall, J. N., Pinsonneault, M. H., Basu, S. and Christensen-Dalsgaard, J., Phys. Rev. Lett. **78**, 171 (1997).
2. Fogli, G. L., Lisi, E., Marrone, A., Montanino, D. and Palazzo, A., Phys. Rev. D **66**, 053010 (2002).
3. Bahcall, J. N. and Pinsonneault, M. H., Phys. Rev. Lett. **92**, 121301 (2004).
4. Bahcall, J. N., Pinsonneault, M. H. and Basu, S., Astrophys. J. **555**, 990 (2001).
5. Bahcall, J. N., Phys. Rev. C **56**, 3391 (1997); Bahcall, J. N. *et al.*, Phys. Rev. C **54**, 411 (1996).
6. Grevesse, N. and Sauval, A. J., Space Sci. Rev. **85**, 161 (1998).
7. Allende Prieto, C., Lambert, D. L. and Asplund, M., Astrophys. J. **573**, L137 (2002); *ibid.* **556**, L63 (2001); Asplund, M., Grevesse, N., Sauval, A. J., Allende Prieto, C. and Kiselman, D., to appear in Astron. Astrophys. astro-ph/0312290; Asplund, M., private communication (2004); see also Asplund, M., Nordlund, A., Trampedach, R. and Stein, R. F., Astron. Astrophys. **359**, 743 (2000); Asplund, M., Astron. Astrophys. **359**, 755 (2000).
8. Bahcall, J. N., Basu, S., Pinsonneault, M. and Serenelli, A. M., Astrophys. J. **618**, 1049 (2005).
9. Junghans, A. R. *et al.*, Phys. Rev. C **68** 065803 (2003); Baby, L. *et al.* (ISOLDE Collaboration), Phys. Rev. Lett. **90**, 022501 (2003); *ibid.* Phys. Rev. C **67**, 065805 (2003); Hammache, F. *et al.*, Phys. Rev. Lett. **80**, 928 (1998); Hammache, F. *et al.*, Phys. Rev. Lett. **86**, 3985 (2001); Hass, M. *et al.* (ISOLDE Collaboration), Phys. Lett. B **462**, 237 (1999); Strieder, F. *et al.*, Nucl. Phys. A **696**, 219 (2001); Junghans, A. R. *et al.*, Phys. Rev. Lett. **88**, 041101 (2002).
10. Park, T. S. *et al.*, Phys. Rev. C **67**, 055206 (2003).
11. Rogers, F. J. and Nayfonov, A., Astrophys. J. **576**, 1064 (2002); Rogers, F. J., Contrib. Plasma Phys. **41**, 179 (2001).
12. Asplund, M., Grevesse, N. and Sauval, A. J., to appear in "Cosmic Abundances as Records of Stellar Evolution and Nucleosynthesis," (eds. F. N. Bash and T. G. Barnes) (ASP Conference Series, 2005), astro-ph/0410214.
13. Adelberger, E. G. *et al.*, Rev. Mod. Phys. **70**, 1265 (1998).
14. Angulo, C. and Descouvemont, P., Nucl. Phys. A **690**, 755 (2001).
15. Formicola, A. *et al.* (LUNA Collaboration), Phys. Lett. B **591**, 61 (2004).
16. Bahcall, J. N., Gonzalez-Garcia, M. C. and Peña-Garay, C., J. High Energy Phys. **02**, 009 (2003).
17. Bahcall, J. N., Fowler, W. A., Iben, I. and Sears, R. L., Astrophys. J. **137**, 344 (1963); Bahcall, J. N., Nucl. Phys. B (Proc. Suppl.) **118**, 77 (2003).
18. Singh, B. S. N., Hass, M., Nir-El, Y. and Haquin, G., nucl-ex/0407017.
19. Park, T. S. *et al.*, Phys. Rev. C. **67**, 055206 (2003).
20. Bahcall, J. N., "Neutrino Astrophysics," (Cambridge University Press, Cambridge 1989).
21. Ahmed, S. N. *et al.* (SNO Collaboration), Phys. Rev. Lett. **92**, 181301 (2004).
22. Bahcall, J. N., Phys. Rev. C **65**, 025801 (2002).
23. Bahcall, J. N., Gonzalez-Garcia, M. C. and Peña-Garay, C., J. High Energy Phys. **08**, 016 (2004).
24. Lodders, K., Astrophys. J. **591**, 1220 (2003).
25. Seaton, M. J. and Badnell, N. R., astro-ph/0404437 (2004).
26. Badnell, N. R., astro-ph/0410744 (2004).
27. Basu, S. and Antia, H. M., Astrophys. J. **606**, L85 (2004).
28. McDonald, A., Suzuki, Y., Suzuki, A. and Gonzalez-Garcia, M. C., (this volume).
29. Gonzalez-Garcia, M. C. and Peña-Garay, C., Phys. Rev. D **68**, 093003 (2003).
30. Murayama, H. and Peña-Garay, C., Phys. Rev. D **69**, 031301 (2003).

Physica Scripta. Vol. T121, 51–56, 2005

Atmospheric Neutrino Fluxes*

Thomas K. Gaisser

Bartol Research Institute, University of Delaware Newark, DE 19716 USA

Received February 13, 2005; accepted February 28, 2005

PACS number: 14.60.Pg

Abstract

Starting with a historical review, I summarize the status of calculations of the flux of atmospheric neutrinos and how they compare to the measurements.

1. Introduction

When cosmic-ray protons and nuclei enter the atmosphere, they interact and produce all kinds of secondary particles, which in turn interact or decay or propagate to the ground, depending on their intrinsic properties and energies. In the GeV energy range, the most abundant particles at the ground are the neutrinos. At production, there are approximately twice as many muon neutrinos as electron neutrinos. This characteristic ratio arises from the pion-muon-electron decay chain in which a muon and one muon-neutrino (or anti-neutrino) are produced when the charged pion decays, while the subsequent muon decay produces an electron neutrino as well as another muon neutrino.

In 1960 Markov [1] suggested using Cherenkov light in a lake or the deep ocean to do neutrino physics with the atmospheric neutrino beam; in particular, to investigate the question whether electron and muon neutrinos are distinct species. About the same time Greisen [2] described a proposal to search for astrophysical sources of neutrinos with a water Cherenkov detector in a deep mine. Atmospheric neutrinos were, however, first detected with electronic detectors–as horizontal neutrino-induced muons in a mine in South Africa [3] and in an iron calorimeter in a deep mine in the Kolar Gold Fields of India [4]. More than a decade passed before a large-scale effort to instrument a large body of water was undertaken by the DUMAND Project [5].

Atmospheric neutrino fluxes were first calculated in the 1960s. One approach [6] is to infer the neutrino fluxes from measurements of closely related muons. More recently this approach has been refined to take account of effects of the geomagnetic field and of the separate contributions of pions and kaons [7]. Most later calculations follow Refs. [8, 9] and calculate both the muon [8] and the neutrino fluxes [9] starting from the primary cosmic-ray spectrum at the top of the atmosphere.

With the advent in the 1980s of large underground detectors to search for proton decay, it became possible to measure increasingly large samples of events induced by neutrinos of both types (and from all directions because the Earth is transparent to neutrinos). Such events were of interest primarily as the background for proton decay. Both water Cherenkov detectors [10, 11] and segmented iron calorimeters [4, 12, 13, 14] were used.

Evidence gradually accumulated that the ratio of the two neutrino flavors was different from the expected value of two. The first hint came from the IMB experiment [15], which reported too few muon decays compared to what was expected from interactions of v_μ inside their detector. The IMB experiment later concluded, however, that their measurement of the ratio of stopping to throughgoing muons was inconsistent with oscillations [16]. The Kamiokande experiment [17] found an anomalously low ratio of v_μ/v_e and suggested neutrino oscillations as a possible explanation. As the IMB experiment accumulated more data [18], they too found a persistently low v_μ/v_e flavor ratio. Meanwhile, however, the Frejus experiment [13] measured a ratio consistent with the expectation. After some six years of running with the shielded Kamiokande II–III detector, enough data was accumulated to see a suggestion of the pathlength dependence expected from oscillations with parameters such that upward v_μ in the multi-GeV energy region oscillate while downward neutrinos do not [19].

By measuring the directions and energies of thousands of neutrinos of both flavors in the past decade, the Super-Kamiokande Collaboration has shown that the anomalous ratio is a consequence of neutrino flavor oscillations during their propagation from the atmosphere to the detector [20]. They have measured the oscillations in the muon neutrino-tau neutrino sector [21] by fitting the energy and angular dependence and other observed features of their data, while ruling out oscillation to sterile neutrinos at 99% confidence level. It is now possible to demonstrate the L/E pathlength/energy dependence expected for oscillations [22]. Best fit parameters in a two-flavor approximation $(v_\mu \leftrightarrow v_\tau)$ are $\delta m^2 = 2.1 \times 10^{-3}\,\mathrm{eV}^2$ and maximal mixing [23]. Results from MACRO [24] and Soudan [14] are consistent with these parameters. Measurements of atmospheric neutrinos are reviewed in Refs. [25, 26, 27].

With the parallel discovery [28, 29, 30] of oscillations of solar electron-neutrinos in a different region of parameter space, a pattern of mixing among the three neutrino types (electron, muon and tau) is beginning to emerge in which two mixing angles are large while one is smaller (though not yet measured). Eventually the nature of the full neutrino mass matrix will be addressed with long-baseline accelerator neutrino beams by experiments now under construction or planned. This will take some time, however. Accordingly, there is interest in refining calculations of the atmospheric neutrino beam to the point where effects of sub-dominant mixing may be resolved. To take full advantage of the power of the Super-Kamiokande experiment requires reducing uncertainties in the absolute normalization of the neutrino flux and in the ratios of electron to muon neutrinos and neutrinos to anti-neutrinos. The normalization depends on the primary cosmic-ray intensity and on the details of production of pions and kaons

---

*Research supported in part by the U.S. Department of Energy under DE-FG02 91ER40626.

in cosmic-ray interactions in the atmosphere. The ratios depend primarily on pion and kaon production. This paper reviews the present level of uncertainties.

2. Calculation

The neutrino flux is a convolution of the primary cosmic-ray spectrum, modified by geomagnetic cutoffs, with the yield of neutrinos per incident cosmic-ray. Schematically,

$$\phi_\nu = \phi_p \otimes R_p \otimes Y_{p \to \nu} + \sum_A \phi_A \otimes R_A \otimes Y_{A \to \nu}. \tag{1}$$

The first term represents the contribution of free protons and the second term the contribution of nucleons bound in nuclei. The two need to be calculated separately even in the superposition approximation in which it is assumed that bound nucleons interact independently as if they were free. This is because the geomagnetic cutoff depends on rigidity (total momentum per unit charge) whereas pion production depends on energy per nucleon. In each term the factor $\phi(E)$ represents the primary cosmic-ray spectrum as a function of energy per nucleon. $R(E, \theta_\oplus, \phi_\oplus, \theta, \phi)$ indicates the cutoff rigidity (more exactly the transmission probability) which depends on latitude and longitude and on the local zenith and azimuth of the arrival direction of each primary cosmic ray. The yield $Y(E_A, E_\nu)$ gives the number of neutrinos with energy E_ν produced per primary of energy E_A and is obtained by calculating the cosmic-ray induced cascades in the atmosphere. A complete review of the calculation of the flux of atmospheric neutrinos is given in Ref. [31]. In this paper I illustrate the considerations involved in calculating the production spectrum of atmospheric neutrinos by comparing results of three recent calculations [32, 33, 34].

All these calculations are three-dimensional, taking account of the deviations of the neutrinos from the directions of the primaries that produce them. In addition, the calculations of Refs. [33] and [34] also account for the effect bending of muons in the geomagnetic field before they decay. There are several other fully three-dimensional calculations, including the recent papers of Refs. [35, 36, 37] which appeared after the review [31]. The challenge of the full three-dimensional calculation is indicated by the five arguments of the cutoff function above. Cascades must be generated for relevant primary energies all over the globe $(\theta_\oplus, \phi_\oplus)$ taking account of the cutoff for primaries from all directions (θ, ϕ). Since most of the created neutrinos miss the detector, the calculation is highly inefficient. Although three-dimensional aspects of a calculation are technically challenging, they are not the most important sources of uncertainty in the calculated fluxes of atmospheric neutrinos. The biggest uncertainties come from the primary cosmic-ray spectrum and the treatment of hadronic interactions.

2.1. *Primary spectrum*

In the past decade new measurements of the primary spectrum with magnetic spectrometers have improved our knowledge of the primary spectrum. In particular, the AMS [38] and BESS [39] detectors give results for the proton spectra up to 100 GeV that agree with each other within 5%. As a consequence, current calculations of atmospheric neutrinos are using fits to the primary spectrum in which these data have been given priority. An example of such a fit [40] is shown by the thin lines in Fig. 1. Here the various nuclear groups are plotted as nucleons per GeV/nucleon, which is the quantity most directly related to secondary particle

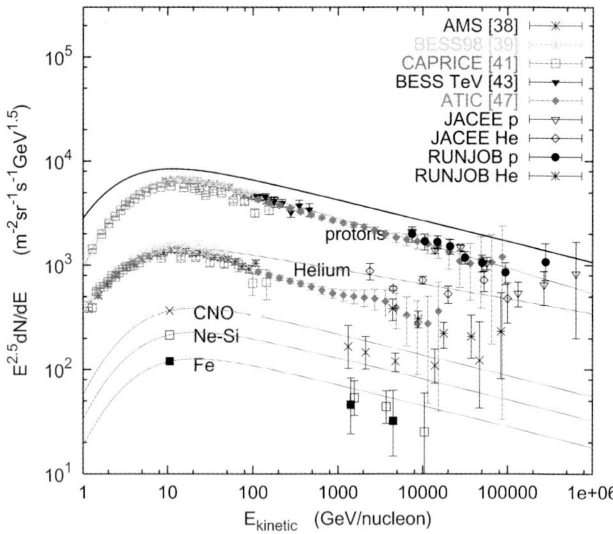

Fig. 1. The flux of nucleons. The heavy black line shows the numerical form of Eq. (2). The lighter lines show extrapolations of fits [40] to measurements of protons, helium and three heavier groups below 100 GeV/nucleon.

production. Free protons make up about 70% of the flux of all nucleons, helium about 20%, and heavier nuclei the rest. Another recent series of spectrometer measurements (CAPRICE [41]) continues to give results for protons about 15% lower than BESS and AMS, which can be taken as an indication of the uncertainty in the absolute normalization of this component of the primary spectrum. It should also be noted that BESS [39] and AMS [42] results for helium do not agree with each other as well as for protons.

Measurements with spectrometers presently extend up to about 500 GeV only [43]. Higher energy data are from balloon-borne ionization calorimeters, which sample the fraction of energy deposited in electromagnetic cascades inside the calorimeter. Systematic errors in assigning primary energy are larger as a consequence. Data from JACEE [44] and RUNJOB [45, 46] are included in Fig. 1. Preliminary data from the ATIC experiment [47] fill in the gap in the 1–100 TeV region. The ATIC data for protons appear consistent with all previous data. Their helium data, if confirmed, indicate a preference for the lower RUNJOB normalization for helium.

The heavy solid line in Fig. 1 is a simple power-law approximation to the spectrum of all nucleons,

$$\phi_N(E) = 1.7 \frac{\text{nucleons}}{\text{cm}^2 \, \text{s} \, \text{sr} \, \text{GeV}} \times \left(\frac{1}{E}\right)^{2.7}, \tag{2}$$

where E is total energy per nucleon. Fig. 2 displays the all-nucleon spectra used to calculate the neutrino flux as a ratio to the power-law form of Eq. (2). Refs. [32, 34] use the fits of

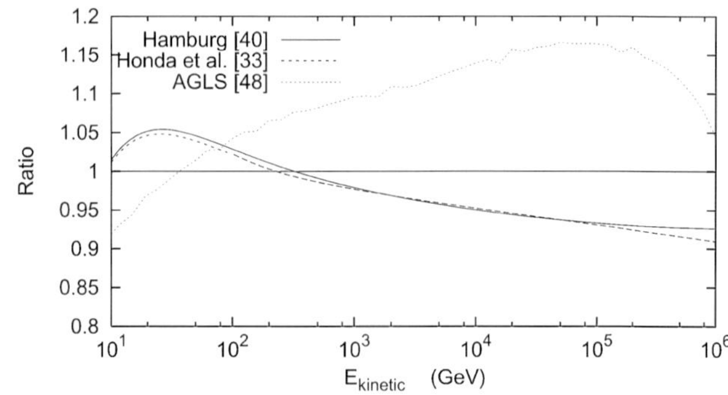

Fig. 2. Ratio of the all-nucleon fluxes used in several calculations to the power-law form of Eq. (2).

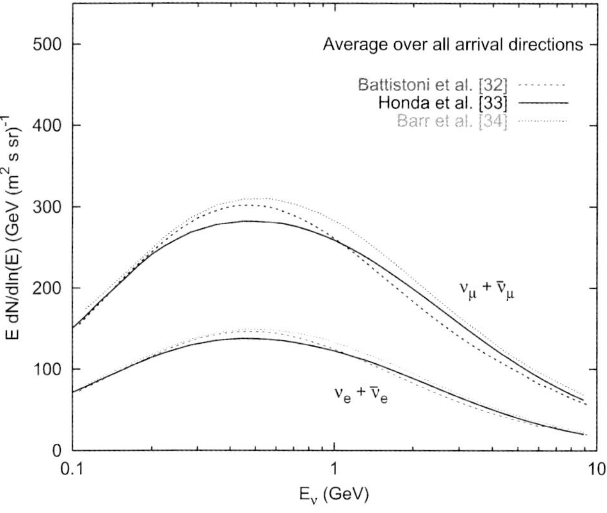
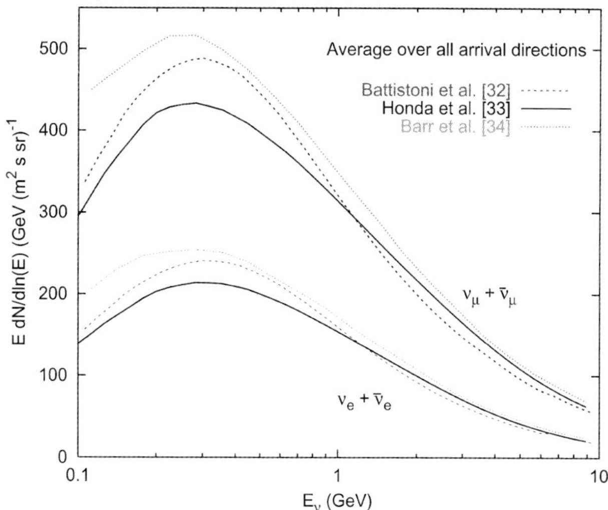

Fig. 3. The production spectrum of $\nu_\mu + \bar{\nu}_\mu$ and $\nu_e + \bar{\nu}_e$ integrated over the atmosphere and averaged over all directions at Super-K (left) and Soudan or SNO (right).

Ref. [40], while Ref. [33] use slightly different parameters for the heavy components at low energy and a slightly flatter power-law extrapolation (-2.71) above 100 GeV. Nevertheless, the primary all-nucleon spectrum of the three calculations [32, 33, 34] are the same to within about 1%. Moreover, simple power law approximation 2 to the all-nucleon spectrum represents these fits to better than 10% over the whole energy region. For reference, Fig. 2 also shows the primary spectrum used in an earlier calculation [48], which is somewhat lower than the new fits below 100 GeV and significantly higher at high energy.

2.2. *Treatment of hadronic interactions*

The representations of hadronic interactions used in Refs. [32, 33, 34] are completely independent. As a consequence, it is not surprising that they give different results. Because the assumptions about primary spectrum are essentially identical, differences among the results of the three calculations reflect the level of uncertainty due to treatment of hadronic interactions. For low energies the fluxes depend also on the geomagnetic cutoffs, so comparison must be made for the same location. Figure 3 shows the atmospheric neutrino fluxes from the three calculations [32, 33, 34]. Differences among the calculations are at the level of 10% at Super-K. The differences are somewhat larger for the sites at high geomagnetic latitude (where the cutoff for downward primaries is negligible), which probably indicates that the low-energy hadronic interactions are less well understood.

The flavor ratios for the three calculations are shown for the low-energy region in Fig. 4, along with the neutrino/antineutrino ratios. Differences are at the level of 3% in this energy range for the flavor ratios and 5% for the $\nu/\bar{\nu}$ ratios.

3. Higher energies

Response functions for various measurements of atmospheric neutrinos are given in Ref. [31] and reproduced here in Fig. 5. Sub-GeV and multi-GeV events involve neutrinos with energies below 2 GeV and from 1 to 20 GeV respectively. Upward stopping muons are produced by muon neutrinos with energies in the range 3–100 GeV, while upward throughgoing muons come from neutrinos with energies in the range 10 GeV to 10 TeV. The relevant primary energies per nucleon are roughly a factor of ten higher, but with rather broader distributions. Thus the normalization of the sub-GeV neutrinos is determined almost entirely by primaries with energies less than 100 GeV/nucleon and hence covered by the most precise magnetic spectrometer measurements. On the other hand, the primary response function for neutrino-induced muons extends to 10 TeV and beyond and hence is subject to somewhat larger uncertainty in normalization. The difference in response function for neutrino-induced upward muons at Super-K from that at AMANDA arises from the higher muon energy threshold in AMANDA, which is of order 100 GeV at the interaction vertex [49].

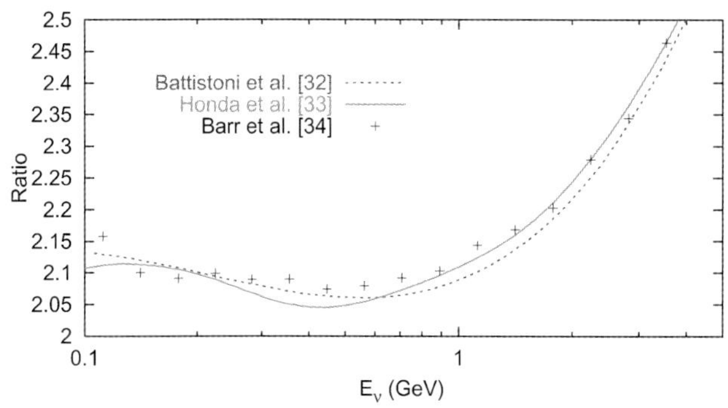
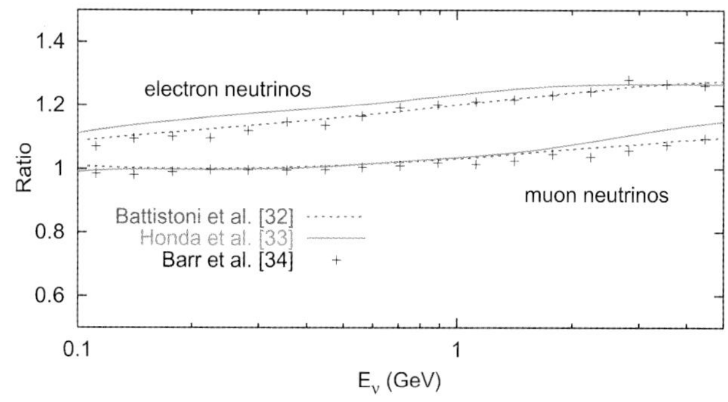

Fig. 4. The ratio $\nu_\mu + \bar{\nu}_\mu$ to $\nu_e + \bar{\nu}_e$ (left panel) and $\nu/\bar{\nu}$ (right panel) at Super-K. The production spectra are integrated over the atmosphere neglecting oscillations and averaged over all directions before taking the ratio.

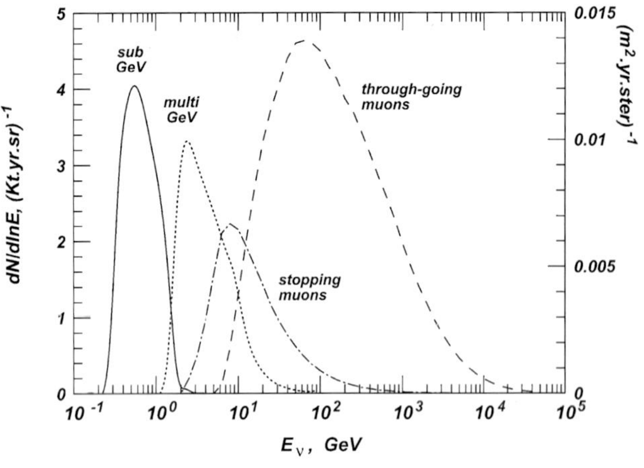

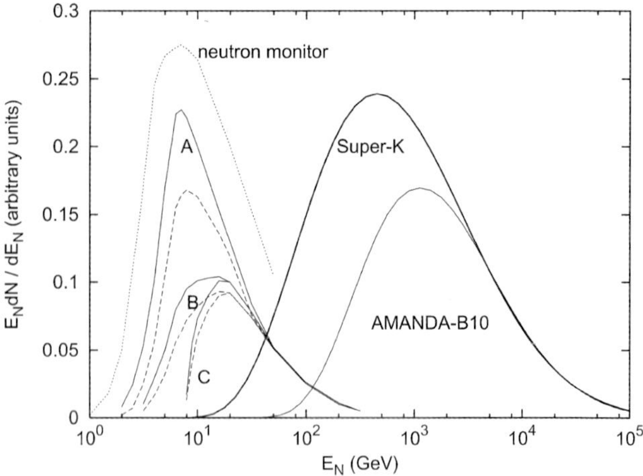

Fig. 5. Distribution of neutrino energies (left) and primary energy per nucleon (right) giving rise to various classes of events.

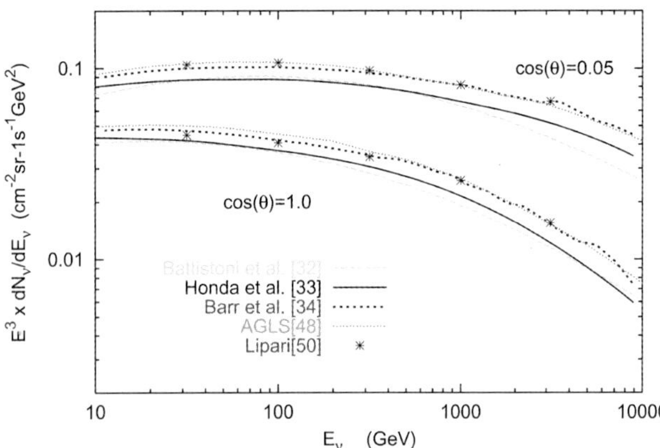

Fig. 6. Comparison of fluxes of muon neutrinos at high energy.

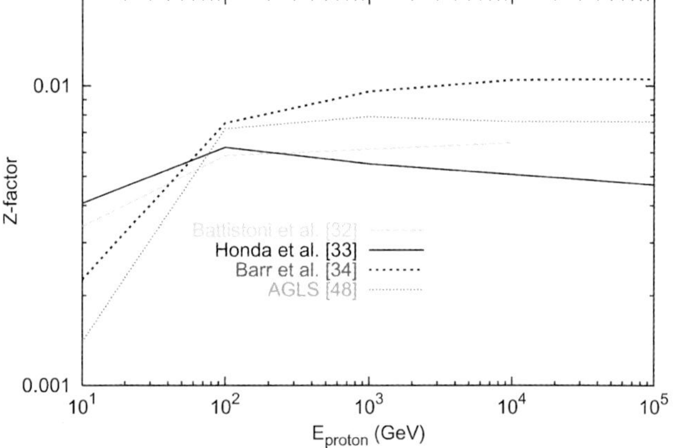

Fig. 7. Spectrum-weighted Z-factor for K^+ production by protons.

Again we can make use of the fact that the assumed primary spectra in the three calculations [32, 33, 34] are almost the same to illustrate differences due to the treatment of hadronic interactions. Fig. 6 compares fluxes of muon neutrinos for these three calculations in the energy range responsible for neutrino-induced muons.

3.1. Analytic approximations

At high energy, numerical integration of the analytic cascade equations gives sufficiently accurate results to be useful for understanding the characteristic features of the results. The asterisks in Fig. 6 show the numerical results of Lipari [50], including the contribution from decay of muons.[1] At high energy most muons reach the ground before decaying and therefore do not contribute to the neutrino flux. For $E_\nu > 100\,\mathrm{GeV}$ for example, less than 15% of muon neutrinos are from decay of muons, and the fraction decreases further as energy increases.

Neglecting neutrinos from muon decay, a simple approximation for the flux of $\nu_\mu + \bar\nu_\mu$ from decay of pions and kaons is

$$\frac{\mathrm{d}N_\nu}{\mathrm{d}E_\nu} = \frac{\phi_N(E_\nu)}{(1 - Z_{NN})(\gamma+1)} \left\{ \left[\frac{Z_{N\pi}(1-r_\pi)^\gamma}{1 + B_{\pi\nu}\cos\theta E_\nu/\epsilon_\pi} \right] \right.$$

$$\left. + 0.635 \left[\frac{Z_{NK}(1-r_K)^\gamma}{1 + B_{K\nu}\cos\theta E_\nu/\epsilon_K} \right] \right\}, \tag{3}$$

where $\phi_N(E_\nu)$ is the primary spectrum of nucleons evaluated at the energy of the neutrino (see Eq. (2)). The constants $r_i = m_\mu^2/m_i^2$ for $i = (\pi, K)$, while the constants B depend on the hadron attenuation lengths as well as decay kinematics. The critical energy for pions is $\epsilon_\pi \approx 115\,\mathrm{GeV}$, while for kaons $\epsilon_K \approx 850\,\mathrm{GeV}$ [51].

Asymptotically, the neutrino spectrum is one power steeper than the primary spectrum. For fixed zenith angle θ the transition occurs first for neutrinos from pion decay because of its lower critical energy. As a consequence, the relative contribution from kaons increases with energy. Kaons become the dominant source of neutrinos for $E_\nu > 100\,\mathrm{GeV}$. (The onset of kaon dominance is somewhat slower at large angle because of the cosine factor in the denominator of the kaon term in Eq. (3)). Important differences among the calculations can be traced to differences in treatment of kaon production, in particular to the channel $p \to \Lambda K^+$. Fig. 7 shows the spectrum-weighted moments [51] for four calculations.

Several features of Fig. 6 can be understood with the help of Eq. (3).

- The general decrease of the vertical to horizontal ratio at large energy, which appears in all calculations, is a consequence of the factor of $\cos(\theta)$ in the denominators.
- The primary spectrum of [34] is significantly lower than that of [48] at high energy, as shown in Fig. 2. The calculation of Ref. [34] is nevertheless high, comparable to that of Ref. [48] because the kaon production assumed in the new calculation is larger, as shown in Fig. 7.

---

[1] The primary spectrum of [50] has the same form as Eq. (2) with a higher normalization.

- The ratio (>1) of the flux of Ref. [34] compared to that of Ref. [33], and its increase with energy, may also be attributed to the differences in the assumptions for kaon production as shown in Fig. 7.

3.2. *Muon fluxes*

Calculation of the flux of atmospheric muons follows the same form as Eq. (1) with the yield of neutrinos replaced by the yield of muons. Because of the close relation between muons and neutrinos, comparison of its predicted muon intensity to measurements is an important check on any calculation of atmospheric muons. Comparison to muon fluxes also serves to check the normalization of the calculation. Because muons lose about $2\,\mathrm{GeV}$ in passing through the atmosphere, a ground level measurement probes high energies and is therefore most relevant to multi-GeV neutrinos and to neutrino-induced upward muons. Because of the kinematics of the $\pi \to \mu$ decay, however, pion decay remains the dominant source of muons at all energies. As a consequence, the leverage of the muon measurements for controlling the neutrino flux is limited above $\sim 100\,\mathrm{GeV}$ where neutrino production is dominated by the contribution from kaons.

Fig. 8 contains a summary of measurements of the spectrum of vertical muons at the ground. All measurements have been corrected to sea level. At very high energy the muon spectrum becomes one power steeper than the parent spectrum of nucleons as a consequence of the extra power of $1/E_\pi$ in the ratio of pion decay length to interaction length, which reflects the decreasing probability of decay relative to re-interaction for charged pions at high energy. For $E < \epsilon_\pi \approx 115\,\mathrm{GeV}$ essentially all pions decay, and the muon production spectrum has the same power behavior as the parent nucleon spectrum ($\alpha \approx 2.7$). At low energy, however, muon energy-loss and decay become important, and the muon spectrum at the ground falls increasingly below the production spectrum. To account for all the complications one generally resorts to Monte Carlo calculations. However, analytic approximations of the effects are also possible [50]. Fig. 8 shows a comparison of one such calculation [52], which uses as input the simple power-law primary spectrum of Eq. (2). Data in Fig. 8 are from many measurements with spectrometers at the surface [53, 54, 55, 56, 57, 58] and underground [59, 60].

The comparison of the analytic calculation with the data in Fig. 8 leads to two remarks. First the good agreement gives a general confirmation of the primary spectrum at high energy. Second, it serves to illustrate the level of systematic differences among the various measurements.

4. Concluding remarks

In their recent detailed paper, [23] the Super-Kamiokande group give an extensive discussion of how the best fit oscillation parameters are determined from a comparison of their data with simulations based on the atmospheric neutrino flux of Ref. [33]. In particular, their Table VII enumerates the assumed uncertainties in various properties of the calculated neutrino flux and the shifts in those properties needed to obtain the best fit. Apart from the overall increase in normalization, the largest shifts relative to what is assumed in Ref. [33] concern the slope of the primary spectrum at high energy and the K/pi ratio. The latter is decreased by 6% overall, while the primary spectrum is significantly harder as compared to the fits of Ref. [40] (from -2.71 to -2.66 above $100\,\mathrm{GeV}$). As noted in the discussion of Figs. 6 and 7, the high energy spectrum could instead be increased by enhancing kaon production preferentially at high energy.

The Super-K fitting procedure uses the atmospheric neutrino data itself to correct the calculated neutrino flux. The neutrino data have the advantage in principle that (apart from oscillations!) the intensity observed at the detector integrates over the whole atmosphere without complications of energy loss and decay that dominate the observed muon flux. For example the required increase in the overall normalization of 11.9% could be interpreted as favoring the higher normalization of Refs. [38, 39] as compared to that of Ref. [41]. The observations depend, however, on properties of neutrino interactions as well as on the neutrino flux. As properties of neutrino interactions become better determined, for example from near detectors of long- baseline experiments, this iterative procedure may become an important way to refine our understanding of the atmospheric neutrino flux at production.

Acknowledgments

I am grateful to G. Battistoni and M. Honda for providing the Z-factors for their hadronic interaction models and to G. Barr, G. Battistoni, M. Honda, T. Kajita, P. Lipari, S. Robbins and T. Stanev for helpful discussions.

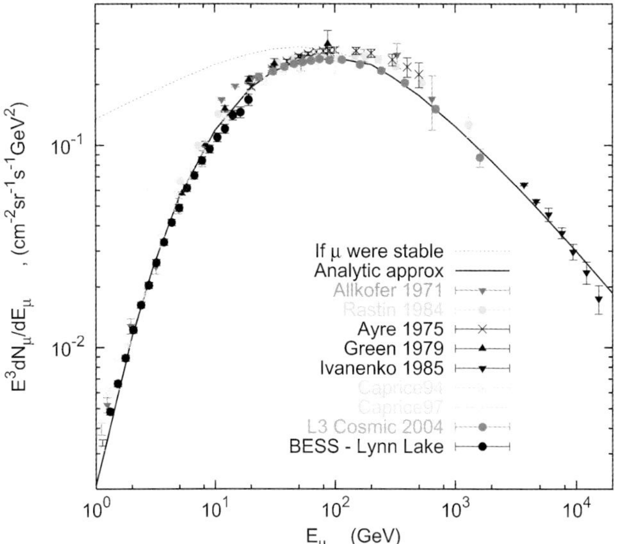

Fig. 8. Summary of measurements of the vertical muon intensity at the ground. The solid line shows an analytic calculation [52]. The dotted line shows the spectrum in the absence of decay and energy loss, or equivalently the muon production spectrum integrated over the atmosphere.

References

1. Markov, M. A., Proc. 1960 Annual International Conference on High Energy Physics at Rochester (ed. E. C. G. Sudarshan, J. H. Tinlot & A. C. Melissinos, Rochester) 578 (1960).
2. Greisen, K., Ann. Rev. Nucl. Sci. **10**, 63 (1960).
3. Reines, F. *et al.*, Phys. Rev. Lett. **15**, 429 (1965).
4. Achar, C. V. *et al.*, Phys. Lett. **18**, 196 (1965).
5. Babson, J. *et al.*, Phys. Rev. D **42**, 3613 (1990).
6. Markov, M. A. and Zheleznykh, I. M., Nucl. Phys. **27**, 385 (1961).
7. Perkins, D. H., Astropart. Phys. **2**, 249 (1994) and references therein.
8. Zatsepin, G. T. and Kuz'min, V. A., Sov. Phys. JETP **12**, 1171 (1961).
9. Zatsepin, G. T. and Kuz'min, V. A., Sov. Phys. JETP **14**, 1294 (1962).
10. Becker-Szendy, R. *et al.*, Phys. Rev. D **42**, 2974 (1990).
11. Hirata, K. S. *et al.*, Phys. Rev. D **38**, 448 (1988).
12. Aglietta, M. *et al.*, Europhys. Lett. **8**, 611 (1989).

13. Berger, Ch. *et al.*, Phys. Lett. B **227**, 489 (1989) and B **245**, (1990). See also Daum, K. *et al.*, Z. Phys. C **66**, 417 (1995).
14. Sanchez, M. *et al.*, Phys. Rev. D **68**, 113004 (2003).
15. Haines, T. J. *et al.*, Phys. Rev. Lett. **57**, 1986 (1986).
16. Becker-Szendy, R. *et al.*, Phys. Rev. Lett. **69**, 1010 (1992).
17. Hirata, K. S. *et al.*, Phys. Lett. B **205**, 416 (1988).
18. Becker-Szendy, R. *et al.*, Phys. Rev. D **46**, 3720 (1992) and references therein.
19. Fukuda, Y. *et al.*, Phys. Lett. B **335**, 237 (1994).
20. Fukuda, Y. *et al.*, Phys. Rev. Lett. **81**, 1562 (1998).
21. Fukuda, S. *et al.*, Phys. Rev. Lett. **85**, 3999 (2000).
22. Ashie, Y. *et al.*, Phys. Rev. Lett. **93**, 101801 (2004).
23. Ashie, Y. *et al.*, Phys. Rev. **D71**, 112005 (2005).
24. Ambrosio, M. *et al.*, Phys. Lett. B**566**, 35 (2003).
25. Kajita, T. and Totsuka, Y., Rev. Mod. Phys. **73**, 85 (2001).
26. Jung, C. K., Kajita, T., Mann, T. and McGrew, C., Ann. Rev. Nucl. Part. Sci. **51**, 451 (2001).
27. Goodman, M., Phys. Lett. B (Proc. Suppl.) **118**, 99 (2002).
28. Ahmad, Q. R. *et al.*, Phys. Rev. Lett. **87**, 071301 (2001).
29. Smy, M. *et al.*, Phys. Rev. D **69**, 011104 (2004).
30. Ahmad, Q. R. *et al.*, Phys. Rev. Lett. **89**, 011301 and 011302 (2002).
31. Gaisser, T. K. and Honda, M., Ann. Rev. Nucl. Part. Sci. **52**, 153 (2002).
32. Battistoni, G. *et al.*, Astropart. Phys. **19**, 269 (2003), (Erratum pp 291–294). http://www.mi.infn.it/battist/neutrino.html
33. Honda, M. *et al.*, Phys. Rev. D **70**, 043008 (2004). http://www.icrr.u-tokyo.ac.jp/mhonda/
34. Barr, G. D. *et al.*, Phys. Rev. D **70**, 023006 (2004). http://www-pnp.physics.ox.ac.uk/barr/fluxfiles/
35. Wentz, J. *et al.*, Phys. Rev. D **67**, 073020 (2003).
36. Liu, Y. *et al.*, Phys. Rev. D **67**, 073022 (2003).
37. Favier, J. *et al.*, Phys. Rev. D **68**, 093006 (2003).
38. Alcarez, J. *et al.*, Phys. Lett. B**490**, 27 (2000).
39. Sanuki, T. *et al.*, Astrophys. J. **545**, 1135 (2000).
40. Gaisser, T. K., Honda, M., Lipari, P. and Stanev, T., Proc. 27th Int. Cosmic Ray Conf., (Hamburg) **5**, 1643 (2001).
41. Boezio, M. *et al.*, Astrophys. J. **518**, 457 (1999).
42. Alcarez, J. *et al.*, Phys. Lett. B **494**, 193 (2000).
43. Haino, S. *et al.*, Phys. Lett. B **594**, 35 (2004).
44. Asakimori, K. *et al.*, Astrophys. J. **502**, 278 (1998).
45. Apanasenko, A. V. *et al.*, Astropart. Phys. **16**, 13 (2001).
46. Furukawa, M. *et al.*, Proc. 28th Int. Cosmic Ray Conf. (Tsukuba) **4**, 1837 (2003).
47. Wefel, J. *et al.*, in Proc. 28th Int. Cosmic Ray Conf. (Tsukuba) **4**, 1849 (2003). Preliminary ATIC data are from the presentation of H. Ahn at COSPAR 2004 (Paris).
48. Agrawal, V., Gaisser, T. K., Lipari, P., and Stanev, T., Phys. Rev. D**53**, 1314 (1996).
49. Ahrens, J. *et al.*, Phys. Rev. D **66**, 012005 (2002).
50. Lipari, P., Astropart. Phys. **1**, 195 (1993).
51. Gaisser, T. K. "Cosmic Rays and Particle Physics" (Cambridge University Press, 1990).
52. Gaisser, T. K., Astropart. Phys. **16**, 285 (2002).
53. Allkofer, O. C., Carstensen, K. and Dau, W. D., Phys. Lett. B**36**, 425 (1971).
54. Rastin, B. C., J. Phys. G **10**, 1609 (1984).
55. Ayre, C. A. *et al.*, J. Phys. G **1**, 584 (1975).
56. Green, P. J. *et al.*, Phys. Rev. D **20**, 1598 (1979).
57. Kremer, J. *et al.*, (CAPRICE) Phys. Rev. Letters **83**, 4241 (1999).
58. Motoki, M. *et al.*, (BESS) Astropart. Phys. **19**, 113 (2003).
59. Achard, P. *et al.*, (L3) Phys. Lett. B **598**, 15 (2004).
60. Ivanenko, I. P. *et al.*, Proc. 19th Int. Cosmic Ray Conf., La Jolla (NASA Conf. Publ. No. 2376) **8**, 210 (1985).

Physica Scripta. Vol. T121, 57–64, 2005

The MSW Effect and Matter Effects in Neutrino Oscillations

A. Yu. Smirnov[1,2]

[1]The Abdus Salam International Centre for Theoretical Physics, I-34100 Trieste, Italy
[2]Institute for Nuclear Research of Russian Academy of Sciences, Moscow 117312, Russia

Received December 8, 2004; accepted in revised form December 29, 2004

PACS number: 14.60.Pq, 26.65.+t, 91.35.Pn, 95.85.Ry, 97.60.Bw

Abstract

The MSW (Mikheyev-Smirnov-Wolfenstein) effect is the adiabatic or partially adiabatic neutrino flavor conversion in media with varying density. The main notions related to the effect, its dynamics and physical picture are reviewed. The large mixing MSW effect is realized inside the Sun providing a solution of the solar neutrino problem. The small mixing MSW effect driven by the 1–3 mixing can be realized for the supernova (SN) neutrinos. Inside collapsing stars new elements of the MSW dynamics may show up: non-oscillatory transition, non-adiabatic conversion, time dependent adiabaticity violation induced by shock waves. Effects of the resonance enhancement and the parametric enhancement of oscillations can be realized for atmospheric and accelerator neutrinos in the Earth. Precise results for neutrino oscillations in low density media with arbitrary density profile are presented and the attenuation effect is described. The area of applications is the solar and SN neutrinos inside the Earth, and the results are crucial for the neutrino oscillation tomography.

1. Introduction

A variety of matter effects [1] on propagation of mixed neutrinos [2, 3] depends on (i) channel of mixing – the type of neutrino states involved: active, sterile, mass eigenstates, *etc.*; (ii) density profile: constant, monotonous, periodic, fluctuating; (iii) properties of medium: polarization, motion, chemical composition (thermal bath and neutrino gases are special cases); (iv) presence of new non-standard neutrino interactions.

The dynamics of the matter effects can be classified by the degrees of freedom involved. In the 2ν case, an arbitrary neutrino state can be expressed in terms of the eigenstates of the instantaneous Hamiltonian, ν_{1m} and ν_{2m}, as

$$\nu(x) = \cos\theta_a \nu_{1m} + \sin\theta_a \nu_{2m} e^{i\Phi_m}, \tag{1}$$

where

- $\theta_a = \theta_a(x)$ determines the admixtures of eigenstates in $\nu(x)$;
- $\Phi_m(x)$ is the phase difference between the two eigenstates (phase of oscillations):

$$\Phi_m(x) = \int_{x_0}^{x} \Delta H dt'. \tag{2}$$

Here $\Delta H \equiv H_{2m} - H_{1m}$ is the difference of eigenvalues. Eq. (2) gives the adiabatic phase, additional contributions can be related to non-adiabaticity and topology.

- The mixing angle in matter, θ_m, defined as $\langle \nu_e|\nu_{1m}\rangle \equiv \cos\theta_m$, $\langle \nu_e|\nu_{2m}\rangle \equiv \sin\theta_m$, *etc.*, determines the flavor contents (or flavors) of the eigenstates. The angle is the function of matter density $n_e(x)$: $\theta_m = \theta_m(n_e(x))$.

Besides these, the effects of absorption and loss of the coherence can change the normalization of the eigenstates or suppress interference.

Different processes are associated with these degrees of freedom. In particular, in pure form the oscillations [2, 3, 4, 1, 5] are an effect of the monotonous phase difference increase Φ_m, which occurs in the uniform medium. In contrast, the MSW

effect [1, 6] is associated to the change of flavors of the eigenstates (change of $\theta_m(x)$) in the nonuniform medium. In general, an interplay of different effects occurs which is induced by simultaneous operation of several degrees of freedom.

In this paper I will discuss only a few effects selected on the following ground: usual massive neutrinos with flavor mixing and standard weak interactions; the effects in the Sun, collapsing stars and the Earth; effects which are realized in Nature or have a good chance to be realized and established in future studies.

2. The MSW effect

The MSW effect [1, 6] is the adiabatic or partially adiabatic neutrino flavor conversion in a medium with varying density. The flavor of a neutrino state follows the density change. Here I review the main notions related to the effect and describe its dynamics.

2.1. Main notions

2.1.1. Refraction. At low energies the *elastic forward scattering* only is relevant in most of applications [1, 7]. It can be described by the potential, V, or equivalently, the refraction index: $n_{ref} - 1 = V/p$.

The difference of potentials (for a single neutrino) in different spatial points $V = V(x)$ leads to the "neutrino optics" phenomena such as reflection, complete internal reflection, bending of neutrino trajectories, focusing by stars, *etc.* The values of refraction index are very close to unity, *e.g.*: $n_{ref} - 1 = 10^{-20}$ inside the Earth, 10^{-18} inside the Sun, 10^{-6} in the neutron stars. So, the "neutrino lenses" should be of astrophysical size.

The difference of potentials for different neutrinos, ν_e and ν_a: $V \equiv V_e - V_a$ influences the evolution of mixed neutrinos (even for constant potentials). In usual matter the difference is due to the charged current scattering of ν_e on electrons ($\nu_e e \rightarrow \nu_e e$) [1]:

$$V \equiv V_e - V_a = \sqrt{2} G_F n_e, \tag{3}$$

where G_F is the Fermi coupling constant and n_e is the number density of electrons. It leads to appearance of an additional phase in the neutrino system: $\Delta\phi_{matter} \equiv (V_e - V_a)t$. The distance over which this "matter" phase equals 2π determines the *refraction length*:

$$l_0 \equiv \frac{2\pi}{V_e - V_a} = \frac{\sqrt{2}\pi}{G_F n_e}. \tag{4}$$

2.1.2. Eigenstates and mixing in matter. In the presence of matter the Hamiltonian of the system changes: $H_0 \rightarrow H = H_0 + V$, where H_0 is the Hamiltonian in vacuum. Correspondingly, the eigenstates and the eigenvalues of H change: $\nu_1, \nu_2 \rightarrow \nu_{1m}, \nu_{2m}$, $m_1^2/2E, m_2^2/2E \rightarrow H_{1m}, H_{2m}$. Here ν_1 and ν_2 are the mass eigenstates with masses m_1 and m_2.

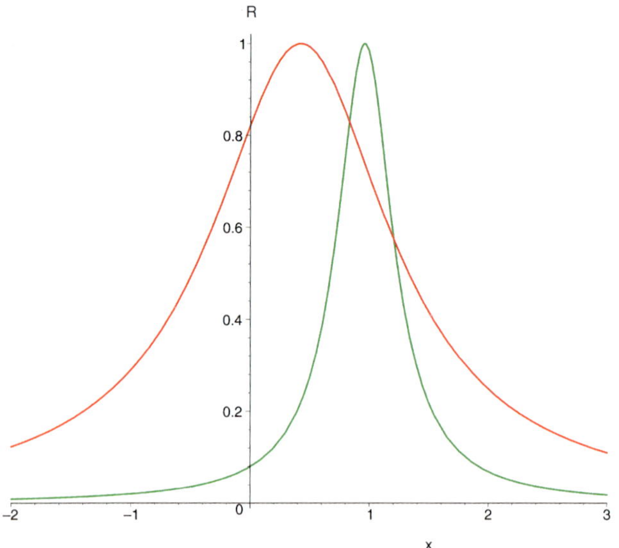

Fig. 1. The dependence of the effective mixing parameter $R \equiv \sin^2 2\theta_m$ on $x \equiv l_\nu/l_0 \propto E n_e$ for two different values of the vacuum mixing: $\sin^2 2\theta = 0.825$ (black, red) which is the LMA mixing and $\tan^2 \theta = 0.08$ (grey, green). The semi-plane $x < 0$ corresponds to the antineutrino channel.

The mixing in matter is defined with respect to the eigenstates ν_{1m} and ν_{2m}. Similarly to the vacuum case, the mixing angle in matter, θ_m, determines the relations between the eigenstates in matter and the flavor states: $\nu_e \equiv \cos\theta_m \nu_{1m} + \sin\theta_m \nu_{2m}$, etc. In matter, both the eigenstates and eigenvalues, and consequently, the mixing angle depend on matter density and neutrino energy.

2.1.3. *Resonance.* The dependence of the effective mixing parameter in matter, $R \equiv \sin^2 2\theta_m$, on the ratio of the vacuum oscillation length, $l_\nu = 4\pi E/\Delta m^2$, and the refraction length: $x \equiv l_\nu/l_0 = 2EV/\Delta m^2 \propto E n_e$ (fig. 1) has a resonance character [6]. At

$$l_\nu = l_0 \cos 2\theta \qquad \text{(resonance condition)} \qquad (5)$$

the mixing becomes maximal: $\sin^2 2\theta_m = 1$. For small θ the condition (5) reads:

Vacuum oscillation length $\approx$ *Refraction length.* $\qquad (6)$

That is, the eigenfrequency which characterizes a system of mixed neutrinos, $1/l_\nu$, coincides with the eigenfrequency of the medium, $1/l_0$. For large vacuum mixing (LMA has $\cos 2\theta = 0.4$) there is a significant deviation from the equality (6). Large mixing corresponds to the case of strongly coupled system for which the shift of frequencies occurs.

The resonance condition (5) determines the resonance density:

$$n_e^R = n_0 \cos 2\theta, \qquad n_0 \equiv \frac{\Delta m^2}{2\sqrt{2} E G_F}. \qquad (7)$$

The width of resonance on the half of height (in the density scale) is given by

$$2\Delta n_e^R = 2n_e^R \tan 2\theta = 2n_0 \sin 2\theta. \qquad (8)$$

When the vacuum mixing approaches the maximal value, $\theta = \pi/4$, the resonance shifts to zero density, $n_e^R \to 0$, and the width of resonance Δn_e^R increases converging to n_0.

In a medium with varying density, the resonance layer is determined by the interval in which the density changes from $n_e^R - \Delta n_e^R$ to $n_e^R + \Delta n_e^R$.

2.1.4. *Adiabaticity.* Since in a non-uniform medium the density changes on the way of neutrinos, $n_e = n_e(x)$, the Hamiltonian of the system depends on time (distance): $H = H(t)$. Therefore, (*i*) the mixing angle changes in the course of propagation: $\theta_m = \theta_m(n_e(x))$; (ii) the eigenstates of the instantaneous Hamiltonian, ν_{1m} and ν_{2m}, are no more "eigenstates" of propagation, and transitions $\nu_{1m} \leftrightarrow \nu_{2m}$ occur.

If the density changes slowly, the system (mixed neutrinos) has time to adjust the change leading to adiabatic evolution [1, 6, 8, 9]. The adiabaticity condition is [9]

$$\gamma = \left| \frac{\dot\theta_m}{H_{2m} - H_{1m}} \right| \ll 1. \qquad (9)$$

As follows from the evolution equation for the neutrino eigenstates [6, 9], $|\dot\theta_m|$ determines the energy of the transition $\nu_{1m} \leftrightarrow \nu_{2m}$, and $|H_{2m} - H_{1m}|$ gives the energy gap between the levels. So, the condition (9) means that the transitions $\nu_{1m} \leftrightarrow \nu_{2m}$ can be neglected and the eigenstates propagate independently ($\theta_a = const$ in (1)).

If θ is small, the adiabaticity is critical in the resonance. It takes the form [6]

$$\Delta r_R \geq l_R, \qquad (10)$$

where $l_R = l_\nu/\sin 2\theta$ is the oscillation length in resonance, and $\Delta r_R = n_e^R \tan 2\theta/(dn_e/dr)_R$ is the spatial width of the resonance layer. The adiabaticity condition has been considered outside the resonance and in the non-resonance channel in [10]. In the case of large vacuum mixing the point of maximal adiabaticity violation, n_e^{ad} [11], is shifted to densities larger than the resonance one: $n_e^{ad} \to n_0 > n^R$.

2.2. *Dynamics of the MSW effect*

2.2.1. *Dynamical features.* In the adiabatic case the features of the effect can be summarized as follows.

- The flavors of eigenstates change according to density change; the flavors are determined by $\theta_m(x) = \theta_m(n_e(x))$.
- The admixtures of the eigenstates in a propagating neutrino state do not change due to the adiabaticity; there are no

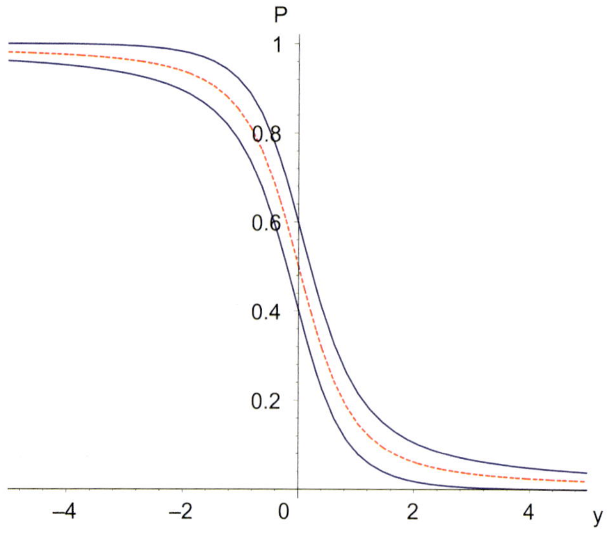

Fig. 2. Dependences of the average probability (dashed line) and the depth of oscillations (P^{max}, P^{min} – solid lines) on y for $y_0 = -5$. The resonance layer corresponds to $y = 0$ and the resonance layer is given by the interval $y = -1 \div 1$. For $\tan^2 \theta = 0.4$ (large mixing MSW solution) the evolution stops at $y_f = 0.47$.

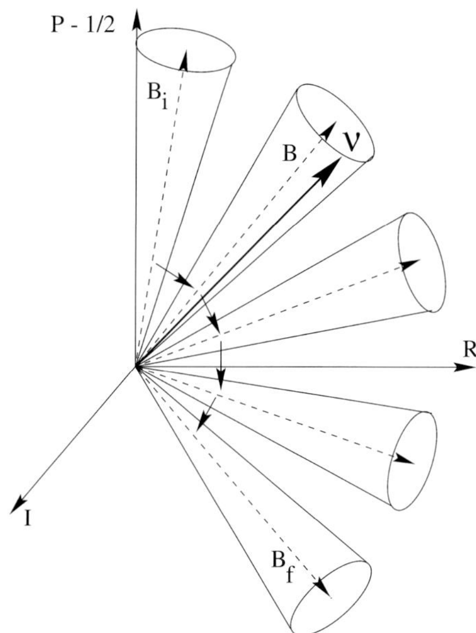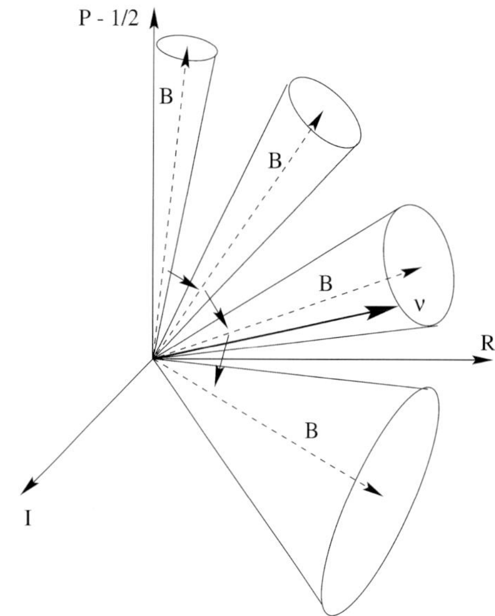

Fig. 3. Graphic representation of the MSW effect. The direction of the cone axis is determined by $2\theta_m$, the cone angle is given by θ_a and the position of the neutrino vector on the surface of the cone is fixed by the phase Φ_m. In the adiabatic case (left panel) the direction of the axis flips according to the $\theta_m(n_e)$ change, and the cone angle is unchanged. In the non-adiabatic case (right panel) also the cone angle changes.

$v_{1m} \leftrightarrow v_{2m}$ transitions. The admixtures are given by the mixing in production point, θ_m^0.

- The phase difference $\Phi_m(x)$ (2) increases leading to oscillations.

Two degrees of freedom are operative: the phase Φ_m and the flavor, θ_m. The MSW effect is driven by the change of flavors of the neutrino eigenstates in matter with varying density. The change of phase produces an oscillation effect on top of the adiabatic conversion.

2.2.2. Spatial picture of the MSW effect.

In fig. 2 are shown dependences of the average probability, $\bar{P}$, and depth of oscillations given by $P^{max} - P^{min}$ on

$$y \equiv \frac{n_e^R - n_e}{\Delta n_e^R}, \tag{11}$$

the distance (in the density scale) from the resonance density in units of the width of the resonance layer [6]. In terms of n the conversion pattern depends on the initial value n_0 only which reflects the *universality* of the adiabatic evolution. The probability is the oscillatory function which is inscribed into the band shown by the solid lines. There is no explicit dependence on the vacuum angle θ. With decrease of y_0, the oscillation band becomes narrower approaching the line of *non-oscillatory conversion*. For zero final density: $y_f = 1/\tan 2\theta$. The smaller the mixing (and therefore, the larger the final y_f) the stronger the transition.

2.2.3. Adiabaticity violation.

If the density changes rapidly, so that condition (9) is not satisfied, the transitions $v_{1m} \leftrightarrow v_{2m}$ become efficient [6, 12]. Therefore admixtures of the eigenstates in a given propagating state change: $\theta_a = \theta_a(x)$. Now all three degrees of freedom – phase, flavor, and admixture – become operative. Typically, adiabaticity breaking leads to weakening of the flavor transition and enhancement of oscillations.

2.2.4. Graphic representation [13].

It is based on the analogy of the neutrino evolution with the behavior of the spin of an electron in the magnetic field. The neutrino evolution equation can be written as

$$\frac{d\vec{v}}{dt} = (\vec{B} \times \vec{v}), \tag{12}$$

where the neutrino vector of length $1/2$ (equivalent of spin) is

$$\vec{v} \equiv (R, I, P - 1/2) = \left(\mathrm{Re}v_e^\dagger v_\mu, \ \mathrm{Im}v_e^\dagger v_\mu, \ v_e^\dagger v_e - 1/2\right), \tag{13}$$

$v_\alpha = v_\alpha(x), (\alpha = \mu, e)$ are neutrino wave functions, and P is the survival probability.[1] See fig. 3. The vector of "magnetic field" equals

$$\vec{B} \equiv \frac{2\pi}{l_m}(\sin 2\theta_m, 0, \cos 2\theta_m), \tag{14}$$

where $l_m = 2\pi/\Delta H$ is the oscillation length in matter. The vector $\vec{v}$ moves on the surface of the cone with axis $\vec{B}$ according to the increase of the oscillation phase, Φ_m.

3. Realizations of the MSW effect

General conditions for the MSW conversion are: (*i*) slow enough density change; (*ii*) crossing the resonance layer; (*iii*) large enough matter width (minimal width condition) [14]. These conditions are satisfied for the solar neutrinos inside the Sun, for the supernova neutrinos inside collapsing stars and can be satisfied for neutrinos in the Early Universe.

3.1. Solar neutrinos. Large angle MSW solution

The large mixing MSW conversion provides a solution of the solar neutrino problem [15]. The best fit values of the oscillation parameters from combined analysis of the solar and KamLAND data (in assumption of the CPT invariance) are [16]:

$$\Delta m^2 = 7.9 \cdot 10^{-5} \, \mathrm{eV}^2, \qquad \tan^2 \theta = 0.40. \tag{15}$$

Analysis of the solar neutrino data alone leads to smaller mass split, $\Delta m^2 = 6.3 \cdot 10^{-5} \mathrm{eV}^2$, which agrees with (15) within 1σ.

---

[1] The elements of this vector are nothing but components of the density matrix.

3.1.1. *Physical picture.* According to LMA, inside the Sun the initially produced electron neutrinos undergo highly adiabatic conversion: $\nu_e \rightarrow \cos\theta_m^0 \nu_1 + \sin\theta_m^0 \nu_2$, where θ_m^0 is the mixing angle in the production point. On the way from the central parts of the Sun the coherence of the neutrino state is lost after several hundred oscillation lengths [17], and incoherent fluxes of the mass states ν_1 and ν_2 arrive at the surface of the Earth. In the matter of the Earth ν_1 and ν_2 oscillate, partially regenerating the ν_e-flux. The averaged survival probability can be written as

$$P = \sin^2\theta + \cos^2\theta_{12}^{m0}\cos 2\theta_{12} - \cos 2\theta_{12}^{m0} f_{reg}, \qquad (16)$$

where the first term corresponds to the non-oscillatory transition (dominates at high energies), the second term is the contribution from the averaged oscillations which increases with decrease of energy, and the third term is the regeneration effect f_{reg}. At low energies P reduces to the vacuum oscillation probability with very small matter corrections.

3.1.2. *Status of LMA.* The solution provides a very good global fit of the solar neutrino data. There is no statistically significant deviation from the description given by the standard solar model (SSM) [18] and the LMA solution.

The key observation which testifies for the MSW (matter) effect in the Sun is stronger than $1/2$ suppression of the signals at SK and SNO. The ν_e-survival probability extracted from the CC/NC ratio at SNO is

$$\langle P_{ee}\rangle = 0.31^{+0.12}_{-0.08} \quad (3\sigma), \qquad (17)$$

that is, $\langle P_{ee}\rangle < 0.50$ at 5σ, whereas the vacuum 2ν oscillations can produce $\langle P_{ee}\rangle \geq 0.5$.

Observations of two other signatures of the solution (*i*) the upturn of the spectrum at low energies ($E < 7$–$8\,$MeV), (*ii*) the day-night asymmetry of signals with larger flux during the night are the main objectives of forthcoming studies.

No viable alternative to the LMA solution exists and possible effects beyond LMA are substantially restricted already.

There is a very good agreement of the results from solar neutrinos and KamLAND which implies CPT conservation. Furthermore, it shows correctness of theory of both vacuum oscillations and conversion in matter.

3.1.3. *Testing the MSW effect in the Sun.* An important way to test the effect is to introduce the free parameter a_{MSW} in the the matter potential

$$V \rightarrow a_{MSW}V, \qquad (18)$$

and to determine a_{MSW} from the data [19]. A global analysis of the solar and reactor (KamLAND + CHOOZ) results with Δm^2_{21}, $\sin^2\theta_{12}$ and a_{MSW} being unconstrained gives [19] $a_{MSW} = 0.8$–3.0 (95% C.L.) with the best fit value $a_{MSW} = 1.6$; zero value of a_{MSW} is excluded at 6σ.

3.1.4. *Precision measurements in solar neutrinos.* Identification of the LMA solution opens new possibilities in [20] (*i*) precise description of the LMA conversion both inside the Sun and in the Earth taking into account various corrections; (*ii*) estimation of the accuracy of the approximation made; (*iii*) obtaining accurate *analytic* expressions for probabilities and observables as functions of the oscillation parameters. There are three small quantities which allow for a very precise expansion.

(1) Smallness of the adiabaticity parameter

$$\gamma(x) = \frac{l_m}{4\pi h(x)} \sim 10^{-3}\text{–}10^{-4}, \qquad (19)$$

where h is the height of the solar density profile, allows to use the adiabatic perturbation theory. The non-adiabatic corrections to the averaged survival probability are of the order γ^2 [20].

(2) Smallness of the ratio $\Delta r_{prod}/h$, where Δr_{prod} is the size of the neutrino production region, allows one to make averaging of the survival probability over the neutrino production region in the analytic form [20].

(3) Smallness of the parameter

$$\epsilon(r) = \frac{2EV_E}{\Delta m^2_{12}} \leq 0.02\text{–}0.04, \qquad (20)$$

where V_E is the matter potential in the Earth, allows one to develop a very precise perturbation theory for the neutrino oscillations inside the Earth [20].

3.2. *MSW effect and Supernova neutrinos*

In supernovae one expects new elements of the MSW dynamics. The SN neutrinos probe whole 3ν level crossing scheme, and the effects of both resonances (due to Δm^2_{12} and Δm^2_{13}) should show up. Various effects associated with 1–3 mixing can be realized, depending on the value of θ_{13} (fig. 4). As follows from fig. 4, the SN neutrinos are sensitive to $\sin^2\theta_{13}$ as small as 10^{-5}. Studies of SN neutrinos will also give information on the type of mass hierarchy [21, 22, 23, 24, 25].

3.2.1. *The small mixing MSW conversion.* This can be realized due to the 1–3 mixing and the "atmospheric" mass split Δm^2_{13}.

3.2.2. *Non-oscillatory adiabatic conversion* [6]. It is expected for $\sin^2\theta_{13} > 10^{-3}$. The density in the production point is extremely large: $n(0) \gg n^R$, and therefore the mixing in the initial state is strongly suppressed. So, the neutrino state coincides with the eigenstate in matter: $\nu_e = \nu_{2m}$. Due to adiabaticity (no $\nu_{im} \leftrightarrow \nu_{jm}$ transition) the neutrino state coincides with this eigenstate during the whole evolution: $\nu(t) = \nu_{2m}$. The interference is negligible since there is simply no second component to interfere with, and consequently, oscillations are absent. The flavor of the neutrino state changes as the flavor of the eigenstate ν_{2m} and the latter follows the density change.

3.2.3. *Adiabaticity violation.* It occurs if the 1–3 mixing is small, $\sin^2\theta_{13} < 10^{-3}$.

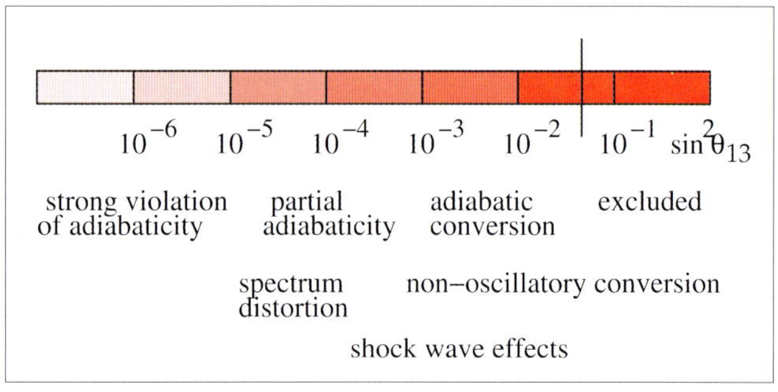

Fig. 4. Scales of the 1–3 mixing probed with supernova neutrinos. Indicated are the regions of various effects in neutrino propagation.

The shock wave can reach the region of the neutrino conversion, $\rho \sim 10^4\,\text{g/cc}$, after $t_s = (3\text{–}5)\,\text{s}$ from the bounce (beginning of the $\nu-$ burst) [26]. Changing suddenly the density profile and therefore breaking the adiabaticity, the shock wave front influences the conversion in the resonance characterized by Δm_{13}^2 and $\sin^2\theta_{13}$, if $\sin^2\theta_{13} > 10^{-4}$.

The following shock wave effects can be seen in neutrinos (antineutrinos) for normal (inverted) hierarchy: (1) change of the total number of events in time [26]; (2) wave of softening of the spectrum which propagates in the energy scale from low energies to high energies [27]; (3) delayed Earth matter effect in the "wrong" channel (*e.g.*, in neutrino channel for normal mass hierarchy) [21]. Modification of the density profile by the shock wave leads to appearance of additional resonances below the front [28]. Reverse shock produces a "double dip" time feature in the average neutrino energy [29, 25]. Monitoring the shock wave with neutrinos can shed some light on the mechanism of explosion.

3.2.4. *Neutrinos from SN1987A.* After confirmation of the LMA MSW solution we can definitely say that some effect of flavor conversion has already been observed in 1987.

In the case of normal mass hierarchy the adiabatic $\bar{\nu}_e \to \bar{\nu}_1$ and $\bar{\nu}_{\mu,\tau} \to \bar{\nu}_2$ transitions occurred inside the star, and then ν_1 and ν_2 oscillated inside the Earth [10, 21]. In terms of the original fluxes of the electron and muon antineutrinos, $F^0(\bar{\nu}_e)$ and $F^0(\bar{\nu}_\mu)$, the $\bar{\nu}_e-$ flux at the detector can be written as

$$F(\bar{\nu}_e) = F^0(\bar{\nu}_e) + \bar{p}\Delta F^0, \tag{21}$$

where $\Delta F^0 \equiv F(\bar{\nu}_\mu) - F(\bar{\nu}_e)$, and $\bar{p}$ is the *permutation* factor which can be calculated precisely: $\bar{p} = \sin^2\theta_{12} - \bar{f}_{reg}$. Due to difference in the distances traveled by neutrinos to Kamiokande, IMB and Baksan detectors inside the Earth, the $\bar{\nu}_e$ regeneration factors $\bar{f}_{reg}$ and therefore $\bar{p}$ differ for these detectors. This can partially explain the difference of the Kamiokande and IMB energy spectra of events [30].

One must take into account the conversion effects in analysis of SN1987A[30] as well as future supernova neutrino data. The conversion can lead to increase of the average energy of the observed events by (30–40)%. Inversely, not taking into account the conversion effect produces errors in determination of the average energy of the original $\bar{\nu}_e$ spectrum up to 40–50% in Kamiokande, and a factor of 2 in IMB.

For the inverted mass hierarchy and $\sin^2\theta_{13} > 10^{-4}$ one would get nearly complete permutation, $\bar{p} \approx 1$, and therefore a harder $\bar{\nu}_e$ spectrum, as well as an absence of the Earth matter effect. This is disfavored by the data [10, 22], see however [23].

4. Matter effects in neutrino oscillations

Pure oscillation effect can be realized in a uniform medium. Mixing is constant, $\theta_m(E, n) = const.$, and therefore

- the flavors of eigenstates do not change;
- the admixtures of eigenstates do not change; there are no $\nu_{1m} \leftrightarrow \nu_{2m}$ transitions: ν_{1m} and ν_{2m} are eigenstates of propagation;
- monotonous increase of Φ_m – a phase difference between the eigenstates occurs.

Only one degree of freedom operates – the phase Φ_m and all others are frozen. This is similar to what happens in vacuum. The depth and length of oscillations are determined by the mixing and energy splitting in matter: $\sin^2 2\theta_m$, $l_m = 2\pi/(H_{2m} - H_{1m})$.

The oscillations are realized in the Earth which can be considered as a multi-layer medium with nearly constant density in each layer. A variety of possibilities exists depending on the neutrino trajectory (zenith angle), neutrino energy and channel of oscillations.

4.1. *Resonance enhancement of oscillations*

For a given (constant) density, the length and depth of oscillations depend on the neutrino energy. This leads to a characteristic distortion of the energy spectrum, $F(E)/F_0(E)$, where $F_0(E)$ and $F(E)$ are the spectra of (*e.g.* ν_e) neutrinos in the source and detector correspondingly (fig. 5).The ratio $F(E)/F_0(E)$ given by the *survival probability* has an oscillatory dependence on the energy. At the resonance energy, E_R, the oscillations proceed with maximal depth; they are enhanced in the resonance range [6]: $E = E_R \pm \Delta E_R$, where $\Delta E_R = \tan 2\theta E_R = \sin 2\theta E_R^0$ and $E_R^0 = \Delta m^2 / 2\sqrt{2} G_F n_e$.

The effect is realized for high energy neutrinos ($E \sim$ GeV) in the mantle of the Earth: constant density is a good first approximation. The effect is expected to be seen in long baseline accelerator experiments [31], and future high statistics atmospheric neutrino studies [32, 33].

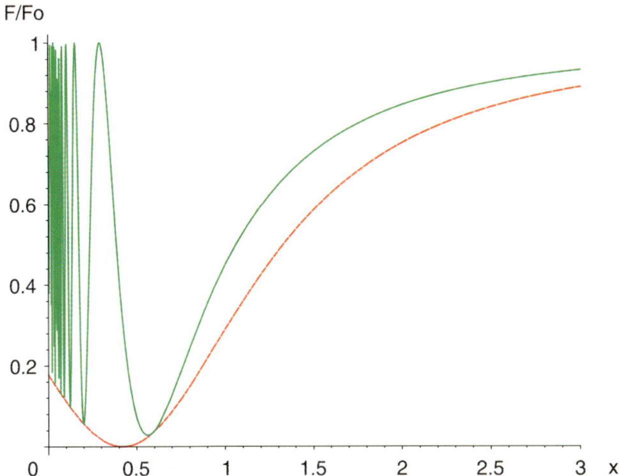

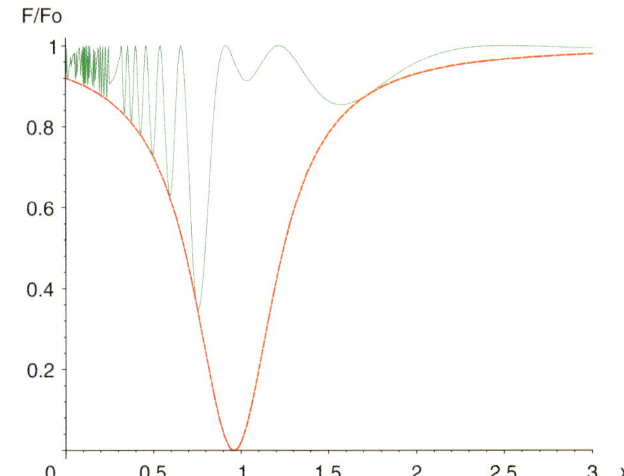

Fig. 5. Resonance enhancement of oscillations in matter with constant density. Shown is the dependence of the ratio of the final and original fluxes, (of *e.g.* ν_e) F/F_0, on energy ($x \equiv l_\nu/l_0 \propto E$) for a thin layer, $L = l_0/\pi$, and $\sin^2 2\theta = 0.824$ (left panel), and a thick layer, $L = 10l_0/\pi$, and $\tan^2\theta = 0.08$ (right panel). The oscillatory curves are inscribed into the resonance curve $(1 - \sin^2 2\theta_m)$.

4.2. *Parametric enhancement of oscillations*

This enhancement is related to a certain condition for the phase of oscillations [34, 35]. It provides another way of getting a strong transition: no matter enhancement or resonance conversion are needed. No large or maximal mixings in vacuum or matter are required.

The simplest case which can be realized in Nature is neutrinos in a castle wall profile. The latter consists of alternate layers with two different densities [35, 36, 37]. Let Φ_1, Φ_2 and θ_1^m, θ_2^m be the phases and mixing angles in the layers 1 and 2. Then under the condition:

$$s_1 c_2 \cos 2\theta_1^m + s_2 c_1 \cos 2\theta_2^m = 0, \tag{22}$$

where $s_1 \equiv \sin \Phi_i/2$, $c_1 \equiv \cos \Phi_i/2$, $(i = 1, 2)$, which is called the parametric resonance condition [38], the flavor transition can be complete. Simple realization of (22) is

$$\Phi_1 = \Phi_2 = \pi \tag{23}$$

which leads to $c_1 = c_2 = 0$. Eq. (22) can be satisfied for neutrinos (the $v_\mu - v_e$ channel with the Δm_{13}^2 and 1–3 mixing) with few GeV energies which cross the core of the Earth [32, 37]. These neutrinos propagate in three layers of matter: mantle-core-mantle. In the approximation of constant densities inside the layers, the profile can be considered as a part of the castle wall profile [36].

For small mixings in both layers, the maximal enhancement occurs when the condition (23) is satisfied (fig. 6, left panel). Apparently for a few more layers that would lead to maximal transition. On the other hand, even three layers are enough to get nearly maximal transition provided that the mixings in matter are not small (fig. 6, right panel).

For $\Delta m_{13}^2 = 2 \cdot 10^{-3}$ eV^2, the strongest parametric effect (fig. 6 right panel) can be observed in a sample of atmospheric neutrinos with $E > 2.5\,\text{MeV}$ (2 times larger than the energies of the multi-GeV sample [32]). A manifestation is the excess of e-like events for the core crossing trajectories. It can be seen as an enhanced up-down asymmetry of the e-like events [32, 33].

4.3. *Neutrino oscillations in the low density medium*

The condition of low density,

$$V(x) \ll \frac{\Delta m^2}{2E}, \tag{24}$$

means that the potential energy is much smaller than the kinetic energy. In this case one can use a small parameter $\epsilon(x)$ (20) to develop the perturbation theory [39].

For the LMA oscillation parameters and the solar and supernova neutrinos: $\epsilon(x) = (1\text{–}3) \cdot 10^{-2}$. The relevant channel of oscillations is the mass-to-flavor transition, $v_2 \rightarrow v_e$, since both the solar and SN neutrinos arrive at the Earth as incoherent fluxes of mass eigenstates. The probability of $v_2 \rightarrow v_e$ can be written as $P_{2e} = \sin^2 \theta + f_{reg}$, and oscillations appear in the first order in ϵ. Using $\epsilon-$ perturbation theory the following expression for the regeneration factor f_{reg} has been obtained [39, 40]

$$f_{reg} = \tfrac{1}{2} \sin^2 2\theta \int_{x_0}^{x_f} \mathrm{d}x\, V(x) \sin \Phi_m(x \rightarrow x_f). \tag{25}$$

Here x_0 and x_f are the initial and final points of propagation correspondingly, $\Phi_m(x \rightarrow x_f)$ is the adiabatic phase (2) acquired between a given point of trajectory, x, and a final point, x_f. The latter feature has the important consequence of leading to the attenuation effect.

In the second order in V, and therefore ϵ, the regeneration factor becomes [41]

$$f_{reg} = \sin^2 2\theta[\sin \Phi_m(x_c \rightarrow x_f)I + \cos 2\theta I^2], \tag{26}$$

where for the symmetric profile

$$I = \int_{x_c}^{x_f} \mathrm{d}x\, V(x) \cos \Phi_m(x_c \rightarrow x). \tag{27}$$

In I the integration proceeds from the central point of trajectory, x_c, to the final point, x_f. The phase is calculated from x_c to a given point x. Essentially, the integral I plays the role of expansion parameter and it can be estimated as $I < 2EV_{max}/\Delta m^2 = \epsilon_{max}$.

The perturbation theory [41] can be improved if the expansion is performed with respect to a certain potential V_0 rather than zero. The effective expansion parameter becomes smaller.

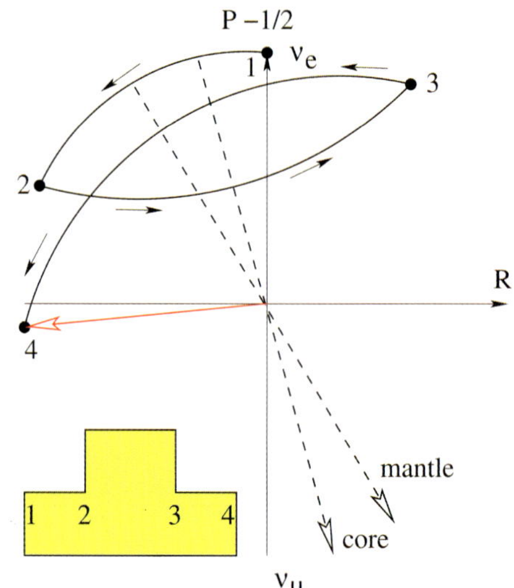

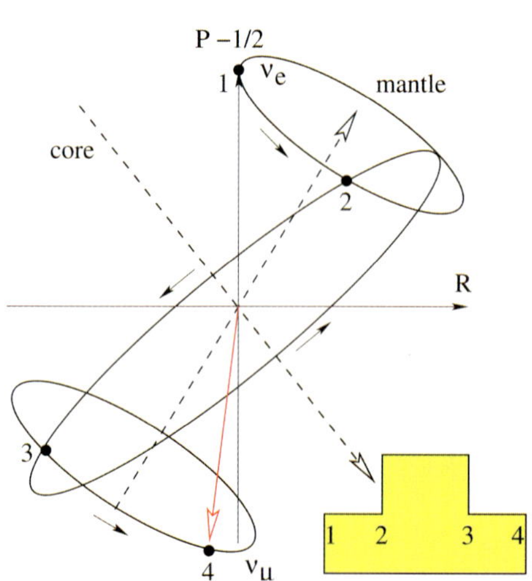

Fig. 6. Graphic representation of the parametric enhancement of neutrino oscillations. The notations are explained in fig. 3 and in Eqs. (13, 14). Left panel: the neutrino energy above the resonance energy in the mantle (matter suppressed mixing). Right panel: the neutrino energy is between the core and mantle resonance energies. The mixing in the core is large; in the mantle $\Phi_1 > \pi/4$ and in the core $\Phi_2 > 2\pi$.

4.4. *Attenuation effect*

Eq. (25) allows one to estimate the sensitivity of the oscillation effects to structures of the density profile [39]. Consider some structure in the point x of the trajectory at the distance $d \equiv x_f - x$ from the detector. According to (25) for the mass-to-flavor transition the potential $V(x)$ is integrated with $\sin \Phi_m(d)$. The larger d, (and therefore, the larger $\Phi_m(d)$), the stronger averaging effect when some integration over the energy is performed. So, weaker sensitivity of the oscillation effects to remote structures of the profile should show up. Integration over energy with the energy resolution function $R(E, E')$,

$$\bar{f}_{reg}(E) = \int dE' R(E, E') f_{reg}(E'), \tag{28}$$

can be expressed as

$$\bar{f}_{reg} = \tfrac{1}{2} \sin^2 2\theta \int_{x_0}^{x_f} dx V(x) F(x_f - x) \sin \Phi_m(x \to x_f), \tag{29}$$

where $F(x_f - x)$ is called the *attenuation* factor [39]. For the box-like function $R(E, E')$ with width (energy resolution) ΔE we obtain

$$F(d) = \frac{1}{z} \sin z, \qquad z = \frac{\pi d \Delta E}{l_v E}. \tag{30}$$

The factor $F(0) = 1$ and it decreases with distance. This means that the contribution of remote structures to the integral (29) is suppressed. The width of the first peak of $F(d)$

$$l_a = l_v \frac{E}{\Delta E} \tag{31}$$

(corresponds to $z = \pi$) determines the *attenuation length*: at $d < l_a$ the effects of structures are not suppressed. The better the energy resolution, the larger the attenuation length, and consequently, deeper structures can be seen by the neutrino "microscope". This explains, *e.g.*, why for the solar neutrinos the zenith angle dependence of the Earth matter effect is flat and there is no enhancement of the regeneration for the core crossing trajectories in spite of 2–3 times larger density. Indeed, for solar neutrinos with $E \sim 10\,\mathrm{MeV}$ and $\Delta E/E = 0.3$, we obtain $l_a = 1000\,\mathrm{km}$ and therefore the contribution of the core is attenuated. On the contrary, small structures ($\sim 10\,\mathrm{km}$) near the surface can produce strong effects. The attenuation length increases with energy. For $E \sim 25\,\mathrm{MeV}$ and $\Delta E/E = 0.2$, $l_a = 4000\,\mathrm{km}$, so that the core of the Earth can be probed by SN neutrinos [42].

Another insight into the phenomena can be obtained using the adiabatic perturbation theory which leads to [20]

$$f_{reg} = \epsilon(R) \sin^2 2\theta \sin^2[0.5\Phi_m(x_0 \to x_f)]$$
$$+ \sin 2\theta \, Re[c(x_0 \to x_f)]. \tag{32}$$

Here $\epsilon(R)$ is the expansion parameter at the surface of the Earth and

$$c(x) = \int_{x_0}^{x} dx' \frac{d\theta_m(x')}{dx'} e^{i\Phi_m(x' \to x)} \tag{33}$$

is the amplitude of transition between the eigenstates in matter (the adiabaticity violation effect). In the adiabatic case, $c(x) = 0$, the second term in (32) is absent. The adiabaticity condition is broken at the borders of shells only. Due to sharp density change we have for the jth border: $d\theta_m(x)/dx|_j = \Delta V_j \delta(x - x_j)$, where ΔV_j is the jump of the potential between the shells. The integration in $c(x)$ is trivial and simple computations give [20]

$$f_{reg} = \frac{2E \sin^2 2\theta}{\Delta m^2} \sin \frac{\Phi_0}{2} \sum_{j=0\ldots n-1} \Delta V_j \sin \frac{\Phi_j}{2}. \tag{34}$$

Here Φ_0 and Φ_j are the phases acquired along the whole trajectory and on the part of the trajectory inside the borders j. This formula corresponds to a symmetric profile with respect to the center of the trajectory. Using (34) one can easily infer the attenuation effect.

5. Conclusions

The large mixing MSW conversion provides the solution of the solar neutrino problem: it leads to determination of Δm_{12}^2 and θ_{12}. Now we have a detailed physical picture of the conversion and its very precise analytical description both inside the Sun and in the Earth.

Interesting relation emerges between θ_{12} determined in the solar neutrinos and the Cabibbo angle: $\theta_{12} + \theta_C = \pi/2$. If not accidental, it has an important implication for the fundamental physics.

The small mixing MSW conversion driven by the 1–3 mixing can be realized for the supernova neutrinos. Study of these neutrinos will give information on the 1–3 mixing and type of mass hierarchy; it opens a unique possibility to perform monitoring of shock waves.

A number of matter effects can be realized for neutrinos propagating inside the Earth: (*i*) resonance enhancement of oscillations; (*ii*) parametric effects in the multi-layer medium; (*iii*) attenuation effect for low energy neutrinos. The first two effects can be seen in experiments with atmospheric and accelerator neutrinos. They will play an important role in determination of the oscillation parameters and establishing the type of neutrino mass hierarchy. The attenuation effect is realized for the solar and supernova neutrinos, it describes the loss of sensitivity to remote structures of the density profile. The effect is crucial for the oscillation tomography of the Earth.

References

1. Wolfenstein, L., Phys. Rev. **D17**, 2369 (1978); in "Neutrino-78," Purdue Univ. C3, (1978), Phys. Rev. **D20**, 2634 (1979).
2. Pontecorvo, B., Zh. Eksp. Theor. Fiz. **33** (1957); ibidem **34**, 247 (1958).
3. Maki, Z., Nakagawa M. and Sakata, S., Prog. Theor. Phys. **28**, 870 (1962).
4. Pontecorvo, B., ZETF **53**, 1771 (1967) [Sov. Phys. JETP **26**, 984 (1968)]; Gribov, V. N. and Pontecorvo, B., Phys. Lett. **28B**, 493 (1969).
5. Barger, V., Whisnant, K., Pakvasa, S. and Phillips, R. J. N., Phys. Rev. **D22**, 2718 (1980); Pakvasa, S., in DUMAND-80, vol. **2**, 457 (1981).
6. Mikheyev, S. P. and Smirnov, A. Yu., Sov. J. Nucl. Phys. **42**, 913 (1985); Nuovo Cim. **C9**, 17 (1986); Mikheev, S. P. and Smirnov, A. Yu., Sov. Phys. JETP **64**, 4 (1986).
7. Langacker, P., Leville, J. P. and Sheiman, J., Phys. Rev. D **27**, 1228 (1983); Semikoz, V. B., Sov. J. Nucl. Phys. **46**, 946 (1987).
8. Bethe, H., Phys. Rev. Lett. **56**, 1305 (1986).
9. Messiah, A., Proc. 6th Moriond Workshop on Massive Neutrinos in Particle Physics and Astrophysics, (eds. O. Fackler and J. Tran Thanh Van), Tignes, France, Jan. (1986), p. 373; Mikheev, S. P. and Smirnov, A. Y., Sov. Phys. JETP **65**, 230 (1987).
10. Smirnov, A. Yu., Spergel, D. N., Bahcall, J. N., Phys. Rev. **D49**, 1389 (1994).
11. Lisi, E., Marrone, A., Montanino, D., Palazzo, A. and Petcov, S. T., Phys. Rev. **D63**, 093002 (2001); Friedland, A., Phys. Rev. **D64**, 013008 (2001), and hep-ph/0106042.
12. Haxton, W. C., Phys. Rev. Lett. **57**, 1271 (1986); Parke, S. J., Phys. Rev. Lett. **57**, 1275 (1986); Rosen, S. P. and Gelb, J. M., Phys. Rev. D **34**, 969 (1986).
13. Mikheyev, S. P. and Smirnov, A. Yu., Proc. 6th Moriond Workshop on massive Neutrinos in Astrophysics and Particle Physics, Tignes, Savoie, France Jan. (1986) (eds. O. Fackler, and J. Tran, Thanh Van), p. 355 (1986); Bouchez, J. *et al.*, Z. Phys. C **32**, 499 (1986); Ermilova, V. K., Tsarev, V. A. and Chechin, V. A., J. Exp. Theor. Phys. Lett. **43**, 453 (1986).

14. Lunardini, C. and Smirnov, A. Yu., Nucl. Phys. B**583**, 260 (2000).
15. McDonald, A., these proceedings; Suzuki, Y., these proceedings.
16. Suzuki, A., these proceedings, KamLAND Collaboration, Eguchi, K. *et al.*, Phys. Rev. Lett. **90**, 021802 (2003); Araki, T. *et al.*, hep-ex/0406035.
17. de Holanda, P. C. and Smirnov, A. Yu., Astropart. Phys. **21**, 287 (2004).
18. Bahcall, J. N. and Pinsonneault, M. H., Phys. Rev. Lett. **92**, 121301 (2004).
19. Fogli, G. and Lisi, E., New J. Phys. **6**, 139 (2004); Fogli, G. L., Lisi, E., Marrone, A. and Palazzo, A., Phys. Lett. B**583**, 149 (2004).
20. de Holanda, P. C., Wei Liao and Smirnov, A. Yu., Nucl. Phys. B**702**, 307 (2004).
21. Dighe, A. S. and Smirnov, A. Yu., Phys. Rev. D**62**, 033007 (2000); Lunardini, C. and Smirnov, A. Yu., JCAP **0306**, 009 (2003).
22. Minakata, H. and Nunokawa, H., Phys. Lett. B **504**, 301 (2001).
23. Barger, V., Marfatia, D. and Wood, B. P., Phys. Lett. B **532**, 19 (2002).
24. Takahashi, K., Sato, K., Burrows, A. and Thompson, T. A., Phys. Rev. D **68**, 113009 (2003).
25. Raffelt, G., these proceedings.
26. Schirato, R. C. and Fuller, G. M., astro-ph/0205390.
27. Takahashi, K., Sato, K., Dalhed, H. E. and Wilson, J. R. Astropart. Phys. **20**, 189 (2003).
28. Fogli, G. L., Lisi, E., Montanino, D. and Mirizzi, A., Phys. Rev. D **68**, 033005 (2003).
29. Tomas, R. *et al.*, astro-ph/0407132.
30. Lunardini, C. and Smirnov, A. Yu., Phys. Rev. D **63**, 073009 (2001); Astropart. Phys. **21**, 703 (2004); Costantini, M. L., Ianni, A. and Vissani, F., Phys. Rev. D **70**, 043006 (2004).
31. Freund, M. and Ohlsson, T., Mod. Phys. Lett. A **15**, 867 (2000); Freund, M., Lindner, M., Petcov, S. T. and Romanino, A., Nucl. Phys. B **578**, 27 (2000); Ohlsson, T. and Snellman, H., Phys. Lett. B **474**, 153 (2000).
32. Akhmedov, E. K., Dighe, A., Lipari, P. and Smirnov, A. Y., Nucl. Phys. B **542**, 3 (1999).
33. Bernabeu, J., Palomares-Ruiz, S., Perez, A. and Petcov, S. T., Phys. Lett. B **531**, 90 (2002); Bernabeu, J., Palomares Ruiz, S. and Petcov, S. T., Nucl. Phys. B **669**, 255 (2003).
34. Ermilova, V. K., Tsarev V. A., Chechin, V. A. and Soob, Kr., Fiz. [Short Notices of the Lebedev Institute] **5**, 26 (1986).
35. Akhmedov, E. Kh., Yad. Fiz. **47**, 475 (1988) [Sov. J. Nucl. Phys. **47**, 301 (1988)].
36. Liu, Q. Y. and Smirnov, A. Yu., Nucl. Phys. B **524**, 505 (1998); Liu, Q. Y., Mikheyev, S. P. and Smirnov, A. Yu., Phys. Lett. B**440**, 319 (1998).
37. Petcov, S. T., Phys. Lett. **B434** (1998) 321; Chizhov, M., Maris, M. and Petcov, S. T., hep-ph/9810501; Chizhov, M. V. and Petcov, S. T., Phys. Rev. Lett. **83**, 1096 (1999); Phys. Rev. D **63**, 073003 (2001).
38. Akhmedov, E. K., Nucl. Phys. B **538**, 25 (1999), [hep-ph/9805272]; hep-ph/9903302; Pramana **54**, 47 (2000), [hep-ph/9907435]. Akhmedov, E. K., Smirnov, A. Yu., Phys. Rev. Lett. **85**, 3978 (2000), and hep-ph/9910433.
39. Ioannisian, A. N. and Smirnov, A. Yu., Phys. Rev. Lett. **93**, 241801 (2004).
40. Akhmedov, E. Kh., Tortola, M. A. and Valle, J. W. F., JHEP **0405**, 057 (2004).
41. Ioannisian, A. N., Kazarian, N. A., Smirnov, A. Yu. and Wyler, D., hep-ph/0407138.
42. Lindner, M., Ohlsson, T., Tomas, R. and Winter, W., Astropart. Phys. **19**, 755 (2003).

Physica Scripta. Vol. T121, 65–71, 2005

Three-Flavour Effects and CP- and T-Violation in Neutrino Oscillations

E. Kh. Akhmedov[*]

AHEP Group, Instituto de Física Corpuscular – C.S.I.C. Universitat de València, Edificio Institutos de Paterna, Apt 22085, E-46071 València, Spain

Received December 1, 2004; accepted in revised form December 17, 2004

PACS numbers: 14.60.Pq, 14.60.Lm, 26.65+t

Abstract

Some theoretical aspects of 3-flavour (3f) neutrino oscillations are reviewed. These include: general properties of 3f oscillation probabilities; matter effects in $\nu_\mu \leftrightarrow \nu_\tau$ oscillations; 3f effects in oscillations of solar, atmospheric, reactor and supernova neutrinos and in accelerator long-baseline experiments; CP and T violation in neutrino oscillations in vacuum and in matter, and the problem of U_{e3}.

1. Introduction

Explanation of the solar, atmospheric, reactor and accelerator neutrino data [1] (with the exception of the LSND result, which still awaits its confirmation) in terms of neutrino oscillations requires at least three neutrino species, and in fact three neutrino species are known to exist $-\nu_e$, ν_μ and ν_τ. Yet, until a few years ago most studies of neutrino oscillations were performed in the 2-flavour framework. There were essentially two reasons for that: (i) simplicity – there are much fewer parameters in the 2-flavour case than in the 3-flavour one, and the expressions for the transition probabilities are much simpler and by far more tractable, and (ii) the hierarchy $\Delta m_{\text{sol}} \ll \Delta m_{\text{atm}}$ and the smallness of the leptonic mixing parameter $|U_{e3}|$, which allow to effectively decouple different oscillation channels. The 2-flavour approach indeed proved to be a good first approximation, which is a consequence of the above point (ii).

However, the increased accuracy of the available and especially of expected neutrino data makes it very important to take even relatively small effects in neutrino oscillations into account. As we shall see, 3-flavour (3f) effects can lead to corrections up to $\sim$10% to 2-flavour oscillation probabilities, which is comparable to the accuracy of the present-day neutrino data. In addition, effects specific to ≥ 3 flavour neutrino oscillations, such as CP and T violation, are of great interest and being widely discussed now. All this makes 3f analyses of neutrino oscillations mandatory.

In my talk some theoretical issues pertaining to 3f neutrino oscillations are reviewed. The topics that are discussed include: general properties of 3f oscillation probabilities; matter effects in $\nu_\mu \leftrightarrow \nu_\tau$ oscillations; 3f effects in oscillations of solar, atmospheric, reactor and supernova neutrinos and in accelerator long-baseline experiments; CP and T violation in neutrino oscillations in vacuum and in matter; the problem of U_{e3}.

2. Leptonic mixing and neutrino oscillations

The leptonic mixing matrix U connects neutrino flavour eigenstates $|\nu_a\rangle$ $(a = e, \mu, \tau)$ with the mass eigenstates $|\nu_i\rangle$ $(i = 1, 2, 3)$:

$$|\nu_a\rangle = \sum_i U_{ai}^* |\nu_i\rangle. \tag{1}$$

In the 3f case the leptonic mixing matrix U is a unitary 3×3 matrix, which in the standard parameterization can be written as

$$U = O_{23}\Gamma_\delta O_{13}\Gamma_\delta^\dagger O_{12}$$

$$= \begin{pmatrix} c_{12}c_{13} & s_{12}c_{13} & s_{13}e^{-i\delta_{CP}} \\ -s_{12}c_{23} - c_{12}s_{13}s_{23}e^{i\delta_{CP}} & c_{12}c_{23} - s_{12}s_{13}s_{23}e^{i\delta_{CP}} & c_{13}s_{23} \\ s_{12}s_{23} - c_{12}s_{13}c_{23}e^{i\delta_{CP}} & -c_{12}s_{23} - s_{12}s_{13}c_{23}e^{i\delta_{CP}} & c_{13}c_{23} \end{pmatrix}. \tag{2}$$

Here O_{ij} is the orthogonal rotation matrix in the ij-plane which depends on the mixing angle θ_{ij}, $\Gamma_\delta = \text{diag}(1, 1, e^{i\delta_{CP}})$, δ_{CP} being the Dirac-type CP-violating phase, $s_{ij} \equiv \sin\theta_{ij}$ and $c_{ij} \equiv \cos\theta_{ij}$. In the 3f case there are also, in general, two Majorana-type CP-violating phases; however, these phases do not affect neutrino oscillations, and I will not discuss them.

Neutrino data allow two different neutrino mass orderings, the normal mass hierarchy and the inverted mass hierarchy (see Figs. 1 and 2). The lines of the matrix U in Eq. (2) represent the neutrino flavour eigenstates in terms of mass eigenstates, whereas its columns give the mass eigenstates in terms of flavour

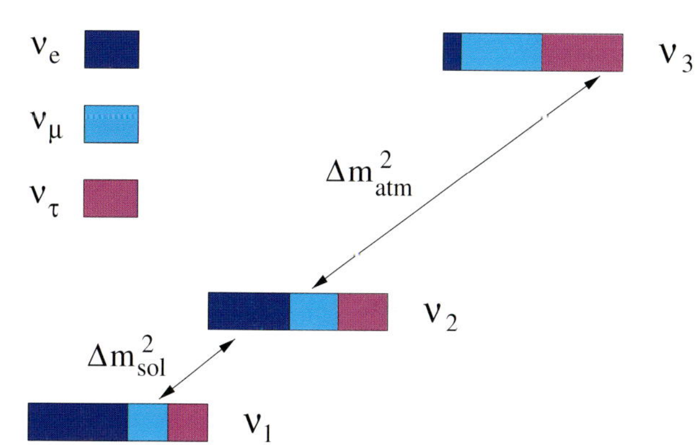

Fig. 1. Normal mass hierarchy.

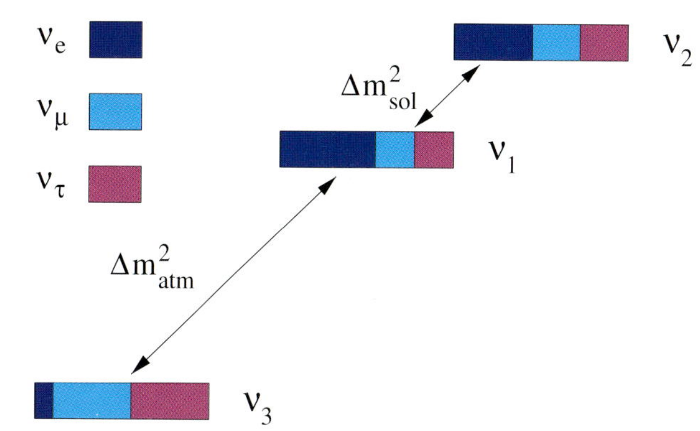

Fig. 2. Inverted mass hierarchy.

[*]On leave from National Research Centre Kurchatov Institute, Moscow 123182, Russia. Present address: ICTP, Strada Costiera 11, 34014 Trieste, Italy. E-mail: akhmedov@ictp.trieste.it

eigenstates. In particular, the value of $|U_{e3}|^2$ is the weight of ν_e in the third mass eigenstate.

3. Three-flavour neutrino oscillations in matter

Neutrino oscillations in matter are described by the Schrödinger-like evolution equation

$$i\frac{d}{dt}\begin{pmatrix} \nu_e \\ \nu_\mu \\ \nu_\tau \end{pmatrix} = \left[U \begin{pmatrix} E_1 & 0 & 0 \\ 0 & E_2 & 0 \\ 0 & 0 & E_3 \end{pmatrix} U^\dagger + \begin{pmatrix} V(t) & 0 & 0 \\ 0 & 0 & 0 \\ 0 & 0 & 0 \end{pmatrix} \right] \begin{pmatrix} \nu_e \\ \nu_\mu \\ \nu_\tau \end{pmatrix}. \quad (3)$$

Here E_i are the neutrino eigenenergies in vacuum, and the effective potential $V = \sqrt{2} G_F N_e$ is due to the charged-current interaction of ν_e with the electrons of the medium [2]. The neutral current induced potentials are omitted from Eq. (3) because they are the same for neutrinos of all three species and therefore do not affect neutrino oscillations. This, however, is only true in leading (tree) order; radiative corrections induce tiny differences between the neutral current potentials of ν_e, ν_μ and ν_τ and, in particular, result in a very small $\nu_\mu - \nu_\tau$ potential difference $V_{\mu\tau} \sim 10^{-5} V$ [3]. This quantity is negligible in most situations but may be important for supernova neutrinos.

For matter of constant density, closed-form solutions of the evolution equation can be found [4]; however, the corresponding expressions are rather complicated and not easily tractable. For a general electron density profile $N_e \neq const$ no closed-form solutions exist. It is therefore desirable to have approximate analytic solutions of the neutrino evolution equation. A number of such solutions were found, most of them based on the expansions in one (or both) of the two small parameters: $\Delta m_{21}^2/\Delta m_{31}^2 = \Delta m_{\rm sol}^2/\Delta m_{\rm atm}^2 \simeq 0.03$, $\quad |U_{e3}| = |\sin\theta_{13}| \lesssim 0.2$ [1]. For a recent discussion and a summary of the earlier results see ref. [5].

In the limits $\Delta m_{21}^2 = 0$ or $U_{e3} = 0$ the transition probabilities acquire an effective 2-flavour (2f) form. When both these parameters vanish, the genuine 2f case is recovered.

4. General properties of 3f oscillation probabilities

In the 3f case, there are nine oscillation (survival and transition) probabilities for neutrinos and the same number for antineutrinos, hence altogether 18 probabilities. How many of them are independent?

Consider first only neutrinos. Unitarity (probability conservation) gives 6 constraints

$$\sum_b P_{ab} = \sum_a P_{ab} = 1 \quad (a, b = e, \mu, \tau), \quad (4)$$

of which 5 are independent; this leaves $9 - 5 = 4$ oscillation probabilities independent. However, recently it has been realized [5] that one can further reduce the number of independent oscillation probabilities. This is possible because the evolution equation (3) has a symmetry related to the fact that the matrix of matter-induced potentials $diag(V(t), 0, 0)$ commutes with O_{23}. Let us define the "θ_{23}-transformed" probabilities

$$\tilde{P}_{ab} = P_{ab}(s_{23}^2 \leftrightarrow c_{23}^2, \sin 2\theta_{23} \rightarrow -\sin 2\theta_{23}) \quad (5)$$

(this transformation can, e.g., be achieved through the shift $\theta_{23} \rightarrow \theta_{23} + \pi/2$). Then, by inspecting the properties of Eq. (3) with respect to the rotation by O_{23}, it is easy to show that

$$P_{e\tau} = \tilde{P}_{e\mu}, \qquad P_{\tau\mu} = \tilde{P}_{\mu\tau}, \qquad P_{\tau\tau} = \tilde{P}_{\mu\mu}. \quad (6)$$

Two of these three relations are independent, which leaves us with $4 - 2 = 2$ independent probabilities. Not any two probabilities would do; one possible choice is $P_{e\mu}$ and $P_{\mu\tau}$.

Since the oscillation probabilities for antineutrinos $P_{\bar{a}\bar{b}}$ are related to those for neutrinos P_{ab} through

$$P_{\bar{a}\bar{b}} = P_{ab}(\delta_{\rm CP} \rightarrow -\delta_{\rm CP}, V \rightarrow -V), \quad (7)$$

one concludes that all 18 neutrino and antineutrino probabilities can be expressed through just two [5].

Analogously, by rotating Eq. (3) with the matrix $O'_{23} = O_{23} \times diag(1, 1, e^{i\delta_{\rm CP}})$, one can study the general dependence of the oscillation probabilities on the CP-violating phase $\delta_{\rm CP}$ [6]. For example, for $P_{e\mu}$ and $P_{\mu\tau}$ one finds

$$P_{e\mu} = A_{e\mu} \cos\delta_{\rm CP} + B_{e\mu} \sin\delta_{\rm CP} + C_{e\mu},$$

$$P_{\mu\tau} = A_{\mu\tau} \cos\delta_{\rm CP} + B_{\mu\tau} \sin\delta_{\rm CP} + C_{\mu\tau} + D_{\mu\tau} \cos 2\delta_{\rm CP}$$

$$+ E_{\mu\tau} \sin 2\delta_{\rm CP}. \quad (8)$$

5. 3f effects in neutrino oscillations

5.1. *Two kinds of three-flavour effects*

There are two kinds 3f effects in neutrino oscillations. First, there are effects which, in a sense, are trivial. These include:

- The existence of new physical oscillation channels – i.e., in addition to $\nu_e \leftrightarrow \nu_\mu$ there are $\nu_e \leftrightarrow \nu_\tau$ and $\nu_\mu \leftrightarrow \nu_\tau$ channels; mutual influence of the channels through unitarity;
- New "parameter channels" for the same physical channel. For example, $\nu_e \leftrightarrow \nu_\mu$ oscillations can be governed by two pairs of parameters, $(\Delta m_{21}^2, \theta_{12})$ and $(\Delta m_{31}^2, \theta_{13})$, corresponding to two ways in which these oscillations can occur.

Second, there are non-trivial effects, i.e. qualitatively new effects that are specific for three (or more) flavours and do not occur in the 2f case:

- Fundamental CP- and T-violation;
- Matter-induced T-violation;
- Interference of different "parameter channels" – specific contributions to oscillation probabilities;
- Matter effects on $\nu_\mu \leftrightarrow \nu_\tau$ oscillations.

A characteristic feature of the non-trivial 3f effects (except for the last one) is that they disappear if at least one mixing angle is 0 or $90°$, or at least one $\Delta m_{ij}^2 = 0$.

I will discuss both types of 3f effects.

5.2. *Matter effects in $\nu_\mu \leftrightarrow \nu_\tau$ oscillations*

Since the matter-induced potentials for ν_μ and ν_τ are the same (neglecting the tiny radiative corrections), in the 2f case the $\nu_\mu \leftrightarrow \nu_\tau$ oscillations are not affected by matter. This, however, is not true in the 3f case; therefore the effect of matter on $\nu_\mu \leftrightarrow \nu_\tau$ oscillations is a pure 3f effect. It vanishes only when both Δm_{21}^2 and U_{e3} vanish. The effects of the Earth's matter on $\nu_\mu \leftrightarrow \nu_\tau$ oscillations can manifest themselves in the long-baseline accelerator as well as in atmospheric neutrino experiments [7, 8]. It has been demonstrated recently that these effects can be rather large [9].

5.3. *Solar neutrinos*

In the 3f case, solar ν_e can in principle oscillate into either ν_μ, or ν_τ, or some their combination. What do they actually oscillate to?

It is easy to answer this question. The smallness of the mixing parameter $|U_{e3}|$ implies that the mass eigenstate v_3, separated by a large mass gap from the other two, is approximately given by

$$v_3 \simeq s_{23}\, v_\mu + c_{23}\, v_\tau \tag{9}$$

and, to first approximation, does not participate in the solar neutrino oscillations. From the unitarity of the leptonic mixing matrix it then follows that the solar neutrino oscillations are the oscillations between v_e and a state v' which is the linear combination of v_μ and v_τ, orthogonal to v_3:

$$v' = c_{23}\, v_\mu - s_{23}\, v_\tau \tag{10}$$

Therefore for solar neutrinos

$$P(v_e \to v_\mu)/P(v_e \to v_\tau) \simeq c_{23}^2/s_{23}^2. \tag{11}$$

Since the mixing angle θ_{23}, responsible for the atmospheric neutrino oscillations, is known to be close to $45°$, Eq. (11) implies that the solar v_e oscillate into a superposition of v_μ and v_τ with equal or almost equal weights. The same argument applies to the long-baseline oscillations of reactor antineutrinos (KamLAND experiment). For reactor experiments with relatively short baselines $L \simeq 1$ km (such as CHOOZ and Palo Verde), the same is true when $\theta_{13} \ll 0.03$. In the opposite limit, $\theta_{13} \gg 0.03$, one finds $P(\bar{v}_e \to \bar{v}_\mu)/P(\bar{v}_e \to \bar{v}_\tau) \simeq s_{23}^2/c_{23}^2$, which is also close to unity. In the intermediate case, $\theta_{13} \sim 0.03$, deviations from unity are possible due to the interference terms in the probabilities $P(\bar{v}_e \to \bar{v}_\mu)$ and $P(\bar{v}_e \to \bar{v}_\tau)$.

What are the 3f effects in the oscillation probabilities of solar neutrinos? Since at low energies v_μ and v_τ are experimentally indistinguishable, all the observables depend on just one probability – the v_e survival probability $P(v_e \to v_e)$. The loss of coherence of the neutrino state in the course of neutrino propagation between the Sun and the Earth leads to an effective averaging over fast oscillations due to the large mass squared difference $\Delta m_{\text{atm}}^2 = \Delta m_{31}^2$, which yields [10]

$$P(v_e \to v_e) \simeq c_{13}^4\, \tilde{P}_{2ee}(\Delta m_{21}^2, \theta_{12}, N_{\text{eff}}) + s_{13}^4. \tag{12}$$

Here $\tilde{P}_{2ee}(\Delta m_{21}^2, \theta_{12}, N_{\text{eff}})$ is the 2f survival probability of v_e in matter with the effective electron density $N_{\text{eff}} = c_{13}^2\, N_e$.

As follows from the CHOOZ data [11], the second term in Eq. (12), s_{13}^4, does not exceed 10^{-3}, i.e. is negligible. At the same time, the coefficient c_{13}^4 of $\tilde{P}_{2ee}$ in the first term may differ from unity by as much as $\sim$5–10 %. Thus, 3f effects may lead to an approximately energy-independent suppression of the v_e survival probability by up to 10 %. With high precision solar data this must be taken into account.

From Eq. (12) it is clear that the fluxes of various components of the solar neutrino spectrum f_i ($i = pp$, ^{7}Be, ^{8}B,...) are always extracted from the charged-current experimental data in the combinations $f_i c_{13}^4$. This leads to an intrinsic uncertainty in their values due to the uncertainty in θ_{13}. In contrast to this, the neutral currents experiments give the fluxes which are free of *both* astrophysics and θ_{13} uncertainties.

While the day-time survival probability of solar v_e (12) scales essentially as c_{13}^4, the day-night signal difference due to the Earth matter effect scales as c_{13}^6 [12, 13]. Unfortunately, the experimental errors of the day-night asymmetry measured by the Super-Kamiokande (SK) and SNO detectors are too large and no useful information on θ_{13} can presently be extracted (the SK value, which has a smaller error, is $A_{ND} = 2.1\% \pm 2.0\%(\text{stat}) \pm 1.3\%(\text{syst})$). However, future large water Cherenkov detectors, such as UNO or Hyper-Kamiokande, may be able to probe θ_{13} [13]. This is illustrated in Fig. 3, which shows the night-day signal asymmetry A_{ND} in future detectors. Fig. 3 allows for possible deviations of the central value of the observed asymmetry from the current SK one within 1σ error of the latter. As one can see from the left panel, if the future central value of measured A_{ND} is higher than the present SK one, the current upper limit on θ_{13} can be improved. If, on the contrary, a lower value of A_{ND} is measured (right panel), the derived upper limit on θ_{13} will be substantially weaker than the current one and thus irrelevant. However, as can be seen from the figure, in that case a *lower* bound on θ_{13} may appear; together with the current upper bound it may actually lead to a rather precise determination of θ_{13} [13]. If the future central value of A_{ND} coincides with the current SK one, no useful information on θ_{13} can be obtained.

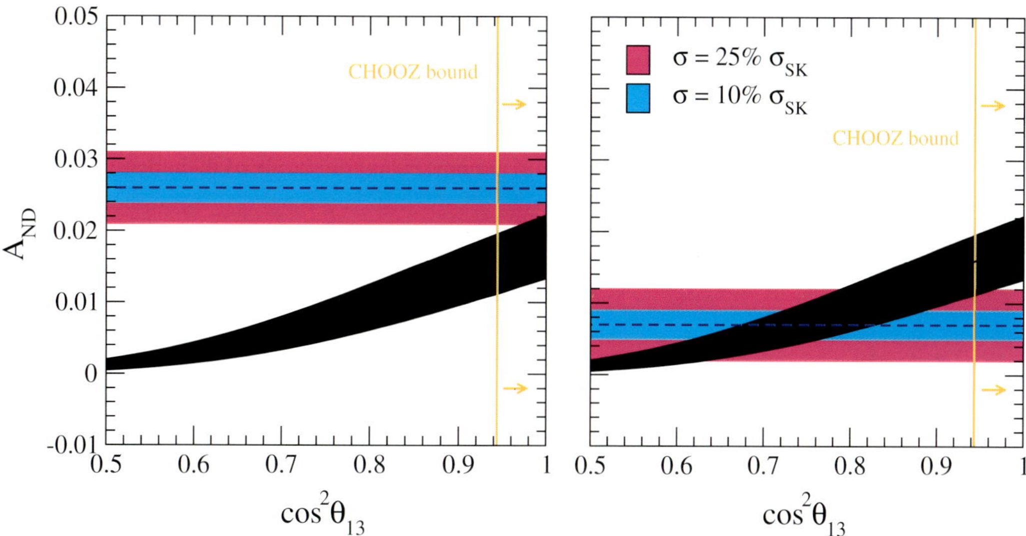

Fig. 3. Expected night-day asymmetry at UNO and Hyper-Kamiokande (horizontal bands) with central values larger (left panel) and smaller (right panel) than the current SK one. Hatched areas are theoretical expectations based on 3σ allowed regions of Δm_{21}^2 and θ_{12} from KamLAND and solar neutrino data. Regions allowed by CHOOZ (3σ) are to the right of vertical lines.

5.4. *Atmospheric neutrinos*

(1) The dominant channel $\nu_\mu \leftrightarrow \nu_\tau$. In the 2f limit, there are no matter effects in this channel (neglecting tiny $V_{\mu\tau}$ caused by radiative corrections). The oscillation probability is independent from the sign of Δm_{31}^2, i.e. cannot differentiate between the normal and inverted neutrino mass hierarchies. The 3f effects result in a sensitivity to matter effects and to the sign of Δm_{31}^2.

(2) The subdominant channels $\nu_e \leftrightarrow \nu_{\mu,\tau}$. Contributions of these oscillation channels to the number of μ – like events are subleading and difficult to observe. For e-like events, one could *a priori* expect significant oscillations effects. However, these effects are in fact strongly suppressed because of the specific composition of the atmospheric neutrino flux and proximity of the mixing angle θ_{23} to $45°$. Indeed, in the 2f limits one finds

$$\frac{F_e - F_e^0}{F_e^0} = \tilde{P}_2(\Delta m_{31}^2, \theta_{13}, V) \cdot (rs_{23}^2 - 1) \tag{13}$$

in the limit $\Delta m_{21}^2 \to 0$ [7], and

$$\frac{F_e - F_e^0}{F_e^0} = \tilde{P}_2(\Delta m_{21}^2, \theta_{12}, V) \cdot (rc_{23}^2 - 1) \tag{14}$$

in the limit $s_{13} \to 0$ [14]. Here F_e^0 and F_e are the ν_e fluxes in the absence and in the presence of the oscillations, respectively, and $r \equiv F_\mu^0/F_e^0$. At low energies $r \simeq 2$; also, we know that $s_{23}^2 \simeq c_{23}^2 \simeq 1/2$. Therefore the factors $(rs_{23}^2 - 1)$ and $(rc_{23}^2 - 1)$ in Eqs. (13) and (14) are very small and strongly suppress the oscillation effects even if the transition probabilities $\tilde{P}_2$ are close to unity. This happens because of the strong cancellations of the transitions from and to the ν_e state.

All this looks as a conspiracy to hide the oscillation effects on the e-like events! This conspiracy is, however, broken by the 3f effects. Keeping both Δm_{21}^2 and s_{13} in leading order yields [15].

$$\frac{F_e - F_e^0}{F_e^0} \simeq \tilde{P}_2(\Delta m_{31}^2, \theta_{13}) \cdot (r\,s_{23}^2 - 1) + \tilde{P}_2(\Delta m_{21}^2, \theta_{12})$$

$$\times (r\,c_{23}^2 - 1) - 2s_{13}\,s_{23}\,c_{23}\,r\,\mathrm{Re}(\tilde{A}_{ee}^*\,\tilde{A}_{\mu e}). \tag{15}$$

Here $\tilde{A}_{ee}$ and $\tilde{A}_{\mu e}$ are the ν_e survival and transition amplitudes in the rotated basis $\tilde{\nu} \approx O_{13}^\dagger O_{23}^\dagger \nu_{fl}$, where ν_{fl} is the neutrino state in the flavour basis. The interference term, which represents the genuinely 3f effects, is not suppressed by the flavour composition of the atmospheric neutrino flux; it can reach a few per cent and may be partially responsible for some excess of the upward-going sub-GeV e-like events observed at Super-Kamiokande. However, this term seems to be insufficient to fully explain the excess, which may be a hint of a deviation of θ_{23} from $45°$ [15].

5.5. *Reactor antineutrinos*

For reactor neutrino experiments, the $\bar{\nu}_e$ survival probability can be written as

$$P_{\bar{e}\bar{e}} \simeq 1 - \sin^2 2\theta_{13} \cdot \sin^2\left(\frac{\Delta m_{31}^2}{4E}L\right)$$

$$- c_{13}^4 \sin^2 2\theta_{12} \cdot \sin^2\left(\frac{\Delta m_{21}^2}{4E}L\right). \tag{16}$$

Since the average energy of reactor $\bar{\nu}_e$'s is $\bar{E} \sim 4\,\mathrm{MeV}$, for intermediate-baseline experiments, such as CHOOZ and Palo Verde ($L \lesssim 1$ km), one has

$$\frac{\Delta m_{31}^2}{4E} L \sim 1, \qquad \frac{\Delta m_{21}^2}{4E} L \ll 1. \tag{17}$$

This justifies the use of the one mass scale dominance approximation, in which the last term in (16) is neglected:

$$P(\bar{\nu}_e \to \bar{\nu}_e) = 1 - \sin^2 2\theta_{13} \cdot \sin^2\left(\frac{\Delta m_{31}^2}{4E}L\right). \tag{18}$$

This is a pure 2f result. Note, however, that disregarding the last term in (16) is only legitimate if θ_{13} is larger than ~ 0.03, which is about the reach of the currently discussed next-generation reactor neutrino experiments.

For KamLAND, which is a very long baseline reactor experiment ($\bar{L} \simeq 170$ km), one has

$$\frac{\Delta m_{31}^2}{4E} L \gg 1, \qquad \frac{\Delta m_{21}^2}{4E} L \gtrsim 1. \tag{19}$$

Averaging over the fast oscillations driven by $\Delta m_{31}^2 = \Delta m_{\mathrm{atm}}^2$ yields

$$P(\bar{\nu}_e \to \bar{\nu}_e) = c_{13}^4 P_{2\bar{e}\bar{e}}(\Delta m_{21}^2, \theta_{12}) + s_{13}^4. \tag{20}$$

This has the same form as Eq. (12). The 2f survival probability $P_{2\bar{e}\bar{e}}$ is, in first approximation, just the corresponding $\bar{\nu}_e$ survival probability in vacuum, which can be obtained from (18) by substituting $\theta_{13} \to \theta_{12}$, $\Delta m_{31}^2 \to \Delta m_{21}^2$. Note, however, that matter effects in KamLAND can reach a few per cent, i.e. can be comparable with the effects of non-zero θ_{13}, and so should be taken into account in 3f analyses. The probability (20) can differ from the 2f probability $P_{2\bar{e}\bar{e}}$ by up to $\sim 10\%$.

5.6. *Long-baseline accelerator experiments*

(1) ν_μ disappearance.

3f effects can result in up to $\sim 10\%$ corrections to the disappearance probability, mainly due to the factor c_{13}^4 in the effective amplitude of the $\nu_\mu \leftrightarrow \nu_\tau$ oscillations, $\sin^2(2\theta_{\mu\tau})_{\mathrm{eff}} \equiv c_{13}^4 \sin^2 2\theta_{23}$. Another manifestation of 3-flavourness are matter effects in $\nu_\mu \leftrightarrow \nu_\tau$ oscillations. The same applies to ν_τ appearance in experiments with the conventional neutrino beams. ν_μ disappearance receives contributions also from the subdominant $\nu_\mu \leftrightarrow \nu_e$ oscillations.

(2) ν_μ appearance at neutrino factories; ν_e appearance at neutrino factories and in experiments with the conventional neutrino beams.

These are driven by the $\nu_e \leftrightarrow \nu_{\mu,\tau}$ oscillations. There are two channels through which these subdominant oscillations can proceed – those governed by the parameters $(\theta_{13}, \Delta m_{31}^2)$ and $(\theta_{12}, \Delta m_{21}^2)$. For typical energies of the long-baseline (LBL) accelerator experiments, a few GeV to tens of GeV, one finds that for θ_{13} in the range $3 \cdot 10^{-3} \lesssim \theta_{13} \lesssim 3 \cdot 10^{-2}$ the two channels compete; otherwise one of them dominates.

Unlike in the case of atmospheric neutrinos, there is no suppression of the oscillation effects on the ν_e flux due to the flavour composition of the original flux.

The dependence of the oscillation probabilities on the CP-violating phase δ_{CP} comes from the interference terms and is a pure 3f effect. The 3f effects will be especially important for the future experiments at neutrino factories which are designed for precision measurements of neutrino parameters.

5.7. *Supernova neutrinos*

In supernovae, matter density varies in a very wide range, and the conditions for three MSW [2] resonances are satisfied (taking into account that due to radiative corrections $V_{\mu\tau} \neq 0$). The hierarchy $\Delta m_{21}^2 \ll \Delta m_{31}^2$ leads to the approximate factorization of the transition dynamics at the resonances, so that the transitions, to first approximation, are effectively 2f ones. However, the observable effects of the supernova neutrino oscillations depend on the transitions between all three neutrino species [16].

Supernova neutrinos can propagate significant distances inside the Earth before reaching the detector. Matter effects on the oscillations of supernova neutrinos inside the Earth depend crucially on the sequence of the neutrino flavour conversions in the supernova which, in turn, depends on the sign of Δm_{31}^2 and is very sensitive to the value of the leptonic mixing parameter U_{e3}. Thus, the Earth matter effects on supernova neutrinos can be used to determine the sign of Δm_{31}^2 and to probe $|U_{e3}|$ down to very small values ($\sim 10^{-3}$) [17]. The transitions due to the $v_\mu - v_\tau$ potential difference $V_{\mu\tau}$ caused by radiative corrections may have observable consequences if the originally produced v_μ and v_τ fluxes are not exactly the same [18].

If neutrinos are Majorana particles, a combination of the MSW effect and resonance spin-flavour precession (RSFP) due to the interaction of neutrino transition magnetic moments μ_v with supernova magnetic fields B can result in the conversion $v_e \rightarrow \bar{v}_e$. Such a conversion would lead to the transformation of the supernova v_e, born in the neutronization process, into their antiparticles. This effect would have a clear experimental signature and its observation would be a smoking gun evidence for the neutrino transition magnetic moments. It would also signify the leptonic mixing parameter $|U_{e3}|$ in excess of 10^{-2}. The conversion mechanism is efficient if $\mu_v B_{res} \gtrsim 10^{-13} \mu_B \cdot 10^9$ G. In the 2f approach, the $v_e \rightarrow \bar{v}_e$ transition is only possible in the case of the inverted neutrino mass hierarchy [19]. However, in the full 3f framework one finds a new RSFP resonance to exist, due to which the $v_e \rightarrow \bar{v}_e$ conversion can occur also for the normal mass hierarchy [20]. Thus, the possibility of $v_e \rightarrow \bar{v}_e$ transitions of supernova neutrinos in the case of the normal neutrino mass hierarchy is a pure 3f effect.

6. CP and T violation in v oscillations in vacuum

The probability of $v_a \rightarrow v_b$ oscillations in vacuum is given by

$$P(v_a, t_0 \rightarrow v_b; t) = \left| \sum_i U_{bi} e^{-iE_i(t-t_0)} U_{ai}^* \right|^2. \tag{21}$$

In the general case of n flavours the leptonic mixing matrix U_{ai} depends on $(n-1)(n-2)/2$ Dirac-type CP-violating phases $\{\delta_{CP}\}$.

Under CP transformation, neutrinos are replaced by their antiparticles ($v_{a,b} \leftrightarrow \bar{v}_{a,b}$), which is equivalent to the complex conjugation of U_{ai}:

$$\text{CP:} \quad v_{a,b} \leftrightarrow \bar{v}_{a,b} \Leftrightarrow U_{ai} \rightarrow U_{ai}^* \quad (\{\delta_{CP}\} \rightarrow -\{\delta_{CP}\}). \tag{22}$$

Time reversal transformation interchanges the initial and final evolution times t_0 and t in Eq. (21), i.e. corresponds to evolution "backwards in time". As follows from Eq. (21), the interchange $t_0 \rightleftarrows t$ is equivalent to the complex conjugation of the exponential factors in the oscillation amplitude. Since the transition probability only depends on the modulus of the amplitude, this is equivalent to the complex conjugation of the factors U_{bi} and U_{ai}^*, which in turn amounts to interchanging $a \rightleftarrows b$. Thus, instead of evolution "backwards in time" one can consider evolution forward in time, but between the interchanged initial and final flavours:

$$\text{T:} \quad t_0 \rightleftarrows t \Leftrightarrow v_a \leftrightarrow v_b \Rightarrow U_{ai} \rightarrow U_{ai}^*(\{\delta_{CP}\} \rightarrow -\{\delta_{CP}\}). \tag{23}$$

Under the combined action of CP and T one has

$$\text{CPT:} \quad v_{a,b} \leftrightarrow \bar{v}_{a,b} \quad \text{and} \quad t_0 \rightleftarrows t \, (v_a \leftrightarrow v_b)$$
$$\Rightarrow P(v_a \rightarrow v_b) \rightarrow P(\bar{v}_b \rightarrow \bar{v}_a). \tag{24}$$

From CPT invariance it follows that CP violation implies T violation and vice versa.

CP and T violation can be characterized by the probability differences

$$\Delta P_{ab}^{CP} \equiv P(v_a \rightarrow v_b) - P(\bar{v}_a \rightarrow \bar{v}_b), \tag{25}$$
$$\Delta P_{ab}^{T} \equiv P(v_a \rightarrow v_b) - P(v_b \rightarrow v_a). \tag{26}$$

From CPT invariance it follows that the CP- and T-violating probability differences coincide, and that the survival probabilities have no CP asymmetry:

$$\Delta P_{ab}^{CP} = \Delta P_{ab}^{T}; \qquad \Delta P_{aa}^{CP} = 0. \tag{27}$$

CP and T violations are absent in the 2f case, so any observable violation of these symmetries in neutrino oscillations in vacuum would be a pure $\geq$3f effect.

In the 3f case, there is only one CP-violating Dirac-type phase δ_{CP} and so only one CP-odd (and T-odd) probability difference:

$$\Delta P_{e\mu}^{CP} = \Delta P_{\mu\tau}^{CP} = \Delta P_{\tau e}^{CP} \equiv \Delta P, \tag{28}$$

where

$$\Delta P = -4 s_{12} c_{12} s_{13} c_{13}^2 s_{23} c_{23} \sin \delta_{CP}$$
$$\times \left[\sin\left(\frac{\Delta m_{12}^2}{2E} L\right) + \sin\left(\frac{\Delta m_{23}^2}{2E} L\right) + \sin\left(\frac{\Delta m_{31}^2}{2E} L\right) \right]. \tag{29}$$

It vanishes

- when at least one $\Delta m_{ij}^2 = 0$
- when at least one $\theta_{ij} = 0$ or $90°$
- when $\delta_{CP} = 0$ or $180°$
- in the averaging regime
- in the limit $L \rightarrow 0$ (as L^3)

Clearly, this quantity is very difficult to observe.

7. CP- and T-odd effects in v oscillations in matter

For neutrino oscillations in matter, CP transformation (substitution $v_a \leftrightarrow \bar{v}_a$) implies not only complex conjugating the leptonic mixing matrix, but also flipping the sign of the matter-induced neutrino potentials:

$$\text{CP:} \quad U_{ai} \rightarrow U_{ai}^*(\{\delta_{CP}\} \rightarrow -\{\delta_{CP}\}), \qquad V(r) \rightarrow -V(r). \tag{30}$$

It can be shown [21] that in matter with an arbitrary density profile, just as well as in vacuum, the action of time reversal on neutrino oscillations is equivalent to interchanging the initial and final neutrino flavours. It is also equivalent to complex conjugating U_{ai} and replacing the matter density profile by the reverse one:

$$\text{T:} \quad U_{ai} \rightarrow U_{ai}^*(\{\delta_{CP}\} \rightarrow -\{\delta_{CP}\}), \qquad V(r) \rightarrow \tilde{V}(r). \tag{31}$$

Here

$$\tilde{V}(r) = \sqrt{2} G_F \tilde{N}(r), \tag{32}$$

$\tilde{N}(r)$ being the reverse profile, i.e. the profile that corresponds to the interchanged positions of the neutrino source and detector. In the case of symmetric matter density profiles (e.g., matter of constant density), $\tilde{N}(r) = N(r)$.

An important point is that the very presence of matter (with unequal numbers of particles and antiparticles) violates C, CP and CPT, leading to CP-odd effects in neutrino oscillations even in the absence of the fundamental CP-violating phases $\{\delta_{CP}\}$. This fake (extrinsic) CP violation may complicate the study of the fundamental (intrinsic) one.

7.1. CP-odd effects in matter

Unlike in vacuum, CP-odd effects in neutrino oscillations in matter exist even in the 2f case (in the case of three or more flavours, even when all $\{\delta_{CP}\} = 0$):

$$P(\nu_a \to \nu_b) \neq P(\bar{\nu}_a \to \bar{\nu}_b). \tag{33}$$

This is actually a well known fact – for example, the MSW effect can enhance the $\nu_e \leftrightarrow \nu_\mu$ oscillations and suppress the $\bar{\nu}_e \leftrightarrow \bar{\nu}_\mu$ ones or vice versa. Moreover, in matter the survival probabilities are not CP-invariant:

$$P(\nu_a \to \nu_a) \neq P(\bar{\nu}_a \to \bar{\nu}_a). \tag{34}$$

To disentangle fundamental CP violation from the matter induced one in the LBL experiments one would need to measure the energy dependence of the oscillated signal or the signals at two baselines, which is a difficult task. The (difficult) alternatives are:

- LBL experiments at relatively low energies and moderate baselines ($E \sim 0.1$–1 GeV, $L \sim 100$–1000 km) [22] – in this case matter effects are negligible.
- Indirect measurements through
  - (A) CP-even terms $\sim\cos\delta_{CP}$ [23];
  - (B) Area of leptonic unitarity triangle [24].

CP-odd effects cannot be studied in the supernova neutrino experiments because of the experimental indistinguishability of low-energy ν_μ and ν_τ.

7.2. T-odd effects in matter

Since CPT is not conserved in matter, CP and T violations are no longer directly connected (although some relations between them still exist [21, 25]). Therefore T-odd effects in neutrino oscillation in matter deserve an independent study. Their characteristic features are:

- Matter does not necessarily induce T-odd effects (only asymmetric matter with $\tilde{N}(r) \neq N(r)$ does).
- There is no T violation (either fundamental or matter induced) in the 2f case. This is a simple consequence of unitarity. For example, for the (ν_e, ν_μ) system one has

$$P_{ee} + P_{e\mu} = 1, \qquad P_{ee} + P_{\mu e} = 1, \tag{35}$$

  from which $P_{e\mu} = P_{\mu e}$.
- In the 3f case there is only one T-odd probability difference for ν's (and one for $\bar{\nu}$'s), irrespective of the matter density profile:

$$\Delta P_{e\mu}^T = \Delta P_{\mu\tau}^T = \Delta P_{\tau e}^T. \tag{36}$$

This is a consequence of 3f unitarity [26].

The matter-induced T-odd effects are very interesting, pure $\geq$3f matter effects, absent in symmetric matter (in particular, in constant-density matter). They do not vanish in the regime of complete averaging of neutrino oscillations [21]. They may fake the fundamental T violation and complicate its study, i.e. the extraction of δ_{CP} from the experiment. The matter-induced T-violating effects vanish when either $U_{e3} = 0$ or $\Delta m_{21}^2 = 0$ (i.e., in the 2f limits) and so are doubly suppressed by both these small parameters. This implies that the perturbation theory can be used to obtain analytic expressions for the T-odd probability differences [21].

In an asymmetric matter, both fundamental and matter-induced T violations contribute to the T-odd probability differences ΔP_{ab}^T. This may hinder the experimental determination of the fundamental CP- and T-violating phase δ_{CP}. In particular, in the accelerator LBL experiments one has to take into account that the Earth's density profile is not perfectly spherically symmetric. To extract the fundamental T violation, strictly speaking one would need to measure

$$P_{\text{dir}}(\nu_a \to \nu_b) - P_{\text{rev}}(\nu_b \to \nu_a), \tag{37}$$

where P_{dir} and P_{rev} correspond to the direct and reverse matter density profiles. (An interesting point is that even the survival probabilities $P_{\mu\mu}$ and $P_{\tau\tau}$ can be used for that [27]).

In practical terms, it would certainly be difficult to measure the quantity in (37): It would not be easy, for example, to move CERN to Gran Sasso and the Gran Sasso Laboratory to CERN. Fortunately, this is not actually necessary – matter-induced T-odd effects due to imperfect sphericity of the Earth's density distribution are very small. They cannot spoil the determination of δ_{CP} if the error in δ_{CP} is $>1\%$ at 99% C.L. [21].

7.3. "CPT in matter"

As was pointed out before, a matter with unequal numbers of particles and antiparticles violates CPT. Is there any relation between CP and T violations in matter which can play a role similar to the CPT relation in vacuum? For symmetric density profiles ($\tilde{V}(r) = V(r)$) such a relation was found in [25]:

$$P(\nu_a \to \nu_b; \delta_{CP}, V(r)) = P(\bar{\nu}_b \to \bar{\nu}_a; \delta_{CP}, -V(r)). \tag{38}$$

It is easy to generalize this to the case of an arbitrary density profile:

$$P(\nu_a \to \nu_b; \delta_{CP}, V(r)) = P(\bar{\nu}_b \to \bar{\nu}_a; \delta_{CP}, -\tilde{V}(r)). \tag{39}$$

Unlike CPT in vacuum, this "CPT in matter" relation does not directly relate observables (there is no anti-Earth), and so is of limited practical use. However, it can be useful for cross-checking theoretical calculations of oscillation probabilities.

8. Why study U_{e3}? (A hymn to U_{e3})

The leptonic mixing parameter U_{e3} plays a very special role in neutrino physics. It is of particular interest for a number of reasons.

First, it is the least known of leptonic mixing parameters: while we have (relatively small) allowed ranges for the other two mixing parameters, we only know an upper bound on $|U_{e3}|$. Its smallness, which looks strange in the light of the fact that the other two mixing parameters, θ_{12} and θ_{23}, are apparently large, remains essentially unexplained. (There are, however, some ideas which relate the smallness of $|U_{e3}|$ to that of $\Delta m_{\text{sol}}^2/\Delta m_{\text{atm}}^2$ [28].)

The smallness of U_{e3} is likely to be the bottleneck for studying the fundamental CP and T violation effects and matter-induced T-odd effects in neutrino oscillations. The same applies to the determination of the sign of Δm_{31}^2 in future LBL experiments, which would allow us to discriminate between the normal and inverted neutrino mass hierarchies. Therefore it would be vitally important to know how small U_{e3} actually is.

The parameter U_{e3} can be efficiently used to discriminate between various neutrino mass models [29]. It is one of the main parameters that drives the subdominant oscillations of atmospheric neutrinos and is important for their study. It also governs the Earth matter effects on supernova neutrino oscillations.

The parameter U_{e3} drives the parametric amplification of oscillations of core-crossing neutrinos inside the Earth, which is an interesting matter effect, different from the MSW resonance enhancement [30].

And finally, U_{e3} apparently provides us with the only opportunity to see the "canonical" MSW effect. While matter effects can be important even in the case of large vacuum mixing angles, the most spectacular phenomenon, strong enhancement of mixing by matter, can only occur if the vacuum mixing angle is small. From what we know now, it seems that the only small leptonic mixing parameter is U_{e3}.

All this makes measuring U_{e3} one of the most important problems in neutrino physics.

9. Conclusions

3f effects in solar, atmospheric, reactor and supernova neutrino oscillations and in LBL accelerator neutrino experiments may be quite important. They can lead to up to $\sim 10\%$ corrections to the oscillation probabilities and also to specific effects, absent in the 2f case. The manifestations of ≥ 3 flavours in neutrino oscillations include fundamental CP violation and T violation, matter-induced T-odd effects, matter effects in $\nu_\mu \leftrightarrow \nu_\tau$ oscillations, and specific CP- and T-conserving interference terms (proportional to the sines of three different mixing angles) in oscillation probabilities. The leptonic mixing parameter U_{e3} plays a very special role and its study is of great interest.

References

1. Gonzalez-Garcia, M. C., these Proceedings, hep-ph/0410030.
2. Mikheev, S. P. and Smirnov, A. Y., Sov. J. Nucl. Phys. **42**, 913 (1985), [Yad. Fiz. **42**, 1441 (1985)]; Wolfenstein, L., Phys. Rev. D **17**, 2369 (1978).
3. Botella, F. J., Lim, C. S. and Marciano, W. J., Phys. Rev. D **35**, 896 (1987).
4. Barger, V. D., Whisnant, K., Pakvasa, S. and Phillips, R. J., Phys. Rev. D **22**, 2718 (1980); Zaglauer, H. W. and Schwarzer, K. H., Z. Phys. C **40**, 273 (1988); Ohlsson, T. and Snellman, H., J. Math. Phys. **41**, 2768 (2000); (Erratum-ibid. **42**, 2345 (2001)); Phys. Lett. B 474 (2000) 153; Kimura, K., Takamura, A. and Yokomakura, H., Phys. Lett. B **537**, 86 (2002).
5. Akhmedov, E. Kh., Johansson, R., Lindner, M., Ohlsson, T. and Schwetz, T., JHEP **0404**, 078 (2004).
6. Yokomakura, H., Kimura, K. and Takamura, A., Phys. Lett. B **544**, 286 (2002).
7. Akhmedov, E. Kh., Dighe, A., Lipari, P. and Smirnov, A. Yu., Nucl. Phys. B **542**, 3 (1999).
8. Akhmedov, E. Kh., Nucl. Phys. Proc. Suppl. **118**, 245 (2003); Phys. Usp. **47**, 114 (2004) [Usp. Fiz. Nauk **47**, 121 (2004)].
9. Gandhi, R., Ghoshal, P., Goswami, S., Mehta, P., Sankar, S. U., hep-ph/0408361.
10. Lim, C.-S., In Proc. BNL Neutrino Workshop on Opportunities for Neutrino Physics, Upton, N. Y., Feb. 5–7, 1987, ed. by M. J. Murtagh, Report No. BNL-52079, C87/02/05, 1987, p. 111; preprint BNL-39675 (scanned version at http://www-lib.kek.jp/cgi-bin/img_index?8710120).
11. The CHOOZ Collaboration, Apollonio, M. *et al.*, Phys. Lett. B **420**, 39 (1998); Phys. Lett. B **466**, 415 (1999).
12. Blennow, M., Ohlsson, T. and Snellman, H., Phys. Rev. D **69**, 073006 (2004).
13. Akhmedov, E. Kh., Tortola, M. A. and Valle, J. W. F., JHEP **0405**, 057 (2004).
14. Peres, O. L. and Smirnov, A. Yu., Phys. Lett. B **456**, 204 (1999).
15. Peres, O. L. G. and Smirnov, A. Y., Nucl. Phys. B **680**, 479 (2004); Gonzalez-Garcia, M. C., Maltoni, M. and Smirnov, A. Y., hep-ph/0408170.
16. Dighe, A. S. and Smirnov, A. Y., Phys. Rev. D **62**, 033007 (2000).
17. Lunardini, C. and Smirnov, A. Yu., Phys. Rev. D **63**, 073009 (2001); Lunardini, C. and Smirnov, A. Yu., Nucl. Phys. B **616**, 307 (2001); Minakata, H. and Nunokawa, H., Phys. Lett. B **504**, 301 (2001); Barger, V., Marfatia, D. and Wood, B. P., Phys. Lett. B **532**, 19 (2002); Lunardini, C. and Smirnov, A. Y., JCAP **0306**, 009 (2003).
18. Akhmedov, E. Kh., Lunardini, C. and Smirnov, A. Yu., Nucl. Phys. B **643**, 339 (2002).
19. Ando, S. and Sato, K., JCAP **0310**, 001 (2003).
20. Akhmedov, E. Kh. and Fukuyama, T., JCAP **0312**, 007 (2003).
21. Akhmedov, E. Kh., Huber, P., Lindner, M. and Ohlsson, T., Nucl. Phys. B **608**, 394 (2001).
22. Koike, M. and Sato, J., Phys. Rev. D **61**, 073012 (2000), (Erratum-ibid. D **62**, 079903 (2000)). Minakata, H. and Nunokawa, H., Phys. Lett. B **495**, 369 (2000); JHEP **0110**, 001 (2001).
23. Lipari, P., Phys. Rev. D **64**, 033002 (2001).
24. Farzan, Y. and Smirnov, A. Yu., Phys. Rev. D **65**, 113001 (2002); Aguilar-Saavedra, J. A. and Branco, G. C., Phys. Rev. D **62**, 096009 (2000); Sato, J., Nucl. Instr. Meth. A **472**, 434 (2000); Fritzsch, H. and Xing, Z. Z., Prog. Part. Nucl. Phys. **45**, 1 (2000).
25. Minakata, H., Nunokawa, H. and Parke, S., Phys. Lett. B **537**, 249 (2002).
26. Krastev, P. I. and Petcov, S. T., Phys. Lett. B **205**, 84 (1988).
27. Fishbane, P. M. and Kaus, P., Phys. Lett. B **506**, 275 (2001).
28. Akhmedov, E. Kh., Branco, G. C. and Rebelo, M. N., Phys. Rev. Lett. **84**, 3535 (2000); King, S. F., JHEP **0209**, 11 (2002); Kuchimanchi, R. and Mohapatra, R. N., Phys. Lett. B **552**, 198 (2003).
29. Barr, S. M. and Dorsner, I., Nucl. Phys. B **585**, 79 (2000); Tanimoto, M., hep-ph/0106064; Feruglio, F., hep-ph/0410131; Joshipura, A. S., hep-ph/0411154; Lindner, M., these Proceedings.
30. See, e.g., Smirnov, A. Yu., these Proceedings, and references therein.

Physica Scripta. Vol. T121, 72–77, 2005

Global Analysis of Neutrino Data

M. C. Gonzalez-Garcia

Theory Division, CERN, CH-1211, Geneva 23, Switzerland, and Y.I.T.P., SUNY at Stony Brook, Stony Brook, NY 11794-3840, USA, and IFIC, Universitat de València - C.S.I.C., Apt 22085, 46071 València, Spain

Received December 14, 2004; accepted in revised form December 16, 2004

PACS numbers: 12.15Ff, 14.60Pq, 14.60St

Abstract

In this talk I review the present status of neutrino masses and mixing and some of their implications for particle physics phenomenology.

1. Introduction: the New Minimal Standard Model

The SM is based on the gauge symmetry $SU(3)_C \times SU(2)_L \times U(1)_Y$ spontaneously broken to $SU(3)_C \times U(1)_{EM}$ by the vacuum expectations value (VEV), v, of a Higgs doublet field ϕ. The SM contains three fermion generations which reside in chiral representations of the gauge group. Right-handed fields are included for charged fermions as they are needed to build the electromagnetic and strong currents. No right-handed neutrino is included in the model since neutrinos are neutral.

In the SM, fermion masses arise from the Yukawa interactions which couple the right-handed fermion singlets to the left-handed fermion doublets and the Higgs doublet. After spontaneous electroweak symmetry breaking these interactions lead to charged fermion masses but leave the neutrinos massless. No Yukawa interaction can be written that would give a tree level mass to the neutrino because no right-handed neutrino field exists in the model.

One could think that neutrino masses could arise from loop corrections if these corrections induced effective terms $\frac{Y_{ij}^v}{v}(\bar{L}_{Li}\tilde{\phi})(\tilde{\phi}^T L_{Lj}^C)$ where L_{Li} are the lepton doublets. This, however, cannot happen because within the SM $G_{SM}^{global} = U(1)_B \times U(1)_e \times U(1)_\mu \times U(1)_\tau$ is an accidental global symmetry. Here $U(1)_B$ is the baryon number symmetry, and $U(1)_{e,\mu,\tau}$ are the three lepton flavor symmetries. Terms of the form above violate G_{SM}^{global} and therefore cannot be induced by loop corrections. Furthermore, they cannot be induced by non-perturbative corrections because the $U(1)_{B-L}$ subgroup of G_{SM}^{global} is non-anomalous.

It follows then that the SM predicts that neutrinos are *strictly* massless. Consequently, there is neither mixing nor CP violation in the leptonic sector.

We now know that this picture cannot be correct. Over several years we have accumulated important experimental evidence that neutrinos are massive particles and there is mixing in the leptonic sector:

- Solar $v_e's$ convert to v_μ or v_τ with confidence level (CL) of more than 7σ [1, 2].
- KamLAND find that reactor $\bar{v}_e$ disappear over distances of about 180 km and they observe a distortion of their energy spectrum. Altogether their evidence has more than 3σ CL [3].
- The evidence of atmospheric (ATM) v_μ disappearing is now at $>15\sigma$, most likely converting to v_τ [1].
- K2K observe the disappearance of accelerator v_μ's at distance of 250 km and find a distortion of their energy spectrum with a CL of 2.5–4 σ [1, 4].
- LSND found evidence for $\overline{v_\mu} \to \overline{v_e}$. This evidence has not been confirm by any other experiment so far and it is being tested by MiniBooNE [4].

These results imply that neutrinos are massive and the Standard Model has to be extended at least to include neutrino masses. This minimal extension is what I call *The New Minimal Standard Model*.

In the New Minimal Standard Model flavour is mixed in the CC interactions of the leptons, and a leptonic mixing matrix appears analogous to the CKM matrix for the quarks. However the discussion of leptonic mixing is complicated by two factors. First the number of massive neutrinos (n) is unknown, since there are no constraints on the number of right-handed, SM-singlet, neutrinos. Second, since neutrinos carry neither color nor electromagnetic charge, they could be Majorana fermions. As a consequence the number of new parameters in the model depends on the number of massive neutrino states and on whether they are Dirac or Majorana particles.

In general, if we denote the neutrino mass eigenstates by v_i, $i = 1, 2, \ldots, n$, and the charged lepton mass eigenstates by $l_i = (e, \mu, \tau)$, in the mass basis, leptonic CC interactions are given by

$$-\mathcal{L}_{CC} = \frac{g}{\sqrt{2}} \overline{l_{iL}} \, \gamma^\mu \, U_{ij} \, v_j \, W_\mu^+ + \text{h.c.} \tag{1}$$

Here U is a $3 \times n$ matrix $U_{ij} = P_{\ell,ii} V_{ik}^{\ell\,\dagger} V_{kj}^v (P_{v,jj})$ where V^ℓ (3×3) and V^v ($n \times n$) are the diagonalizing matrix of the charged leptons and neutrino mass matrix respectively $V^{\ell\dagger} M_\ell M_\ell^\dagger V^\ell = \text{diag}(m_e^2, m_\mu^2, m_\tau^2)$ and $V^{v\dagger} M_v^\dagger M_v V^v = \text{diag}(m_1^2, m_2^2, m_3^2, \ldots, m_n^2)$. P_ℓ is a diagonal 3×3 phase matrix, that is conventionally used to reduce by three the number of phases in U. P_v is a diagonal matrix with additional arbitrary phases (chosen to reduce the number of phases in U) only for Dirac states. For Majorana neutrinos, this matrix is simply a unit matrix, the reason being that if one rotates a Majorana neutrino by a phase, this phase will appear in its mass term which will no longer be real. Thus, the number of phases that can be absorbed by redefining the mass eigenstates depends on whether the neutrinos are Dirac or Majorana particles. In particular, if there are only three Majorana (Dirac) neutrinos, U is a 3×3 matrix analogous to the CKM matrix for the quarks but due to the Majorana (Dirac) nature of the neutrinos it depends on six (four) independent parameters: three mixing angles and three (one) phases.

A consequence of the presence of the leptonic mixing is the possibility of flavour oscillations of the neutrinos. In this symposium we had a beautiful historical introduction to neutrino oscillations by S. Bilenky and two very interesting talks on the phenomenology of oscillations in matter by A. Smirnov and on three neutrino mixing effects by E. Akhmedov. So I will only briefly summarize here the elements which are relevant for the phenomenological analysis that I will present.

Neutrino oscillations appear because a neutrino of energy E produced in a CC interaction with a charged lepton l_α can be detected via a CC interaction with a charged lepton l_β with a probability which presents an oscillatory behaviour, with oscillation lengths $L_{0,ij}^{\mathrm{osc}} = \frac{4\pi E}{\Delta m_{ij}^2}$ and amplitude that is proportional to elements in the mixing matrix. Neutrino oscillations are only sensitive to mass squared differences. Also, the Majorana phases cancel out and only the Dirac phase is observable. Experimental information on absolute neutrino masses can be obtained from Tritium β decay experiments [5] and from its effect on the cosmic microwave background radiation and large structure formation data [6]. Also if neutrinos are Majorana particles their mass and also additional phases can be determined in ν-less $\beta\beta$ decay experiments [7].

When neutrinos travel through regions of dense matter, they can undergo forward scattering with the particles in the medium. These interactions are, in general, flavour dependent and as a consequence the oscillation pattern described above is modified but it still depends only on the mass squared differences and it is independent of the Majorana phases.

The neutrino experiments described above have measured some non-vanishing $P_{\alpha\beta}$ and from these measurements we have inferred all the positive evidence that we have on the non-vanishing values of neutrino masses and mixing. In the following I will derive the allowed ranges for the mass and mixing parameters when the bulk of data is consistently combined.

2. Orthodox fits

I denote by *Orthodox* fits those which try to explain the evidences from solar, KamLAND, ATM and K2K experiments and assume that the LSND evidence will not be confirmed by MiniBoone.

2.1. Analysis of Solar and KamLAND

Fig. 1 shows the results from our latest analysis [8] of KamLAND $\bar{\nu}_e$ disappearance data, solar ν_e data and their combination under the hypothesis of CPT symmetry. The main new ingredient is the inclusion of the new results from KamLAND presented in ν-2004 conference in June. We have also taken into account the new gallium measurement which leads to the average value 68.1 ± 3.75 SNU. The main features of these results are:

- In the analysis of solar data, only LMA is allowed at more than 3σ and maximal mixing is rejected by the solar analysis at more than 5σ. This is so since the release of the SNO salt-data (SNOII) in Sep. 2003.
- In the analysis of the new KamLAND data the 3σ region does not extend to mass values larger than $\Delta m_{21}^2 = 2 \times 10^{-4} \, \mathrm{eV}^2$ because for larger Δm_{21}^2 values, the predicted spectral distortions are too small to fit the new KamLAND data.
- the combined analysis allows only the lowest LMA region at 3σ with best-fit point and 1σ ranges:

$$\Delta m^2 = [8.2_{-0.3}^{+0.3}] \times 10^{-5} \, \mathrm{eV}^2, \qquad \tan^2\theta = 0.39_{-0.04}^{+0.05}. \tag{2}$$

These results are in agreement with those reported in the several state-of-the-art analysis of solar and KamLAND data which exist in the literature.

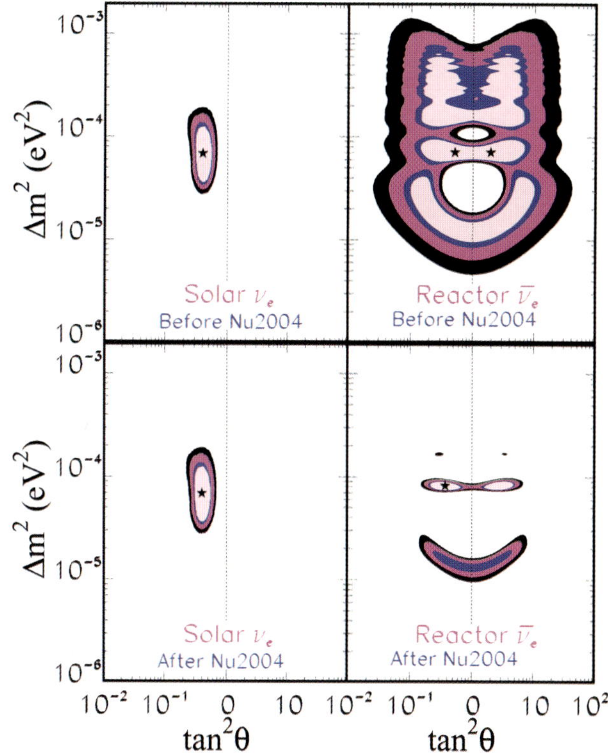

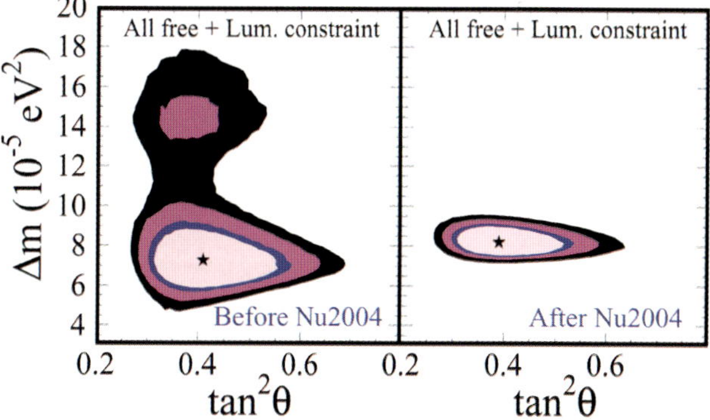

Fig. 1. Allowed regions for 2-ν oscillations of solar ν_e and KamLAND $\bar{\nu}_e$ (upper four panels), and for the combination of KamLAND and solar data under the hypothesis of CPT conservation (lower two panels). The different contours correspond to the allowed regions at 90%, 95%, 99% and 3σ CL.

2.2. Atmospheric and K2K Neutrinos

Fig. 2 show the results of our latest analysis of the ATM neutrino data [9], which includes the full data set of Super-Kamiokande phase I (SK1) as well as:

- use of new three-dimensional fluxes from Honda;
- improved interaction cross sections;
- some improvements in the Monte-Carlo which lead to some changes in the actual values of the data points.

Our results show good quantitative agreement with those of the SK collaboration. In particular we find that after inclusion of the above effects, the allowed region is shifted to lower Δm^2 with best-fit point and 1σ ranges:

$$\Delta m^2 = [2.2_{-0.4}^{+0.6}] \times 10^{-3} \, \mathrm{eV}^2, \qquad \tan^2\theta = 1_{-0.26}^{+0.35}. \tag{3}$$

At this point I would like to raise a word of caution. In all present analysis of ATM data, two main sources of theoretical flux uncertainties are included: an energy independent normalization error and a "tilt" error which parametrizes the uncertainty in the $E^{-\gamma}$ dependence of the flux. Some additional uncertainties in the

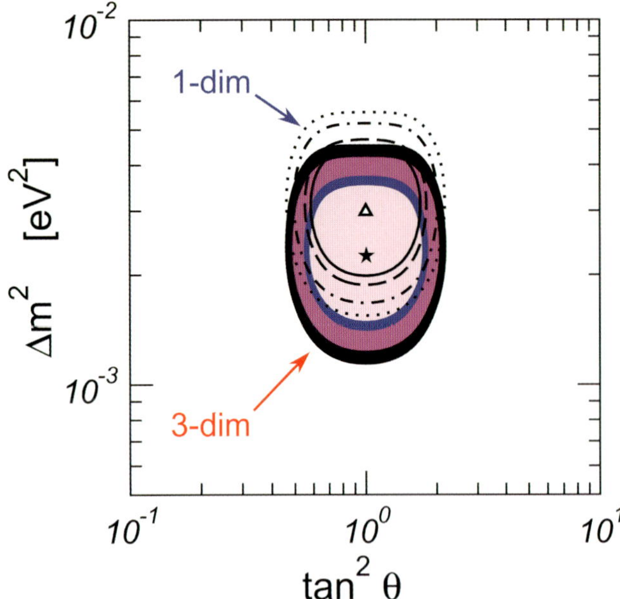

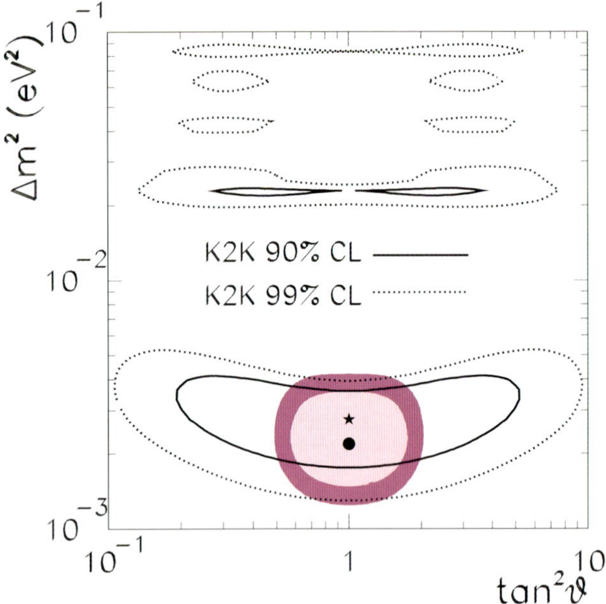

Fig. 2. Upper: Allowed regions from the analysis of ATM data using the new (full regions labeled "3-dim") and old (empty curves labeled as "1-dim") SK1 data and ATM fluxes. The different contours correspond to 90%, 95%, 99% and 3σ CL. *Lower:* Allowed regions from the analysis of K2K data at 90% (full line) and 99% (dashed line) confidence level. For comparison we also show the corresponding allowed regions from ATM neutrinos at the same CL.

ratios of the samples at different energies are also allowed as well as uncertainties in the zenith dependences. However, we still lack a well established range of theoretical flux uncertainties within a given ATM flux calculation, in a similar fashion to what it is provided for the solar neutrino fluxes by the SSM. In the absence of these, we cannot be sure that we are accounting for the most general characterization of the energy dependence of the ATM neutrino flux uncertainties. Given the large amount of data points provided by the SK experiment, this is becoming an important issue in the ATM neutrino analysis. There is a chance that the ATM fluxes may be still too "rigid", even when allowed to change within the presently considered uncertainties. As a consequence, we may be over-constraining the oscillation parameters.

The evidence of oscillation of ATM ν_μ has been now confirmed by the long-baseline (LBL) K2K experiment which has observed

not only a deficit of ν_μ's at a distance of 250 km but it has also measured the distortion of their energy spectrum. The lower panel of Fig. 1 show the results of our preliminary analysis of the K2K data which graphically illustrates this agreement.

2.3. *Three-Neutrino Oscillations*

The minimum joint description of ATM, K2K, solar and reactor data requires that all the three known neutrinos take part in the oscillations. The mixing parameters are encoded in the 3×3 lepton mixing matrix which can be conveniently parametrized in the standard form

$$U = \begin{pmatrix} 1 & 0 & 0 \\ 0 & c_{23} & s_{23} \\ 0 & -s_{23} & c_{23} \end{pmatrix} \begin{pmatrix} c_{13} & 0 & s_{13}e^{i\Delta} \\ 0 & 1 & 0 \\ -s_{13}e^{-i\Delta} & 0 & c_{13} \end{pmatrix} \begin{pmatrix} c_{21} & s_{12} & 0 \\ -s_{12} & c_{12} & 0 \\ 0 & 0 & 1 \end{pmatrix}$$

where $c_{ij} \equiv \cos \theta_{ij}$ and $s_{ij} \equiv \sin \theta_{ij}$. The angles θ_{ij} can be taken without loss of generality to lie in the first quadrant, $\theta_{ij} \in [0, \pi/2]$.

There are two possible mass orderings, which we denote as *Normal* and *Inverted*. In the normal scheme $m_1 < m_2 < m_3$ while in the inverted one $m_3 < m_1 < m_2$.

In total the 3-ν oscillation analysis involves seven parameters: 2 mass differences, 3 mixing angles, the CP phase and the mass ordering. Generic 3-ν oscillation effects include:

– coupled oscillations with two different wavelengths;

– CP violating effects;

– difference between Normal and Inverted schemes.

The strength of these effects is controlled by the values of the ratio of mass differences $\alpha \equiv \Delta m_{21}^2/|\Delta m_{31}^2|$, by the mixing angle θ_{13} and by the CP phase Δ.

For solar and ATM oscillations,

$$\Delta m_\odot^2 = \Delta m_{21}^2 \ll |\Delta m_{31}^2| \simeq |\Delta m_{32}^2| = \Delta m_{\text{atm}}^2, \tag{4}$$

and the joint 3-ν analysis simplifies as follows:

– for solar and KamLAND neutrinos, the oscillations with the ATM oscillation length are completely averaged and the survival probability takes the form:

$$P_{ee}^{3\nu} = \sin^4 \theta_{13} + \cos^4 \theta_{13} P_{ee}^{2\nu} \tag{5}$$

where in the Sun $P_{ee}^{2\nu}$ is obtained with the modified sun density $N_e \to \cos^2 \theta_{13} N_e$. So the analyses of solar data constrain three of the seven parameters: Δm_{21}^2, θ_{12} and θ_{13}. The effect of θ_{13} is to decrease the energy dependence of the survival probability;

– for ATM and K2K neutrinos, the solar wavelength is too long and the corresponding oscillating phase is negligible. As a consequence, the ATM data analysis restricts $\Delta m_{31}^2 \simeq \Delta m_{32}^2$, θ_{23} and θ_{13}, the latter being the only parameter common to both solar and ATM neutrino oscillations and which may potentially allow for some mutual influence. The effect of θ_{13} is to add a $\nu_\mu \to \nu_e$ contribution to the ATM oscillations;

– at CHOOZ the solar wavelength is unobservable and the relevant survival probability oscillates with wavelength determined by Δm_{31}^2 and amplitude determined by θ_{13}.

In this approximation, the CP phase is unobservable. In principle there is a dependence on the Normal versus Inverted orderings due to matter effects in the Earth for ATM neutrinos. However, this effect is controlled by the mixing angle θ_{13}. Presently all data favour small θ_{13} with best fit point very near $\theta_{13} = 0$. The dominant constraint arises from the combined

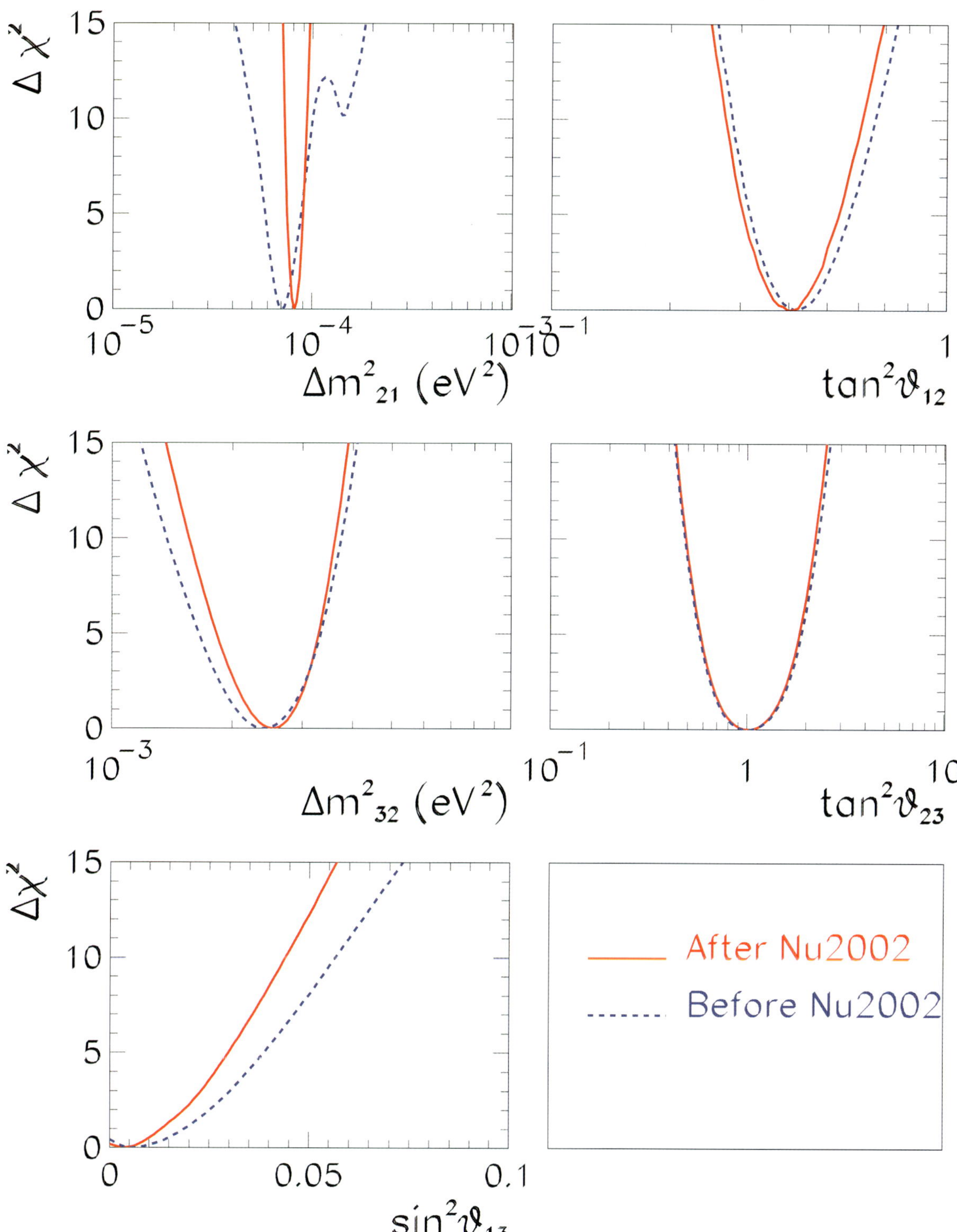

Fig. 3. Global 3ν oscillation analysis: Dependence of $\Delta\chi^2$ on each of the five parameters from the global analysis compared to the bound prior to the inclusion of the new KamLAND and K2K data.

analysis of CHOOZ reactor and ATM data and it is further limited by the solar and KamLAND results. As a consequence, this effect is too small to be statistically meaningful in the present analysis.

Fig. 3 shows the individual bounds on each of the five parameters derived from the global analysis. To illustrate the effect of the new KamLAND and K2K data we also show the results before their inclusion. In each panel the displayed $\Delta\chi^2$ has been marginalized with respect to the undisplayed parameters. As seen in the figure the main effect of the new data is a better determination of the mass differences. A secondary effect is the

slight improvement of upper bound on θ_{13}. This is mainly driven by K2K which favours the higher mass part of the ATM neutrino region for which the corresponding bound from CHOOZ is tighter.

Altogether we find the following 3σ ranges:

$$7.3 \leq \frac{\Delta m^2_{21}}{10^{-5}\ \text{eV}^2} \leq 9.3, \qquad 0.28 \leq \tan^2\theta_{12} \leq 0.60,$$

$$1.6 \leq \frac{\Delta m^2_{32}}{10^{-3}\ \text{eV}^2} \leq 3.6, \qquad 0.5 \leq \tan^2\theta_{23} \leq 2.1, \tag{6}$$

$$\sin^2\theta_{13} \leq 0.041.$$

These results can be translated into our present knowledge of the moduli of the mixing matrix U:

$$|U| = \begin{pmatrix} 0.79 - 0.88 & 0.47 - 0.61 & < 0.20 \\ 0.19 - 0.52 & 0.42 - 0.73 & 0.58 - 0.82 \\ 0.20 - 0.53 & 0.44 - 0.74 & 0.56 - 0.81 \end{pmatrix}. \qquad (7)$$

To finish this section I would like to discuss the possible observability of Δm_{21}^2 oscillations in ATM neutrinos. These effects although small can, in principle, be visible, mostly in the low energy ν_e events provided that the mixing angle θ_{23} deviates from maximal (see Ref. [10] and references in therein). As a matter of fact, they can lead to an increase or a decrease of the sub-GeV e-like events depending on the octant of the angle θ_{23} and they can be our best observable to detect both deviations of θ_{23} from maximal as well as the sign of the deviation.

The present data may already give some hint of deviation of the 2–3 mixing from maximal. Indeed, there is some excess of the $e-$like events in the sub-GeV range. The excess increases with decrease of energy within the sample as expected from a Δm_{21}^2 effect. To illustrate this I show in Fig. 4 the results of the global analysis of ATM and CHOOZ data in the framework of 3ν oscillations taking into account also the effect of Δm_{21}^2 oscillations [10].

From the figure we see that, even with the present uncertainties, the ATM data has some sensitivity to Δm_{21}^2 oscillation effects and that these effects break the symmetry in θ_{23} around maximal mixing. Although statistically not very significant, this preference for non-maximal 2–3 mixing is a physical effect on the present neutrino data, induced by the fact than an excess of events is observed in sub-GeV electrons but not in sub-GeV muons nor, in the same amount, in the multi-GeV electrons. As a consequence, this excess cannot be fully explained by a combination of a global rescaling and a "tilt", of the fluxes within the assumed uncertainties.

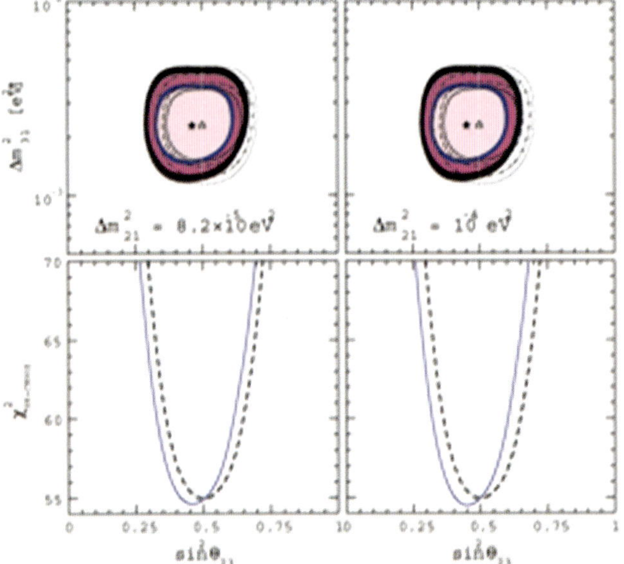

Fig. 4. Effect of Δm_{21}^2 oscillations on the allowed regions of the oscillation parameters Δm_{31}^2 and $\sin^2 \theta_{23}$ from the combined analysis of all the ATM and CHOOZ data samples. In the lower panels we show the dependence of the χ^2 function on θ_{23}, marginalized with respect to Δm_{31}^2. Hollow and dashed black lines are for $\Delta m_{21}^2 = 0$.

3. Unorthodox fits

I denote by *Unorthodox* fits those in which either the matter contents of the SM has been extended to include new light sterile neutrinos or the basic symmetries of the model are violated, in most cases with the aim of accommodating the LSND signal.

3.1. *LSND and Sterile Neutrinos*

The LSND experiment found evidence of $\bar\nu_\mu \to \bar\nu_e$ neutrino conversion with $\Delta m^2 \geq 0.1 \ \mathrm{eV}^2$. This result can be accommodated together with those from solar, reactor, ATM and LBL experiments into a single neutrino oscillation framework only if there are at least three different scales of neutrino mass-squared differences.

As a first attempt to generate the required scales one can invoke the existence of a fourth light neutrino, which must be *sterile* in order not to affect the invisible Z^0 decay width, precisely measured at LEP. There are six possible four-neutrino schemes which, in principle, can do the job. They can be divided in two classes: $(3 + 1)$ and $(2 + 2)$. In the $(3 + 1)$ schemes, there is a group of three close-by neutrino masses that is separated from the fourth one by a gap of the order of $1\,\mathrm{eV}^2$, which is responsible for the SBL oscillations observed in the LSND experiment. In $(2 + 2)$ schemes, there are two pairs of close masses separated by the LSND gap. The main difference between these two classes is that in $(2 + 2)$-spectra the transition into the sterile neutrino is a solution of either the solar or the ATM neutrino problem, or the sterile neutrino takes part in both. This is not the case for a $(3 + 1)$-spectrum, where the sterile neutrino could be only slightly mixed with the active ones and mainly provide a description of the LSND result.

I show in Fig. 5 the latest results of the analysis of neutrino data in these scenarios. The phenomenological situation at present is that none of the four-neutrino scenarios are favored by the data. $(3 + 1)$-spectra are disfavored by the incompatibility between the LSND signal and the negative results found by other short-baseline (SBL) laboratory experiments. There is also a constraint on the possible value of the heavier neutrino mass in this scenario from their contribution to the energy density in the Universe which is presently constrained by cosmic microwave background radiation and large scale structure formation data [6]. This is illustrated in the upper panel of Fig. 5 taken from Ref. [11] where they find that after the inclusion of the cosmological bound there is only a marginal overlap at 99% CL between the allowed LSND region and the excluded region from SBL + ATM experiments. Concerning (2+2)-spectra, they are ruled out by the existing constraints from the sterile oscillations in solar and ATM data as illustrated in the lower panel which shows that the lower bound on the sterile component from the analysis of ATM data and the upper bound from the analysis of solar data do not overlap at more than 4σ.

3.2. *Neutrinos as tests of fundamental symmetries*

Alternative explanations to the LSND result include the possibility of CPT [12] violation, which implies that the masses and mixing angles of neutrinos may be different from those of antineutrinos. To test this possibility, in Ref. [13] we performed an analysis of the existing data from solar, ATM, LBL, reactor and SBL experiments in the framework of CPT violating oscillations. The outcome of the analysis is that, presently, the hypothesis of CPT violation is not supported by the data. This arises from two main

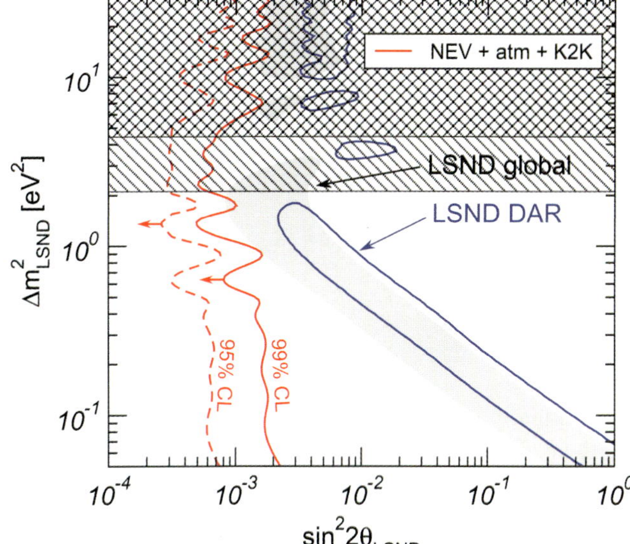

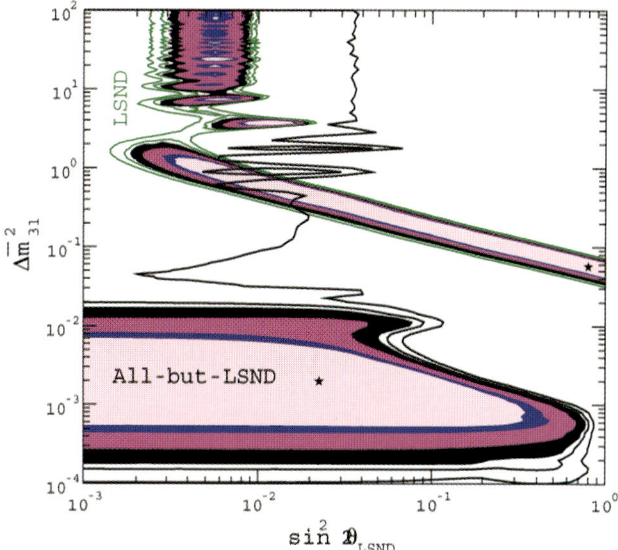

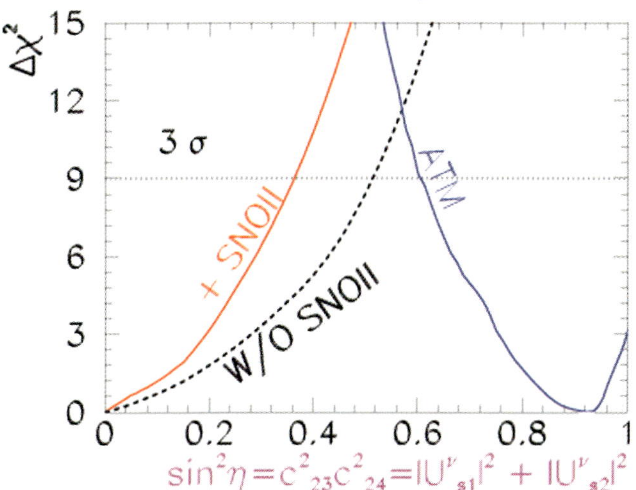

Fig. 5. *Upper*: Status of the 3 + 1 oscillation scenarios. *Lower*: Present status of the bounds on the active-sterile admixture from solar and ATM neutrino data in $(2 + 2)$-models.

Fig. 6. 90%, 95%, 99%, and 3σ CL allowed regions (filled) in required to explain the LSND signal together with the corresponding allowed regions from our global analysis of all-but-LSND data. The contour lines correspond to $\Delta\chi^2 = 13$ and 16 (3.2σ and 3.6σ, respectively).

ATM neutrino events extend over several decades in energy. As a consequence they can test the presence of this effect even at the subdominant level. In Ref. [9] we have performed an analysis of ATM and LBL neutrino data in terms of neutrino mass oscillations plus these new physics effects and we have concluded that the determination of mass and mixing parameters is robust under the presence of these unknown forms of new physics. Conversely, the analysis permits to impose strong constraints on the violations of these symmetries. For instance we find that at 90% CL the possible VLI a and VEP are limited to

$$\frac{|\Delta c|}{c} \leq 8.1 \times 10^{-25}, \qquad |\phi\,\Delta\gamma| \leq 4.0 \times 10^{-25} \tag{8}$$

which constitute the strongest constraints on the violation of these fundamental symmetries.

Acknowledgments

This work was supported in part by the National Science Foundation grant PHY-0354776. M.C.G.-G. is also supported by Spanish Grants No FPA-2001-3031 and GRUPOS03/013.

facts: (*i*) KamLand finds that reactor $\bar{\nu}_e$ oscillate with wavelength and amplitude in good agreement with the expectations from the LMA solution of the solar ν_e; (*ii*) both ATM neutrinos and antineutrinos have to oscillate with similar wavelengths and amplitudes to explain the ATM data. In general, as a result of these effects, the best fit to the data is very near CPT conservation and in particular this rules out this scenario as explanation of the LSND anomaly. This is illustrated in Fig. 6, which shows clearly that there is no overlap below the 3σ level between the LSND and the all-but-LSND allowed regions.

Using the good description of neutrino data in terms of neutrino matter oscillations, it is also possible to constraint other exotic forms of new physics such as the violation of Lorentz Invariance (VLI) [14] induced by different asymptotic values of the velocity of the neutrinos, $c_1 \neq c_2$, or the violation of the equivalence principle (VEP) [15] due to non universal coupling of the neutrinos, $\gamma_1 \neq \gamma_2$ to the local gravitational potential, among others. These forms of new physics, if non-universal, can also induce neutrino flavour oscillations whose main differentiating characteristic is a different energy dependence of the oscillation wavelength. For example for both VLI and VEP the oscillation wavelength decreases with energy unlike for mass oscillations.

References

1. See talk by Y. Suzuki in these proceedings.
2. See talk by A. McDonald in these proceedings.
3. See talk by A. Suzuki in these proceedings.
4. See talk by J. Conrad in these proceedings.
5. See talk by C. Weinheimer in these proceedings.
6. See talk by M. Tegmark in these proceedings.
7. See talks by S. Petcov and E. Fiorini in these proceedings.
8. Bahcall, J. N. *et al.*, J. High. En. Phys. **0408**, 016 (2004).
9. Gonzalez-Garcia, M. C. *et al.*, Phys. Rev. D **70**, 033010 (2004).
10. Gonzalez-Garcia, M. C. *et al.*, hep-ph/0408170.
11. Maltoni, M. *et al.*, New J. Phys. **6**, 122 (2004).
12. Murayama, H. and Yanagida, T., Phys. Lett. B **520**, 263 (2001); Barenboim, G. *et al.*, J. High. En. Phys. **0210**, 001 (2002) .
13. Gonzalez-Garcia, M. C. *et al.*, Phys. Rev. D **68**, 053007 (2003).
14. Coleman, S. and Glashow, S. L., Phys. Lett. B **405**, 249 (1997); Colladay, D. and Kostelecky, V. A., Phys. Rev. **D55**, 6760 (1997).
15. Gasperini, M., Phys. Rev. D **38**, 2635 (1988); Phys. Rev. D **39**, 3606 (1989).

Physica Scripta. Vol. T121, 78–85, 2005

Future Precision Neutrino Oscillation Experiments and Theoretical Implications

M. Lindner[*]

Physik Department, Technische Universität München, D-85748 Garching bei München, Germany

Received February 16, 2005; accepted in revised form February 21, 2005

PACS numbers: 12.15.Ff, 14.60.Lm, 14.60.Pq, 14.60.St

Abstract

Future neutrino oscillation experiments will lead to precision measurements of neutrino mass splittings and mixings. The flavour structure of the lepton sector will therefore at some point become better known than that of the quark sector. This article discusses the realistic potential of future oscillation experiments on the basis of detailed simulations with an emphasis on experiments which can be done in about ten years. In addition, some theoretical implications for neutrino mass models will be briefly discussed.

1. Introduction

The observation of atmospheric neutrino oscillations by the SuperKamiokande experiment [1] triggered a remarkable discovery phase. The initial evidence turned into a solid proof of neutrino flavour conversions as well as of the L/E dependence as required by oscillations. The solar neutrino problem has also been resolved in the last years. The Gallex experiment [2] detected initially a rate effect which implied flavour conversion on the basis of solar models. The SNO experiment proved then model independent neutrino flavour transitions [3, 4]. The initially allowed parameter islands were cleaned up by the KamLAND experiment, which demonstrated finally with reactor anti-neutrinos [5, 6] that the so-called LMA-solution is correct. Altogether the existing experimental results fit now very nicely into a picture with three massive neutrinos, which corresponds to the simplest scenario for three generations. The only exemption is the disputed LSND result [7], which would have far reaching consequences if it were confirmed, but this possibility will be ignored here. The oscillations of three neutrino generations involve then two mass-squared differences $\Delta m_{12}^2 \simeq \Delta m_{sol.}^2$ and $\Delta m_{23}^2 \simeq \Delta m_{atm.}^2$, three mixing angles, θ_{12}, θ_{23}, and θ_{13}, and a CP-violating phase δ. Atmospheric neutrino data [8] and the first results from the K2K long-baseline accelerator experiment [8] determine $\Delta m_{23}^2 = (2.2^{+0.6}_{-0.4}) \times 10^{-3}\,\mathrm{eV}^2$ and $\theta_{23} \approx 45°$ [8, 9], whereas solar neutrino data [10, 11], combined with the results from the KamLAND reactor experiment [5] lead to $\Delta m_{12}^2 = (8.2^{+0.3}_{-0.3}) \times 10^{-5}\,\mathrm{eV}^2$ and $\tan^2 \theta_{12} = 0.39^{+0.05}_{-0.04}$ [9]. The results can now approximatively be summarized by two independent two flavour oscillations where the solar and atmospheric Δm^2 values are roughly now known.

The key parameter for genuine three flavour effects is the mixing angle θ_{13} which is so far only known to be small from the CHOOZ [12, 13] and Palo Verde [14] experiments. The current bound for θ_{13} depends on the value of the atmospheric mass squared difference and it gets rather weak for $\Delta m_{31}^2 \lesssim 2 \times 10^{-3}\,\mathrm{eV}^2$. However, in that region an additional constraint on θ_{13} from global solar neutrino data becomes important [15].

At the current best fit value of $\Delta m_{31}^2 = 2.2 \times 10^{-3}\,\mathrm{eV}^2$ we have at 3σ the bound $\sin^2 \theta_{13} \leq 0.041$ [9]. There is no reason why θ_{13} should vanish and one should expect therefore θ_{13} to be finite.

One might think that neutrino oscillations are in future less interesting, since it will lead only to parameter improvements of the leading 2×2 oscillations and maybe a finite parameter value of θ_{13}. However, such a view misses completely the fact that the neutrino sector is, unlike the quark sector, not obscured by hadronic uncertainties. The precision to which the underlying flavour information is determined will therefore only be limited by the ultimate experimental precision. If high precision measurements are possible, then they will be very sensitive tests of flavour models and related topics, like the unitarity of three flavours. Genuine three flavour oscillation effects occur only for a finite value of θ_{13} and establishing a finite value of θ_{13} is therefore one of the next milestones in neutrino physics. Leptonic CP violation is another three flavour effect which can only be tested if θ_{13} is finite. The usual see-saw scenario includes besides δ in addition two further Majorana CP phases in the light neutrino sector, as well as other CP phases in the heavy Majorana sector, which are involved in leptogenesis. In general the heavy and light CP phases are not connected, but most flavour models create relations between these two sectors, relating thus low energy leptonic CP violation to leptogenesis and mass models. Precision measurements of neutrino oscillations test therefore very interesting questions of particle physics which are connected to the origin of flavour and to phenomenological consequences of flavour. There is thus a very strong motivation to establish first in the next generation of experiments a finite value of θ_{13} in order to aim in the long run at a measurement of leptonic CP violation [16, 17, 18, 19, 20].

2. Three neutrino oscillation in matter

An effective two flavour treatment is insufficient for future oscillation experiments and matter effects must be included in addition. The generalization of the oscillation formulae in vacuum to N neutrinos leads to the probabilities for flavour transitions $\nu_{f_l} \rightarrow \nu_{f_m}$ given by

$$P(\nu_{f_l} \rightarrow \nu_{f_m})$$
$$= \delta_{lm} - 4 \underbrace{\sum_{i>j} \mathrm{Re}\, J_{ij}^{f_l f_m} \sin^2 \Delta_{ij}}_{P_{CP}} - 2 \underbrace{\sum_{i>j} \mathrm{Im}\, J_{ij}^{f_l f_m} \sin 2\Delta_{ij}}_{P_{\overline{CP}}} \quad (1)$$

where the shorthands $J_{ij}^{f_l f_m} := U_{li} U_{lj}^* U_{mi}^* U_{mj}$ and $\Delta_{ij} := \frac{\Delta m_{ij}^2 L}{4E}$ have been used. These generalized vacuum transition probabilities depend on all combinations of quadratic mass differences

---

*Email address: lindner@ph.tum.de

$\Delta m_{ij}^2 = m_i^2 - m_j^2$ as well as on different products of elements of the leptonic mixing matrix U. We will assume a three neutrino framework, *i.e.* $1 \le i, j \le 3$ and U is a 3×3 mixing matrix parameterized in the standard way

$$U = \begin{pmatrix} c_{12}c_{13} & s_{12}c_{13} & s_{13}e^{-i\delta} \\ -s_{12}c_{23} - c_{12}s_{23}s_{13}e^{i\delta} & c_{12}c_{23} - s_{12}s_{23}s_{13}e^{i\delta} & s_{23}c_{13} \\ s_{12}s_{23} - c_{12}c_{23}s_{13}e^{i\delta} & -c_{12}s_{23} - s_{12}c_{23}s_{13}e^{i\delta} & c_{23}c_{13} \end{pmatrix}, \tag{2}$$

where $c_{ij} = \cos(\theta_{ij})$ and $s_{ij} = \sin(\theta_{ij})$. U contains three leptonic mixing angles and one Dirac-like leptonic CP phase δ. Note that the most general mixing matrix for three Majorana neutrinos contains two further Majorana-like CP phases, but it can easily be seen that these extra phases do not enter in the above oscillation formulae. Disappearance probabilities, *i.e.* the transitions $v_{f_l} \to v_{f_l}$, do not even depend on δ, since $J_{ij}^{f_l f_l}$ is only a function of the modulus of elements of U. Appearance probabilities, like $v_e \to v_\mu$ are therefore the place where leptonic CP violation can be studied. Eq. (1) contains a CP conserving part P_{CP} and a CP violating part $P_{\cancel{CP}}$, and both terms depend on the CP phase δ. An obvious extraction strategy for CP-violation would thus be to look at CP asymmetries [21]. Note, however, that the beams of a long baseline experiment traverse the Earth and the presence of matter violates CP by itself. This implies modifications of eq. (1) and it makes a measurement of leptonic CP violation more involved.

The general oscillation formulae in vacuum, eq. (1), lead to well known, but rather lengthy trigonometric expressions for the oscillation probabilities in vacuum. These expressions become even longer and do not exist in closed form when arbitrary matter corrections are taken into account. For effectively constant matter densities, which is often a good assumption, the problem simplifies somewhat, but the general oscillation probabilities are still very lengthy. The Hamiltonian describing three neutrino oscillation in matter can then be written in flavour basis as

$$H = \frac{1}{2E_v} U \begin{pmatrix} m_1^2 & 0 & 0 \\ 0 & m_2^2 & 0 \\ 0 & 0 & m_3^2 \end{pmatrix} U^T + \frac{1}{2E_v} \begin{pmatrix} A + A' & 0 & 0 \\ 0 & A' & 0 \\ 0 & 0 & A' \end{pmatrix}. \tag{3}$$

The first term describes oscillations in vacuum in flavour basis. The quantities A and A' in the second term are given by the charged current and neutral current contributions to coherent forward scattering in matter. The charged current contribution is given by

$$A = \pm \frac{2\sqrt{2}G_F Y \rho E_v}{m_n} = 2VE_v, \tag{4}$$

where G_F is Fermi's constant, Y is the number of electrons per nucleon, m_n is the nucleon mass and ρ is the matter density. A is positive for neutrinos in matter and anti-neutrinos in anti-matter, while it is negative for anti-neutrinos in matter and neutrinos in anti-matter. The flavour universal neutral current contributions A' lead to an overall phase which does not enter the transition probabilities. The over-all neutrino mass scale m_1^2 can be written as a term proportional to the unit matrix and can similarly be removed, such that only Δm_{21}^2 and Δm_{31}^2 remain in the first term of eq. (3). After re-diagonalization of the Hamiltonian in constant matter density one finds that matter effects lead in a very good approximation to an A-dependent parameter mapping in the 1–3

subspace which can be written as [22]

$$\sin^2 2\theta_{13,m} = \frac{\sin^2 2\theta_{13}}{C_\pm^2}, \tag{5}$$

$$\Delta m_{31,m}^2 = \Delta m_{31}^2 C_\pm, \tag{6}$$

$$\Delta m_{32,m}^2 = \frac{\Delta m_{31}^2 (C_\pm + 1) + A}{2}, \tag{7}$$

$$\Delta m_{21,m}^2 = \frac{\Delta m_{31}^2 (C_\pm - 1) - A}{2}. \tag{8}$$

The index m denotes effective quantities in matter, where

$$C_\pm^2 = \left(\frac{A}{\Delta m_{31}^2} - \cos 2\theta \right)^2 + \sin^2 2\theta. \tag{9}$$

Note that A in $C_\pm$ can change its sign and the mappings for neutrinos and anti-neutrinos are therefore different, resulting in different effective mixings and masses. This is an important effect, which will allow detailed tests of coherent forward scattering of neutrinos in matter. Note that oscillations in matter depend unlike vacuum oscillations on the sign of Δm_{31}^2. This allows to extract the $sign(\Delta m_{31}^2)$ via matter effects.

Inserting the parameter mappings eqs. (5)–(8) into the full oscillation formulae leads still to quite lengthy expressions for the oscillation probabilities in matter, where it is not easy to oversee all effects. It is therefore instructive to simplify the problem further such that a qualitative analytic understanding of all effects becomes possible, while quantitative statements should be evaluated numerically with the full expressions. The key for further simplification is to expand the oscillation probabilities in small quantities. These expansion parameters are $\alpha = \Delta m_{21}^2 / \Delta m_{31}^2 = O(10^{-2})$ and $\sin^2 2\theta_{13} \le 0.16$. The matter effects can be parameterized by the dimensionless quantity $\hat{A} = A/\Delta m_{31}^2 = 2VE/\Delta m_{31}^2$, where $V = \sqrt{2}G_F n_e$. The oscillation probabilities for all channels can in this way be significantly simplified [23]. Using $\Delta \equiv \Delta_{31}$, the leading terms for $P(v_\mu \to v_\mu)$ and $P(v_e \to v_\mu)$ can, for example, be written as [17, 24, 22]

$$P(v_\mu \to v_\mu) \approx 1 - \cos^2 \theta_{13} \sin^2 2\theta_{23} \sin^2 \Delta + 2\alpha \cos^2 \theta_{13} \cos^2 \theta_{12} \sin^2 2\theta_{23} \Delta \cos \Delta, \tag{10}$$

$$P(v_e \to v_\mu) \approx \sin^2 2\theta_{13} \sin^2 \theta_{23} \frac{\sin^2((1 - \hat{A})\Delta)}{(1 - \hat{A})^2}$$
$$\pm \sin \delta \cdot \sin 2\theta_{13}\, \alpha\, \sin 2\theta_{12} \cos \theta_{13} \sin 2\theta_{23} \sin(\Delta)$$
$$\times \frac{\sin(\hat{A}\Delta) \sin((1 - \hat{A})\Delta)}{\hat{A}(1 - \hat{A})} + \cos \delta \cdot \sin 2\theta_{13} \alpha \sin 2\theta_{12}$$
$$\times \cos \theta_{13} \sin 2\theta_{23} \cos(\Delta) \frac{\sin(\hat{A}\Delta) \sin((1 - \hat{A})\Delta)}{\hat{A}(1 - \hat{A})}$$
$$+ \alpha^2\, \sin^2 2\theta_{12} \cos^2 \theta_{23} \frac{\sin^2(\hat{A}\Delta)}{\hat{A}^2}, \tag{11}$$

where "+" in eq. (11) stands for neutrinos and "−" for anti-neutrinos. The most important feature of eq. (11) is that all interesting effects in the $v_e \to v_\mu$ transition depend crucially on θ_{13}. The size of $\sin^2 2\theta_{13}$ determines thus if the total transition rate, matter effects, effects due to the sign of Δm_{31}^2 and CP violating effects are measurable. This is the reason why the size of θ_{13} is one of the most important questions for future oscillation experiments.

Before we discuss some features of eqs. (10) and (11) in more detail, we would like to comment on the underlying assumptions and the reliability of these equations. First eqs. (10) and (11) are an expansion in terms of the small quantities α and $\sin 2\theta_{13}$. Higher order terms are suppressed at least by another power of one of these small parameters and these corrections are thus typically at the percent level. Note that the expansion in α is actually an expansion in the solar and not the atmospheric frequency. The expansion does therefore not break down at the first atmospheric oscillation maximum, *i.e.* at $\Delta \simeq 1$, but at much larger baselines before the first (sub-dominant) solar oscillation maximum, *i.e.* at $\alpha\Delta \simeq 1$. The latter condition gives an upper bound for the baseline where eqs. (10) and (11) are good approximations

$$L \lesssim 8000\,\text{km} \left(\frac{E_\nu}{\text{GeV}}\right)\left(\frac{10^{-4}\,\text{eV}^2}{\Delta m_{21}^2}\right), \tag{12}$$

while the first oscillation maximum sits at $\alpha \cdot L \simeq L/30$. Eqs. (10) and (11) are therefore excellent approximations at and well beyond the first oscillation maximum of long baseline experiments. The matter corrections in eqs. (10) and (11) are derived for constant average matter density which is a good approximation.

Note that all quantitative results which will be presented are based on numerical simulations in matter. The results are therefore not affected by any approximation. Eqs. (10) and (11) will only be used to understand the problem analytically, which is extremely helpful in order to oversee the multi-dimensional parameter space.

In addition to long baseline experiments, reactor experiments with identical near and far detectors have an excellent potential for precise measurements. The near detector is used to eliminate many common systematical errors and the far detector is located typically at a baseline of a few kilometer. For these short baselines matter effects can be ignored and one finds to second order in the small quantities $\sin 2\theta_{13}$ and α for the oscillation probability

$$1 - P_{\bar{e}\bar{e}} = \sin^2 2\theta_{13}\sin^2 \Delta_{31} + \alpha^2 \Delta_{31}^2 \cos^4 \theta_{13}\sin^2 2\theta_{12}. \tag{13}$$

At the first atmospheric oscillation maximum, Δ_{31} is approximately $\pi/2$ and $\sin^2 \Delta_{31}$ is close to one, which means that the second term on the right-hand side of this equation can be neglected for $\sin^2 2\theta_{13} \gtrsim 10^{-3}$. The reactor measurement is dominated in this case at short baselines by the product of $\sin^2 2\theta_{13}$ and $\sin^2 \Delta_{31}$, which must be measured as deviation from one. Eq. (13) implies that correlations and degeneracies play essentially no role in reactor experiments. The behavior in the $\sin^2 2\theta_{13} - \Delta m_{21}^2$-plane will also be different since eq. (13) is essentially independent of Δm_{21}^2. A reactor experiment will improve the global parameter determination in two ways. First, a direct, essentially uncorrelated and clean measurement for θ_{13} [25] can be obtained which can be used to disentangle the long baseline results. Secondly, the reactor measurement can replace the cross-section suppressed anti-neutrino running of the accelerator experiments, leading to statistical improvements in the neutrino measurements [26].

3. Correlations and degeneracies

Eqs. (10) and (11) exhibit certain parameter correlations and degeneracies, which play an important role in the analysis of long baseline experiments, and which would be hard to understand in a purely numerical analysis of the high dimensional parameter space. The most important properties are:

- Eqs. (10) and (11) depend only on the product $\alpha \cdot \sin 2\theta_{12}$ or equivalently $\Delta m_{21}^2 \cdot \sin 2\theta_{12}$. This are the parameters related to solar oscillations which will be taken as external input. Note that the product is better determined than the product of the individual measurements of Δm_{21}^2 and $\sin 2\theta_{12}$.

- Next we observe in eq. (11) that the second and third term contain both a factor $\sin(\hat{A}\Delta)$, while the last term contains a factor $\sin^2(\hat{A}\Delta)$. Since $\hat{A}\Delta = VL/2$, we find that these factors depend only on L, resulting in a "magic baseline" [27] when $VL_{magic} = 2\pi$, where $\sin(\hat{A}\Delta)$ vanishes. At this magic baseline only the first term in eq. (11) survives and $P(\nu_e \to \nu_\mu)$ does no longer depend on δ, α and $\sin 2\theta_{12}$. This is in principle very important, since it implies that $\sin^2 2\theta_{13}$ can be determined at the magic baseline from the first term of eq. (11) whatever the values and errors of δ, α and $\sin 2\theta_{12}$ are. For the given matter density of the Earth we find $L_{magic} = 2\pi/V \simeq 8100$ km which fits nicely into the Earth. This value is quite amazing, since V is given in terms of completely unrelated constants of nature like G_F.

- Next we observe that only the second and third term of eq. (11) depend on the CP phase δ, and both terms contain a factor $\sin 2\theta_{13} \cdot \alpha$, while the first and fourth term of eq. (11) do not depend on the CP phase δ and contain factors of $\sin^2 2\theta_{13}$ and α^2, respectively. The extraction of CP violation is thus always suppressed by the product $\sin 2\theta_{13} \cdot \alpha$ and the CP violating terms are obscured by large CP independent terms if either $\sin^2 2\theta_{13} \ll \alpha^2$ or $\sin^2 2\theta_{13} \gg \alpha^2$. The relative contribution of the CP phase δ to the probability is thus largest for $\sin^2 2\theta_{13} \simeq 4\theta_{13}^2 \simeq \alpha^2$.

- Another observation is that the last term in eq. (11), which is proportional to $\alpha^2 = (\Delta m_{21}^2)^2/(\Delta m_{31}^2)^2$, dominates in the limit of tiny $\sin^2 2\theta_{13}$. The error of Δm_{21}^2 limits therefore for small $\sin^2 2\theta_{13}$ the parameter extraction. This last term implies a finite transition probability even for $\theta_{13} = 0$. Observing $\nu_e \to \nu_\mu$ or $\nu_\mu \to \nu_e$ appearance transitions does therefore not necessarily establish a finite value of $\theta_{13} = 0$ in a three flavour framework.

- Eqs. (10) and (11) suggest that transformations exist which leave these equations invariant. Therefore degeneracies, *i.e.* parameter sets having identical oscillation probabilities for a fixed L/E are expected. An example of such an invariance is given by a simultaneous replacement of neutrinos by anti-neutrinos and $\Delta m_{31}^2 \to -\Delta m_{31}^2$. This is equivalent to changing the sign of the second term of eq. (11) and replacing $\alpha \to -\alpha$ and $\Delta \to -\Delta$, while $\hat{A} \to \hat{A}$. It is easy to see that eqs. (10) and (11) are unchanged, but this is not a degeneracy, since neutrinos and anti-neutrinos can be distinguished experimentally.

- The first real degeneracy [28, 29] can be seen in the disappearance probability eq. (10), which is invariant under the replacement $\theta_{23} \to \pi/2 - \theta_{23}$. Note that the second and third term in eq. (11) are not really invariant under this transformation, but this change in the sub-leading appearance probability can approximately be compensated by tiny parameter shifts. This implies that the degeneracy can in principle be lifted with high precision measurements in the disappearance channels.

- The second degeneracy can be found in the appearance probability eq. (11) in the $(\delta - \theta_{13})$-plane [30, 31]. In terms of θ_{13} (which is small) and δ the four terms of eq. (11) have the structure

$$P(\nu_e \to \nu_\mu) \approx \theta_{13}^2 \cdot F_1 + \theta_{13} \cdot (\pm \sin \delta F_2 + \cos \delta F_3) + F_4, \tag{14}$$

where the quantities F_i, $i = 1, \ldots, 4$ contain all the other parameters. The requirement $P(v_e \to v_\mu) = const.$ leads for both neutrinos and anti-neutrinos to parameter manifolds of degen-erate or correlated solutions. Having both neutrino and anti-neutrino beams, the two channels can be used independently, which is equivalent to considering simultaneously eq. (14) for $F_2 \equiv 0$ and $F_3 \equiv 0$. The requirement that these probabilities are now independently constant, *i.e.* $P(v_e \to v_\mu) = const.$ for $F_2 \equiv 0$ and $F_3 \equiv 0$, leads to more constraint manifolds in the $(\delta - \theta_{13})$-plane, but some degeneracies still survive.

- The third degeneracy [32] is given by the fact that a change in sign of Δm_{31}^2 can essentially be compensated by an offset in δ. Therefore we note again that the transformation $\Delta m_{31}^2 \to -\Delta m_{31}^2$ leads to $\alpha \to -\alpha$, $\Delta \to -\Delta$ and $\hat{A} \to -\hat{A}$. All terms of the disappearance probability, eq. (10), are invariant under this transformation. The first and fourth term in the appearance probability eq. (10), which do not depend on the CP phase δ, are also invariant. The second and third term of eq. (10) depend on the CP phase and change by the transformation $\Delta m_{31}^2 \to -\Delta m_{31}^2$. The fact that these changes can be compensated by an offset in the CP phase δ is the third degeneracy.

- Altogether there exists an eight-fold degeneracy [29], as long as only the $v_\mu \to v_\mu$, $\bar{v}_\mu \to \bar{v}_\mu$, $v \to v_\mu$ and $\bar{v}_e \to \bar{v}_\mu$ channels and one fixed L/E are considered. However, eqs. (10) and (11) also imply that the degeneracies can be broken by using in a suitable way information from different L/E values. This can be achieved in total event rates by changing or combining different L or E [33, 34, 35] but it can in principle also be done by using information in the event rate spectrum of a single baseline L, which requires detectors with very good energy resolution [17]. Another strategy to break the degeneracies is to include further oscillation channels in the analysis ("silver channels") [33, 36].

The discussion of this section shows the strength of the analytic approximations, which allow to understand the complicated parameter interdependence. It also helps to optimally plan experimental setups and to find strategies to resolve the degeneracies.

4. The potential of future neutrino oscillation experiments

Triggered by the spectacular results in neutrino physics during the last ten years, several new experimental projects are under way in this field. It is therefore interesting to investigate where we should stand in the determination of neutrino oscillation parameters in ten years from now. It is also interesting to look further and to estimate the ultimate precision which could be obtained.

The precision of quantities like $\sin^2 2\theta_{13}$ which is found form the simulation of experiments will be presented in a way shown in fig. 1. The bands show how the initial value, which is given by statistics alone (left edge of blue/dark grey band) deteriorates by systematic errors, by parameter correlations (e.g. with the unknown or partly known CP phase) and parameter degeneracies (due to trigonometric ambiguities). It is important to note that a given experiment (or combination of experiments) typically measures some parameter combination with a precision which is considerably better than the final limit. This precision of the experiment is shown in fig. 1 as the right edge of the blue/dark grey band. This precision might be called $(\sin^2 2\theta_{13})_{eff}$, since it expresses the precision if all other unknown parameters

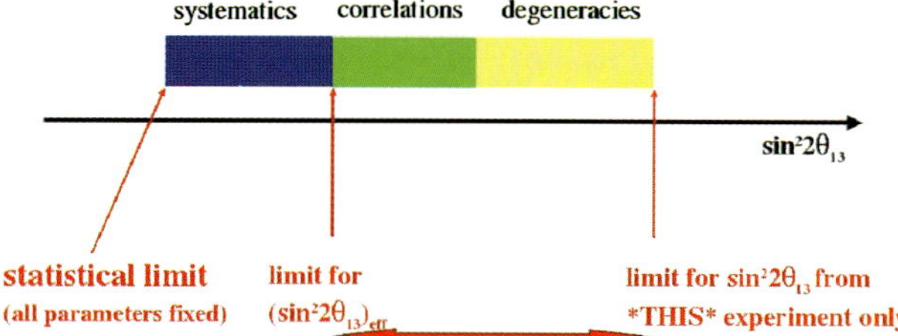

Fig. 1. The precision for $\sin^2 2\theta_{13}$ is shown in colored bands, where the left edge of the blue/dark grey band shows the initial value which is obtained if only statistics is considered. The right edge of the blue/dark grey band is the result after the systematic errors are included. This is the principal precision of the experiment. However, the sensitivity for $\sin^2 2\theta_{13}$ deteriorates further due to parameter correlations and parameter degeneracies. The final value is the right edge of the yellow/light grey band. The range covered by the green and yellow bands can lead to remarkable synergies when this experiment is combined with another experiment with similar precision, but different parameter dependence.

are fixed and no errors are included. However, if one properly extracts a limit of $\sin^2 2\theta_{13}$ with all unknowns properly taken into account, then one ends up at the right edge of the yellow/light grey band. Distinguishing in this way between the precision and the sensitivity is quite useful, since it also shows the room for improvement by combinations with other similarly precise experiments with other parameter dependence.

4.1. Next generation experiments

Future oscillation experiments can be grouped according to their time scale of operation. The K2K experiment is already running and it tests the leading atmospheric oscillation already now. Next come the MINOS and CNGS projects which are already operating or under construction, respectively. Therefore we include in our study the conventional beam experiments MINOS [37], and the CNGS experiments ICARUS [38] and OPERA [39]. We include also the subsequent superbeam experiments J-PARC to SuperKamiokande (T2K) [40] and NuMI off-axis (NOvA) [41], as well as new reactor neutrino experiments [42] with a near and far detector. The main characteristics of these experiments are summarized in tab. I. For the reactor experiments we use the Double-CHOOZ proposal (D-CHOOZ) [43] as initial stage setup with roughly 6×10^4 events, and an optimized setup called Reactor-II, with a slightly longer baseline and 6×10^5 events. Such a configuration could be realized at several other sites under discussion [42]. The results presented in the following are based on Ref. [44], where more details on the analysis can be found. The simulations of the experiments as well as the statistical analysis is performed with the GLoBES software package [45].

Table I. *Characteristics of the considered experiments.*

Label	L [km]	$\langle E_v \rangle$	t_{run}	channel
Conventional beam experiments:				
MINOS	735	3 GeV	5 yr	$v_\mu \to v_{\mu,e}$
ICARUS	732	17 GeV	5 yr	$v_\mu \to v_{e,\mu,\tau}$
OPERA	732	17 GeV	5 yr	$v_\mu \to v_{e,\mu,\tau}$
Off-axis superbeams:				
T2K	295	0.76 GeV	5 yr	$v_\mu \to v_{e,\mu}$
NOvA	812	2.22 GeV	5 yr	$v_\mu \to v_{e,\mu}$
Reactor experiments:				
D-CHOOZ	1.05	~4 MeV	3 yr	$v_e \to v_e$
Reactor-II	1.70	~4 MeV	5 yr	$v_e \to v_e$

Table II. *Precision for* $|\Delta m^2_{31}|$
and $\sin^2\theta_{23}$ *at* 3σ *for the
values* $\Delta m^2_{31} = 2 \times 10^{-3}\,eV^2$,
$\sin^2\theta_{23} = 0.5$.

	$|\Delta m^2_{13}|$	$\sin^2\theta_{23}$
current	88%	79%
MINOS+CNGS	26%	78%
T2K	12%	46%
NOνA	25%	86%
Combination	9%	42%

Tab. II depends on the value of Δm^2_{31} which is shown in fig. 2. The sensitivity suffers for all experiments for low values of Δm^2_{31}. T2K will provide a precise determination of Δm^2_{31} at the level of a few percent for $\Delta m^2_{31} \gtrsim 2 \times 10^{-3}\,\mathrm{eV}^2$. Although NOνA can put a comparable lower bound on Δm^2_{31}, the upper bound is significantly weaker, and similar to the bound from MINOS. The reason for this is a strong correlation between Δm^2_{31} and θ_{23}, which disappears only for $\Delta m^2_{31} \gtrsim 3 \times 10^{-3}\,\mathrm{eV}^2$. From the right panel of fig. 2 one can see that for $\Delta m^2_{31} \sim 2 \times 10^{-3}\,\mathrm{eV}^2$ only T2K is able to improve the current bound on $\sin^2\theta_{23}$. One reason for the rather poor performance on $\sin^2\theta_{23}$ is the fact that these experiments are sensitive mainly to $\sin^2 2\theta_{23}$. This implies that for $\theta_{23} \approx \pi/4$ it is very hard to achieve a good accuracy on $\sin^2\theta_{23}$, although $\sin^2 2\theta_{23}$ can be measured with relatively high precision [46].

Another interesting question is if the next generation long baseline experiments which will operate in the next years will be able to test the three flavourness of the oscillations. The sensitivity to a finite value of the key parameter θ_{13} is shown in fig. 3 for MINOS, OPERA and ICARUS. It can be seen that these experiments have only a modest potential for improvements of the existing θ_{13} limit.

The combined sensitivity to $\sin^2 2\theta_{13}$ of the next-to-next generation experiments is compared in the left panel of fig. 4 with new reactor experiments, T2K (JPARC-SK) and NOνA (NuMI). It can be seen that the $\sin^2 2\theta_{13}$-limits from beam experiments are strongly affected by parameter correlations and degeneracies, whereas reactor experiments provide a "clean" measurement of $\sin^2 2\theta_{13}$, dominated by statistics and systematics [48].

The dependence of the $\sin^2 2\theta_{13}$-limit on the value of Δm^2_{31} is shown in the right panel of fig. 4, where the sensitivity of all experiments gets again rather poor for low values of Δm^2_{31}. For $\Delta m^2_{31} \sim 2 \times 10^{-3}\,\mathrm{eV}^2$ we find roughly an improvement by a factor 2 from conventional beam experiments (MINOS + ICARUS + OPERA combined), a factor 4 from D-CHOOZ, and a factor 6 from the superbeams T2K and NOνA with respect to the current bound from global data [15]. Note that an optimized reactor experiment such as Reactor-II has the potential for even better $\sin^2 2\theta_{13}$-sensitivities than the superbeams (c.f. left panel of fig. 4).

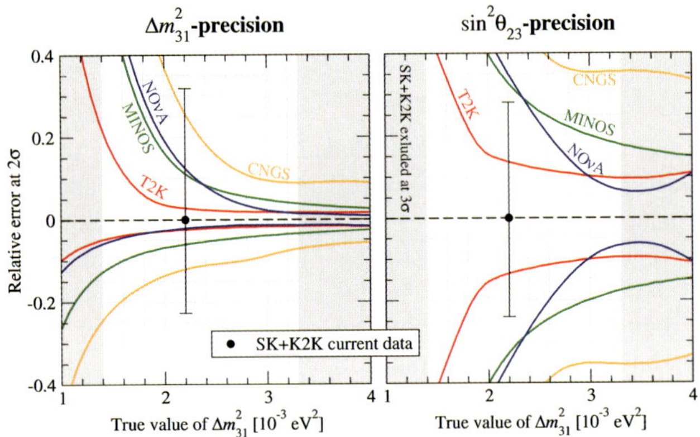

Fig. 2. The precision of Δm^2_{31} (left panel) and $\sin^2\theta_{23}$ (right panel) as a function of the "true value" of Δm^2_{31} for $\theta_{23}^{\mathrm{true}} = \pi/4$ (from Ref. [47]).

A first interesting question concerns improvements of Δm^2_{31} and $\sin^2\theta_{23}$. In tab. II we show the precision which can be obtained in the future in comparison to the current precision, as obtained from a global fit to SuperKamiokande (SK) atmospheric and K2K long-baseline data [15]. The last row is the precision which can be obtained by combining all experiments. We observe from these numbers, that the accuracy on Δm^2_{31} can be improved by one order of magnitude, whereas the accuracy on $\sin^2\theta_{23}$ will be improved only by a factor two.

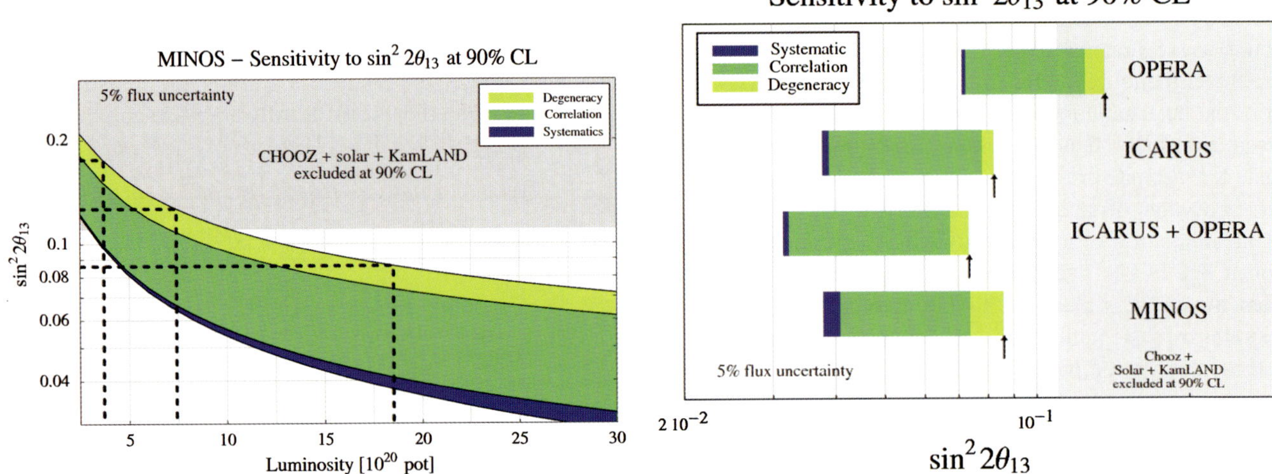

Fig. 3. Left plot: The sensitivity of the MINOS experiment to θ_{13} as a function of the protons on target (pot) assuming a 5% flux uncertainty. The dashed lines represent what 1, 2 and 5 years of operation might achieve (from left to right). Right plot: Comparison of 5 years of operation for the MINOS and CNGS experiments. The grey area for large $\sin^2 2\theta_{13}$ indicates in all cases the current limit from the CHOOZ experiment. The color code of the error bars is explained in fig. 1. Further details can be found in [44].

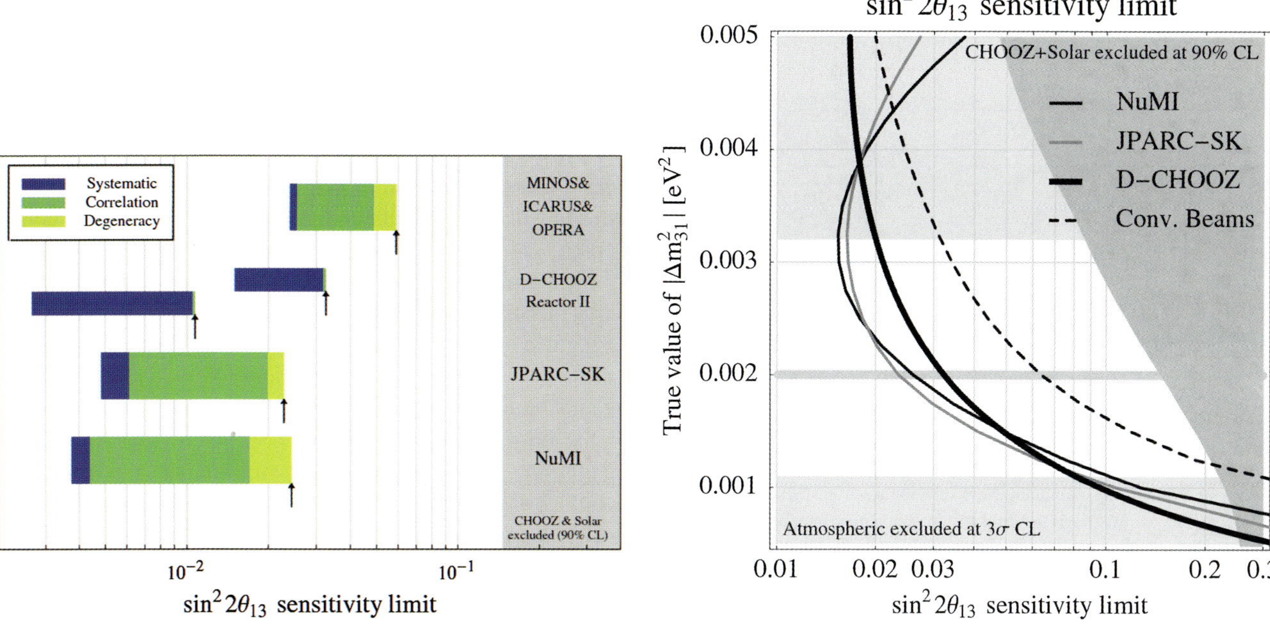

Fig. 4. Left plot: Sensitivity to $\sin^2 2\theta_{13}$ at 90% CL for $\Delta m_{31}^2 = 2 \times 10^{-3}$ eV^2, $\Delta m_{21}^2 = 7 \times 10^{-5}$ eV^2. Right plot: Sensitivity to $\sin^2 2\theta_{13}$ at 90% CL as a function of the "true value" of Δm_{31}^2.

4.2. Synergies

The previous discussion shows that competing plans with similar potential might be realized at the same time scale. This allows to combine the statistics of similar experiments leading to improved global fits. However, it is also possible to utilize synergies between experiments which are more than the simple addition of statistics. The point is that individual experiments measure a certain parameter combinations which contain different degeneracies and correlations. Experiments with similar sensitivities and different correlations and degeneracies allow to separate the parameters partly or fully. An example of such a discussion is given by combining the T2K and NOνA experiments for a fixed time of operation in the best possible way. T2K is essentially insensitive to matter effects, while matter effects play already some role for the longer NOVνA baselines. Both experiments could run partly with neutrino and partly with anti-neutrino beams. The cross-sections for anti-neutrinos are, however, smaller, leading to fewer events for the same running period. An anti-neutrino running is moreover in many aspects like a different experiment, but it is clear that anti-neutrino information is crucial in order to resolve the parameters. A comparable reactor neutrino experiment would be very useful here. It could provide the required information such that both T2K and NOνA could initially run fully in neutrino mode.

Such a synergetic combination would be especially interesting if $\sin^2 2\theta_{13}$ would be close to the current bound. In order to demonstrate these synergy effects we assume that $\sin^2 2\theta_{13} = 0.1$ and investigate what we could be learned about the CP-phase δ and the neutrino mass ordering. First we note that T2K, NOνA and the reactor experiment will all be able to establish the non-zero value of $\sin^2 2\theta_{13}$. However, depending on the unknown value of δ different values of $\sin^2 2\theta_{13}$ will be allowed. This can be seen as allowed regions in the θ_{13}-δ-plane shown in Figs. 8 and 9 of Ref. [44]. None of the experiments on their own can give any information on the CP-phase δ and on the mass hierarchy. The determination of $\sin^2 2\theta_{13}$ from beam experiments is strongly affected by the correlations with δ, and especially for NOνA

also correlations with other parameters are important. Moreover, the inability to rule out the wrong mass hierarchy leads to a further ambiguity in the determination of $\sin^2 2\theta_{13}$. In contrast, since the $\bar{\nu}_e$-survival probability does not depend on δ, Reactor-II provides a clean determination of $\sin^2 2\theta_{13}$ at the level of 20% at 90% CL. If all experiments are combined the complementarity of reactor and beam experiments allows to exclude up to 40% of all possible values of the CP-phase for a given hierarchy. The wrong hierarchy can be ruled out at a modest confidence level with $\Delta \chi^2 \simeq 3$ due to matter effects in NOνA. However, at high confidence levels all values of δ are allowed, and moreover, even for a given hierarchy CP-conserving and CP-violating values of δ cannot be distinguished at 90% CL. These results depend also to some extent on the value of δ.

So far we have considered only neutrino running for the superbeams, since it is unlikely that significant data can be collected with anti-neutrinos within ten years from now. Nevertheless, it might be interesting to investigate the potential of a neutrino-antineutrino comparison. In fig. 5 we show the results from T2K + NOνA with 3 years of neutrinos plus 3 years of anti-neutrinos each (left), in comparison with the case where the antineutrino running is replaced by Reactor-II (right). We find that antineutrino data at that level does neither solve the problems related to the CP-phase nor to the hierarchy. Still CP-violating and CP-conserving values cannot be distinguished at 90% CL. Moreover, the determination of $\sin^2 2\theta_{13}$ is less precise than from the reactor measurement. To benefit from antineutrino measurements a significantly longer measurement period would be necessary, to obtain large enough data samples.

4.3. Long term perspectives

Beyond the discussed accelerator and reactor based oscillation experiments exist more ambitious projects like the JHF-HyperKamiokande project, beta beams and neutrino factories. Such experiments clearly require further R&D before they can be built. However, assuming current knowledge, we believe that such setups are possible in the long run. The potential of the

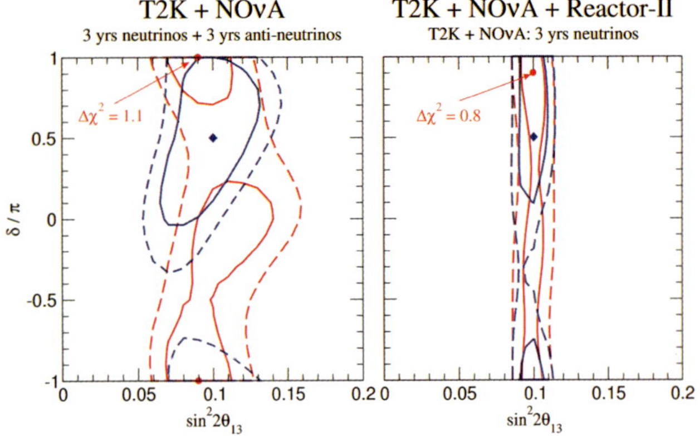

Fig. 5. Antineutrino running vs Reactor-II. We show the 90% CL (solid curves) and 3σ (dashed curves) allowed regions in the $\sin^2 2\theta_{13}$-δ-plane for the assumed values $\sin^2 2\theta_{13} = 0.1$ and $\delta = \pi/2$. The blue/dark curves refer to the allowed regions for the normal mass hierarchy, whereas the red/light curves refer to the sgn(Δm_{31}^2)-degenerate solution (inverted hierarchy), where the projections of the minima onto the $\sin^2 2\theta_{13}$-δ-plane are shown as diamonds (normal hierarchy) and dots (inverted hierarchy). For the latter, the $\Delta\chi^2$-value with respect to the best-fit point is also given.

JHF-HyperKamiokande experiment and a neutrino factory are compared in fig. 6 to T2K (JHF-SK) and NOvA (NuMI). It can be seen that the existing limits can be improved by a few orders of magnitude compared to now.

5. Theoretical implications and conclusions

One of the most interesting unsolved topics is the origin of of flavour and fermion masses. There exist apparent regularities in the fermionic field content which make it very tempting to introduce right-handed neutrino fields leading to both Dirac and Majorana mass terms for neutrinos. Diagonalization of the resulting mass matrices yields Majorana mass eigenstates and due to the see-saw mechanism very small neutrino masses. This can also be nicely realized in embeddings of the SM into GUTs with larger symmetries, such as SO(10). Before the discovery of large leptonic mixing, many theorists expected the leptonic mixings to be similar to quark mixing, characterized by small mixing angles. Experiment led theory in showing the

striking results that $\sin^2 2\theta_{23} \simeq 1$ and $\tan^2 \theta_{12} \simeq 0.39$, while θ_{13} is small. By finding two large mixing angles, neutrino physics has already provided surprising and very valuable information which severely constrains models of neutrino masses. Future neutrino oscillation experiments will lead to very precise measurements which are also very valuable, since they will provide precision tests of the flavour sector. The level of precision will confirm or rule out many ideas about the origin of flavour and connected topics.

One topic is the small value of $\sin^2 2\theta_{13}$. Since there are two large mixing angles, there is no particular reason to expect the third angle, θ_{13}, to be extremely small or even zero. A small value of $\sin^2 2\theta_{13} \simeq 0.1$ could be a numerical coincidence in a framework which predicts generically large or sizable mixings. However, if the limit on $\sin^2 2\theta_{13}$ would become smaller by an order of magnitude then some protective mechanism like a "symmetry" would be required. This can be seen in neutrino mass models which are able to predict the large values for θ_{12} and θ_{23}. Such models have a certain tendency to predict a sizable value of θ_{13} as can be seen, for example, in [49, 50, 51, 52, 53, 54]. The conclusion is that a value of θ_{13} close to the CHOOZ bound would be quite natural, while much smaller values are less likely or hard to understand.

Future precision measurements can also test if relations like $\theta_{23} = \pi/4$ or $\theta_{12} + \theta_C = \pi/4$ which are currently fulfilled within experimental errors still hold at the percent level or below. If so, then this would provide strong constrains on the origin of the flavour structure. Precision is also valuable, even if such special relations are not found. The point is that it is generally not easy to predict a set of very precisely known masses and mixings in a certain model or class of models.

Neutrino masses and mixing parameters are also subject to quantum corrections between low scales, where measurements are performed, and high scales where some theory typically predicts the masses and mixings. This has interesting implications, since it implies that certain deviations from special relations are expected due to quantum corrections (renormalization group or RGE effects). Suppose, for example, that some theory were able to predict $\theta_{13} \equiv 0$. Then RGE effects would still predict a tiny, but finite value at low energy. Strictly speaking, $\theta_{13} = 0$ cannot be

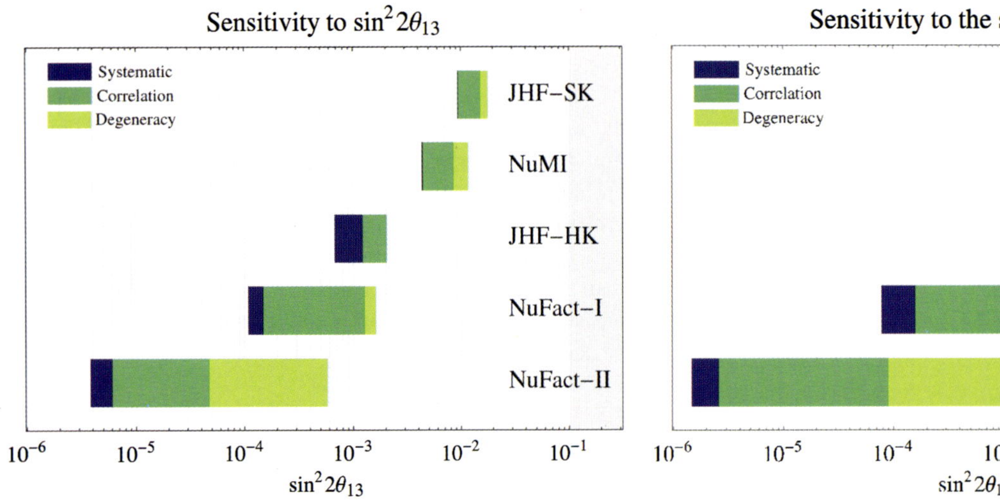

Fig. 6. Left plot: The θ_{13} sensitivity of different future accelerator based neutrino oscillation experiments [20]. Right plot: The θ_{13} values for which sensitivity to matter effects, i.e. $sign(\Delta m_{31}^2)$ exists. The shown bands are again the reduction of sensitivity from a purely statistical limit (left end of the dark grey/blue range) by systematics (right end of dark grey/blue), correlations (medium grey/green) and degeneracies (light grey/yellow). The right end of the light grey(yellow) band represents the final 90%CL limit. The grey area for large $\sin^2 2\theta_{13}$ indicates the current limit from the CHOOZ experiment.

excluded completely by this argument, as the high-energy value could be just as large as the change due to running and of opposite sign. However, a complete cancellation of this kind would be a miraculous fine-tuning, since the physics generating the value at high energy is not directly related to quantum corrections at lower energies. The strength of the running of θ_{13} depends on the neutrino mass spectrum and whether or not supersymmetry is realized. For the Minimal Supersymmetric Standard Model one finds a shift $\Delta \sin^2 2\theta_{13} > 0.01$ for a considerable parameter range, i.e. one would expect to measure a finite value of θ_{13} [55]. Conversely, limits on model parameters would be obtained if an experiment were to set an upper bound on $\sin^2 2\theta_{13}$ in the range of 0.01. In any case, it should be clear that a precision of the order of quantum corrections to neutrino masses and mixings is very interesting in a number of ways.

Precision measurements would also allow very interesting tests of many other topics, like MSW matter effects, three neutrino unitarity, neutrino decay, de-coherence and NSI effects. There exists also an interesting interplay with theories beyond the Standard Models, flavour models as well as astroparticle physics (leptogenesis, supernovae, nucleosynthesis, structure formation). In summary, future precision oscillation experiments will provide the best window into the so far un-understood flavour sector. It may give us a glimpse on the origin of flavour, but it may also lead to unexpected results and insights as it happened before in neutrino physics.

Acknowledgments

Work supported by the "Sonderforschungsbereich 375 für Astro-Teilchenphysik der Deutschen Forschungsgemeinschaft".

References

1. Toshito, T. (SuperKamiokande Collab.), proceedings of 36th Rencontres de Moriond on Electroweak Interactions and Unified Theories, Les Arcs, France, 10–17 Mar (2001), hep-ex/0105023.
2. Hampel, W. *et al.* (GALLEX Collab.), Phys. Lett. B **447**, 127 (1999).
3. McDonald, A., these proceedings.
4. Ahmad, Q. R. *et al.* (SNO Collab.), Phys. Rev. Lett. **89**, 011301 (2002), nucl-ex/0204008.
5. Suzuki, A., these proceedings.
6. Eguchi, K. *et al.* (KamLAND Collab.), Phys. Rev. Lett. **90**, 021802 (2003), hep-ex/0212021.
7. Church, E. D., Eitel, K., Mills, G. B. and Steidl, M., Phys. Rev. D **66**, 013001 (2002), hep-ex/0203023.
8. Suzuki, Y., these proceedings.
9. Gonzalez-Garcia, M. C., these proceedings.
10. Fukuda, S. *et al.*, Phys. Lett. B **539**, 179 (2002); Cleveland, B. T. *et al.*, Astrophys. J. **496**, 505 (1998); Abdurashitov, D. N. *et al.*, Phys. Rev. **C60**, 055801 (1999); Hampel, W. *et al.*, Phys. Lett. **B447**, 127 (1999); Cattadori, C., Nucl. Phys. B (Proc. Suppl.) **110**, 311 (2002); Ahmad, Q. R. *et al.*, Phys. Rev. Lett. **89**, 011301 (2002); Phys. Rev. Lett. **89**, 011302 (2002).
11. SNO collaboration, submitted to Phys. Rev. Lett. (2003), nucl-ex/0309004.
12. Apollonio, M. *et al.*, Phys. Lett. B **420**, 397 (1998).
13. Apollonio, M. *et al.*, Eur. Phys. J. C **27**, 331 (2003).
14. Boehm, F. *et al.*, Phys. Rev. Lett. **84**, 3764 (2000).
15. Maltoni, M., Schwetz, T., Tortola, M. A. and Valle, J. W. F., New J. Phys. **6**, 122 (2004), hep-ph/0405172; Phys. Rev. D **68**, 113010 (2003), hep-ph/0309130.
16. Freund, M., Huber, P. and Lindner, M., Nucl. Phys. B **585**, 105 (2000), hep-ph/0004085.
17. Freund, M., Huber, P. and Lindner, M., Nucl. Phys. B **615**, 331 (2001), hep-ph/0105071.
18. Huber, P., Lindner, M. and Winter, W., Nucl. Phys. B **654**, 3 (2003), hep-ph/0211300.
19. Lindner, M., in "Neutrino mass", Springer tracts in modern physics, (ed. by G. Altarelli and K. Winter), hep-ph/0209083.
20. Huber, P., Lindner, M. and Winter, W., Nucl. Phys. B **645**, 3 (2002), hep-ph/0204352.
21. Dick, K., Freund, M., Lindner, M. and Romanino, A., Nucl. Phys. B **562**, 29 (1999), hep-ph/9903308.
22. Freund, M., Phys. Rev. **D64**, 053003 (2001), hep-ph/0103300.
23. Akhmedov, E. K., Johansson, R., Lindner, M., Ohlsson, T. and Schwetz, T., J. High En. Phys. **0404**, 078 (2004), hep-ph/0402175.
24. Cervera, A. *et al.*, Nucl. Phys. **B579**, 17 (2000), erratum ibid., Nucl. Phys. **B593**, 731 (2001).
25. Minakata, H., Sugiyama, H., Yasuda, O., Inoue, K. and Suekane, F., Phys. Rev. D **68**, 033017 (2003), hep-ph/0211111.
26. Huber, P., Lindner, M., Schwetz, T. and Winter, W., Nucl. Phys. B **665**, 487 (2003), hep-ph/0303232.
27. Huber, P. and Winter, W., Phys. Rev. D **68**, 037301 (2003), hep-ph/0301257.
28. Fogli, G. L. and Lisi, E., Phys. Rev. D **54**, 3667 (1996), hep-ph/9604415.
29. Barger, V., Marfatia, D. and Whisnant, K., Phys. Rev. **D65**, 073023 (2002), hep-ph/0112119.
30. Koike, M., Ota, T. and Sato, J., Phys. Rev. D **65**, 053015 (2002), hep-ph/0011387.
31. Burguet-Castell, J., Gavela, M. B., Gomez-Cadenas, J. J., Hernandez, P. and Mena, O., Nucl. Phys. **B608**, 301 (2001), hep-ph/0103258.
32. Minakata, H. and Nunokawa, H., J. High En. Phys. **10**, 001 (2001), hep-ph/0108085.
33. Burguet-Castell, J., Gavela, M. B., Gomez-Cadenas, J. J., Hernandez, P. and Mena, O., Nucl. Phys. B **646**, 301 (2002), hep-ph/0207080.
34. Barger, V., Marfatia, D. and Whisnant, K., Phys. Rev. **D66**, 053007 (2002), hep-ph/0206038.
35. Huber, P., Maltoni, M. and Schwetz, T., hep-ph/0501037.
36. Donini, A., Meloni, D. and Migliozzi, P., Nucl. Phys. B **646**, 321 (2002), hep-ph/0206034.
37. Ables, E. *et al.*, (MINOS) FERMILAB-PROPOSAL-P-875.
38. Aprili, P. *et al.*, (ICARUS) CERN-SPSC-2002-027.
39. Duchesneau, D., (OPERA), hep-ex/0209082.
40. Itow, Y. *et al.*, hep-ex/0106019.
41. Ayres, D. *et al.*, hep-ex/0210005,
42. Anderson, K. *et al.*, hep-ex/0402041.
43. Ardellier, F. *et al.*, hep-ex/0405032.
44. Huber, P., Lindner, M., Rolinec, M., Schwetz, T. and Winter, W., Phys. Rev. D **70**, 073014 (2004), hep-ph/0403068.
45. Huber, P., Lindner, M. and Winter, W., GLoBES (Global Long Baseline Experiment Simulator), hep-ph/0407333, to appear in Comp. Phys. Comm., http://www.ph.tum.de/˜globes.
46. Minakata, H., Sonoyama, M. and Sugiyama, H., Phys. Rev. D **70**, 113012 (2004), hep-ph/0406073.
47. Huber, P., Lindner, M., Rolinec, M., Schwetz, T. and Winter, W., hep-ph/0412133.
48. Huber, P., Lindner, M., Schwetz, T. and Winter, W., Nucl. Phys. B **665**, 487 (2003), hep-ph/0303232.
49. Barr, S. M. and Dorsner, I., Nucl. Phys. B **585**, 79 (2000), hep-ph/0003058.
50. Altarelli, G. and Feruglio, F., in "Neutrino Mass", Springer Tracts in Modern Physics, (ed. by G. Altarelli and K. Winter), hep-ph/0206077.
51. Barbieri, R., Hambye, T. and Romanino, A., J. High En. Phys. **0303**, 017 (2003), hep-ph/0302118.
52. Chen, M. C. and Mahanthappa, K. T., Int. J. Mod. Phys. **A18**, 5819 (2003), hep-ph/0305088.
53. King, S. F., Rept. Prog. Phys. **67**, 107 (2004), hep-ph/0310204.
54. Goh, H. S., Mohapatra, R. N. and Ng, S. P., Phys. Rev. D **68**, 115008 (2003), hep-ph/0308197.
55. Antusch, S., Kersten, J., Lindner, M. and Ratz, M., Nucl. Phys. B **674**, 401 (2003), hep-ph/0305273.

Physica Scripta. Vol. T121, 86–93, 2005

Experimental Prospects of Neutrinoless Double Beta Decay

Ettore Fiorini[1]

Dipartimento di Fisica G.P.S. Occhialini dell' Universita' di Milano-Bicocca e Sezione di Milano dell' INFN, Piazza della Scienza 3, 20126 Milano, Italy

Received October 4, 2004; accepted in revised form February 15, 2005

PACS numbers: 11.30.Fs, 14.60.Lm, 14.60.Pq

Abstract

The results obtained so far in searches for double beta decay (DBD) are reviewed with special emphasis on the neutrinoless channel which could indicate a non-zero effective neutrino mass under the hypothesis that this particle is of Majorana type. The connection of the value of this mass with the indications coming from the recent results of experiments on neutrino oscillation is underlined. The difficulties and uncertainties are emphasized of the calculation of the nuclear matrix elements relevant for DBD. It is stressed that, as a consequence, different nuclear candidates for double beta decay should be investigated. There is an additional need to do so: a peak indicative of neutrinoless double beta decay could be mimicked by some radioactive contamination. Detection of the different peaks expected for DBD in two or more candidate nuclei becomes therefore mandatory.

The experimental methods to search for DBD, particularly in its neutrinoless channel, are compared and discussed together with the results obtained so far. The two most constraining experiments presently running (NEMO 3 and CUORICINO) are described in some detail and their preliminary limits discussed. The report is concluded with a description of the presently proposed experiments aiming to reach a sensitivity of a few tens of milli-electronvolts as required by the results coming from oscillation experiments. Special emphasis is given to possible improvements in increasing active mass and resolution, and decreasing background. This comparison refers in particular to the role played with respect to the classical methods of detection (scintillation and ionisation counters, TPC and tracking chambers, large arrays of diodes) by new detecting methods.

1. Introduction

As it is well known the rare process of double beta decay (DBD) can occur substantially in three channels:

$$(A, Z) \rightarrow (A, Z+2) + 2e^- + 2\bar{\nu}, \tag{1}$$

$$(A, Z) \rightarrow (A, Z+2) + 2e^- + 1, 2, 3 \ldots \chi, \tag{2}$$

$$(A, Z) \rightarrow (A, Z+2) + 2e^-. \tag{3}$$

Both channels (2) and (3) violate lepton number, while channel (1) (*two neutrino* DBD) (Fig. 1a) is allowed by the standard model. Our attention here will be concerned with channel (3), normally called *neutrinoless* DBD (fig. 1b), even if channel (2), where χ is a massless Goldston boson, is also neutrinoless. The presence of neutrinoless DBD would allow in principle to determine the absolute value of the mass and test the Majorana nature of the neutrino [1].

From the experimental point of view, phase space would strongly favour neutrinoless over two-neutrino DBD and as a consequence searches for DBD would be capable to detect even a tiny violation of the lepton number. In addition, and that is even more appealing for an experimentalist, neutrinoless DBD would produce a peak in the spectrum of the sum of the electron energies (Fig. 2).

Searches for neutrinoless DBD have been stimulated by the recent results on neutrino oscillations [2, 3] which clearly indicate that the difference of the squared masses of two neutrinos of

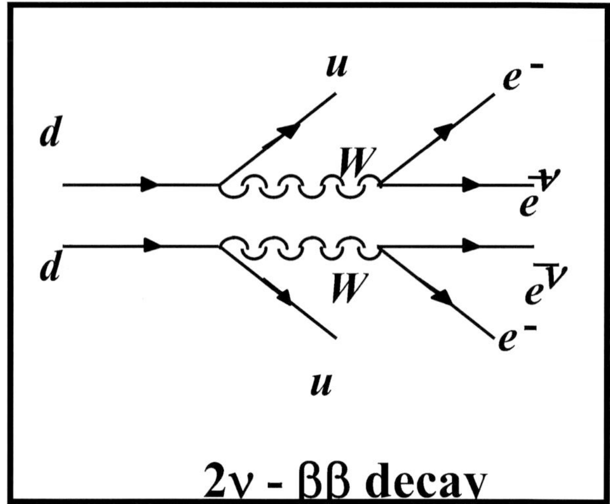

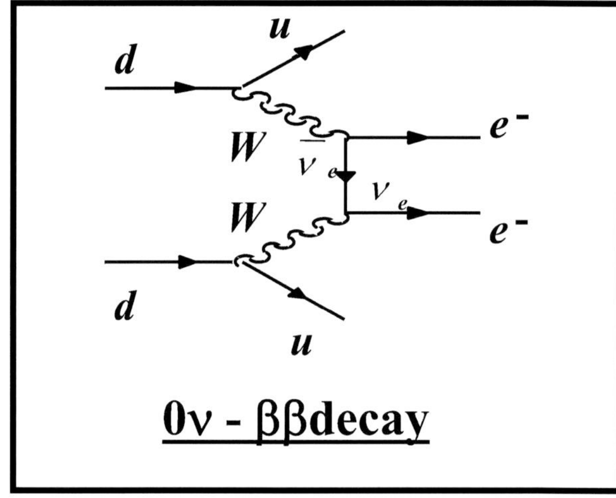

Fig. 1. **a.** Two-neutrino DBD; **b.** Neutrinoless DBD.

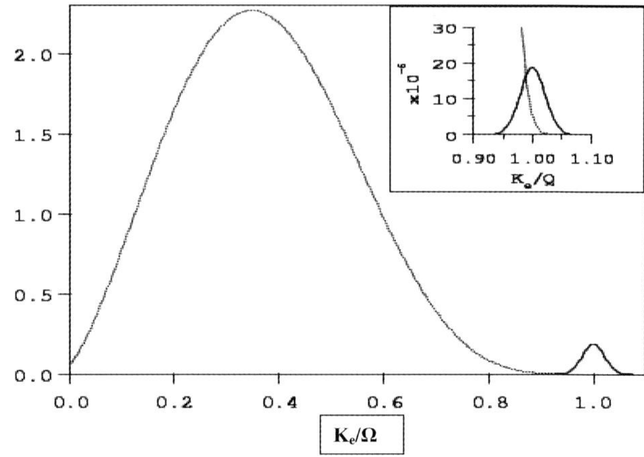

Fig. 2. Two-neutrino and neutrinoless DBD.

[1] Ettore.Fiorini@mib.infn.it

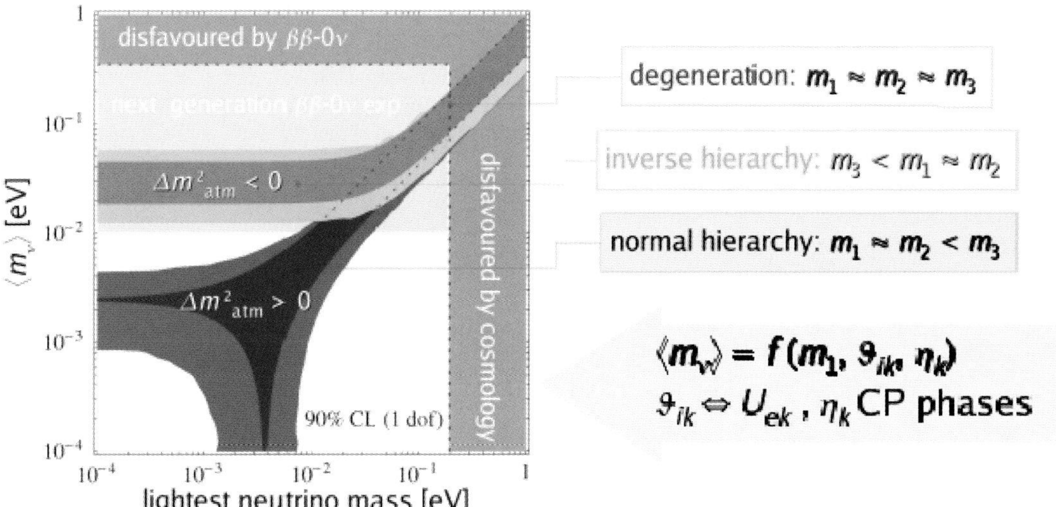

Fig. 3. Expected values of effective neutrino mass form oscillation experiments.

different flavours is different from zero. This would imply that the absolute mass of at least one neutrino has a finite value. As shown in Fig. 3, values of absolute masses of a few meV or of a few tens of meV are expected under the hypothesis of direct or inverted mass hierarchies, respectively [1].

2. Experimental approach

Experiments on DBD can and have been carried out with two totally different approaches: indirect and direct [5, 6].

Indirect experiments are basically of two types: geochemical and radiochemical. In the former, a rock containing the (A, Z) nucleus is searched for an isotopic excess in $(A, Z + 2)$. Positive results have been obtained on DBD of the following nuclei

$$^{82}\mathrm{Se} \rightarrow {}^{82}\mathrm{Kr}, {}^{96}\mathrm{Zr} \rightarrow {}^{96}\mathrm{Mo}, {}^{128}\mathrm{Te} \rightarrow {}^{128}\mathrm{Xe}, {}^{130}\mathrm{Te} \rightarrow {}^{130}\mathrm{Xe}.$$

All these decays have been confirmed by direct experiments with the exception of DBD of ^{128}Te.

In radiochemical experiments a DBD active massive source is stored for a long time underground and successively tested for the decay of the $(A, Z + 2)$ nucleus produced in the stored material. Evidence for the decay

$$^{238}\mathrm{U} \rightarrow {}^{238}\mathrm{Pu}$$

followed by α decay of the daughter nuclei has been obtained.

The limitation of both methods is due to the fact that they measure only the lifetime of DBD and cannot therefore distinguish among the various DBD channels and even between decays to the ground and excited states.

Direct experiments aiming to actually detect DBD processes can be divided into two types: experiments with the DBD source inserted, normally in form of thin layers of material, among detection planes (*Source $\neq$ Detector approach*) and experiments where the detector is made of a DBD active substance (*Source = Detector approach*).

Many conventional detecting approaches, like ionization and scintillation detectors or time projection or tracking chambers, have been employed so far. More recently a new type of detector has been implemented in DBD Source = Detector experiments: the Low Temperature Thermal detector, or bolometer [7, 8]. The scheme of these detectors is shown in Fig. 4. When a crystal of a diamagnetic and dielectric material is kept at low temperature, its

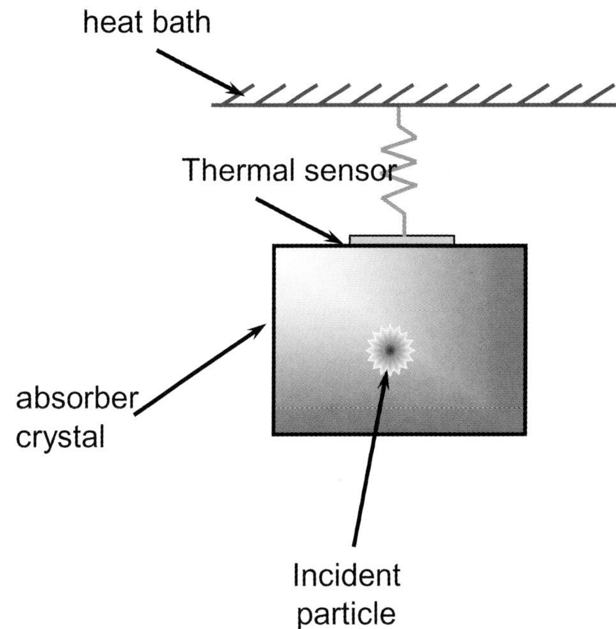

Fig. 4. Scheme of a thermal detector.

heat capacity is proportional to the cube of the temperature and can therefore become extremely small. As a consequence, the tiny increase of temperature due to the energy loss by a particle in this "absorber" can be revealed and measured by means of a suitable thermal sensor.

These detectors, even if still in their infancy, have proved to yield excellent resolutions. For X-rays of a few keV, like those of ^{55}Mn, the energy resolution has reached 4 eV, more than an order of magnitude better compared to any other detector. For energies of a few MeV, their resolution is comparable (and for α particles even superior) to the one reachable with Ge diodes. In the case of experiments on DBD they offer a perhaps more important advantage. The absorber can be made from different compounds allowing an ample choice of DBD candidate nuclei.

3. Present experimental results

The results of the most recent experiments on two-neutrino DBD are reported in Table I.The most constraining results on

Table I. *Summary of the most recent two-neutrino DBD experiments.*

Isotope	$\tau_{1/2}^{2\nu}$ (y) (measured)	$\tau_{1/2}^{2\nu}$ (y) (calculated)
^{48}Ca	$(4.2^{+2.1}_{-1.0}) \times 10^{19}$	6×10^{18}–5×10^{20}
^{76}Ge	$(1.42^{+0.09}_{-0.07}) \times 10^{21}$	7×10^{19}–6×10^{22}
^{82}Se	$(0.9 \pm 0.1) \times 10^{20}$	3×10^{18}–6×10^{21}
^{96}Zr	$(2.1^{+0.8}_{-0.4}) \times 10^{19}$	3×10^{17}–6×10^{20}
^{100}Mo	$(8.0 \pm 0.7) \times 10^{18}$	1×10^{17}–2×10^{22}
^{100}Mo(0^{+*})	$(6.8 \pm 1.2) \times 10^{20}$	5×10^{19}–2×10^{21}
^{116}Cd	$(3.3^{+0.4}_{-0.3}) \times 10^{19}$	3×10^{18}–2×10^{21}
^{128}Te	$(2.5 \pm 0.4) \times 10^{24}$	9×10^{22}–3×10^{25}
^{130}Te	$(0.9 \pm 0.15) \times 10^{21}$	2×10^{19}–7×10^{20}
^{150}Nd	$(7.0 \pm 1.7) \times 10^{18}$	6×10^{16}–4×10^{20}
^{238}U	$(2.0 \pm 0.6) \times 10^{21}$	1.2×10^{19}

Table II. *Results on neutrinoless DBD since one year ago.*

Experiment	Isotope	$\tau_{1/2}^{0\nu}$ (y)	m^*_{ee} (eV)	Range m_{ee}
Heidelberg – Moscow 2001	^{76}Ge	1.9×10^{25}	<0.35	<0.3–2.5
IGEX 2002		1.6×10^{25}	<0.38	<0.3–2.5
Mi DBD – ν 2002	^{130}Te	2.1×10^{23}	<1.5	<0.9–2.1
Bernatowicz *et al.* 1993	^{128}Te^{geo}	7.7×10^{24}	<1.0	<1.0–4.4
Belli *et al.* 2003	^{136}Xe	1.7×10^{23}	<1.7	<0.8–2.4
Bizzeti *et al.* 2003	^{116}Cd	1.7×10^{23}	<4.8	<1.6–5.5
Ejiri *et al.* 2001	^{100}Mo	5.5×10^{22}	<6.0	1.4–256

neutrinoless DBD obtained up to one year ago are reported in Table II. They are all negative with exception of an indication for neutrinoless DBD of ^{76}Ge [9] with a lifetime of 1.12×10^{25} y and a value of 0.46 eV for $\langle m_\nu \rangle$. This result is however widely contested [10].

Two new experiments on DBD are presently running; one is based on the source $\neq$ detector and the other on the source $=$ detector approach.

NEMO 3 (Fig. 5), presently operating in the Frejus Underground Laboratory, is a detector with drift wire tracking chambers operating in the Geiger mode (6180 cells) while the calorimetry is performed by means of plastic scintillators coupled to low radioactivity PMT. Up to 10 kg of DBD isotopes are inserted in the detector in form of thin sheets. The preliminary results on two neutrino and neutrinoless DBD [11] are reported in Table III.

CUORICINO is a cryogenic detector searching for DBD of ^{130}Te. This nucleus has the unique advantage of a good natural isotopic abundance (33.87%) which allows the use of

Table III. *Preliminary results from NEMO 3.*

	$\tau_{2\nu}$	$\tau_{0\nu}$	$\langle m_\nu \rangle$
^{100}Mo	$7.72 \pm .02 \pm .54 \times 10^{18}$ y	$>3.5 \times 10^{23}$	0.7–1.2
^{82}Se	$10.3 \pm .2 \pm 1.0 \times 10^{19}$ y	1.9×10^{23}	1.3–3.6
^{116}Cd	$3.0 \pm .1 \pm .3 \times 10^{19}$ y		
^{150}Nd	$9.7 \pm .7 \pm 1.0 \times 10^{18}$ y		
^{96}Zr	$2.0 \pm .3 \pm .2 \times 10^{19}$ y		

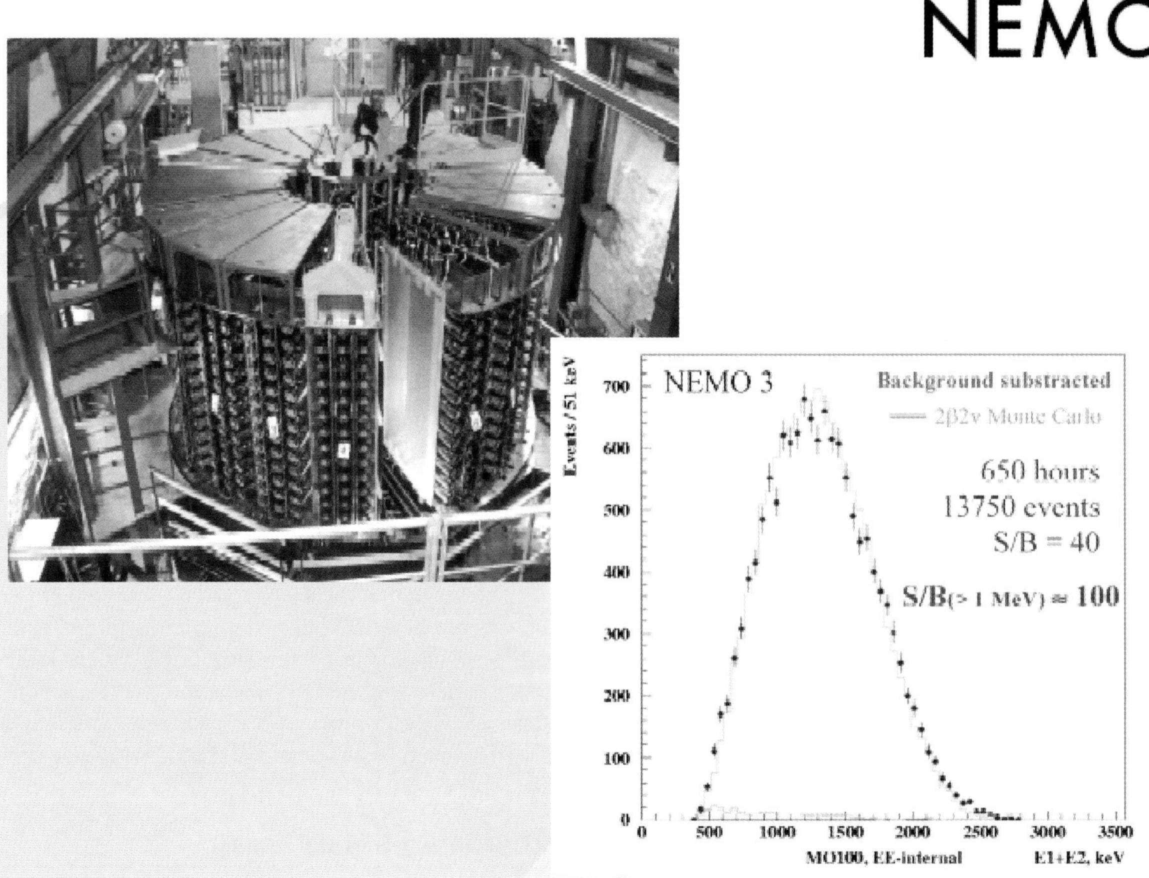

Fig. 5. NEMO 3.

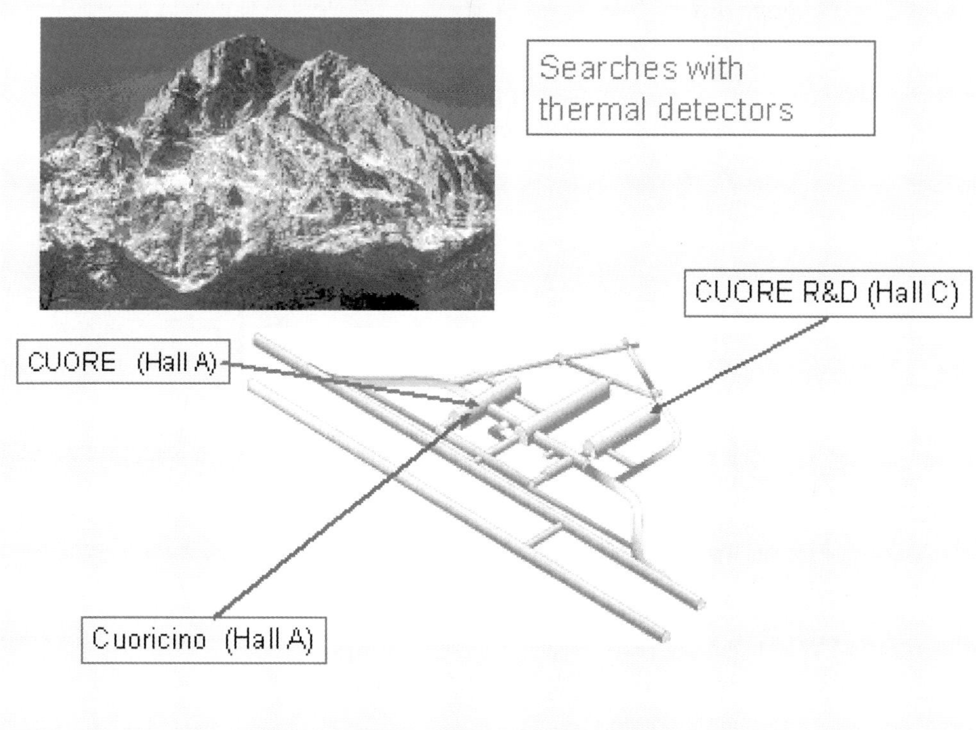

Fig. 6. Location of CUORICINO, CUORE and R&D for CUORE in Gran Sasso Lab.

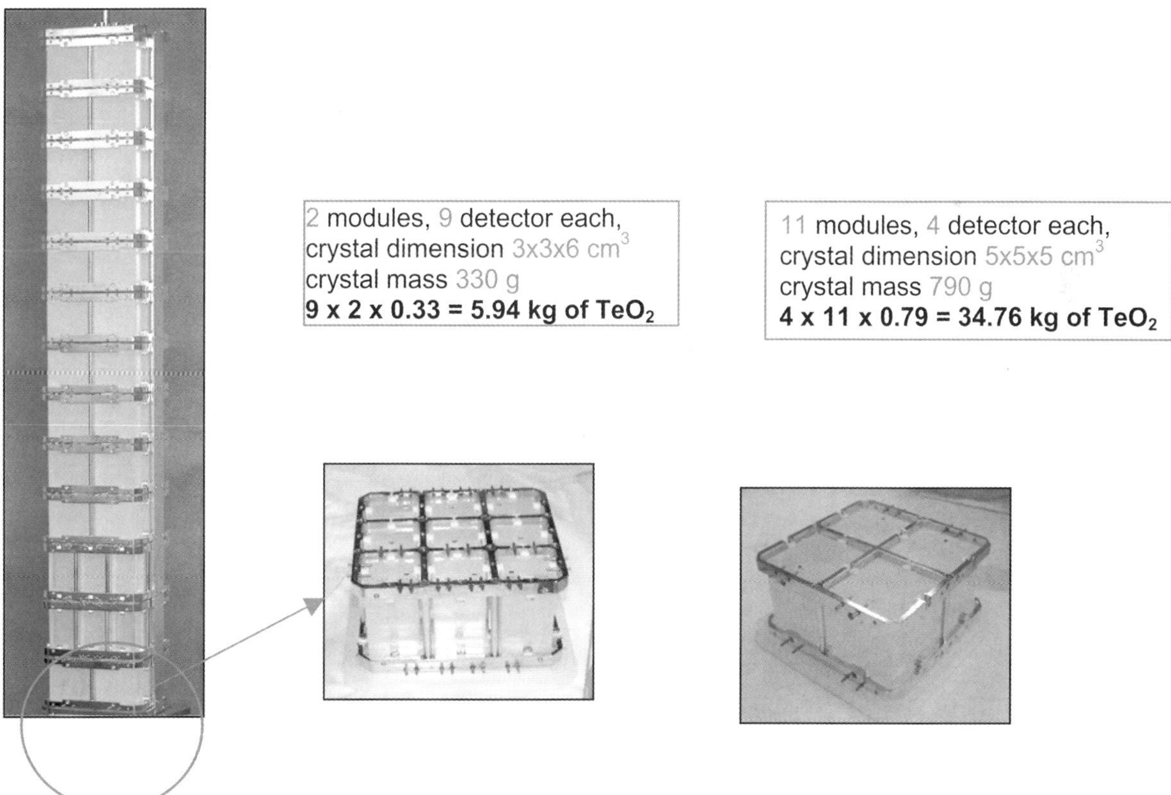

Fig. 7. The structure of CUORICINO.

natural tellurium avoiding large cost and possible enrichment problems. It has a high transition energy $(2528.8 \pm 1.3)\,\text{keV}$ and its DBD has been observed with geo-chemical techniques $(\tau_{1/2\,\text{incl}} = (0.7 - 2.7) \times 10^{21}\,\text{y})$. **CUORICINO** is running in the Gran Sasso Underground Laboratory, where a series of cryogenic experiments on DBD has been performed and the location of the final experiment CUORE has been already been allocated (Fig. 6).

As shown in Fig. 7, CUORICINO is an array of 44 and 18 crystals of TeO$_2$ of $5 \times 5 \times 5$ and $3 \times 3 \times 6\,\text{cm}^3$, respectively. The total active mass is $40.7\,\text{kg}$, more than an order of magnitude

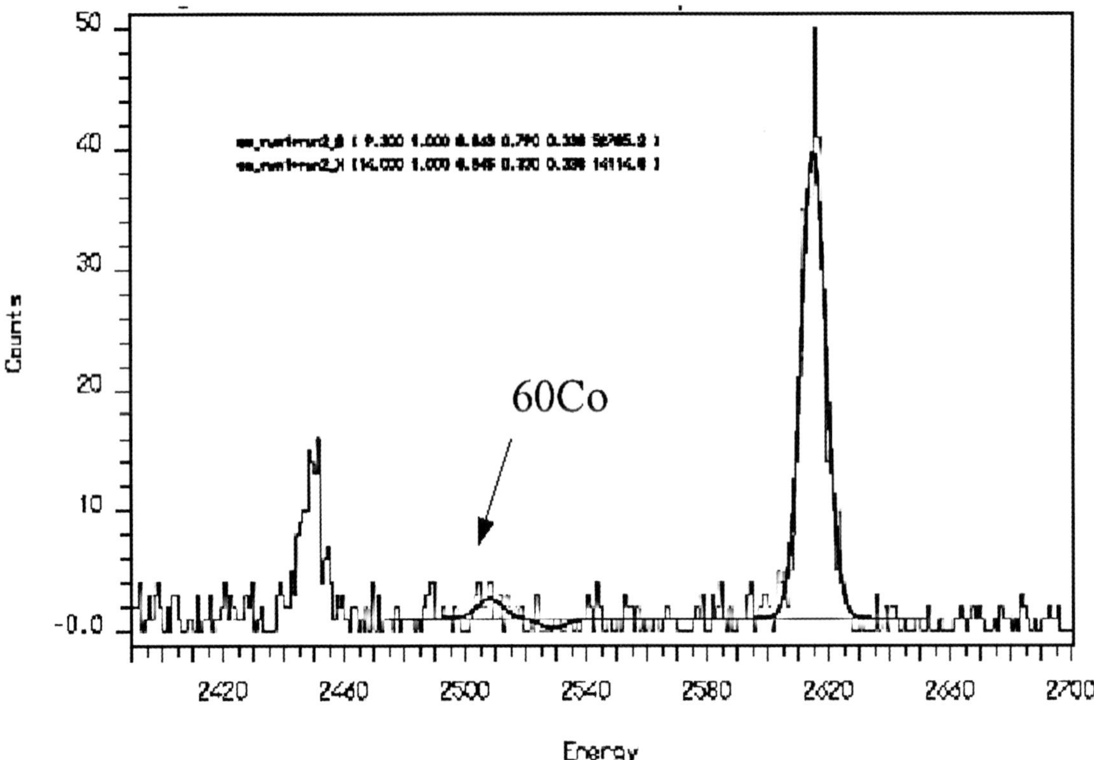

Fig. 8. Spectrum of CUORICINO in the region of neutrinoless DBD.

larger than any other cryogenic detector. After a few months of running time no evidence (fig. 8) was found for neutrinoless DBD of ^{130}Te with a 90% c. l. lower limit on the lifetime of 1.0×10^{24} years. Taking into account the large uncertainties on the nuclear matrix elements, this translates into a constraint on the effective neutrino mass ranging from 0.26 to 1.4 eV. This span already partially covers the indication by Klapdor *et al.* [9] (0.1–1.4) eV, when evaluated taking into account the same uncertainty.

4. The future

Various experiments on DBD based on different techniques have been proposed or are being studied (Table IV). I will report here only a few of them.

CUORE (for Cryogenic Underground Observatory of Rare Decays) (Fig. 9) is the only future experiment for which a definite location has already been decided (Fig. 6). It will consist of 19 columns made by 13 planes of 4 TeO$_2$ crystals of $5 \times 5 \times 54$ cm^3. The total mass will be $\sim$760 kg, corresponding to $\sim$203 kg of ^{130}Te. Crystals are separated by a few mm, only, with little material among them. The array will be shielded (from the interior to the exterior) by layers of Roman, low radioactivity and modern lead with ^{210}Pb contamination of <4 mBq/kg, 16 ± 4 Bq/kg and $\sim$160 Bq/kg, respectively.

MAJORANA (Fig. 10) is an extension of the previous experiments on DBD of ^{76}Ge. It will be located in a not yet decided underground laboratory in Canada or USA. It consists in a fragmented array of Ge diodes enriched in ^{76}Ge with a total mass of 500 kg. A strong background suppression is expected for pulse shape discrimination.

MOON (Fig. 11), planned to be operated in Japan, is an experiment based on the source $\neq$ detector approach. The setup will be made of molybdenum foils of a mass up to 750 kg with

Fig. 9. CUORE.

plastic scintillators for energy measurement and optical fibers for position readout.

EXO (for Enriched Xenon Observatory) (Fig. 12) is an experiment based on a totally new technique aiming to detect DBD of ^{136}Xe into ^{136}Ba. The principle is to detect the daughter ^{136}Ba$^+$ ions by tagging with lasers. The gas option (a liquid option is also considered) is based on 40 m of xenon at 40

Table IV. *Proposed second generation DBD experiments.*

Experiment	Nucleus	Detector	$T^{0\nu}$ (y)	$<m_\nu>$
NEMO III	^{100}Mo *et al.*	10 kg of enrich. Isotopes-tracking	4×10^{24}	.23–.98
Cuoricino	^{130}Te+others	40 kg of TeO$_2$ bolometers (nat)	1.5×10^{25}	.09–.63
CUORE	^{130}Te+others	770 kg of TeO$_2$ bolometers (nat)	7×10^{26}	.014–.091
EXO	^{100}Mo	1 t Xe TPC	8×10^{26}	.055–.12
GENIUS	^{76}Ge	1 t Ge diodes in LN	1×10^{28}	.013–.050
Majorana	^{76}Ge	1 t Ge diodes	4×10^{27}	.021–.070
MOON	^{100}Mo	34 t of nat. Mo sheets in plastic sc.	1×10^{27}	.014–.057
DCBA	^{150}Nd	20 kg Nd-tracking	2×10^{25}	.035–.055
CAMEO	^{116}Cd	1 t CdWO$_4$ in liquid scintillators	$> 10^{26}$	.053–.24
COBRA	^{116}Cd, ^{130}Te	10 kg of CdTe semiconductors	1×10^{24}	.5–2.
Candles	^{48}Ca	Tons of CaF$_2$ in liquid scintillators	1×10^{26}	.15–.26
GSO	^{116}Cd	2 t Gd$_2$SiO$_5$: Ce scintill. in liquid sc.	2×10^{26}	.038–.172
Xe	^{136}Xe	1.56 Xenon in liquid scintillator	5×10^{26}	.066–.157
Xmass	^{136}Xe	1 t of liquid Xe	3×10^{26}	.086–.252

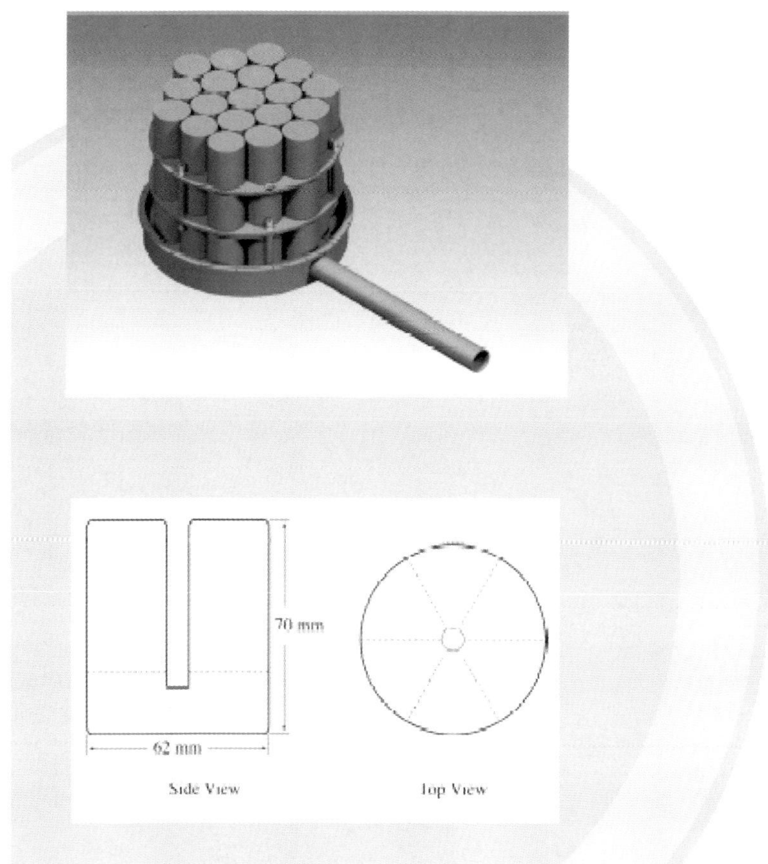
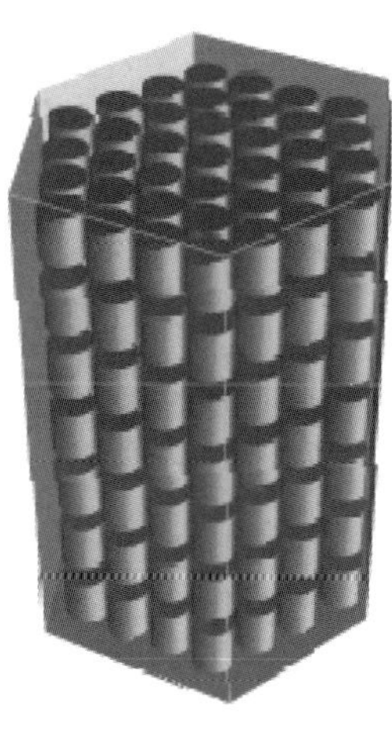

Fig. 10. Majorana.

atmospheres corresponding to 2000 kg, enriched to 80% in ^{136}Xe. The experiment is likely to be located in the DOE WIPP site in Carlsbad (New Mexico).

5. Conclusions

Present constraints (or the indication) on $\langle m_\nu \rangle$ are at the level of a fraction of eV while neutrino oscillations demand sensitivities of a few tens of meV, requiring therefore an improvement of an order of magnitude. We note that:

– The sensitivity on $\langle m_\nu \rangle$ depends linearly on the nuclear matrix element and on the square root of the inverse of the limit on half-life.

– If no evidence is observed, the limit on lifetime depends linearly on isotopic abundance, and on the square root of effective mass, measurement time and inverse of the energy

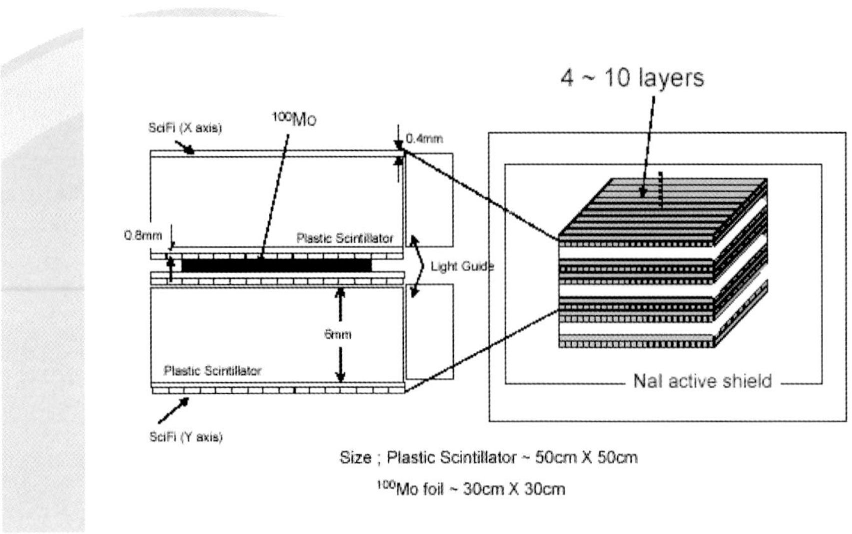

Fig. 11. MOON.

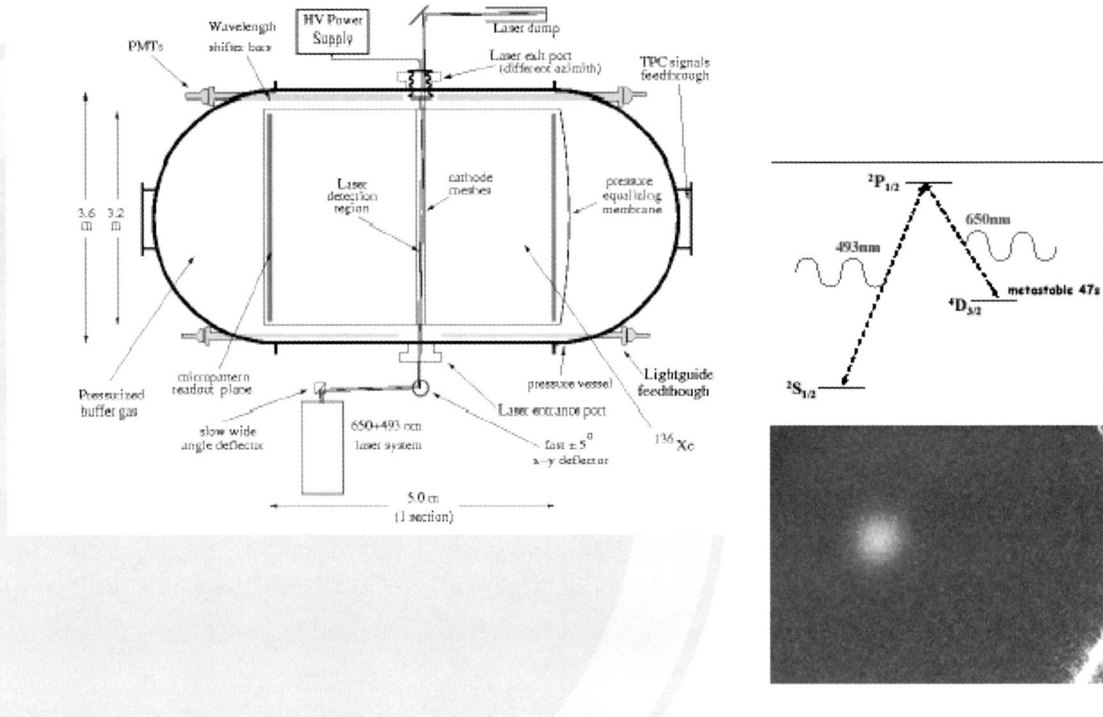

Fig. 12. EXO.

resolution and of the background. As a consequence one needs to:

- Increase the isotopic abundance (expensive, slow, risky for the background).
- Increase the mass (expensive and technically complicated).
- Increase the time of measurement (tedious).

- Increase the energy resolution (already at the level of a few keV).
- Decrease the background (already at $\sim$0.1 counts^{-1} keV^{-1} kg^{-1} y^{-1}).

Despite all these difficulties, the discovery potential of searches for neutrinoless DBD is such as to induce more and more experimental physicists to enter in this multidisciplinary challenge which involves astroparticle and nuclear physics, material science, cosmo-chronology, low radioactivity etc.

References

1. Petkov, S., Theoretical prospects of neutrino double beta decay, in Neutrino Physics, Proc. of a Nobel Symposium, Enköping (Sweden), August 19–24, (2004).
2. Suzuki, Y., Super-Kamiokande results, Proc. of a Nobel Symposium, Enköping (Sweden), August 19–24, (2004).
3. McDonald, A., SNO results, Proc. of a Nobel Symposium, Enköping (Sweden), August 19–24, (2004).
4. Suzuki, A., KamLand results, Proc. of a Nobel Symposium, Enköping (Sweden), August 19–24, (2004).
5. Tretyak, V. and Zdesenko, Y., At. Nucl. Data Tables **80**, 83 (2002).
6. Elliot, S. and Engel, J., J. Phys. G: Nucl. Part. Phys. **30**, R 183 (2004).
7. Fiorini, E. and Niinikoski, T., Nucl. Instr. Meth. **224**, 83 (1984).
8. For recent results of thermal detectors see the reports to the Workshop LTD10 on Low temperature detectors, Nucl. Instr. Meth. **A 520** (2004).
9. Klapdor-Kleingrothaus, H. V., Krivosheina, I. V., Dietz, A. and Chkvoretz, O., Phys. Lett. **B 586**, 198 (2004).
10. Zdesenko, Yu. G., Danevich, F. A. and Tretyak, V. I., Phys. Lett. **B 546**, 206 (2002) and references therein.
11. Zarazin, X., NEMO 3: first results, report to the International Conference "Neutrino 2004", Paris, June (2004), in press.
12. Fiorini, E. (on behalf of the CUORE collaboration): Results of CUORICINO and Perspective for CUORE, report to the International Conference "Neutrino 2004", Paris, June (2004), in press.

Physica Scripta. Vol. T121, 94–101, 2005

Theoretical Prospects of Neutrinoless Double Beta Decay

S. T. Petcov[1]*

[1] Scuola Internazionale Superiore di Studi Avanzati and INFN,I-34014 Trieste, Italy

Received February 10, 2005; accepted in revised form February 18, 2005

PACS numbers: 14.60.Pq, 23.40.−s

Abstract

The compelling experimental evidences for oscillations of solar and atmospheric neutrinos imply the existence of 3-neutrino mixing in vacuum. We briefly review the phenomenology of 3-ν mixing, and the current data on the 3-neutrino mixing parameters. The open questions and the main goals of future research in the field of neutrino mixing and oscillations are outlined. The predictions for the effective Majorana mass $|\langle m \rangle|$ in $(\beta\beta)_{0\nu}$-decay in the case of 3-ν mixing and massive Majorana neutrinos are reviewed. The physics potential of the experiments, searching for $(\beta\beta)_{0\nu}$-decay and having sensitivity to $|\langle m \rangle| \gtrsim 0.01\,\mathrm{eV}$, for providing information on the type of the neutrino mass spectrum, on the absolute scale of neutrino masses and on the Majorana CP-violation phases in the PMNS neutrino mixing matrix, is discussed.

1. Introduction

There has been a remarkable progress in the studies of neutrino oscillations in the last several years. The experiments with solar, atmospheric and reactor neutrinos [1, 2, 3, 4, 5, 6, 7] have provided compelling evidences for the existence of neutrino oscillations driven by nonzero neutrino masses and neutrino mixing. Evidences for oscillations of neutrinos were obtained also in the first long baseline accelerator neutrino experiment K2K [8].

The idea of neutrino oscillations was formulated in [9]. It was predicted in 1967 [10] that the existence of ν_e oscillations would cause a "disappearance" of solar ν_e on the way to the Earth. The hypothesis of solar ν_e oscillations, which (in one variety or another) were considered from ~ 1970 on as the most natural explanation of the observed [1] solar ν_e deficit (see, e.g., refs. [11, 12, 13, 14]), has received a convincing confirmation from the measurement of the solar neutrino flux through the neutral current reaction on deuterium by the SNO experiment [5], and by the first results of the KamLAND (KL) experiment [7]. The combined analysis of the solar neutrino data obtained by Homestake, SAGE, GALLEX/GNO, Super-Kamiokande (SK) and SNO experiments, and of the KL reactor $\bar{\nu}_e$ data [7], established the large mixing angle (LMA) MSW oscillations/transitions [12] as the dominant mechanism at the origin of the observed solar ν_e deficit (see, e.g., [15]). The Kamiokande experiment [16] provided the first evidences for oscillations of atmospheric ν_μ and $\bar{\nu}_\mu$, while the data of the SK experiment made the case of atmospheric neutrino oscillations convincing [3, 6]. Indications for ν-oscillations were reported by the LSND collaboration [17].

The latest contributions to these magnificent progress are the new SK data on the L/E-dependence of the μ-like atmospheric neutrino events [18], L and E being the distance traveled by neutrinos and the neutrino energy, and the new spectrum data

of the KL and K2K experiments [19, 20]. For the first time the data exhibit directly the effects of the oscillatory dependence on L/E and E of the probabilities of ν-oscillations in vacuum [21]. As a result of these developments, the oscillations of solar ν_e, atmospheric ν_μ and $\bar{\nu}_\mu$, accelerator ν_μ (at $L \sim 250\,\mathrm{km}$) and reactor $\bar{\nu}_e$ (at $L \sim 180\,\mathrm{km}$), driven by nonzero ν-masses and ν-mixing, can be considered as practically established.

2. The neutrino mixing parameters and $(\beta\beta)_{0\nu}$-decay

The SK atmospheric neutrino and K2K data are best described in terms of dominant 2-neutrino $\nu_\mu \to \nu_\tau$ ($\bar{\nu}_\mu \to \bar{\nu}_\tau$) vacuum oscillations. The best fit values and the 99.73% C.L. allowed ranges of the atmospheric neutrino oscillation parameters read [3]:

$$|\Delta m^2_{\mathrm{A}}| = 2.1 \times 10^{-3}\,\mathrm{eV}^2, \quad \sin^2 2\theta_{\mathrm{A}} = 1.0;$$

$$|\Delta m^2_{\mathrm{A}}| = (1.3\text{–}4.2) \times 10^{-3}\,\mathrm{eV}^2, \quad \sin^2 2\theta_{\mathrm{A}} \geq 0.85. \tag{1}$$

It should be noted that the signs of Δm^2_{A} and of $\cos 2\theta_{\mathrm{A}}$, if $\sin^2 2\theta_{\mathrm{A}} \neq 1.0$, cannot be determined using the existing data.

Combined 2-neutrino oscillation analyses of the solar neutrino and the new KL spectrum data show [19, 22] that the $\nu_\odot$-oscillation parameters lie in the low-LMA region: $\Delta m^2_\odot = (7.9^{+0.6}_{-0.5}) \times 10^{-5}\,\mathrm{eV}^2$, $\tan^2 \theta_\odot = (0.40^{+0.09}_{-0.07})$. The high-LMA solution (see, e.g., [15]) is excluded at $\sim 3.3\sigma$. Maximal $\nu_\odot$-mixing is ruled out at $\sim 6\sigma$; at 95% C.L. one finds $\cos 2\theta_\odot \geq 0.28$ [22], which has important implications (see further). One also has: $\Delta m^2_\odot / |\Delta m^2_{\mathrm{A}}| \sim 0.04 \ll 1$.

The evidences for ν-oscillations obtained in the solar and atmospheric neutrino and KL and K2K experiments imply the existence of 3-ν mixing in the weak charged lepton current:

$$\nu_{l\mathrm{L}} = \sum_{j=1}^{3} U_{lj}\, \nu_{j\mathrm{L}}, \qquad l = \mathrm{e}, \mu, \tau, \tag{2}$$

where $\nu_{l L}$ are the flavour neutrino fields, $\nu_{j\mathrm{L}}$ is the field of neutrino ν_j having a mass m_j and U is the Pontecorvo-Maki-Nakagawa-Sakata (PMNS) mixing matrix [9, 23], $U \equiv U_{\mathrm{PMNS}}$. All existing ν-oscillation data, except the data of LSND experiment [17], can be described assuming 3-ν mixing in vacuum and we will consider this possibility in what follows.[1]

The PMNS matrix can be parametrized by 3 angles, and, depending on whether the massive neutrinos ν_j are Dirac or

---

*Also at: Institute of Nuclear Research and Nuclear Energy, Bulgarian Academy of Sciences, 1784 Sofia, Bulgaria.

---

[1] In the LSND experiment indications for $\bar{\nu}_\mu \to \bar{\nu}_e$ oscillations with $(\Delta m^2)_{\mathrm{LSND}} \simeq 1\,\mathrm{eV}^2$ were obtained. The minimal 4-ν mixing scheme which could incorporate the LSND indications for $\bar{\nu}_\mu$ oscillations is strongly disfavored by the data [24]. The ν-oscillation explanation of the LSND results is possible assuming 5-ν mixing [25]. The LSND results are being tested in the MiniBooNE experiment [25].

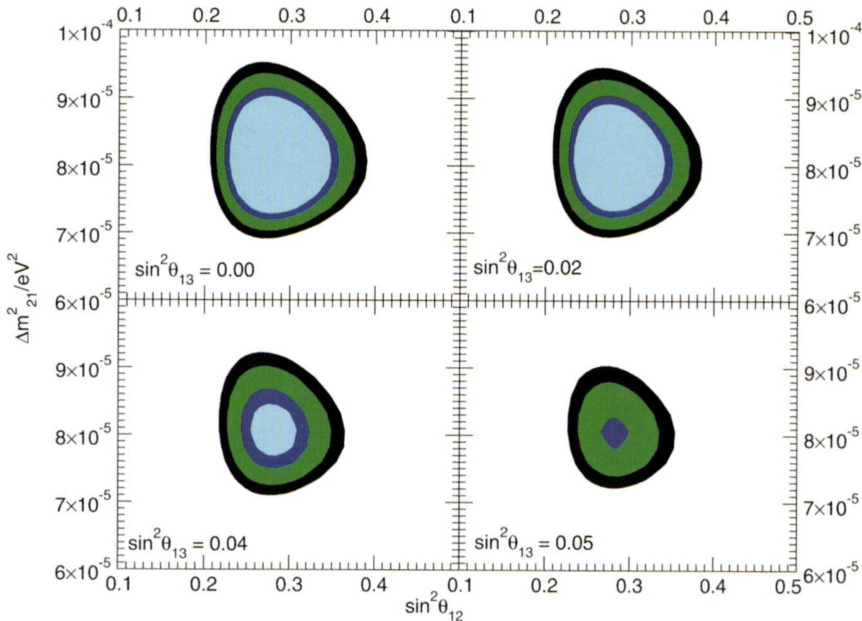

Fig. 1. The 90%, 95%, 99% and 99.73% C.L. allowed regions in the $\Delta m^2_{21} - \sin^2\theta_{12}$ plane, obtained in a 3-ν oscillation analysis of the solar neutrino, KL and CHOOZ data [22].

Majorana particles, by 1 or 3 CP-violation (*CPV*) phases [26, 27]. In the standardly used parameterization (see, e.g., [28]), U_{PMNS} has the form:

$$U_{\mathrm{PMNS}}$$

$$= \begin{pmatrix} c_{12}c_{13} & s_{12}c_{13} & s_{13} \\ -s_{12}c_{23} - c_{12}s_{23}s_{13}e^{i\delta} & c_{12}c_{23} - s_{12}s_{23}s_{13}e^{i\delta} & s_{23}c_{13}e^{i\delta} \\ s_{12}s_{23} - c_{12}c_{23}s_{13}e^{i\delta} & -c_{12}s_{23} - s_{12}c_{23}s_{13}e^{i\delta} & c_{23}c_{13}e^{i\delta} \end{pmatrix}$$

$$\times \mathrm{diag}(1, e^{i\frac{\alpha_{21}}{2}}, e^{i\frac{\alpha_{31}}{2}}), \qquad (3)$$

where $c_{ij} = \cos\theta_{ij}$, $s_{ij} = \sin\theta_{ij}$, the angles $\theta_{ij} = [0, \pi/2]$, $\delta = [0, 2\pi]$ is the Dirac *CPV* phase and α_{21}, α_{31} are two Majorana *CPV* phases [26, 27]. One can identify $\Delta m^2_\odot = \Delta m^2_{21} > 0$. In this case $|\Delta m^2_A| = |\Delta m^2_{31}| \cong |\Delta m^2_{32}| \gg \Delta m^2_{21}$, $\theta_{12} = \theta_\odot$, $\theta_{23} = \theta_A$. The angle θ_{13} is limited by the data from the CHOOZ and Palo Verde experiments [29]. The presently existing atmospheric neutrino data are essentially insensitive to θ_{13} satisfying the CHOOZ limit [3]. The probabilities of survival of solar ν_e and reactor $\bar\nu_e$, relevant for the interpretation of the solar neutrino, KL and CHOOZ data, depend in the case of interest, $|\Delta m^2_{31}| \gg \Delta m^2_{21}$, on θ_{13}:

$$P^{3\nu}_{\mathrm{KL}} \cong \sin^4\theta_{13} + \cos^4\theta_{13}\left[1 - \sin^2 2\theta_{12}\sin^2\frac{\Delta m^2_{21}L}{4E}\right],$$

$$P^{3\nu}_{\mathrm{CHOOZ}} \cong 1 - \sin^2 2\theta_{13}\sin^2\frac{\Delta m^2_{31}L}{4E},$$

$$P^{3\nu}_\odot \cong \sin^4\theta_{13} + \cos^4\theta_{13}\, P^{2\nu}_\odot(\Delta m^2_{21}, \theta_{12}; N_e \cos^2\theta_{13}),$$

where $P^{2\nu}_\odot$ is the solar ν_e survival probability [30] corresponding to 2-ν oscillations driven by Δm^2_{21} and θ_{12}, in which the solar e^- number density N_e is replaced by $N_e \cos^2\theta_{13}$ [31], $P^{2\nu}_\odot = \bar{P}^{2\nu}_\odot + P^{2\nu}_{\odot\,\mathrm{osc}}$, $P^{2\nu}_{\odot\,\mathrm{osc}}$ being an oscillating term [30] and

$$\bar{P}^{2\nu}_\odot = \frac{1}{2} + (\frac{1}{2} - P')\cos 2\theta^m_{12}(t_0)\cos 2\theta_{12},$$

$$P' = \frac{e^{-2\pi r_0 \frac{\Delta m^2_{21}}{2E}\sin^2\theta_{12}} - e^{-2\pi r_0 \frac{\Delta m^2_{21}}{2E}}}{1 - e^{-2\pi r_0 \frac{\Delta m^2_{21}}{2E}}}. \qquad (4)$$

Here $\bar{P}^{2\nu}_\odot$ is the average probability [30, 32], P' is the "double exponential" jump probability [30] and r_0 is the "running" scale-height of the change of N_e along the ν-trajectory in the Sun^2 [30, 33, 34]. In the LMA solution region $P^{2\nu}_{\odot\,\mathrm{osc}} \cong 0$ [34]. Using the expressions for $P^{3\nu}_{\mathrm{KL}}$, $P^{3\nu}_{\mathrm{CHOOZ}}$ and $P^{3\nu}_\odot$ given above, the 3σ allowed range of $|\Delta m^2_A|$ from [3], and performing a combined analysis of the solar neutrino, CHOOZ and KL data, one finds [22]:

$$\sin^2\theta_{13} < 0.055, \qquad 99.73\%\ \mathrm{C.L.} \qquad (5)$$

Similar constraint is obtained from a global 3-ν oscillation analysis of solar, atmospheric and reactor neutrino data [24, 36]. A combined 3-ν oscillation analysis of the solar neutrino, CHOOZ and KL data show also [22] that for $\sin^2\theta_{13} \lesssim 0.02$ the allowed ranges of Δm^2_{21} and $\sin^2\theta_{21}$ do not differ substantially from those derived in the 2-ν oscillation analyzes (see, e.g., ref. [19]). The best fit values, e.g., read^3 [22]:

$$\Delta m^2_{21} = 8.0 \times 10^{-5}\,\mathrm{eV}^2, \qquad \sin^2\theta_{21} = 0.28. \qquad (6)$$

In Fig. 1 we show the allowed regions in the $\Delta m^2_{21} - \sin^2\theta_{12}$ plane, obtained in a 3-ν oscillation analysis of the solar neutrino, KL and CHOOZ data for few fixed values of $\sin^2\theta_{13}$.

As we have seen, the fundamental parameters characterizing the 3-neutrino mixing are: *i*) the 3 angles θ_{12}, θ_{23}, θ_{13}, *ii*) depending on the nature of $\nu_j - 1$ Dirac (δ), or 1 Dirac + 2 Majorana ($\delta, \alpha_{21}, \alpha_{31}$), *CPV* phases, and *iii*) the 3 neutrino masses, m_1, m_2, m_3. It is convenient to express the two larger masses in terms of the third mass and the measured $\Delta m^2_\odot = \Delta m^2_{21} > 0$ and

---

^{2}The analyses and the extensive numerical studies performed in [33, 34] show that expression (4) for $\bar{P}^{2\nu}_\odot$ provides a high precision description of the average solar ν_e survival probability in the Sun for any values of Δm^2_{21} and θ_{12} (the relevant error does not exceed $\sim$(2–3)%), including the values from the LMA region. The results obtained recently in [35] imply actually that the use of the double exponential expression for P' for description of the LMA transitions brings an imprecision in $\bar{P}^{2\nu}_\odot$ which does not exceed $\sim 10^{-6}$.

^{3}The best fit value of $\sin^2\theta_{13} = 0.004$ is different from zero but not at statistically significant level [22].

$\Delta m_{\rm A}^2$. We have remarked earlier that the atmospheric neutrino and K2K data do not allow one to determine the sign of $\Delta m_{\rm A}^2$. This implies that if we identify $\Delta m_{\rm A}^2$ with $\Delta m_{31(2)}^2$ in the case of 3-neutrino mixing, one can have $\Delta m_{31(2)}^2 > 0$ or $\Delta m_{31(2)}^2 < 0$. The two possible signs of $\Delta m_{\rm A}^2$ correspond to two types of ν-mass spectrum:

– *with normal hierarchy*, $m_1 < m_2 < m_3$, $\Delta m_{\rm A}^2 = \Delta m_{31}^2 > 0$, $m_{2(3)} = (m_1^2 + \Delta m_{21(31)}^2)^{\frac{1}{2}}$, and

– *with inverted hierarchy*[4], $m_3 < m_1 < m_2$, $\Delta m_{\rm A}^2 = \Delta m_{32}^2 < 0$, $m_2 = (m_3^2 - \Delta m_{32}^2)^{\frac{1}{2}}$, etc.

The neutrino mass spectrum can also be

– *Normal Hierarchical (NH)*: $m_1 \ll m_2 \ll m_3$, $m_2 \cong (\Delta m_\odot^2)^{\frac{1}{2}}$ $\sim 0.009\,{\rm eV}$, $m_3 \cong |\Delta m_{\rm A}^2|^{\frac{1}{2}}$; or

– *Inverted Hierarchical (IH)*: $m_3 \ll m_1 < m_2$, with $m_{1,2} \cong |\Delta m_{\rm A}^2|^{\frac{1}{2}} \sim 0.045\,{\rm eV}$; or

– *Quasi-Degenerate (QD)*: $m_1 \cong m_2 \cong m_3 \cong m_0$, $m_j^2 \gg |\Delta m_{\rm A}^2|$, $m_0 \gtrsim 0.20\,{\rm eV}$.

Neutrino oscillation experiments allow to determine differences of squares of neutrino masses, but not the absolute values of the masses, or $min(m_j)$. Information on the absolute scale of ν-masses can be derived in ${}^3{\rm H}$ β-decay experiments [39, 40, 41] and from cosmological and astrophysical data (see, e.g., ref. [42]). The currently existing most stringent upper bounds on the $\bar\nu_e$ mass were obtained in the Troitzk [40] and Mainz [41] experiments:

$$m_{\bar\nu_e} < 2.3\,{\rm eV} \quad (95\%\,{\rm C.L.}). \tag{7}$$

We have $m_{\bar\nu_e} \cong m_{1,2,3}$ in the case of QD ν-mass spectrum. The KATRIN experiment [41] is planned to reach a sensitivity to $m_{\bar\nu_e} \sim 0.20\,{\rm eV}$, i.e., to probe the region of the QD spectrum. The CMB data of the WMAP experiment were used to obtain the upper limit [43]:

$$\sum_j m_j < (0.7\text{–}2.0)\,{\rm eV} \quad (95\%\,{\rm C.L.}), \tag{8}$$

where we have included a conservative estimate of the uncertainty in the upper limit (see, e.g., [42]). The WMAP and future PLANCK experiments can be sensitive to $\sum_j m_j \cong 0.4\,{\rm eV}$. Data on weak lensing of galaxies by large scale structure, combined with data from the WMAP and PLANCK experiments may allow one to determine $(m_1 + m_2 + m_3)$ with an uncertainty of $\delta \sim (0.04\text{–}0.10)\,{\rm eV}$ (see [42] and the references quoted therein).

The type of neutrino mass spectrum, i.e., $sgn(\Delta m_{31}^2)$, can be determined by studying oscillations of neutrinos and antineutrinos, say, $\nu_\mu \leftrightarrow \nu_e$ and $\bar\nu_\mu \leftrightarrow \bar\nu_e$, in which matter effects are sufficiently large. This can be done in long base-line ν-oscillation experiments (see, e.g., [44]). If $\sin^2 2\theta_{13} \gtrsim 0.05$ and $\sin^2 \theta_{23} \gtrsim 0.50$, information on $sgn(\Delta m_{31}^2)$ might be obtained in atmospheric neutrino experiments by investigating the effects of the subdominant transitions $\nu_{\mu(e)} \to \nu_{e(\mu)}$ and $\bar\nu_{\mu(e)} \to \bar\nu_{e(\mu)}$ of atmospheric neutrinos which traverse the Earth [45]. For $\nu_{\mu(e)}$ (or $\bar\nu_{\mu(e)}$) crossing the Earth core, new type of resonance-like enhancement of the indicated transitions takes place due to the

(Earth) mantle-core constructive interference effect (neutrino oscillation length resonance (NOLR)) [46][5]. As a consequence of this effect[6] the corresponding $\nu_{\mu(e)}$ (*or* $\bar\nu_{\mu(e)}$) transition probabilities can be maximal [47]. For $\Delta m_{31}^2 > 0$, the neutrino transitions $\nu_{\mu(e)} \to \nu_{e(\mu)}$ are enhanced, while for $\Delta m_{31}^2 < 0$ the enhancement of antineutrino transitions $\bar\nu_{\mu(e)} \to \bar\nu_{e(\mu)}$ takes place, which might allow to determine $sgn(\Delta m_{31}^2)$.

After the spectacular experimental progress made in the studies of ν-oscillations, further understanding of the structure of neutrino mixing, of its origins and of the status of CP-symmetry in the lepton sector, requires a large and challenging program of research to be pursued in neutrino physics. The main goals of this research program should include [50]:

– High precision measurement of neutrino mixing parameters which control the solar and the dominant atmospheric neutrino oscillations, Δm_{21}^2, θ_{21}, and Δm_{31}^2, θ_{23}.

– Measurement of, or improving by at least a factor of (5–10) the existing upper limit on, θ_{13} – the only small mixing angle in $U_{\rm PMNS}$.

– Determination of the $sgn(\Delta m_{31}^2)$ and of the type of ν-mass spectrum (NH, IH, QD, etc.).

– Determination or obtaining significant constraints on the absolute scale of ν-masses.

– Determination of the nature – Dirac or Majorana, of massive neutrinos ν_j.

– Establishing whether the CP-symmetry is violated in the lepton sector a) due to the Dirac phase δ, and/or b) due to the Majorana phases α_{21} and α_{31} if ν_j are Majorana particles.

– Searching with increased sensitivity for possible manifestations, other than flavour neutrino oscillations, of the non-conservation of the individual lepton charges L_l, $l = {\rm e}, \mu, \tau$, such as $\mu \to {\rm e} + \gamma$, $\tau \to \mu + \gamma$, etc. decays.

– Understanding at fundamental level the mechanism giving rise to neutrino masses and mixing and to L_l−non-conservation. This includes understanding the origin of the patterns of ν-mixing and ν-masses suggested by the data. Are the observed patterns of ν-mixing and of $\Delta m_{21,31}^2$ related to the existence of new symmetry of particle interactions? Is there any relations between quark mixing and neutrino mixing? Is $\theta_{23} = \pi/4$, or $\theta_{23} > \pi/4$ or else $\theta_{23} < \pi/4$? Is there any correlation between the values of CPV phases and of mixing angles in $U_{\rm PMNS}$? Progress in the theory of neutrino mixing might also lead, in particular, to a better understanding of the origin of baryon asymmetry of the Universe [57].

The mixing angles, θ_{21}, θ_{23} and θ_{13}, the Dirac CPV phase δ and Δm_{21}^2 and Δm_{31}^2 can, in principle, be measured with a sufficiently high precision in ν-oscillation experiments (see, e.g., [44, 51]). These experiments, however, cannot provide information on the ν-mass scale and on the nature of massive neutrinos ν_j, they are insensitive to the Majorana CPV phases $\alpha_{21,31}$ [26, 52]. Establishing whether ν_j have distinct antiparticles

---

[4]In the convention we use (called A), the neutrino masses are not ordered in magnitude according to their index number: $\Delta m_{31}^2 < 0$ corresponds to $m_3 < m_1 < m_2$. We can also always number the neutrinos with definite mass in such a way that [37] $m_1 < m_2 < m_3$. In this convention (called B), we have in the case of inverted hierarchy spectrum: $\Delta m_\odot^2 = \Delta m_{32}^2$, $\Delta m_{\rm A}^2 = \Delta m_{31}^2$. Convention B is used, e.g., in [28, 38].

[5]For the precise conditions of the mantle-core (NOLR) enhancement see [46, 47].
[6]The Earth mantle-core (NOLR) enhancement of neutrino transitions differs [46] from the MSW one. It also differs [46, 47] from the parametric resonance mechanisms of enhancement discussed in [48]: the conditions of enhancement found in [48] are not realized for the neutrino transitions in the Earth. In [49] it is erroneously concluded that the $\nu_{\mu(e)} \leftrightarrow \nu_{e(\mu)}$ transitions of atmospheric neutrinos crossing the Earth core cannot be enhanced by the interplay of the transitions in the Earth mantle and of those in the Earth core.

(Dirac fermions) or not (Majorana fermions) is of fundamental importance for understanding the underlying symmetries of particle interactions and the origin of v-masses. If v_j are Majorana fermions, getting experimental information about the Majorana CPV phases in U_{PMNS} would be a remarkably challenging problem [28, 37, 53, 54, 55]. The phases $\alpha_{21,31}$ can affect significantly the predictions for the rates of (LFV) decays $\mu \to e + \gamma$, $\tau \to \mu + \gamma$, etc. in a large class of supersymmetric theories with see-saw mechanism of v-mass generation (see, e.g., [56]). Majorana CPV phases might be at the origin of the baryon asymmetry of the Universe [57].

In the present article we will review the potential contribution the studies of neutrinoless double beta $((\beta\beta)_{0v}$-) decay of even-even nuclei, $(A, Z) \to (A, Z + 2) + e^- + e^-$, can make to the program of research outlined above. The $(\beta\beta)_{0v}$-decay is allowed if the neutrinos with definite mass are Majorana particles (for reviews see, e.g., [13, 58, 59, 60]). Let us recall that the nature – Dirac or Majorana, of the massive neutrinos v_j, is related to the symmetries of particle interactions. Neutrinos v_j will be Dirac fermions if the particle interactions conserve some lepton charge, e.g., the total lepton charge L. If there does not exist any conserved lepton charge, neutrinos with definite mass can be Majorana particles. As is well-known, the massive neutrinos are predicted to be of Majorana nature by the see-saw mechanism of neutrino mass generation (see, e.g., [61]), which also provides a very attractive explanation of the smallness of the neutrino masses and – through the leptogenesis theory [57], of the observed baryon asymmetry of the Universe.

If the massive neutrinos v_j are Majorana fermions, processes in which the total lepton charge L is not conserved and changes by two units, such as $K^+ \to \pi^- + \mu^+ + \mu^+$, $\mu^+ + (A, Z) \to (A, Z + 2) + \mu^-$, etc., should exist. The process most sensitive to the possible Majorana nature of neutrinos v_j is the $(\beta\beta)_{0v}$-decay (see, e.g., [13, 62]). If the $A(\beta\beta)_{0v}$-decay is generated *only by the (V-A) charged current weak interaction via the exchange of the three Majorana neutrinos* v_j $(m_j \lesssim 1\,\mathrm{eV})$, which will be assumed in what follows, the dependence of the $(\beta\beta)_{0v}$-decay amplitude $A(\beta\beta)_{0v}$ on the neutrino mass and mixing parameters factorizes in the effective Majorana mass $\langle m \rangle$ (see, e.g., [13, 59]):

$$A(\beta\beta)_{0v} \sim \langle m \rangle M, \tag{9}$$

where M is the corresponding nuclear matrix element (NME) and

$$|\langle m \rangle| = |m_1|U_{e1}|^2 + m_2|U_{e2}|^2\,\mathrm{e}^{i\alpha_{21}} + m_3|U_{e3}|^2\,\mathrm{e}^{i\alpha_{31}}|. \tag{10}$$

If CP-invariance holds[7], one has [63] $\alpha_{21} = k\pi$, $\alpha_{31} = k'\pi$, where $k, k' = 0, 1, 2, \ldots$, and

$$\eta_{21} \equiv \mathrm{e}^{i\alpha_{21}} = \pm 1, \qquad \eta_{31} \equiv \mathrm{e}^{i\alpha_{31}} = \pm 1 \tag{11}$$

represent the relative CP-parities of Majorana neutrinos v_1 and v_2, and v_1 and v_3, respectively. It follows from eq. (10) that the measurement of $|\langle m \rangle|$ will provide information, in particular, on m_j. As eq. (9) indicates, the observation of $(\beta\beta)_{0v}$-decay of a given nucleus and the measurement of the corresponding half-life, would allow to determine $|\langle m \rangle|$ only if the value of the relevant NME is known with a relatively small uncertainty.

The experimental searches for $(\beta\beta)_{0v}$-decay have a long history (see, e.g., [58, 59]).The best sensitivity was achieved in the

Heidelberg-Moscow ^{76}Ge experiment [64]:

$$|\langle m \rangle| < (0.35\text{–}1.05)\,\mathrm{eV}, \quad 90\%\ \mathrm{C.L.} \tag{12}$$

where a factor of 3 uncertainty associated with the calculation of the relevant nuclear matrix element [59] is taken into account. A positive signal at $>3\sigma$, corresponding to $|\langle m \rangle| = (0.1\text{–}0.9)\,\mathrm{eV}$ at 99.73% C.L., is claimed to be observed in [65]. This result will be checked in the currently running and future $(\beta\beta)_{0v}$-decay experiments. However, it may take a long time before comprehensive checks could be completed. Two experiments, NEMO3 (with ^{100}Mo and ^{82}Se) [66] and CUORICINO (with ^{130}Te) [67], designed to reach sensitivity to $|\langle m \rangle| \sim (0.2\text{–}0.3)\,\mathrm{eV}$, are taking data. Their first results read, respectively (90% C.L.):

$$|\langle m \rangle| < (0.7\text{–}1.2)\,\mathrm{eV}\ [66], \qquad |\langle m \rangle| < (0.3\text{–}1.6)\,\mathrm{eV}\ [67], \tag{13}$$

where estimated uncertainties in the NME are accounted for. A number of projects aim to reach sensitivity to $|\langle m \rangle| \sim (0.01\text{–}0.05)\,\mathrm{eV}$ [62, 67]: CUORE (^{130}Te), GENIUS (^{76}Ge), EXO (^{136}Xe), MAJORANA (^{76}Ge), MOON (^{100}Mo), XMASS (^{136}Xe), etc. These experiments can probe the region of IH and QD spectra and test the positive result claimed in [65].

The $(\beta\beta)_{0v}$-decay experiments are presently the only feasible experiments capable of establishing the Majorana nature of massive neutrinos [13, 59, 62]. As we will discuss in what follows, a measurement of a nonzero value of $|\langle m \rangle| \gtrsim 10^{-2}\,\mathrm{eV}$:

- Can give also information on neutrino mass spectrum [38, 68, 69] (see also [70]).

- Can provide unique information on the absolute scale of neutrino masses (see, e.g., [68]).

- With additional information from other sources (^{3}H β-decay experiments and/or cosmological/astrophysical data) on the absolute v-mass scale, can provide unique information on the Majorana CPV phases α_{21} and/or α_{31} [28, 37, 55, 68, 71].

3. Properties of Majorana neutrinos: brief summary

The properties of Majorana fields (particles) are very different from those of Dirac fields (particles). A massive Majorana neutrino χ_k can be described (in local quantum field theory) by 4-component complex spin 1/2 field $\chi_k(x)$ which satisfies the Majorana condition:

$$C\,(\bar{\chi}_k(x))^{\mathrm{T}} = \xi_k \chi_k(x), \qquad |\xi_k|^2 = 1. \tag{14}$$

where C is the charge conjugation matrix. The Majorana condition is invariant under *proper* Lorentz transformations. It reduces by 2 the number of independent components in $\chi_k(x)$.

The condition (14) is invariant with respect to $U(1)$ global gauge transformations of the field $\chi_k(x)$ carrying a $U(1)$ charge Q, $\chi_k(x) \to \mathrm{e}^{i\alpha Q}\chi_k(x)$, only if $Q = 0$. As a result, *i*) χ_k cannot carry nonzero additive quantum numbers (lepton charge, etc.), and *ii*) the field $\chi_k(x)$ cannot "absorb" phases[8]. Thus, $\chi_k(x)$ describes 2 spin states of a spin 1/2, *absolutely neutral particle*, which is identical with its antiparticle, $\chi_k \equiv \bar{\chi}_k$. If CP-invariance holds, Majorana neutrinos have definite CP-parity $\eta_{CP}(\chi_k) = \pm i$:

$$U_{CP}\,\chi_k(x)\,U_{CP}^{-1} = \eta_{CP}(\chi_k)\,\gamma_0\,\chi_k(x'), \qquad \eta_{CP}(\chi_k) = \pm i. \tag{15}$$

It follows from the Majorana condition that the currents $\bar{\chi}_k(x)O^i\chi_k(x) \equiv 0$, for $O^i = \gamma_\alpha$; $\sigma_{\alpha\beta}$; $\sigma_{\alpha\beta}\gamma_5$. This means that

---

[7]We assume that $m_j > 0$ and that the fields of the Majorana neutrinos v_j satisfy the Majorana condition: $C(\bar{v}_j)^T = v_j$, $j = 1, 2, 3$, where C is the charge conjugation matrix.

[8]This is the reason why the *PMNS* matrix contains two additional CPV phases in the case of massive Majorana neutrinos [26].

Majorana fermions (neutrinos) cannot have nonzero $U(1)$ charges and intrinsic magnetic and electric dipole moments.

Finally, if $\Psi(x)$ is a Dirac field and we define the standard propagator of $\Psi(x)$ as

$$\langle 0|T(\Psi_\alpha(x)\bar{\Psi}_\beta(y))|0\rangle = S^F_{\alpha\beta}(x-y), \tag{16}$$

one has

$$\langle 0|T(\Psi_\alpha(x)\Psi_\beta(y))|0\rangle = 0, \qquad \langle 0|T(\bar{\Psi}_\alpha(x)\bar{\Psi}_\beta(y))|0\rangle = 0. \tag{17}$$

In contrast, a Majorana neutrino field $\chi_k(x)$ has, in addition to the standard propagator

$$\langle 0|T(\chi_{k\alpha}(x)\bar{\chi}_{k\beta}(y))|0\rangle = S^{Fk}_{\alpha\beta}(x-y), \tag{18}$$

two non-trivial *non-standard (Majorana)* propagators[9]

$$\langle 0|T(\chi_{k\alpha}(x)\chi_{k\beta}(y))|0\rangle = -\xi^* S^{Fk}_{\alpha\delta}(x-y)C_{\delta\beta},$$

$$\langle 0|T(\bar{\chi}_{k\alpha}(x)\bar{\chi}_{k\beta}(y))|0\rangle = \xi\, C^{-1}_{\alpha\delta} S^{Fk}_{\delta\beta}(x-y). \tag{19}$$

This result implies that if $v_j(x)$ in eq. (2) are massive Majorana neutrinos, $(\beta\beta)_{0\nu}$-decay can proceed by exchange of virtual neutrinos v_j since $\langle 0|T(v_{j\alpha}(x)v_{j\beta}(y))|0\rangle \neq 0$.

4. Predictions for the effective Majorana mass

The predicted value of $|\langle m\rangle|$ depends in the case of $3-v$ mixing on [72] (see also [28, 70]): *i)* $\Delta m^2_A = \Delta m^2_{31(32)}$, *ii)* $\theta_\odot = \theta_{12}$ and $\Delta m^2_\odot = \Delta m^2_{21}$, *iii)* the lightest neutrino mass, $min(m_j)$ and on *iv)* the mixing angle θ_{13}. In the convention (A) employed by us, one has $|U_{e1}|^2 = \cos^2\theta_\odot(1-|U_{e3}|^2)$, $|U_{e2}|^2 = \sin^2\theta_\odot(1-|U_{e3}|^2)$, and $|U_{e3}|^2 \equiv \sin^2\theta_{13}$.

Given $\Delta m^2_\odot$, Δm^2_A, $\theta_\odot$ and $\sin^2\theta_{13}$, the value of $|\langle m\rangle|$ depends strongly on the type of the neutrino mass spectrum as well as on the values of the two Majorana CPV phases of the PMNS matrix, $\alpha_{21,31}$ (see eq. (10)). In the case of QD spectrum ($m_1 \cong m_2 \cong m_3 = m_0$, $m_0^2 \gg |\Delta m^2_A|, \Delta m^2_{21}$), $|\langle m\rangle|$ is essentially independent on Δm^2_A and $\Delta m^2_\odot$, and the two possibilities, $\Delta m^2_A > 0$ and $\Delta m^2_A < 0$, lead *effectively* to the same predictions for $|\langle m\rangle|$.

4.1. Normal hierarchical neutrino mass spectrum

In this case one has [28]

$$|\langle m\rangle| = \left|(m_1\cos^2\theta_\odot + e^{i\alpha_{21}}\sqrt{\Delta m^2_\odot}\sin^2\theta_\odot)\cos^2\theta_{13}\right.$$

$$\left.+ \sqrt{\Delta m^2_A}\sin^2\theta_{13}\, e^{i\alpha_{31}}\right| \tag{20}$$

$$\simeq \left|\sqrt{\Delta m^2_\odot}\sin^2\theta_\odot\cos^2\theta_{13} + \sqrt{\Delta m^2_A}\sin^2\theta_{13}e^{i(\alpha_{31}-\alpha_{21})}\right| \tag{21}$$

where we have neglected the term $\sim m_1$ in eq. (21). Although the neutrino v_1 effectively "decouples" and does not contribute to $|\langle m\rangle|$, eq. (21), the value of $|\langle m\rangle|$ still depends on the Majorana CPV phase $\alpha_{32} = \alpha_{31} - \alpha_{21}$. This reflects the fact that in contrast to the case of massive Dirac neutrinos (or quarks), CP-violation can take place in the mixing of only two massive Majorana neutrinos [26]. Further, since [22] $\sqrt{\Delta m^2_\odot} \lesssim 9.5 \times 10^{-3}\,$eV, $\sin^2\theta_\odot \lesssim 0.36$, $\sqrt{\Delta m^2_A} \lesssim 5.4 \times 10^{-2}\,$eV [24] (at 90% C.L.), and the largest neutrino mass enters into the expression for $|\langle m\rangle|$ with the factor $\sin^2\theta_{13} < 0.055$, the predicted value of $|\langle m\rangle|$ is typically $\sim few \times 10^{-3}\,$eV: for $\sin^2\theta_{13} = 0.04\ (0.02)$

---

one finds $|\langle m\rangle| \lesssim 0.005\ (0.004)\,$eV. Using the best fit values of the indicated parameters (see eqs. (1) and (6)) we get $|\langle m\rangle| \lesssim 0.0044\ (0.0035)\,$eV. It follows from eq. (20) and the allowed ranges of values of $\Delta m^2_\odot$, Δm^2_A, $\sin^2\theta_\odot$, $\sin^2\theta_{13}$ as well as of the lightest neutrino mass m_1 and CPV phases $\alpha_{21,31}$ that in the case of spectrum with *normal hierarchy* there can be a complete cancellation between the three terms in eq. (20) and one can have [68] $|\langle m\rangle| = 0$.

4.2. Inverted hierarchical spectrum

For *IH* neutrino mass spectrum (see, e.g., [28]) $m_3 \ll m_1 \cong m_2 \cong \sqrt{|\Delta m^2_A|} = \sqrt{\Delta m^2_{23}}$. Neglecting $m_3\sin^2\theta_{13}$ in eq. (10), we find [37]:

$$|\langle m\rangle| \cong \sqrt{|\Delta m^2_A|}\cos^2\theta_{13}\sqrt{1 - \sin^2 2\theta_\odot\sin^2\frac{\alpha_{21}}{2}}. \tag{22}$$

Even though one of the three massive Majorana neutrinos "decouples", the value of $|\langle m\rangle|$ depends on the Majorana CP-violating phase α_{21}. Obviously,

$$\sqrt{|\Delta m^2_A|}\,|\cos 2\theta_\odot|\,\cos^2\theta_{13} \leq |\langle m\rangle| \leq \sqrt{|\Delta m^2_A|}\cos^2\theta_{13}. \tag{23}$$

The upper and the lower limits correspond respectively to the CP-conserving cases. Most remarkably, since according to the solar neutrino and KamLAND data $\cos 2\theta_\odot \sim 0.40$, we get a significant lower limit on $|\langle m\rangle|$, typically exceeding $10^{-2}\,$eV, in this case [38, 68]. Using, e.g., the best fit values of $|\Delta m^2_A|$ and $\sin^2\theta_\odot$ one finds: $|\langle m\rangle| \gtrsim 0.02\,$eV. The maximal value of $|\langle m\rangle|$ is determined by $|\Delta m^2_A|$ and can reach, as it follows from eqs. (1) and (5), $|\langle m\rangle| \sim 0.060\,$eV. The indicated values of $|\langle m\rangle|$ are within the range of sensitivity of the next generation of $(\beta\beta)_{0\nu}$-decay experiments.

The expression for $|\langle m\rangle|$, eq. (22), permits to relate the value of $\sin^2\alpha_{21}/2$ to the experimentally measured quantities [28, 37] $|\langle m\rangle|$, Δm^2_A and $\sin^2 2\theta_\odot$:

$$\sin^2\frac{\alpha_{21}}{2} \cong \left(1 - \frac{|\langle m\rangle|^2}{|\Delta m^2_A|\cos^4\theta_{13}}\right)\frac{1}{\sin^2 2\theta_\odot}. \tag{24}$$

A sufficiently accurate measurement of $|\langle m\rangle|$ and of $|\Delta m^2_A|$ and $\theta_\odot$, could allow to get information about the value of α_{21}, provided the neutrino mass spectrum is of the *IH* type.

4.3. Three quasi-degenerate neutrinos

In this case $m_0 \equiv m_1 \cong m_2 \cong m_3$, $m_0^2 \gg |\Delta m^2_A|$, $m_0 \gtrsim 0.20\,$eV. The mass m_0 effectively coincides with the $\bar{v}_e$ mass $m_{\bar{v}_e}$ measured in the ^{3}H β-decay experiments: $m_0 = m_{\bar{v}_e}$. Thus, $m_0 < 2.3\,$eV, or if we use a conservative cosmological upper limit [42], $m_0 < 0.7\,$eV. The QD v-mass spectrum is realized for values of m_0, which can be measured in the ^{3}H β-decay experiment KATRIN [41].

The effective Majorana mass $|\langle m\rangle|$ is given by

$$|\langle m\rangle| \cong m_0 \left|(\cos^2\theta_\odot + \sin^2\theta_\odot e^{i\alpha_{21}})\cos^2\theta_{13} + e^{i\alpha_{31}}\sin^2\theta_{13}\right| \tag{25}$$

$$\cong m_0 \left|\cos^2\theta_\odot + \sin^2\theta_\odot e^{i\alpha_{21}}\right| = m_0\sqrt{1 - \sin^2 2\theta_\odot\sin^2\frac{\alpha_{21}}{2}}. \tag{26}$$

Similarly to the case of *IH* spectrum, one has:

$$m_0\,|\cos 2\theta_\odot| \lesssim |\langle m\rangle| \lesssim m_0. \tag{27}$$

---

[9]For further detailed discussion of the properties of Majorana neutrinos (fermions) see, e.g., [13].

For $\cos 2\theta_\odot \sim 0.40$, favored by the data, one finds a non-trivial lower limit on $|\langle m \rangle|$, $|\langle m \rangle| \gtrsim 0.08\,\text{eV}$. For the 90% C.L. allowed ranges of values of the parameters one has $|\langle m \rangle| \gtrsim 0.06\,\text{eV}$. Using the conservative cosmological upper bound on $\sum_j m_j$ we get $|\langle m \rangle| < 0.70\,\text{eV}$. Also in this case one can obtain, in principle, direct information on one CPV phase from the measurement of $|\langle m \rangle|$, m_0 and $\sin^2 2\theta_\odot$:

$$\sin^2 \frac{\alpha_{21}}{2} \cong \left(1 - \frac{|\langle m \rangle|^2}{m_0^2} \right) \frac{1}{\sin^2 2\theta_\odot}. \tag{28}$$

The specific features of the predictions for $|\langle m \rangle|$ in the cases of the three types of neutrino mass spectrum discussed above are evident in Fig. 2, where the dependence of $|\langle m \rangle|$ on $min(m_j)$ for the LMA solution is shown. If the spectrum is with normal hierarchy, $|\langle m \rangle|$ can lie anywhere between 0 and the presently existing upper limits, eqs. (12) and (13). This conclusion does not change even under the most favorable conditions for the determination of $|\langle m \rangle|$, namely, even when $|\Delta m_A^2|$, $\Delta m_\odot^2$, $\theta_\odot$ and θ_{13} are known with negligible uncertainty. If the results in [65] implying $|\langle m \rangle| = (0.1\text{--}0.9)\,\text{eV}$ are confirmed, this would mean, in particular, that the neutrino mass spectrum is of the QD type.

5. Implications of measuring $|\langle m \rangle| \neq 0$

If the $(\beta\beta)_{0\nu}$-decay of a given nucleus would be observed, it would be possible to determine the value of $|\langle m \rangle|$ from the measurement of the associated half-life of the decay. This would require the knowledge of the nuclear matrix element of the process.

5.1. *On the NME uncertainties*

At present there exist large uncertainties in the calculation of the $(\beta\beta)_{0\nu}$-decay nuclear matrix elements (see, e.g., [59]). This is reflected, in particular, in the factor of ~ 3 uncertainty in the upper limit on $|\langle m \rangle|$, which is extracted from the experimental lower limits on the $(\beta\beta)_{0\nu}$-decay half-life of ^{76}Ge. Recently, encouraging results in what regards the problem of the calculation of the nuclear matrix elements have been obtained in [73]. The observation of a $(\beta\beta)_{0\nu}$-decay of one nucleus is likely to lead to search and eventually to observation of the decay of other nuclei. One can expect that such a progress, in particular, will help to solve the problem of the sufficiently precise calculation of the nuclear matrix elements for the $(\beta\beta)_{0\nu}$-decay [75].

5.2. *Constraining the lightest neutrino mass*

As Fig. 2 indicates, a measurement of $|\langle m \rangle| \gtrsim 0.01\,\text{eV}$ would either *i*) determine a relatively narrow interval of possible values of the lightest ν-mass $min(m_j) \equiv m_{\text{MIN}}$, or *ii*) would establish an upper limit on m_{MIN}. If an upper limit on $|\langle m \rangle|$ is experimentally obtained below 0.01 eV, this would lead to a significant upper limit on m_{MIN} and would imply $\Delta m_A^2 > 0$ for massive Majorana neutrinos.

A measurement of $|\langle m \rangle| = (|\langle m \rangle|)_{exp} \gtrsim 0.02\,\text{eV}$ if $\Delta m_A^2 \equiv \Delta m_{31}^2 > 0$, and of $|\langle m \rangle| = (|\langle m \rangle|)_{exp} \gtrsim \sqrt{|\Delta m_A^2|} \cos^2 \theta_{13}$ in the case of $\Delta m_A^2 \equiv \Delta m_{32}^2 < 0$, for instance, would imply that $m_{\text{MIN}} \gtrsim 0.02\,\text{eV}$ and $m_{\text{MIN}} \gtrsim 0.05\,\text{eV}$, respectively, and thus a ν-mass spectrum with *partial hierarchy* or of QD type [28]. The mass m_{MIN} will be constrained to lie in a rather narrow interval [68] (Fig. 2). If the measured value of $|\langle m \rangle|$, $(|\langle m \rangle|)_{exp}$, lies between

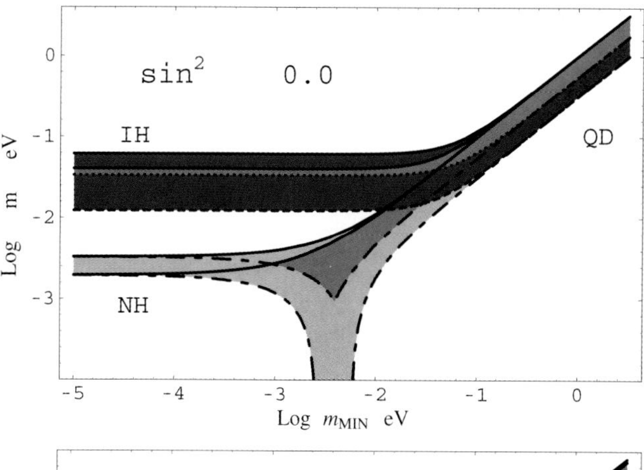

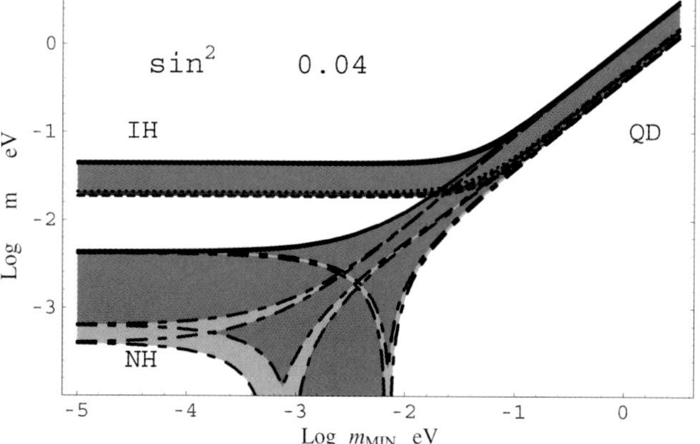

Fig. 2. The value of $|\langle m \rangle|$ as function of $min(m_j)$ for $\sin^2 \theta_{13} = 0.0; 0.04$ and 90% C.L. allowed ranges [22] of Δm_A^2, $\Delta m_\odot^2$ and $\theta_\odot$ (updated version of Fig. 2 from [38]).

the $min(|\langle m \rangle|)$ and $max(|\langle m \rangle|)$, predicted in the case of IH spectrum,

$$|\langle m \rangle|_\pm = \left| \sqrt{|\Delta m_A^2| - \Delta m_\odot^2} \cos^2 \theta_\odot \pm \sqrt{|\Delta m_A^2|} \sin^2 \theta_\odot \right| \cos^2 \theta, \tag{29}$$

m_{MIN} would be limited from above, but $min(m_{\text{MIN}}) = 0$ (Fig. 2). If $(|\langle m \rangle|)_{exp} < (|\langle m \rangle|)_{max}$, where, e.g., in the case of QD spectrum $(|\langle m \rangle|)_{max} \cong m_{\text{MIN}} \cong m_{\bar{\nu}_e}$, this would imply that at least one of the two CPV phases is different from zero: $\alpha_{21} \neq 0$ and/or $\alpha_{31} \neq 0$[10].

A measured value of $m_{\bar{\nu}_e}$, $(m_{\bar{\nu}_e})_{exp} \gtrsim 0.20\,\text{eV}$, satisfying $(m_{\bar{\nu}_e})_{exp} > (m_{\text{MIN}})_{max}$, where $(m_{\text{MIN}})_{max}$ is determined from the upper limit on $|\langle m \rangle|$ in case the $(\beta\beta)_{0\nu}$-decay is not observed, might imply that the massive neutrinos are Dirac particles. If $(\beta\beta)_{0\nu}$-decay has been observed and $|\langle m \rangle|$ measured, the inequality $(m_{\bar{\nu}_e})_{exp} > (m_{\text{MIN}})_{max}$, with $(m_{\text{MIN}})_{max}$ determined from the measured value of $|\langle m \rangle|$, would lead to the conclusion that there exist contribution(s) to the $(\beta\beta)_{0\nu}$-decay rate other than due to the light Majorana neutrino exchange that partially cancels the contribution from the Majorana neutrino exchange.

5.3. *Determining the type of neutrino mass spectrum*

The existence of significant lower bounds on $|\langle m \rangle|$ in the cases of IH and QD spectra [38], which lie either partially (IH spectrum) or completely (QD spectrum) within the range of sensitivity of the next generation of $(\beta\beta)_{0\nu}$-decay experiments, is one of the

---

[10]In general, the knowledge of the value of $|\langle m \rangle|$ alone will not allow to distinguish the case of CP-conservation from that of CP-violation.

most important features of the predictions of $|\langle m \rangle|$. These minimal values are given, up to small corrections, by $\sqrt{|\Delta m_{\rm A}^2|} \cos 2\theta_\odot$ and $m_0 \cos 2\theta_\odot$. According to the combined analysis of the solar and reactor neutrino data [22] *i*) $\cos 2\theta_\odot = 0$ is excluded at $\sim 6\sigma$, *ii*) the best fit value of $\cos 2\theta_\odot$ is $\cos 2\theta_\odot = 0.44$, and *iii*) at 95% C.L. one has for $\sin^2 \theta_{13} = 0(0.02)$, $\cos 2\theta_\odot \gtrsim 0.28(0.30)$. The quoted results on $\cos 2\theta_\odot$ together with the range of possible values of $|\Delta m_{\rm A}^2|$ and m_0 [3, 24, 41, 42], lead to the conclusion about the existence of significant and robust lower bounds on $|\langle m \rangle|$ in the cases of *IH* and *QD* spectrum [38, 74]. At the same time, as Fig. 2 indicates, $|\langle m \rangle|$ does not exceed $\sim 0.006\,{\rm eV}$ for *NH* spectrum. This implies that $max(|\langle m \rangle|)$ in the case of *NH* spectrum is considerably smaller than $min(|\langle m \rangle|)$ for the *IH* and *QD* spectrum. This opens the possibility of obtaining information about the type of ν-mass spectrum from a measurement of $|\langle m \rangle| \neq 0$. In particular, a positive result in the future generation of $(\beta\beta)_{0\nu}$-decay experiments showing that $|\langle m \rangle| > 0.02\,{\rm eV}$ would imply that the *NH* spectrum is strongly disfavored (if not excluded). The uncertainty in the relevant NME and prospective experimental errors in the values of the oscillation parameters and in $|\langle m \rangle|$ can weaken, but do not invalidate, these results (see, e.g., ref. [69]).

5.4. *Constraining the Majorana CPV phases*

The possibility of establishing *CP*-violation due to Majorana *CPV* phases has been studied in [68] and in greater detail in [55]. It was found that it is very challenging[11]: it requires quite accurate measurements of $|\langle m \rangle|$ and of m_0, and holds only for a limited range of values of the relevant parameters. For *IH* and *QD* spectra, which are of interest, the "just CP-violation" region [28] – an experimental point in this region would signal unambiguously CP-violation associated with Majorana neutrinos, is larger for smaller values of $\cos 2\theta_\odot$. More specifically, proving that *CP*-violation associated with Majorana neutrinos takes place requires, in particular, a relative experimental error on the measured value of $|\langle m \rangle|$ smaller than $\sim 15\%$, a "theoretical uncertainty" in the value of $|\langle m \rangle|$ due to an imprecise knowledge of the corresponding *NME* smaller than a factor of 2, a value of $\tan^2 \theta_\odot \gtrsim 0.55$, and values of the relevant Majorana *CPV* phases typically within the ranges of $\sim (\pi/4 - 3\pi/4)$ and $\sim (5\pi/4 - 7\pi/4)$.

6. Conclusions

Future $(\beta\beta)_{0\nu}$-decay experiments have a remarkable physics potential. They can establish the Majorana nature of neutrinos with definite mass ν_j. If the latter are Majorana particles, the $(\beta\beta)_{0\nu}$-decay experiments can provide information on the type of the neutrino mass spectrum and on the absolute scale of neutrino masses. They can also provide unique information on the Majorana *CP*-violation phases present in the PMNS neutrino mixing matrix. The knowledge of the values of the relevant $(\beta\beta)_{0\nu}$-decay nuclear matrix elements with a sufficiently small uncertainty is crucial for obtaining quantitative information on the neutrino mass and mixing parameters from a measurement of $(\beta\beta)_{0\nu}$-decay half-life.

---

[11]Pessimistic conclusion about the prospects to establish *CP*-violation due to Majorana *CPV* phases from a measurement of $|\langle m \rangle|$ and, e.g., of m_0, was reached in [53].

References

1. Cleveland, B. T. *et al.*, Astrophys. J. **496**, 505 (1998); Fukuda, Y. *et al.* [Kamiokande Collaboration], Phys. Rev. Lett. **77**, 1683 (1996); Abdurashitov, J. N. *et al.* [SAGE Collaboration], J. Exp. Theor. Phys. **95**, 181 (2002); Kirsten, T. *et al.* [GALLEX and GNO Collaborations], Nucl. Phys. B (Proc. Suppl.) **118**, 33 (2003); Cattadori, C., Talk given at ν'04 International Conference, June 14–19, (2004), Paris, France.
2. Fukuda, S. *et al.* [Super-Kamiokande Collaboration], Phys. Lett. **B539**, 179 (2002).
3. Suzuki, Y., [Super-Kamiokande Collaboration], these Proceedings.
4. Ahmad, Q. R. *et al.* [SNO Collaboration], Phys. Rev. Lett. **87**, 071301 (2001); *ibid.* **89**, 011301 and 011302 (2002), Ahmed, S. N. *et al.*, Phys. Rev. Lett. **92**, 181301 (2004).
5. McDonald, A., [SNO Collaboration], these Proceedings.
6. Fukuda, Y. *et al.* [Super-Kamiokande Collaboration], Phys. Rev. Lett. **81**, 1562 (1998).
7. Eguchi, K. *et al.* [KamLAND Collaboration], Phys. Rev. Lett. **90**, 021802 (2003).
8. Ahn, M. H. *et al.* [K2K Collaboration], Phys. Rev. Lett. **90**, 041801 (2003).
9. Pontecorvo, B., Zh. Eksp. Teor. Fiz. (JETP) **33**, 549 (1957) and **34**, 247 (1958).
10. Pontecorvo, B., Zh. Eksp. Teor. Fiz. **53**, 1717 (1967) [Sov. Phys. JETP **26**, 984 (1968)].
11. Bilenky, S. M. and Pontecorvo, B., Phys. Rep. **41**, 225 (1978).
12. Wolfenstein, L., Phys. Rev. **D17**, 2369 (1978); Mikheev, S. P. and Smirnov, A. Y., Sov. J. Nucl. Phys. **42**, 913 (1985).
13. Bilenky, S. M. and Petcov, S. T., Rev. Mod. Phys. **59**, 671 (1987).
14. Petcov, S. T., Lecture Notes in Physics **512** (eds. H. Gausterer, and C. B. Lang), (Springer, 1998), p. 281; Gonzalez-Garcia, M. C. and Nir, Y., Rev. Mod. Phys. **75**, 345 (2003).
15. Bandyopadhyay, A. *et al.*, Phys. Lett. **B583**, 134 (2004).
16. Fukuda, Y. *et al.* [Kamiokande Collaboration], Phys. Rev. Lett. **77**, 1683 (1996).
17. Athanassopoulos, C. *et al.*, Phys. Rev. Lett. **81**, 1774 (1998).
18. Ashie, Y. *et al.* [Super-Kamiokande Collaboration], Phys. Rev. Lett. **93**, 101801 (2004).
19. Suzuki, A., [KamLAND Collaboration], these Proceedings.
20. Aliu, E. *et al.* [K2K Collaboration], hep-ex/0411038.
21. Gribov, V. and Pontecorvo, B., Phys. Lett. **B28**, 493 (1969).
22. Bandyopadhyay, A. *et al.*, hep-ph/0406328 (to be published in Phys. Lett. B).
23. Maki, Z., Nakagawa, M. and Sakata, S., Prog. Theor. Phys. **28**, 870 (1962).
24. Maltoni, M. *et al.*, hep-ph/0405172.
25. Conrad, J., these Proceedings; Sorel, M., Conrad, J. and Shaevitz, M., hep-ph/0305255.
26. Bilenky, S. M., Hosek, J. and Petcov, S. T., Phys. Lett. **B94**, 495 (1980).
27. Schechter, J. and Valle, J. W. F., Phys. Rev. **D23**, 2227 (1980); Doi, M. *et al.*, Phys. Lett. **B102**, 323 (1981).
28. Bilenky, S. M., Pascoli, S. and Petcov, S. T., Phys. Rev. **D64**, 053010 (2001).
29. Apollonio, M. *et al.*, Phys. Lett. **B466**, 415 (1999); Boehm, F. *et al.*, Phys. Rev. Lett. **84**, 3764 (2000).
30. Petcov, S. T., Phys. Lett. **B200**, 373 (1988); *ibid.* **B214**, 139 (1988) and **B406**, 355 (1997).
31. Petcov, S. T., Phys. Lett. **B214**, 259 (1988).
32. Haxton, W. C., Phys. Rev. Lett. **57**, 1271 (1986); Parke, S. J., *ibid.* **57**, 1275 (1986).
33. Krastev, P. I. and Petcov, S. T., Phys. Lett. **B207**, 64 (1988); Proc. Moriond Workshop on Neutrinos and Exotic Phenomena, Les Arcs, France, January (1988) (ed. J. Tran Thanh Van, Editions Frontières, Gif-sur-Yvette), p. 173; Bruggen, M., Haxton, W. C. and Quian, Y.-Z., Phys. Rev. **D51**, 4028 (1995).
34. Petcov, S. T. and Rich, J., Phys. Lett. **B224**, 401 (1989); Lisi, E. *et al.*, Phys. Rev. **D63**, 093002 (2000).
35. de Holanda, P. C., Liao, Wei. and Smirnov, A. Yu., Nucl. Phys. **B702**, 307 (2004).
36. Bahcall, J. N., Gonzalez-Garcia, M. C. and Peña-Garay, C., hep-ph/0406294.
37. Bilenky, S. M. *et al.*, Phys. Rev. **D56**, 4432 (1996).
38. Pascoli, S. and Petcov, S. T., Phys. Lett. **B544**, 239 (2002); *ibid.* **B580**, 280 (2004).
39. Perrin, F., Comptes Rendus **197**, 868 (1933); Fermi, E., Nuovo Cim. **11**, 1 (1934).

40. Lobashev, V. *et al.*, Nucl. Phys. B (Proc. Suppl.) **91**, 280 (2001).
41. Weinheimer, C. *et al.*, Nucl. Phys. Proc. Suppl. **118**, 279 (2003) and these Proceedings.
42. Tegmark, M., these Proceedings.
43. Spergel, D. N. *et al.* (WMAP Coll.), Astrophys. J. Suppl. **148**, 175 (2003).
44. Lindner, M., these Proceedings; Blondel, A., these Proceedings.
45. Chizhov, M. V., Maris, M. and Petcov, S. T., hep-ph/9810501; Bernabéu, J., Palomares-Ruiz, S. and Petcov, S. T., Nucl. Phys. B **669**, 255 (2003); Palomares-Ruiz, S. and Petcov, S. T., hep-ph/0406096.
46. Petcov, S. T., Phys. Lett. B **434**, 321 (1998), (E) *ibid.* **B444**, 584 (1998).
47. Chizhov, M. V. and Petcov, S. T., Phys. Rev. Lett. **83**, 1096 (1999); Phys. Rev. Lett. **85**, 3979 (2000); Phys. Rev. **D63**, 073003 (2001).
48. Ermilova, V. K. *et al.*, Short Notices Lebedev Institute **5**, 26 (1986); Akhmedov, E. Kh. and Fiz. Y., **47**, 475 (1988); Krastev, P. I. and Smirnov, A. Yu., Phys. Lett. **B226**, 341 (1989).
49. Liu, Q. Y. and Smirnov, A. Yu., Nucl. Phys. **B524**, 505 (1998); Liu, Q. Y., Mikheyev, S. P. and Smirnov, A. Yu., Phys. Lett. **B440**, 319 (1998).
50. Petcov, S. T., hep-ph/0412410.
51. Choubey, S. and Petcov, S. T., Phys. Lett. **B594**, 333 (2004); Bandyopadhyay, A. *et al.*, hep-ph/0410283; Bandyopadhyay, A. *et al.*, Phys. Rev. **D67**, 113011.
52. Langacker, P. *et al.*, Nucl. Phys. **B282**, 589 (1987).
53. Barger, V. *et al.*, Phys. Lett. **B540**, 247 (2002).
54. de Gouvea, A., Kayser, B. and Mohapatra, R., Phys. Rev. **D67**, 053004 (2003).
55. Pascoli, S., Petcov, S. T. and Rodejohann, W., Phys. Lett. **B549**, 177 (2002).
56. Pascoli, S., Petcov, S. T. and Yaguna, C. E., Phys. Lett. **B564** 241 (2003).
57. Yanagida, T., these Proceedings.
58. Morales, A. and Morales, J., Nucl. Phys. Proc. Suppl. **114**, 141 (2003).
59. Elliot, S. R. and Vogel, P., Ann. Rev. Nucl. Part. Sci. **52**, (2002).
60. Petcov, S. T., New J. Phys. **6**, 109 (2004); Pascoli, S. and Petcov, S. T., hep-ph/0308034.
61. Ramond, P., these Proceedings; Mohapatra, R., these Proceedings.
62. Aalseth, C. *et al.*, hep-ph/0412300.
63. Wolfenstein, L., Phys. Lett. **B107**, 77 (1981); Bilenky, S. M., Nedelcheva, N. P. and Petcov, S. T., Nucl. Phys. **B247**, 61 (1984); Kayser, B., Phys. Rev. **D30**, 1023 (1984).
64. Klapdor-Kleingrothaus, H. V. *et al.*, Nucl. Phys. Proc. Suppl. **100**, 309 (2001).
65. Klapdor-Kleingrothaus, H. V. *et al.*, Phys. Lett. **B586**, 198 (2004).
66. Barabash, A. *et al.*, J. Exp. Theor. Phys. Lett. **80**, 377 (2004).
67. Fiorini, E., these Proceedings.
68. Pascoli, S., Petcov, S. T. and Wolfenstein, L., Phys. Lett. **B524**, 319 (2002); Pascoli, S. and Petcov, S. T., hep-ph/0111203.
69. Pascoli, S., Petcov, S. T. and Rodejohann, W., Phys. Lett. **B558**, 141 (2003).
70. Bilenky, S. M. *et al.*, Phys. Lett. **B465**, 193 (1999); Barger, V. and Whisnant, K., Phys. Lett. **B456**, 194 (1999); Vissani, F., J. High. En. Phys. **06**, 022 (1999); Czakon, M. *et al.*, hep-ph/0003161; Klapdor-Kleingrothaus, H. V., Päs, H. and Smirnov, A. Yu., Phys. Rev. **D63**, 073005 (2001).
71. Rodejohann, W., Nucl. Phys. **B597**, 110 (2001) and hep-ph/0203214; Matsuda, K. *et al.*, Phys. Rev. **D63**, 077301 (2001); Fukuyama, T. *et al.*, Phys. Rev. **D64**, 013001 (2001).
72. Petcov, S. T. and Smirnov, A. Yu., Phys. Lett. **B322**, 109 (1994).
73. Rodin, V. A. *et al.*, Phys. Rev. **C68**, 044302 (2003).
74. Murayama, H. and Peña-Garay, C., Phys. Rev. **D69**, 031301 (2004).
75. Bilenky, S. M. and Petcov, S. T., hep-ph/0405237.

Physica Scripta. Vol. T121, 102–105, 2005

Supernova Neutrino Oscillations

Georg G. Raffelt

Max-Planck-Institut für Physik (Werner-Heisenberg-Institut), Föhringer Ring 6, 80805 München, Germany

Received January 7, 2005; accepted January 11, 2005

PACS numbers: 14.60.Lm, 14.60.Pq, 97.60.Bw

Abstract

Observing a high-statistics neutrino signal from a galactic supernova (SN) would allow one to test the standard delayed explosion scenario and may allow one to distinguish between the normal and inverted neutrino mass ordering due to the effects of flavor oscillations in the SN envelope. One may even observe a signature of SN shock-wave propagation in the detailed time-evolution of the neutrino spectra. A clear identification of flavor oscillation effects in a water Cherenkov detector probably requires a megatonne-class experiment.

1. Introduction

While galactic supernovae are rare, the proliferation of existing or proposed large neutrino detectors has considerably increased the confidence that a high-statistics supernova (SN) neutrino signal will eventually be observed. The scientific harvest would be immense. Most importantly for particle physics, the detailed features of the neutrino signal may reveal the nature of the neutrino mass ordering that is extremely difficult to determine experimentally (e.g. Ref. [1] and references therein). On the other hand, a detailed measurement of the neutrino signal from a galactic SN could yield important clues on the SN explosion mechanism. Neutrinos undoubtedly play a crucial role for the SN dynamics and in particular neutrino energy deposition behind the SN shock is able to initiate and power the SN explosion [2, 3]. However, it is still unclear whether this energy deposition is sufficiently strong because current state-of-the-art models still have problems to produce robust explosions [4]. Empirical constraints on the physics deep inside the SN core would be extremely useful because neutrinos are the only way for a direct access [5] besides gravitational waves [6].

The neutrinos emitted by the collapsed SN core will pass through the mantle and envelope of the progenitor star and on the way encounter a vast range of matter densities ρ from nearly nuclear at the neutrinosphere to that of interstellar space. The Wolfenstein effect [7] causes a resonance in neutrino oscillations [8] when $\Delta m_v^2 \cos 2\theta / 2E_v = \pm \sqrt{2} G_F Y_e \rho$, where the plus and minus sign refers to neutrinos v and antineutrinos $\bar{v}$, respectively. Therefore, depending on the sign of Δm_v^2, the resonance occurs in the v or $\bar{v}$ channel [9]. For the "solar" neutrino mass-squared difference of $\Delta m_{21}^2 \approx 79 \, \mathrm{meV}^2$ [10] one refers to the "L-resonance" (low density) while for the "atmospheric" one of $|\Delta m_{32}^2| \approx 2300 \, \mathrm{meV}^2$ [11] to the "H-resonance" (high density). The resonance is particularly important for 13-oscillations because the 13-mixing angle is known to be small so that the classic MSW enhancement of flavor conversion in an adiabatic density gradient is a crucial feature [12].

Effects of neutrino flavor oscillations will be observable only if the fluxes and/or spectra emitted at the source depend on the neutrino species. Therefore, I will summarize in Sec. 2 the current understanding of SNe as flavor-dependent neutrino sources. In Sec. 3 the effect of neutrino flavor oscillations in different mixing scenarios will be summarized. Sec. 4 is devoted to experimentally observable signatures before turning to conclusions in Sec. 5.

2. Core-collapse Supernovae as neutrino sources

The collapsed core of a SN is essentially a blackbody source for neutrinos of all flavors with a temperature of several MeV, corresponding to the temperature of the medium near the proto-neutron star's surface. In detail, however, the fluxes and spectra differ between the different species because of their different interaction channels. Electron neutrinos and anti-neutrinos interact only by charged-current processes on nucleons while the other species interact primarily by neutral-current reactions because the relevant energies and densities are too low to produce muons or tau-leptons. Therefore, v_μ, $\bar{v}_\mu$, v_τ, and $\bar{v}_\tau$ will be emitted with identical fluxes and spectra and are thus collectively referred to as v_x.

The standard delayed-explosion scenario of SN evolution and thus the neutrino emission is characterized by four distinct phases, schematically indicated in Fig. 1: (1) Core collapse and bounce. (2) Shock propagation and breakout. (3) Accretion and mantle cooling while the shock wave stagnates. (4) Kelvin-Helmholtz cooling of the neutron star after the explosion. Actually, for a very close star the neutrinos emitted during the silicon-burning phase for the last few days before collapse may be observable [13], a possibility that I will not further consider here.

The v_e "light curve" shows a distinct peak, the "prompt deleptonization burst," that occurs when the shock-wave breaks through the neutrino sphere, dissociates iron and thus allows for the quick $e^- + p \rightarrow n + v_e$ conversion and thus for the

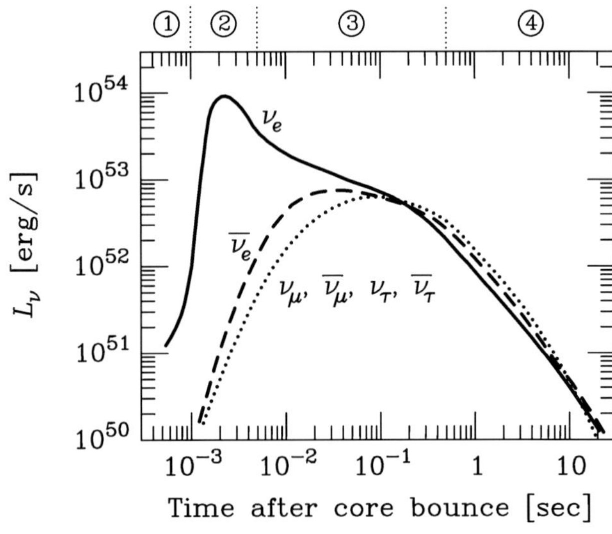

Fig. 1. Schematic SN neutrino "light curves" for different neutrino flavors [14].

prompt deleptonization of the outermost core layers. The flux and spectral characteristics of this burst are probably the least model-dependent features of SN neutrino emission [15, 16]. During the phase of shock-wave stagnation that may last for several hundred milliseconds, the neutrino emission is largely powered by the accretion of matter. After the shock wave has been rejuvenated by neutrino energy deposition and has ejected the stellar mantle and envelope, the neutron star remnant will cool, i.e. the neutrino emission is powered by energy stored deep in the core that diffuses to the surface on a time scale of several seconds.

In the literature it was often assumed that the neutrino emission during the accretion and Kelvin-Helmholtz phases was characterized by an approximate equipartition of the flavor-dependent neutrino luminosities $L_{\nu_e} \approx L_{\bar\nu_e} \approx L_{\nu_x}$ and by a strong hierarchy of average energies $\langle E_{\nu_e} \rangle \ll \langle E_{\bar\nu_e} \rangle \ll \langle E_{\nu_x} \rangle$. This behavior is borne out, for example, by the numerical simulations of the Livermore group [5]. However, traditional treatments of neutrino transport involve a number of physical and numerical simplifications that were justified for the purposes of those calculations but are not accurate enough to judge the flavor dependence of the neutrino fluxes and spectra. In a series of papers we have studied this problem and have concluded that the differences between $\langle E_{\bar\nu_e} \rangle$ and $\langle E_{\bar\nu_x} \rangle$ are probably not more than about 20% or less [17, 18, 19, 20]. On the other hand, the equipartition of energy among the flavors is certainly not exact – the luminosities can differ perhaps by up to a factor of 2. Moreover, it appears that during the accretion phase $L_{\bar\nu_x} < L_{\bar\nu_e}$ while $L_{\bar\nu_e} < L_{\bar\nu_x}$ during the cooling phase [19, 20].

While several recent numerical simulations with a modern treatment of neutrino transfer confirm this picture for the accretion phase, no up-to-date simulations have treated the long-term evolution of the neutrino signal during the cooling phase. Since current numerical models to not produce robust explosions and since a full Boltzmann treatment of neutrino transport is very CPU intensive, numerical studies of the flavor-dependence of the late neutrino signal are not available at the present time. Therefore, forecasting possible effects of neutrino oscillations in the signal from a future galactic SN is unavoidably schematic and must rely on generic assumptions about the flavor-dependent neutrino fluxes and spectra.

3. Impact of flavor oscillations

The neutrino or anti-neutrino fluxes arriving at Earth from a SN are determined by the primary spectra at the source as well as neutrino oscillation effects. For ν_e, the flux arriving at Earth may be written in terms of the energy-dependent "survival probability" $p(E)$ as

$$F_{\nu_e}(E) = p(E)F^0_{\nu_e}(E) + [1 - p(E)]F^0_{\nu_x}(E), \qquad (1)$$

where the superscript zero denotes the primary fluxes and ν_x stands for either ν_μ or ν_τ. An analogous expression pertains to $\bar\nu_e$ with the survival probability $\bar p(E)$.

When neutrinos propagate through the SN mantle and envelope, the MSW resonance corresponding to the solar mixing angle is adiabatic and is always in the neutrino channel. On the other hand, the H-resonance, corresponding to 13-oscillations, is in the neutrino channel for normal neutrino mass ordering while it is in the anti-neutrino channel in the inverted case. This resonance is adiabatic for all relevant energies if the mixing angle θ_{13} is "large" in the sense of $\sin^2 \theta_{13} \gtrsim 10^{-3}$. It is non-adiabatic if the mixing angle is "small" in the sense of $\sin^2 \theta_{13} \lesssim 10^{-5}$.

Table I. *Survival probabilities for neutrinos, p, and anti-neutrinos, $\bar p$, in various mixing scenarios. The channels where one expects Earth effects, shock-wave propagation effects, and where the full ν_e burst is present or absent are indicated.*

Scenario	Hierarchy	$\sin^2 \theta_{13}$	p	$\bar p$	Earth effects	Shock wave	ν_e burst
A	Normal	$\gtrsim 10^{-3}$	0	$\cos^2 \theta_\odot$	$\bar\nu_e$	ν_e	absent
B	Inverted	$\gtrsim 10^{-3}$	$\sin^2 \theta_\odot$	0	ν_e	$\bar\nu_e$	present
C	Any	$\lesssim 10^{-5}$	$\sin^2 \theta_\odot$	$\cos^2 \theta_\odot$	ν_e and $\bar\nu_e$	–	present

For intermediate values, the adiabaticity depends on energy and the situation is more complicated. The survival probabilities thus depend on the magnitude of the 13-mixing angle and on the nature of the mass ordering so that, in principle, the observation of a SN neutrino signal can distinguish between different mixing scenarios [12]. Table I summarizes the survival probabilities for different mixing scenarios where $\theta_\odot$ refers to the "solar" mixing angle.

The indicated survival probabilities pertain to a situation where neutrino oscillations occur only in the SN mantle and envelope. If the SN is shadowed by the Earth for the relevant detector, additional modifications of the survival probabilities arise due to Earth matter effects, causing an energy-dependent modulation of $p(E)$ or $\bar p(E)$. Table I indicates the channel where Earth effects would arise for different mixing scenarios.

The SN shock wave will pass the density region corresponding to the H-resonance a few seconds after core bounce, breaking the adiabaticity of the 13-resonance. The resulting transient modification of the survival probabilities may be observable [21]. They arise in the channels indicated in Table I.

4. Experimental signatures

The most distinct effects of flavor oscillations arise in the ν_e channel because it sports the prompt deleptonization burst and even during the accretion and cooling phases, the expected spectral differences are much larger between ν_e and ν_x than between $\bar\nu_e$ and $\bar\nu_x$. On the other hand, the existing and realistically expected large detectors, notably Super-Kamiokande, IceCube, and a future megatonne-class water Cherenkov detector, are mostly sensitive to the inverse beta decay $\bar\nu_e + p \to n + e^+$, although the ν_e channel can be measured by the elastic scattering reaction $\nu_e + e^- \to e^- + \nu_e$. The largest existing ν_e detector is the Sudbury Neutrino Observatory (SNO) that will be shut down after completing its solar neutrino programme and thus will not be available for a long-term SN watch. Of course, if an efficient ν_e detector such as as a large liquid Argon TPC should become available, it would have unique capabilities for studying SN neutrino oscillations [22, 23].

In a water Cherenkov detector, the ν_e signal can be identified by the directionality of the $\nu_e - e^-$ elastic scattering reaction. Moreover, if future water Cherenkov detectors are doped with enough gadolinium to tag the neutron from $\bar\nu_e + p \to n + e^+$ [24], the identification of the prompt SN ν_e burst will become a realistic possibility. However, in Super-Kamiokande the total number of expected events from the ν_e burst of a "fiducial SN" at a distance of 10 kpc is so small (about a dozen) that a clear identification of the presence or absence of the burst is not likely. In a megatonne detector, on the other hand, the full ν_e burst will produce about 200 events. Therefore, the time structure of the SN signal during the first few tens of milliseconds can provide a clean indication if the full v_e burst is present or absent [16] and therefore allows one to distinguish between different mixing scenarios as indicated by the last column of Table I. For example, if the mass ordering is normal and the 13-mixing angle is large, the v_e burst will fully oscillate into v_x. If the 13-mixing angle is measured in the laboratory to be large, for example by one of the forthcoming reactor experiments [25], then one may distinguish between the normal and inverted mass ordering. On the other hand, if the mixing scenario is independently identified by laboratory experiments, these observations allow one to determine the distance to the SN with a precision of about 5%, even in the likely case that it is optically obscured [16].

A megatonne water Cherenkov detector is a realistic future possibility in view of the world-wide drive towards precision long-baseline oscillation experiments, but it is less clear if a gadolinium-doped detector of such size will ever come into existence. Therefore, I now turn to the main $\bar{v}_e$ detection channel of an ordinary water Cherenkov detector. In this channel the main problem is that one can not rely on theoretical predictions of the flavor dependence of the source fluxes and spectra. Therefore, one has to rely on model-independent signatures. A clear positive detection of such a signature would indicate a specific mixing scenario while the absence would be ambiguous: It could either imply a different mixing scenario or could imply that the flavor-dependent source spectra and fluxes were too similar to provide significant oscillation effects.

One unequivocal indication of oscillation effects would be the energy-dependent modulation of the survival probability $\bar{p}(E)$ caused by Earth matter effects. Even without an identification of the SN location in the sky in the electromagnetic spectrum, its direction can be established with sufficient accuracy by the $v_e - e^-$ elastic scattering signal alone [26, 27]. With a megatonne-class detector, the modulation signal could be well established unless the flavor-dependent flux differences are surprisingly small [28, 29]. The Earth effect would show up in the $\bar{v}_e$ channel for the normal mass hierarchy, assuming that θ_{13} is large (Table I).

Another possibility to establish the presence of Earth effects is to use the signal from two detectors if one of them sees the SN shadowed by the Earth and the other not. As one needs to compare the two signals on the level of a few percent, sufficiently large detectors are needed. The statistical fluctuations may be too large in Super-Kamiokande so that, again, a megatonne-class detector is needed, perhaps in combination with IceCube [30]. IceCube at the South Pole and Hyper-Kamiokande in Japan would be geographically complementary in that about 70% of the sky would be seen directly in one detector and shadowed by the Earth in the other.

At the time of the next galactic SN, the neutrino mixing parameters may already be known from long-baseline laboratory experiments. In that case it may become possible to use the SN neutrino signal in reverse to perform a "tomography" of the Earth's density profile [31].

A few seconds after core bounce, the SN shock wave will pass the density region in the stellar envelope corresponding to the H-resonance and thus break the adiabaticity of 13-oscillations, causing a transient modification of the survival probability and thus a time-dependent signature in the neutrino signal [21, 32, 33, 34, 35, 36]. Probably the most significant signature is the time variation of the average energy of the detected positrons from the $\bar{v}_e + p \rightarrow n + e^+$ reaction. It would show a characteristic dip when the shock wave passes, or a double-dip feature if a reverse shock occurs [36]. This signature may be observable in Super-Kamiokande and almost certainly can be seen in a megatonne-class detector. Of course, apart from identifying the neutrino mixing scenario, such observations would test our theoretical understanding of the core-collapse SN phenomenon.

The shock-wave propagation signature arises in the complementary channel to the Earth effects (Table I), i.e. it shows up in the $\bar{v}_e$ channel for large θ_{13} and the inverted mass hierarchy. Therefore, unless the flavor-dependent source spectra are unexpectedly similar, a megatonne water Cherenkov detector is assured to see either shock-wave propagation effects or Earth effects, although in the latter case it is required, of course, that the SN is shadowed by the Earth. Conversely, if neither Earth effects nor shock-wave propagation effects are seen one has a serious empirical handle on the flavor dependence of the source spectra in that they must be unexpectedly similar.

5. Conclusions

Observing neutrinos from the next galactic SN would provide invaluable information on the astrophysics of the core-collapse explosion phenomenon and on neutrino mixing parameters. Existing or near-future detectors such as Super-Kamiokande and IceCube are sufficient to measure a precise $\bar{v}_e$ light curve and thus can establish, for example, the duration of the accretion phase or perhaps detect unexpected new features. However, these detectors are probably too small to detect unambiguous evidence for neutrino oscillations. On the other hand, a megatonne-class water Cherenkov detector is assured to detect either Earth effects (assuming the SN is shadowed by the Earth) or shock-wave propagation effects, depending on the neutrino mixing scenario. Such observations will help to identify the neutrino mixing scenario and test detailed aspects of theoretical SN physics. Depending on the neutrino mixing information that will be established by laboratory experiments alone and depending on theoretical progress in our understanding of core-collapse SNe, one or the other aspect of these observations will be of greater interest at the time of the next galactic SN. Either way, the importance of such an observation as a probe of neutrino physics and of the astrophysics of supernovae can not be overstated.

Acknowledgments

This work was supported by the European Science Foundation (ESF) under the Network Grant No. 86 Neutrino Astrophysics and by the Deutsche Forschungsgemeinschaft (DFG) under grant No. SFB-375.

References

1. Huber, P., Lindner, M., Rolinec, M., Schwetz, T. and Winter, W., Phys. Rev. D **70**, 073014 (2004) [hep-ph/0403068].
2. Wilson, J. R., "Supernovae and post-collapse behavior", in "Numerical Astrophysics", (edited by J. M. Centrella, J. M. LeBlanc, R. L. Bowers and J. A. Wheeler), (Jones and Bartlett, Boston, 1985), p. 422.
3. Bethe, H. A. and Wilson, J. R., Astrophys. J. **295**, 14 (1985).
4. Buras, R., Rampp, M., Janka, H. T. and Kifonidis, K., Phys. Rev. Lett. **90**, 241101 (2003) [astro-ph/0303171].
5. Totani, T., Sato, K. H., Dalhed, E. and Wilson, J. R., Astrophys. J. **496**, 216 (1998) [astro-ph/9710203].
6. Müller, E., Rampp, M., Buras, R., Janka, H.-T. and Shoemaker, D. H., Astrophys. J. **603**, 221 (2004) [astro-ph/0309833].
7. Wolfenstein, L., Phys. Rev. D **17**, 2369 (1978).

8. Mikheev, S. P. and Smirnov, A. Y., Sov. J. Nucl. Phys. **42**, 913 (1985) [Yad. Fiz. **42**, 1441 (1985)].
9. Langacker, P., Leveille, J. P. and Sheiman, J., Phys. Rev. D **27**, 1228 (1983).
10. Araki, T. *et al.* [KamLAND Collaboration], "Measurement of neutrino oscillation with KamLAND: Evidence of spectral distortion", hep-ex/0406035.
11. Maltoni, M., Schwetz, T., Tórtola, M. A. and Valle, J. W. F., New J. Phys. **6**, 122 (2004) [hep-ph/0405172].
12. Dighe, A. S. and Smirnov, A. Y., Phys. Rev. D **62**, 033007 (2000) [hep-ph/9907423].
13. Odrzywolek, A., Misiaszek, M. and Kutschera, M., Astropart. Phys. **21**, 303 (2004) [astro-ph/0311012].
14. Raffelt, G. G., "Stars as laboratories for fundamental physics", (University of Chicago Press, 1996).
15. Takahashi, K., Sato, K., Burrows, A. and Thompson, T. A., Phys. Rev. D **68**, 113009 (2003) [hep-ph/0306056].
16. Kachelriess, M. *et al.*, "Exploiting the neutronization burst of a galactic supernova", astro-ph/0412082.
17. Raffelt, G. G., Astrophys. J. **561**, 890 (2001) [astro-ph/0105250].
18. Buras, R., Janka, H.-T., Keil, M. T., Raffelt, G. G. and Rampp, M., Astrophys. J. **587**, 320 (2003) [astro-ph/0205006].
19. Keil, M. T., Raffelt, G. G. and Janka, H.-T., Astrophys. J. **590**, 971 (2003) [astro-ph/0208035].
20. Raffelt, G. G., Keil, M. T., Buras, R., Janka, H.-T. and Rampp, M., "Supernova neutrinos: Flavor-dependent fluxes and spectra", Proc. of the 4th Workshop on Neutrino Oscillations and their Origin: NooN 2003 (10–14 February 2003, Kanazawa, Japan), (eds. Y. Suzuki, M. Nakahata, Y. Itow, M. Shiozawa and Y. Obayashi), (World Scientific, Singapore, 2004), pp. 380–387 [astro-ph/0303226].
21. Schirato, R. C. and Fuller, G. M. "Connection between supernova shocks, flavor transformation, and the neutrino signal", astro-ph/0205390.
22. Gil-Botella, I. and Rubbia, A., J. Cosmol. Astropart. Phys. **0310**, 009 (2003) [hep-ph/0307244].
23. Gil-Botella, I. and Rubbia, A., J. Cosmol. Astropart. Phys. **0408**, 001 (2004) [hep-ph/0404151].
24. Beacom, J. F. and Vagins, M. R., Phys. Rev. Lett. **93**, 171101 (2004) [hep-ph/0309300].
25. Anderson, K. *et al.*, "White paper report on using nuclear reactors to search for a value of theta(13)", hep-ex/0402041.
26. Beacom, J. F. and Vogel, P., Phys. Rev. D **60**, 033007 (1999) [astro-ph/9811350].
27. Tomàs, R., Semikoz, D., Raffelt, G. G., Kachelriess, M. and Dighe, A. S., Phys. Rev. D **68**, 093013 (2003) [hep-ph/0307050].
28. Dighe, A. S., Keil, M. T. and Raffelt, G. G., J. Cosmol. Astropart. Phys. **0306**, 006 (2003) [hep-ph/0304150].
29. Dighe, A. S., Kachelriess, M., Raffelt, G. G. and Tomàs, R., J. Cosmol. Astropart. Phys. **0401**, 004 (2004) [hep-ph/0311172].
30. Dighe, A. S., Keil, M. T. and Raffelt, G. G., J. Cosmol. Astropart. Phys. **0306**, 005 (2003) [hep-ph/0303210].
31. Lindner, M., Ohlsson, T., Tomàs, R. and Winter, W., Astropart. Phys. **19**, 755 (2003) [hep-ph/0207238].
32. Takahashi, K., Sato, K., Dalhed, H. E. and Wilson, J. R., Astropart. Phys. **20**, 189 (2003) [astro-ph/0212195].
33. Lunardini, C. and Smirnov, A. Y., J. Cosmol. Astropart. Phys. **0306**, 009 (2003) [hep-ph/0302033].
34. Fogli, G. L., Lisi, E., Montanino, D. and Mirizzi, A., Phys. Rev. D **68**, 033005 (2003) [hep-ph/0304056].
35. Fogli, G. L., Lisi, E., Mirizzi, A. and Montanino, D., "Probing supernova shock waves and neutrino flavor transitions in next-generation water-Cherenkov detectors," hep-ph/0412046.
36. Tomàs, R. *et al.*, J. Cosmol. Astropart. Phys. **0409**, 015 (2004) [astro-ph/0407132].

Physica Scripta. Vol. T121, 106–111, 2005

High-Energy Neutrino Astronomy

Francis Halzen

Department of Physics, University of Wisconsin, Madison, WI, 53706, USA

Received December 2, 2004; accepted February 18, 2005

PACS number: 95.85 Ry

Abstract

Kilometer-scale neutrino detectors such as IceCube are discovery instruments covering nuclear and particle physics, cosmology and astronomy. Examples of their multidisciplinary missions include the search for the particle nature of dark matter and for additional small dimensions of space. In the end, their conceptual design is very much anchored to the observational fact that Nature accelerates protons and photons to energies in excess of 10^{20} and 10^{13} eV, respectively. The cosmic ray connection sets the scale of cosmic neutrino fluxes. In this context, we discuss the first results of the completed AMANDA detector and the reach of its extension, IceCube. Similar experiments are under construction in the Mediterranean. Neutrino astronomy is also expanding in new directions with efforts to detect air showers, acoustic and radio signals initiated by neutrinos with energies similar to those of the highest energy cosmic rays.

1. Neutrinos associated with the highest energy cosmic rays

The flux of cosmic rays is summarized in Fig. 1a,b [1]. The energy spectrum follows a broken power law. The two power laws are separated by a feature dubbed the "knee"; see Fig. 1a. Circumstantial evidence exists that cosmic rays, up to EeV energy, originate in galactic supernova remnants. Any association with our galaxy disappears however in the vicinity of a second feature in the spectrum referred to as the "ankle". Above the ankle, the gyroradius of a proton in the galactic magnetic field exceeds the size of the galaxy and it is generally assumed that we are witnessing the onset of an extragalactic component in the spectrum that extends to energies beyond 100 EeV. Experiments indicate that the highest energy cosmic rays are predominantly protons or, possibly, nuclei. Above a threshold of 50 EeV these protons interact with cosmic microwave photons and lose their energy to pions before reaching our detectors. This is the Greissen-Zatsepin-Kuzmin cutoff that limits the sources to our supercluster of galaxies.

Models for the origin of the highest energy cosmic rays fall into two categories, top-down and bottom-up. In top-down models it is assumed that the cosmic rays are the decay products of cosmological remnants with Grand Unified energy scale $M_{GUT} \sim 10^{24}$ eV. These models predict neutrino fluxes most likely within reach of first-generation telescopes such as AMANDA, and certainly detectable by future kilometer-scale neutrino observatories [2].

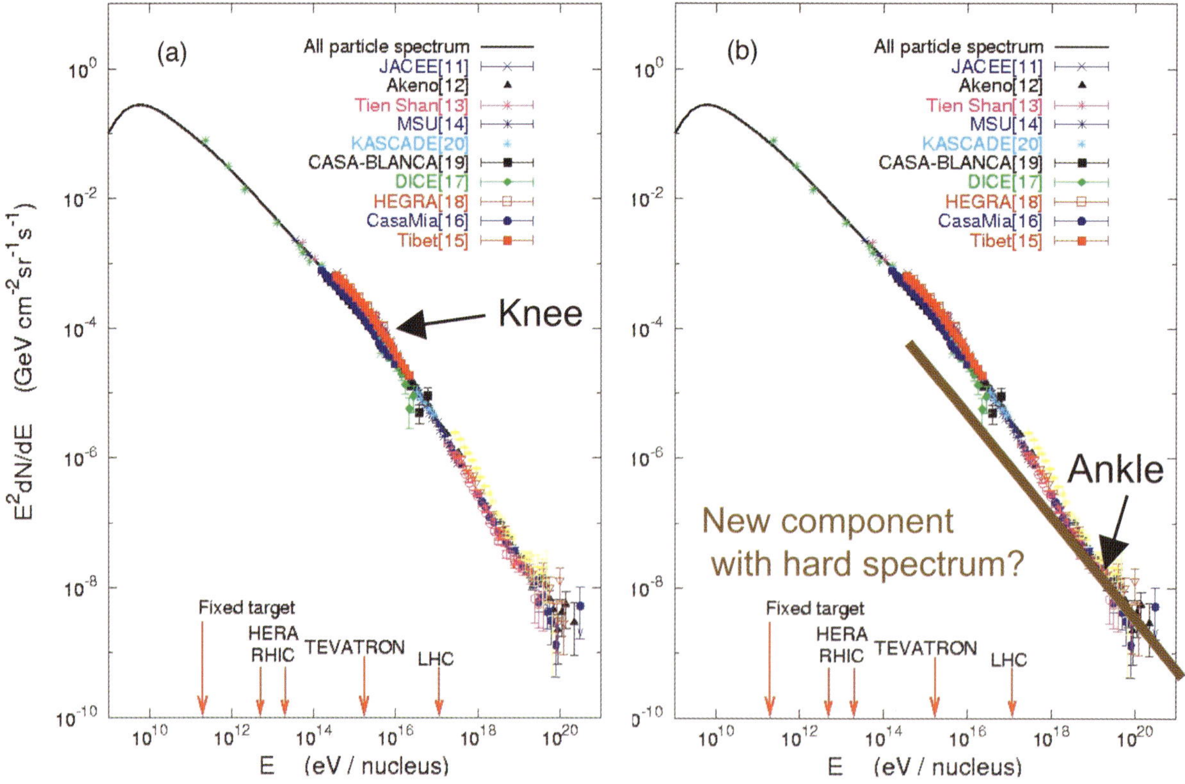

Fig. 1. At the energies of interest here, the cosmic ray spectrum consists of a sequence of 3 power laws. The first two are separated by the "knee" (left panel), the second and third by the "ankle". There is evidence that the cosmic rays beyond the ankle are a new population of particles produced in extragalactic sources; see right panel.

In bottom-up scenarios it is assumed that cosmic rays originate in cosmic accelerators. Accelerating particles to TeV energy and above requires massive bulk flows of relativistic charged particles. These are likely to originate from the exceptional gravitational forces in the vicinity of black holes. Examples include the dense cores of exploding stars, inflows onto supermassive black holes at the centers of active galaxies and annihilating black holes or neutron stars. Before leaving the source, accelerated particles pass through intense radiation fields or dense clouds of gas surrounding the black hole. This results in interactions producing pions decaying into secondary photons and neutrinos that accompany the primary cosmic ray beam as illustrated in Fig. 2.

How many neutrinos are produced in association with the cosmic ray beam? The answer to this question, among many others [2], provides the rationale for building kilometer-scale neutrino detectors. We first consider a neutrino beam produced at an accelerator laboratory; see Fig. 2. Here the target absorbs all parent protons as well as the secondary electromagnetic and hadronic showers. Only neutrinos exit the dump. If nature constructed such a "hidden source" in the heavens, conventional astronomy will not reveal it. It cannot be the source of the cosmic rays, however, because in this case the dump must be transparent to protons.

A more generic "transparent" source can be imagined as follows: protons are accelerated in a region of high magnetic fields where they interact with photons via the processes $p + \gamma \rightarrow \Delta \rightarrow \pi^0 + p, p + \gamma \rightarrow \Delta \rightarrow \pi^+ + n$. While the secondary protons may remain trapped in the acceleration region, equal numbers of neutrons, neutral and charged pions escape. The energy escaping the source is therefore equally distributed between cosmic rays, gamma rays and neutrinos produced by the decay of neutrons and neutral and charged pions, respectively. The neutrino flux from a generic transparent cosmic ray source is often referred to as the Waxman-Bahcall flux [3]. It is easy to calculate and the derivation is revealing.

Figure 1b shows a fit to the observed spectrum above the "ankle" that can be used to derive the total energy in extragalactic cosmic rays. The energy content of this component is $\sim 3 \times 10^{-19}\,\mathrm{erg\,cm^{-3}}$, assuming an E^{-2} energy spectrum with a GZK

cutoff. The power required for a population of sources to generate this energy density over the Hubble time of 10^{10} years is $\sim 3 \times 10^{37}\,\mathrm{erg\,s^{-1}}$ per $(\mathrm{Mpc})^3$ or, as often quoted in the literature, $\sim 5 \times 10^{44}\,\mathrm{TeV}$ per year per $(\mathrm{Mpc})^3$. This works out to [4]

- $\sim 3 \times 10^{39}\,\mathrm{erg\,s^{-1}}$ per galaxy,
- $\sim 3 \times 10^{42}\,\mathrm{erg\,s^{-1}}$ per cluster of galaxies,
- $\sim 2 \times 10^{44}\,\mathrm{erg\,s^{-1}}$ per active galaxy, or
- $\sim 2 \times 10^{52}\,\mathrm{erg}$ per cosmological gamma ray burst.

The coincidence between these numbers and the observed output in electromagnetic energy of these sources explains why they have emerged as the leading candidates for the cosmic ray accelerators. The coincidence is consistent with the relationship between cosmic rays and photons built into the "transparent" source. In the photoproduction processes roughly equal energy goes into the secondary neutrons, neutral and charged pions whose energy ends up in cosmic rays, gamma rays and neutrinos, respectively.

We therefore assume that the same energy density of $\rho_E \sim 3 \times 10^{-19}\,\mathrm{erg\,cm^{-3}}$, observed in cosmic rays and electromagnetic energy, ends up in neutrinos with a spectrum $E_\nu\,dN/dE_\nu \sim E^{-\gamma}\,\mathrm{cm^{-2}\,s^{-1}\,sr^{-1}}$ that continues up to a maximum energy $E_{\max}$. The neutrino flux follows from the relation $\int E_\nu\,dN/dE_\nu = c\rho_E/4\pi$. For $\gamma = 1$ and $E_{\max} = 10^8\,\mathrm{GeV}$, the generic source of the highest energy cosmic rays produces a flux of $E_\nu^2\,dN/dE_\nu \sim 5 \times 10^{-8}\,\mathrm{GeV\,cm^{-2}\,s^{-1}\,sr^{-1}}$.

There are several ways to modify this simple prediction:

- The derivation fails to take into account the fact that there are more cosmic rays in the universe producing neutrinos than observed at earth because of the GZK-effect and neglects evolution of the sources with redshift. This increases the neutrino flux by a factor ~ 3, possibly more.
- For proton-γ interactions muon neutrinos receive only 1/4 of the energy of the charged pion in the decay chain $\pi^+ \rightarrow \mu^+ + \nu_\mu \rightarrow e^+ + \nu_e + \bar{\nu}_\mu + \nu_\mu$ assuming that the energy is equally shared between the 4 leptons and taking into account that oscillations over cosmic distances distribute the neutrino energy equally among the 3 flavors.

The corrections approximately cancel.

Studying specific models of cosmic ray accelerators one finds that the energy supplied by the black hole to cosmic rays usually exceeds that transferred to pions, for instance by a factor 5 in the case of gamma ray bursts. We therefore estimate that the muon-neutrino flux associated with the sources of the highest energy cosmic rays is loosely confined to the range $E_\nu^2\,dN/dE_\nu = 1 \sim 5 \times 10^{-8}\,\mathrm{GeV\,cm^{-2}\,s^{-1}\,sr^{-1}}$ yielding $10 \sim 50$ detected muon neutrinos per km^2 per year. This number depends weakly on $E_{\max}$ and the spectral slope γ. The observed event rate is obtained by folding the predicted flux with the probability that the neutrino is actually detected in a high energy neutrino telescope; the latter is given by the ratio of the muon and neutrino interaction lengths in the detector medium, λ_μ/λ_ν [2].

This flux has to be compared with the sensitivity of $\sim 10^{-7}\,\mathrm{GeV\,cm^{-2}\,s^{-1}\,sr^{-1}}$ reached during the first 4 years of operation of the completed AMANDA detector in 2000–2003 [5]. The analysis of the data has not been completed, but a preliminary limit of $2.9 \times 10^{-7}\,\mathrm{GeV\,cm^{-2}\,s^{-1}\,sr^{-1}}$ has been obtained with a single year of data [6]. On the other hand, after three years of operation IceCube will reach a diffuse flux limit of $E_\nu^2\,dN/dE_\nu = 2 \sim 7 \times 10^{-9}\,\mathrm{GeV\,cm^{-2}\,s^{-1}\,sr^{-1}}$ in 3 years of operation. The exact value depends on the magnitude of the dominant high energy atmospheric neutrino background from the prompt decay

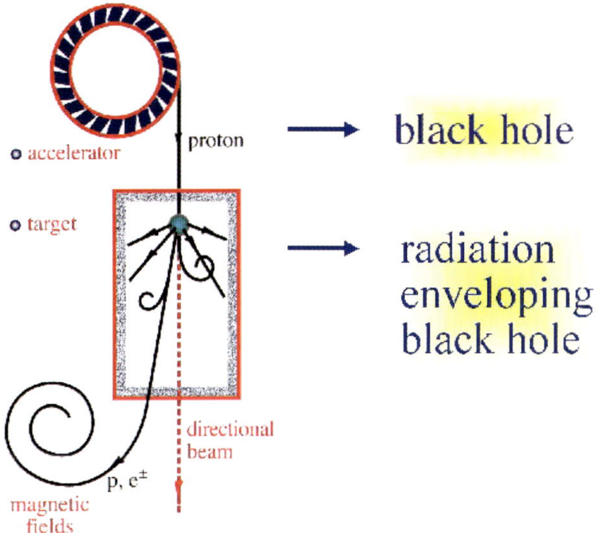

Fig. 2. Cosmic beam dump: sketch of cosmic ray accelerator producing photons and neutrinos.

of atmospheric charmed particles [7]. The level of this background is difficult to anticipate. A cosmic flux at the "Waxman-Bahcall" level will result in the observation of several hundred neutrinos in IceCube [7].

2. Kilometer-Scale detectors

Arguing that a generic cosmic accelerator produces equal energies in cosmic rays, photons and neutrinos, we derived the "Waxman-Bahcall" flux. A kilometer-scale detector is required to detect the roughly $10 \sim 50$ events per km^2 year. Model calculations assuming that active galaxies or gamma-ray bursts are the actual sources of cosmic rays yield similar, or even smaller event rates.

The case for kilometer-scale detectors also emerges from the consideration of "guaranteed" cosmic fluxes. Neutrino fluxes are guaranteed when both the accelerator and the pion producing target material can be identified. Examples include:

- The extragalactic cosmic rays produce $0.1 \sim$ a few events per km^2 year in interactions with cosmic microwave photons. Furthermore, these cosmic rays are magnetically trapped in galaxy clusters and may produce additional neutrinos on the X-ray emitting gas in the cluster.
- Galactic cosmic rays interact with hydrogen in the disk producing an observable neutrino flux in a kilometer-scale detector.
- Air shower arrays have observed a "directional" flux of cosmic rays from the galactic plane, unlikely to be protons whose directions are scrambled in the magnetic field. The flux appears only in a narrow energy range from $1 \sim 3$ EeV, the energy where neutrons reach typical galactic kiloparsec distances within their lifetime of minutes. Both the directionality and the characteristic energy make a compelling case for electrically neutral neutron primaries. For every neutron reaching earth, a calculable number decays into electron antineutrinos before reaching us. Their flux should be observable in neutrino telescopes [8]: from the Cygnus region at the South Pole and from the galactic center for a Mediterranean detector.

In conclusion, observation of "guaranteed" sources also requires kilometer-size neutrino detectors, preferably operated over many years.

Finally and most importantly, with recent observations [9] of the supernova remnant RX J1713.7-3946 using the H.E.S.S. atmospheric Cherenkov telescope array, gamma-ray astronomy may have pointed at a truly guaranteed source of cosmic neutrinos [10]. The observations of TeV-gamma rays from the supernova remnant may have identified the first site where protons are accelerated to energies typical of the main component of the galactic cosmic rays [9]. Although the resolved image of the source (the first ever at TeV energies!) reveals TeV emission from the whole supernova remnant, it shows a clear increase of the flux in the directions of known molecular clouds. This naturally suggests the possibility that protons, shock accelerated in the supernova remnant, interact with the dense clouds to produce neutral pions that are the source of the observed increase of the TeV signal. Furthermore, the high statistics H.E.S.S. data for the flux are power-law behaved over a large range of energies without any signature of a cutoff characteristic of synchrotron or inverse-Compton sources. Other interpretations are not ruled out [11] but, fortunately, higher statistics data is forthcoming.

If future data confirms that a fraction of the TeV flux of RX J1713.7-3946 is of neutral pion origin, then the accompanying charged pions will produce a guaranteed neutrino flux of at least 20 muon-type neutrinos per kilometer-squared per year [10]. From a variety of such sources we can therefore expect event rates of cosmic neutrinos of galactic origin similar to those estimated for extragalactic neutrinos in the previous section. Supernovae associated with molecular clouds are a common feature of the OB associations that exist throughout the galactic plane. They have been suspected to be the sources of the galactic cosmic rays for some time.

It is important to realize that the relation between the neutrino and gamma flux is robust [10]. The $\nu_\mu + \bar{\nu}_\mu$ neutrino flux $(\mathrm{d}N_\nu/\mathrm{d}E_\nu)$ produced by the decay of charged pions in the source can be derived from the observed gamma ray flux by imposing energy conservation:

$$\int_{E_\gamma^{\min}}^{E_\gamma^{\max}} E_\gamma \frac{\mathrm{d}N_\gamma}{\mathrm{d}E_\gamma} \mathrm{d}E_\gamma = K \int_{E_\nu^{\min}}^{E_\nu^{\max}} E_\nu \frac{\mathrm{d}N_\nu}{\mathrm{d}E_\nu} \mathrm{d}E_\nu \tag{1}$$

where $E_\gamma^{\min}$ ($E_\gamma^{\max}$) is the minimum (maximum) energy of the photons that have a hadronic origin. $E_\nu^{\min}$ and $E_\nu^{\max}$ are the corresponding minimum and maximum energy of the neutrinos. The factor K depends on whether the π^0's are of pp or $p\gamma$ origin. Its value can be obtained from routine particle physics. In pp interactions $1/3$ of the proton energy goes into each pion flavor on average. In the pion-to-muon-to-electron decay chain 2 muon-neutrinos are produced with energy $E_\pi/4$ for every photon with energy $E_\pi/2$ (on average). Therefore the energy in neutrinos matches the energy in photons and $K = 1$. This flux has to be reduced by a factor 2 because of oscillations. The estimate should be considered a lower limit because the photon flux to which the calculation is normalized, may be partially absorbed in the source or in the interstellar medium.

3. Neutrino telescopes: first "Light"

While it has been realized for many decades that the case for neutrino astronomy is compelling, the challenge has been to develop a reliable, expandable and affordable detector technology to build the kilometer-scale telescopes required to do the science. Conceptually, the technique is simple. In the case of a high-energy muon neutrino, for instance, the neutrino interacts with a hydrogen or oxygen nucleus in deep ocean water and produces a muon travelling in nearly the same direction as the neutrino. The Cerenkov light emitted along the muon's kilometer-long trajectory is detected by a lattice of photomultiplier tubes deployed on strings at depth shielded from radiation. The orientation of the Cerenkov cone reveals the roughly collinear muon and neutrino direction.

The AMANDA detector, using natural 1 mile-deep Antarctic ice as a Cerenkov detector, has operated for more than 4 years in its final configuration of 667 optical modules on 19 strings. The detector is in steady operation collecting roughly $7 \sim 10$ neutrinos per day using fast on-line analysis software. The lower number will yield a background-free sample all the way to the horizon. AMANDA's performance has been calibrated by reconstructing muons produced by atmospheric muon neutrinos in the 50 GeV to 500 TeV energy range [12].

Using the first 4 years of AMANDA II data, the AMANDA collaboration is performing a search for the emission of muon neutrinos from spatially localized directions in the northern sky. Only the year 2000 data have been published [13]. The skyplot for the 4 years of data is shown in Fig. 3. A 90% upper

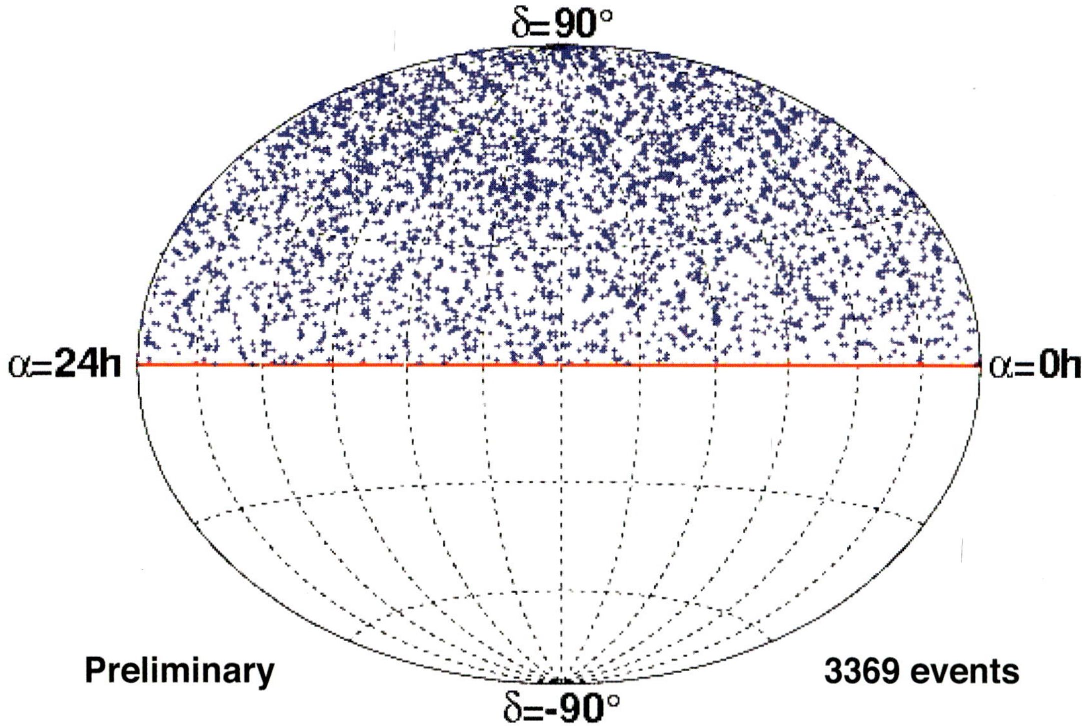

Fig. 3. Skymap showing declination and right ascension of neutrinos detected by the AMANDA II detector during four Antarctic winters of operation in 2000–2003.

limit on the neutrino fluency of point sources is at the level of $6 \times 10^{-8}\,\mathrm{GeV\,cm^{-2}\,s^{-1}}$ or $10^{-10}\,\mathrm{erg\,cm^{-2}\,s^{-1}}$, averaged over declination. This corresponds to a flux of $6 \times 10^{-9}\,\mathrm{cm^{-2}\,s^{-1}}$ integrated above 10 GeV assuming an E^{-2} energy spectrum typical for shock acceleration of particles in high energy sources. In a search for 33 preselected sources, the most significant excess is from the Crab with 10 events where 5 are expected, not significant given the number of trials. IceCube is needed to make conclusive observations of sources. There are several ways to further improve the detector's reach, for instance by including time information on the sources provided by observations at other (photon) wavelengths.

The AMANDA II detector has reached a high-energy effective telescope area of $25{,}000 \sim 40{,}000\,\mathrm{m^2}$, depending on declination. This represents an interesting milestone: known TeV gamma ray sources, such as the active galaxies Markarian 501 and 421, should be observed in neutrinos if the number of gamma rays and neutrinos emitted are roughly equal as expected from cosmic ray accelerators producing pions [10]. Therefore AMANDA must detect the observed TeV photon sources soon, or, its observations will exclude them as the sources of cosmic rays.

4. Mediterranean telescopes

Below PeV energy, South Pole neutrino telescopes do not cover the Southern sky, which is obscured by the large flux of cosmic ray muons and neutrinos. This and the obvious need for more than one telescope – accelerator experiments have clearly demonstrated the value of multiple detectors – provide compelling arguments for deploying northern detectors. With the first observation of neutrinos by a detector in Lake Baikal with a telescope area of $2500\,\mathrm{m^2}$ for TeV muons [14] and after extensive R&D efforts by both the ANTARES [15] and NESTOR [16] collaborations in the Mediterranean, there is optimism that the technological challenges to build neutrino telescopes in deep sea water have been met. Both Mediterranean collaborations have demonstrated

their capability to deploy and retrieve optical sensors, and have reconstructed down-going muons with optical modules deployed for R&D tests.

The ANTARES neutrino telescope is under construction at a 2400 m deep Mediterranean site off Toulon, France. It will consist of 12 strings, each equipped with 75 optical sensors mounted in 25 triplets. The detector performance has been fully simulated [15] with the following results: a sensitivity after one year to point sources of 0.4–$5 \times 10^{-15}\,\mathrm{cm^{-2}\,s^{-1}}$ (note that this is the flux of secondary muons, not neutrinos) and to a diffuse flux of $0.9 \times 10^{-7}\,\mathrm{GeV\,cm^{-2}\,s^{-1}}$ above 50 TeV. As usual, an E^{-2} spectrum has been assumed for the signal. AMANDA II data have reached similar point source limits $(0.6 \times 10^{-15}\,\mathrm{cm^{-2}\,s^{-1}\,sr^{-1}})$ using 4 Antarctic winters, or about 1000 days, of data [13]). This value depends weakly on declination. Also the diffuse limits reached in the absence of a signal are comparable [5]. We have summarized the sensitivity of both experiments in Table 1, where they are also compared to the sensitivity of IceCube.

Given that AMANDA and ANTARES operate at similar depths and have similar total photocathode area (AMANDA II is actually a factor of 2 smaller with 667 8-inch versus 900 10-inch photomultipliers for Antares), the above comparison provides us with a first glimpse at the complex question of the relative merits of water and ice as a Cherenkov medium. The conclusion seems to be that, despite differences in optics and in the background counting rates of the photomultipliers, the telescope sensitivity is approximately the same for equal photocathode area. The comparison is summarized in Table I where we have tabulated the sensitivity of AMANDA and Antares to points sources and to a diffuse flux of neutrinos. At this time, in the absence of a discovery, it is the sensitivity and not the area or angular resolution that represents the relevant quantity of merit. In the same context, the NEMO collaboration has done the interesting exercise of simulating the IceCube detector (augmented from 4800 to 5600 optical modules; see next section) in water rather than ice. One finds a slightly reduced sensitivity in water, probably

Table I. *Comparison of Sensitivities of neutrino experiments.*

	IceCube	AMANDA-II*	ANTARES
# of PMTs	4800/10 inch	600/8 inch	900/10 inch
Point source sensitivity	$6 \times 10^{-17}\,\mathrm{cm}^{-2}\,\mathrm{s}^{-1}$	$1.6 \times 10^{-15}\,\mathrm{cm}^{-2}\,\mathrm{s}^{-1}$	$0.4\text{--}5 \times 10^{-15}\,\mathrm{cm}^{-2}\,\mathrm{s}^{-1}$
(v_μ per year)		weakly dependent on oscillations	depending on declination
diffuse limit[†] (v_μ per year)	$3\text{--}12 \times 10^{-9}\,\mathrm{GeV\,cm}^{-2}\,\mathrm{s}^{-1}\,\mathrm{sr}^{-1}$	$2 \times 10^{-7}\,\mathrm{GeV\,cm}^{-2}\,\mathrm{s}^{-1}\,\mathrm{sr}^{-1}$	$0.8 \times 10^{-7}\,\mathrm{GeV\,cm}^{-2}\,\mathrm{s}^{-1}\,\mathrm{sr}^{-1}$

*includes systematic errors.

[†]depends on assumption for background from atmospheric neutrinos from charm.

not significant within errors and at no energy larger than 50% [17]. Notice that in several years of operation a kilometer-scale detector like IceCube can improve the sensitivity of first-generation telescopes by two orders of magnitude.

5. Kilometer-scale neutrino observatories

The baseline design of kilometer-scale neutrino detectors maximizes sensitivity to v_μ-induced muons with energy above hundreds of GeV, where the acceptance is enhanced by the increasing neutrino cross section and muon range and the Earth is still largely transparent to neutrinos. The mean-free path of a v_μ becomes smaller than the diameter of the earth above 70 TeV – above this energy neutrinos can only reach the detector from angles closer to the horizon. Good identification of other neutrino flavors becomes a priority, especially because v_τ are not absorbed by the earth. Good angular resolution is required to distinguish possible point sources from background, while energy resolution

is needed to enhance the signal from astrophysical sources, which are expected to have flatter energy spectra than the background atmospheric neutrinos.

Overall, AMANDA represents a proof of concept for the kilometer-scale neutrino observatory, IceCube [7], now under construction. IceCube will consist of 80 kilometer-length strings, each instrumented with 60 10-inch photomultipliers spaced by 17 m. The deepest module is 2.4 km below the surface. The strings are arranged at the apexes of equilateral triangles 125 m on a side. The instrumented (not effective!) detector volume is a cubic kilometer. A surface air shower detector, IceTop, consisting of 160 Auger-style Cherenkov detectors deployed over 1 km^2 above IceCube, augments the deep-ice component by providing a tool for calibration, background rejection and air-shower physics, as illustrated in Fig. 4.

The transmission of analogue photomultiplier signals from the deep ice to the surface, used in AMANDA, has been abandoned. The photomultiplier signals will be captured and digitized inside the optical module. The digitized signals are given a global time

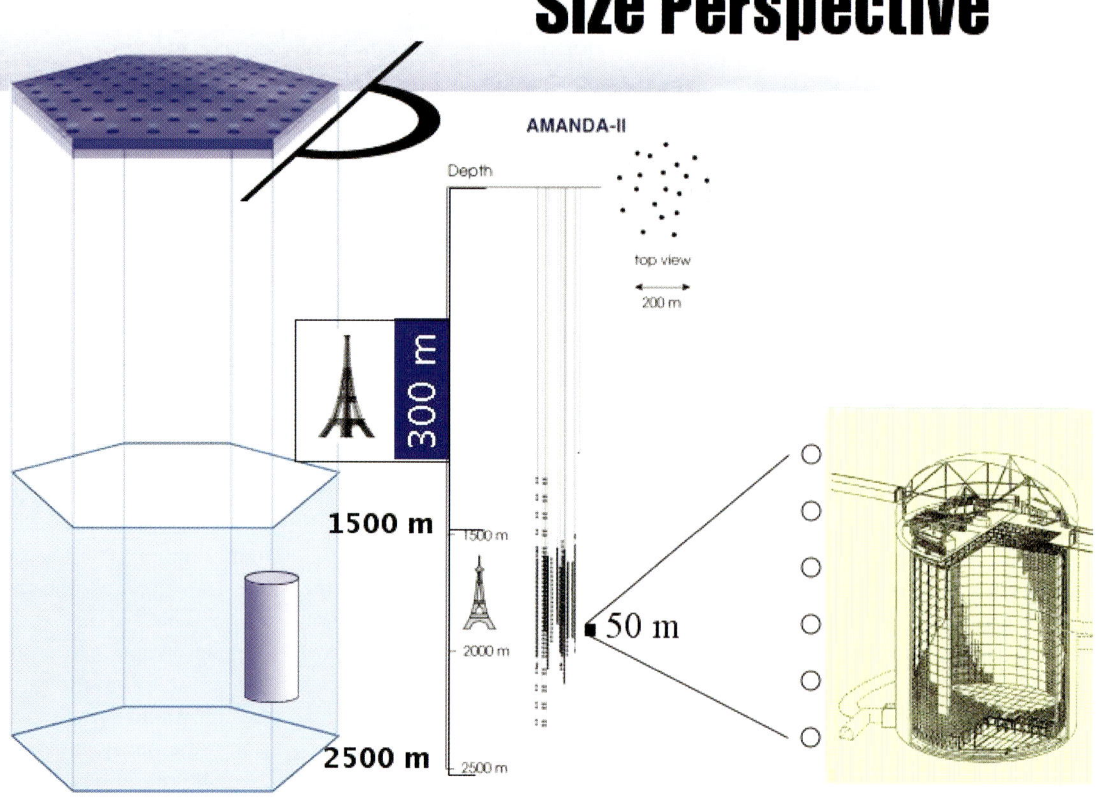

Fig. 4. Relative sizes of the IceCube, AMANDA, and Superkamiokande neutrino detectors. AMANDA will be operated as a lower threshold subsystem of IceCube. As the size of the detector grows, so does the threshold energy of neutrinos detected.

stamp with a precision of <10 ns and transmitted to the surface. The digital messages are sent to a string processor, a global event trigger and an event builder.

Construction of the detector commences in the Austral summer of 2004/2005 and continues for 6 years, possibly less. The growing detector will take data during construction, with each string coming online within days of deployment. The data streams of IceCube, and AMANDA II, embedded inside IceCube, will be merged off-line using GPS timestamps.

IceCube will offer advantages over AMANDA II beyond its larger size: it will have a higher efficiency and superior angular resolution in reconstructing tracks, map showers from electron- and tau-neutrinos (events where both the production and decay of a τ produced by a ν_τ can be identified) and, most importantly, measure neutrino energy. Simulations, benchmarked by AMANDA data, indicate that the direction of muons can be determined with sub-degree accuracy and their energy measured to better than 30% in the logarithm of the energy. The direction of showers will be reconstructed to better than $10°$ above 10 TeV and the response in energy is linear and better than 20%. Energy resolution is critical because, once one establishes that the energy exceeds 1 PeV, there is no atmospheric muon or neutrino background in a kilometer-square detector and full sky coverage of the telescope is achieved. The background counting rate of IceCube signals is expected to be less than 0.5 kHz per optical sensor. In this low background environment, IceCube can detect the excess of MeV anti-ν_e events from a galactic supernova.

NEMO, an INFN R&D project in Italy, has been mapping Mediterranean sites and studying novel mechanical structures, data transfer systems as well as low power electronics, with the goal to deploy a next-generation detector similar to IceCube. A concept has been developed with 81 strings spaced by 140 m. Each consists of 18 bars that are 20 m long and spaced by 40 m. A bar holds a pair of photomultipliers at each end, one looking down and one horizontally. As already mentioned, the simulated performance [18] is, not unexpectedly, similar to that of IceCube with a similar total photocathode area as the NEMO concept.

Recently, a wide array of projects have been initiated to detect neutrinos of the highest energies, typically above a threshold of 10 EeV, exploring other experimental signatures: horizontal air showers and acoustic or radio emission from neutrino-induced showers. Some of these experiments, such as the Radio Ice Cerenkov Experiment [19] and an acoustic array in the Caribbean [20], have taken data; others are under construction,

such as the Antarctic Impulsive Transient Antenna [21]. The more ambitious EUSO/OWL project aims to detect the fluorescence of high energy cosmic rays and neutrinos from a detector attached to the International Space Stations.

Acknowledgments

I thank my AMANDA/IceCube collaborators and Teresa Montaruli for discussions. This research was supported in part by the National Science Foundation under Grant No. OPP-0236449, in part by the U.S. Department of Energy under Grant No. DE-FG02-95ER40896, and in part by the University of Wisconsin Research Committee with funds granted by the Wisconsin Alumni Research Foundation.

References

1. Gaisser, T. K., Proc 31st International Conference on High Energy Physics, Amsterdam, The Netherlands, July (2002).
2. Gaisser, T. K., Halzen, F. and Stanev, T., Phys. Rept. **258**, 173 (1995) [Erratum **271**, 355 (1995)], hep-ph/9410384; Learned, J. G. and Mannheim, K., Ann. Rev. Nucl. Part. Sci. **50**, 679 (2000); Halzen, F. and Hooper, D., Rept. Prog. Phys. **65**, 1025 (2002), astro-ph/0204527.
3. Bahcall, J. N. and Waxman, E., Phys. Rev. D **64**, 023002 (2001).
4. Halzen, F. and Zas, E., Astrophys. J. **488**, 699 (1997), astro-ph/9702193; Gaisser, T. K., OECD Megascience Forum, Taormina, Italy (1997), astro-ph/9707283; Halzen, F. and Zas, E., Astrophysics J. **488**, 669 (1997), astro-ph/9702193.
5. Hill, G. C. *et al.* (AMANDA collaboration), Proc. 28th International Cosmic Ray Conference, Tsukuba, Japan (2003).
6. Ackermann, M. *et al.* (AMANDA Collaboration), astro-ph/0405218.
7. Ahrens, J. *et al.* (IceCube Collaboration), Astropart. Phys. **20**, 507 (2004), astro-ph/0305196 and http://icecube.wisc.edu/science/sci-tech-docs/.
8. Anchordoqui, L. A. *et al.*, Phys. Lett. **B 593**, 42 (2004), astro-ph/0310417.
9. Collaboration, H. E. S. S., Nature **432**, 75 (2004), astro-ph/0411533.
10. Alvarez-Muniz, J. and Halzen, F., Ap. J. **576**, L33 (2002).
11. Hiraga, J. S. *et al.*, astro-ph/0407401.
12. Andres, E. *et al.* (AMANDA Collaboration), Nature **410**, 441 (2001); Phys. Rev. D **66**, 012005 (2002), astro-ph/0205109.
13. Ahrens, J. *et al.* (AMANDA Collaboration), Phys. Rev. Lett. **92**, 171102 (2004), astro-ph/0309585.
14. Balkanov, V. A. *et al.* (Baikal Collaboration), Nucl. Phys. Proc. Suppl. **118**, 363 (2003).
15. Montaruli, T. *et al.* (ANTARES Collaboration), Proc. 28th International Cosmic Ray Conference, Tsukuba, Japan, (2003).
16. http://www.nestor.org.gr/.
17. Migneco, E. CRIS04 Workshop, Catania, Italy, (2004).
18. Coniglione, R. *et al.* (NEMO Collaboration), http://nemoweb.lns.infn.it/publication.htm
19. Frichter, G. M. *et al.*, Phys. Rev. D **53**, 1684 (1996), astro-ph/9507078.
20. Lehtinen, N. G., *et al*, Astropart Phys. **17**, 272 (2002), astro-ph/0104033.
21. Gorham, P. 2002 Aspen Winter Conference on Ultra High Energy Particles from Space, http://astro.uchicago.edu/home/web/olinto/aspen/astroweb.

Physica Scripta. Vol. T121, 112–118, 2005

Neutrino Astrophysics in the Cold: Amanda, Baikal and IceCube

Christian Spiering[*]

DESY, Platanenallee 6, D-15738 Zeuthen, Germany

Received February 8, 2005; accepted February 15, 2005

PACS numbers: 95.55Vj, 95.85 Ry, 96.40 Ty, 95.35+d

Abstract

This talk reviews status and results from the two presently operating underwater/ice neutrino telescopes, NT-200 in Lake Baikal and AMANDA-II at the South Pole. It also gives a description of the design and the expected performance of IceCube, the next-generation neutrino telescope at South Pole.

1. Introduction

More than four decades after deep water Cherenkov telescopes for high energy neutrinos have been proposed [1], two detectors of this type are successfully taking data: NT-200 in Lake Baikal, and AMANDA-II at the South Pole. First components of AMANDA's follow-up project of cubic-kilometer size, IceCube, have been deployed in January 2005.

The science topics of these projects include the search for steady and variable sources of high energy neutrinos like Active Galactic Nuclei (AGN), Supernova Remnants (SNR) or microquasars, as well as the search for neutrinos from burst-like sources like Gamma Ray Bursts (GRB) [2]. Underwater/ice telescopes can also be used to tackle a series of questions besides high energy neutrino astronomy. These include the search for neutrinos from the decay of dark matter particles (WIMPs) and the search for magnetic monopoles or other exotic particles – like strange quark matter or Q-balls (see for reviews [3, 4]). All these topics are addressed not only by Baikal/NT-200, AMANDA and IceCube, but also by the Mediterranean projects reviewed in the talk of J. J. Aubert at this conference [5].

For deep ice detectors there are two modes of operation which are not – or nearly not – possible for detectors in natural water. Firstly, due to the low light activity of the surrounding medium, the photomultiplier (PMT) count rate in ice is only about 1 kHz. This enables the detection of the feeble increase of individual PMT rates as caused, for instance, by multiple interactions of Supernova burst neutrinos over time intervals of a few seconds. Amanda-II is actually monitoring about 90% of the Galaxy for MeV neutrinos from Supernova explosions. Secondly, deep ice arrays can be operated in coincidence with surface air shower arrays [6]. This allows to study questions like the mass composition of cosmic rays up to 10^{18} eV, to calibrate the neutrino telescope, and to use the surface detector as a veto for background rejection.

This paper is organized as follows: In section 2, the design and performance of the two running large neutrino telescopes are sketched, NT-200 in Lake Baikal and AMANDA-II at South Pole. Section 3 is a synopsis of the physics results obtained so far by both detectors (for recent summaries see the talks at the Neutrino 2004 conference [7, 8, 9]. Section 4 desribes design and expected performance of IceCube).

*Email address: christian.spiering@desy.de

2. Design of NT-200 and AMANDA

Underwater/ice neutrino telescopes consist of a lattice of photomultipliers (PMs) housed in transparent glass spheres which are spaced over a large volume in the Ocean, in lakes (like the Baikal telescope) or in ice (as AMANDA and IceCube). The PMs record arrival time and amplitude of the Cherenkov light emitted by muons or particle cascades.

The Baikal Neutrino Telescope NT-200 is operated in Lake Baikal, Siberia, at a depth of 1.1 km. An umbrella-like frame carries 8 strings, each with 24 optical modules (OMs) arranged in pairs. The pressure glass spheres of the OMs contain 37-cm diameter PMs. The instrumented volume forms a cylinder of 70 m height and 42 m diameter. The angular resolution for through-going muons is about $4°$.

The present, final configuration of the AMANDA neutrino telescope is named AMANDA-II. It consists of 677 OMs arranged along 19 vertical strings buried in the glacial ice at the South Pole, with most of the OMs at depths between 1500 and 2000 m. The geometric shape of the array is a cylinder $\sim$500 m high and $\sim$200 m in diameter. Each AMANDA OM houses an 8-inch Hamamatsu photomultiplier. The present angular resolution for muons tracks is about $2.5°$. Figure 1 sketches the configuration of both detectors.

NT-200 takes data since April 1998. It has been deployed in subsequent stages, starting in 1993 with NT-36, the pioneering first stationary underwater array [10]. Components are deployed from an natural fixed platform: the thick ice layer which covers Lake Baikal in February and March, when outside temperatures reach down to $-25°$ (see for more technical details [12]). Figure 2 (left) shows a textbook neutrino event (an upward moving muon track) recorded with the early 4-string configuration of 1996 [11].

The absorption length of deep Baikal water varies between 20 and 24 meters. Since light scattering is small, NT-200 can monitor a volume exceeding its own geometric volume by an order of magnitude. Actually, this is the reason that the sensitivity of NT-200 with respect to high energy, diffuse-flux phenomena is not dwarfed by that of the much larger AMANDA-II (AMANDA is embedded in ice where light scattering is strong and diffuses light from distant sources). In 2005, NT-200 is going to be upgraded with three sparsely instrumented outer strings (NT-200+, see Fig. 1). The three strings will allow a dramatically improved vertex reconstruction of high energy cascades within this volume and will increase the sensitivity to diffuse fluxes by a factor of four (see section 3).

Rather than water, AMANDA uses ice as detection medium. The glacial ice is extremely transparent for Cherenkov light with wavelengths near the peak sensitivity of the OMs: At 400 nm, the average absorption length is 110 m. Scattering, however, is much stronger than in water – the average effective scattering length is only 20 m. Below a depth of 1500 m, both scattering and

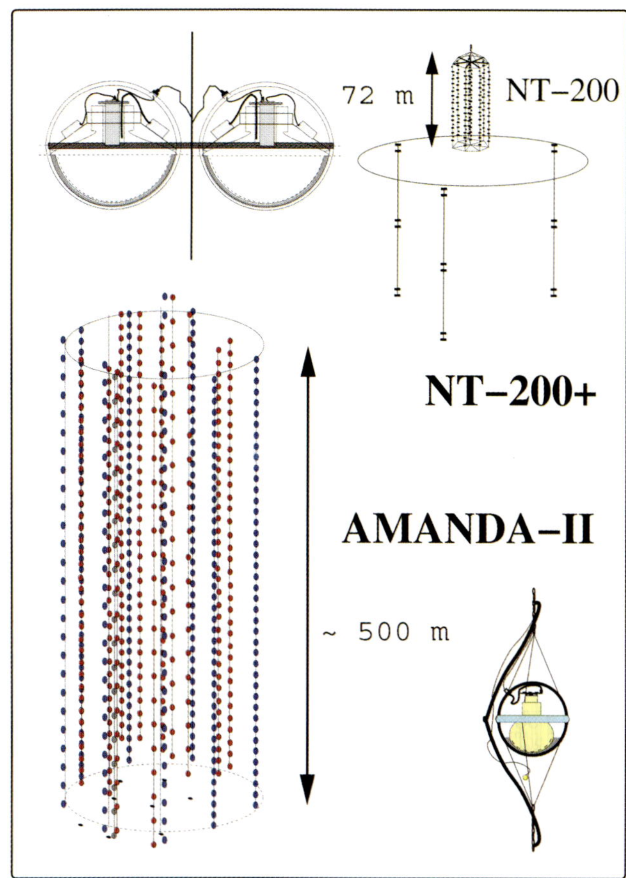

Fig. 1. AMANDA-II and NT-200 (NT200+). For AMANDA-II, only the central part between 1500 and 2000 meters is shown, covering about 90% of the optical modules. Detectors are shown to scale. Also sketched are the optical modules. AMANDA PMs have 20 cm, Baikal PMs 37 cm diameter.

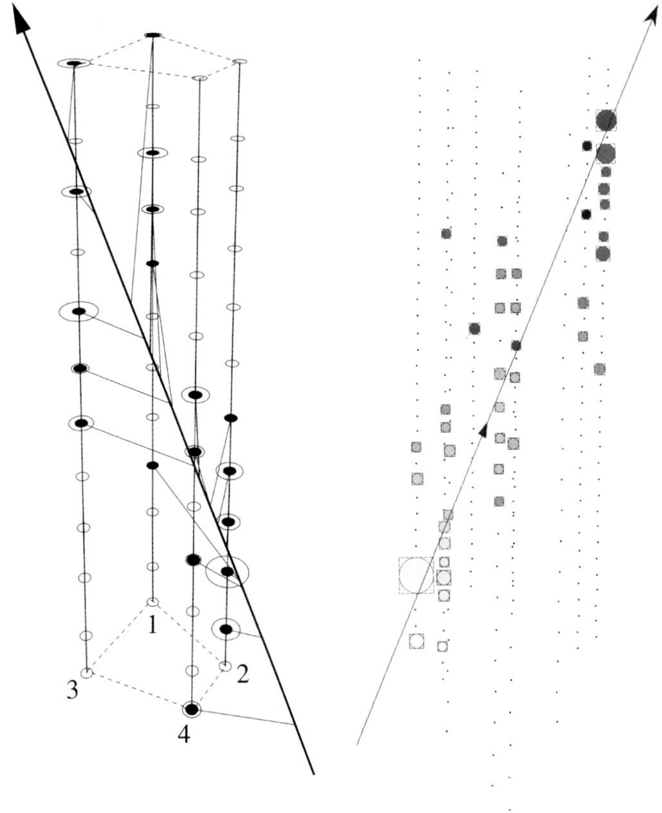

Fig. 2. Left: one of the first clearly upward moving muons recorded with the 1996 four-string-stage of the Baikal detector. Black dots denote PMs. Hit PMs are encircled, with the size of the disc proportional to the recorded amplitude. The arrow line represents the reconstructed muon track, the thin lines the photon paths. Right: Upward muon recorded by the 1997 stage of AMANDA. Dots denote the PMs arranged at ten strings. Hit PMs are highlighted, with the degree of shadowing indicating the arrival time (dark being late), and the size of the symbols the measured amplitude. Note the different vertical scales for these events: 70 m for NT-200, 500 m for AMANDA.

absorption are dominated by dust, and the optical properties vary with dust concentration.

To deploy AMANDA, holes were melted with hot water, and strings with OMs frozen into the ice. Similar to the Baikal telescope, AMANDA was also deployed step by step. An earlier 10-string stage, called AMANDA-B10, was installed between 1997 and 1999 [13]. First neutrinos have been recorded with the inner four strings deployed in 1996. AMANDA-II takes data since 2000.

3. Physics results from NT-200 and AMANDA

3.1. *Proof of principle: atmospheric neutrinos*

For most analysis channels, both AMANDA and NT-200 use the Earth as a filter and separate up-going muons steming from interactions of neutrinos having crossed the Earth. The main class of background are down-going atmospheric muons that are misreconstructed as up-going. After their rejection, basically up-going muons from interactions of neutrinos generated in the atmosphere remain. Atmospheric neutrinos not only constitute the main background when searching for extraterrestrial neutrinos, but are also a natural calibration source. Figure 3 shows the energy spectrum for up-going neutrinos based on AMANDA-II data taken in 2000. It has been obtained by a neural net energy reconstruction, followed by regularized unfolding. This is the first atmospheric neutrino spectrum above a few TeV, and it extends up to 300 TeV.

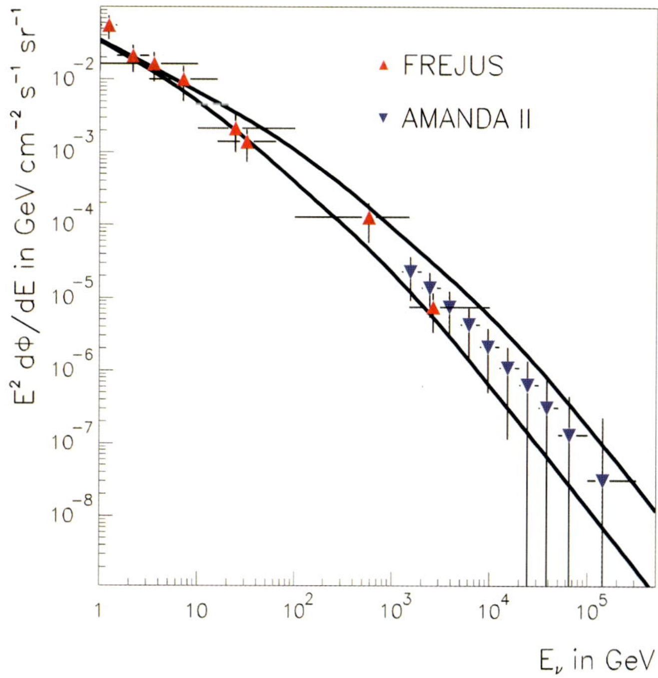

Fig. 3. Atmospheric neutrino energy spectrum (preliminary) from regularized unfolding of AMANDA data, compared to the Frejus spectrum [14] at lower energies. The two solid curves indicate model predictions [15] for the horizontal (upper) and vertical (lower) flux.

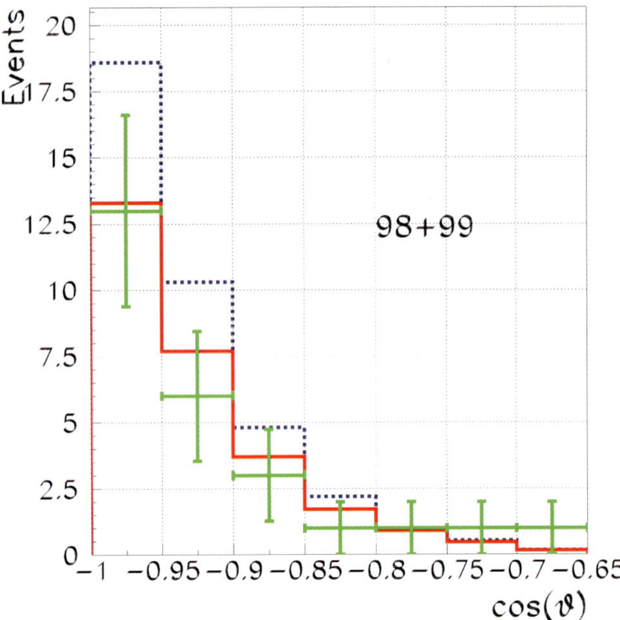

Fig. 4. Angular spectrum of nearly vertical atmospheric neutrinos recorded with NT-200 in 1998/99. $\cos\theta = -1$ corresponds to vertically upward moving tracks. Data are compared to simulated distributions including (full line) and excluding (dotted line) the effect of oscillations.

AMANDA, with its 60 GeV threshold for the standard event selection, is not sensitive to neutrino oscillations. The threshold of the Baikal telescope NT-200 is much lower and, for tight selection procedures, can be reduced down to 10 GeV. Figure 4 shows the angular spectrum for upward tracks close to the opposite zenit. The 20% deficit close to the vertical is well compatible with the the the oscillation parameters $\delta m^2 \approx 2.5 \cdot 10^{-3}$ and $\sin^2\theta \approx 1$. This effect has to be taken into account when using these data as standard signal for detector calibration.

3.2. *Search for a diffuse flux of cosmic neutrinos*

The primary destination of neutrino telescopes is the identification of individual, point-like sources of high-energy neutrinos. Should individual sources be too weak to produce an unambiguous directional signal in the array, the integrated neutrino flux from all sources could still produce a detectable diffuse signal. This flux could be revealed by an high energy excess on top of the omnipresent background of atmospheric neutrinos. The data of both experiments have been searched for such a diffuse signal using complementary techniques in different energy regimes. Event selection is typically optimized to maximize the sensitivity to an E^{-2} signal spectrum. Experimental limits given below include statistical as well as systematic errors (the latter being typically between 20 and 40 percent).

3.2.1. *Upward moving muons:*. The atmospheric neutrino spectrum recorded with AMANDA-II (Fig. 3) was used to set an upper limit on a diffuse E^{-2} flux of extraterrestrial muon neutrinos for the energy range covered by the highest bin, 100–300 TeV, by calculating the maximal non-atmospheric contribution to the flux in that bin given its statistical uncertainty. However, the bins in the unfolded spectrum are correlated and the uncertainty in the last bin can not a priori be assumed to be Poissonian. The statistics in the bin was therefore determined with many Monte Carlo samples used to construct confidence belts. Given the (fractional) unfolded number of experimental events in the bin, a preliminary 90%

C.L. upper limit of $E^2\Phi_{\nu_\mu}(E) < 2.6 \times 10^{-7}\,\mathrm{cm}^{-2}\,\mathrm{s}^{-1}\,\mathrm{sr}^{-1}\,\mathrm{GeV}$ is derived for $100\,\mathrm{TeV} < E_\nu < 300\,\mathrm{TeV}$.

3.2.2. *Cascades.* Apart from muon tracks, *cascades* can be detected. With a typical length of 5–10 m and a diameter of 10 cm, cascades can be considered as quasi point-like compared to the spacing of OMs. All three neutrino flavors contribute to this signature – cascades stem from the leptonic vertex of electron and tau neutrino charged current interactions, and hadronic vertex cascades from all-flavor neutral current interactions.

The AMANDA analysis focuses on contained events (allowing good energy reconstruction), the Baikal/NT-200 survey to bright cascades produced at the neutrino interaction vertex in a large volume around the neutrino telescope. (As pointed out in the introduction, lack of significant light scattering allows to monitor a volume exceeding the geometrical volume of NT-200 by an order of magnitude.)

Both analyses did not yield any excess of candidate events over background: 1 observed event (vs. 0.9 background events) in the case of the year-2000 AMANDA data, no event (vs. 0.4 background events) in the case of the 1998–2000 NT-200 data. The corresponding upper limits with respect to the flux of all three flavors are $\Phi_{\nu_e+\nu_\mu+\nu_\tau}E^2 = 8.6 \cdot 10^{-7}\,\mathrm{cm}^{-2}\,\mathrm{s}^{-1}\,\mathrm{sr}^{-1}\,\mathrm{GeV}$ (AMANDA, $50\,\mathrm{TeV} < E_\nu < 5\,\mathrm{PeV}$ [18]) and $\Phi_{\nu_e+\nu_\mu+\nu_\tau}E^2 = 10.0 \cdot 10^{-7}\,\mathrm{cm}^{-2}\,\mathrm{s}^{-1}\,\mathrm{sr}^{-1}\,\mathrm{GeV}$ (NT-200, $10\,\mathrm{TeV} < E_\nu < 10\,\mathrm{PeV}$ [19]), calculated under the assumption that neutrinos arrive with a ratio $\nu_e : \nu_\mu : \nu_\tau = 1 : 1 : 1$.

3.2.3. *Ultra High Energy neutrinos.* At ultra-high energies (UHE), above 1 PeV, the Earth is opaque to electron- and muon-neutrinos. The AMANDA search for extraterrestrial UHE neutrinos is therefore concentrated on events close to the horizon and even from above. The latter is possible since the atmospheric muon background is low at these high energies due to the steeply falling energy spectrum. The search for UHE events in 1997 AMANDA-B10 data relies on parameters that are sensitive to the expected characteristics of an UHE signal: bright events, long tracks (for muons), low fraction of single photoelectron hits. No excess above background is observed and a 90% C.L. limit on an E^{-2} flux of neutrinos of all flavors is derived, assuming a $1 : 1 : 1$ flavor ratio at Earth [20]: $\Phi_{\nu_e+\nu_\mu+\nu_\tau}E^2 < 9.9 \times 10^{-7}\,\mathrm{cm}^{-2}\,\mathrm{s}^{-1}\,\mathrm{sr}^{-1}\,\mathrm{GeV}$ ($1\,\mathrm{PeV} < E_\nu < 3\,\mathrm{EeV}$).

3.2.4. *Summary of diffuse searches.* Using different analysis techniques, AMANDA and NT-200 yield limits on the diffuse flux of neutrinos with extraterrestrial origin for neutrino energies from 6 TeV up to a few EeV. Figure 5 summarizes the flux predictions, upper bounds derived from observed fluxes of charged cosmic rays and gamma rays, and existing best limits of AMANDA and NT-200 as well as extrapolations to several years of AMANDA data and to IceCube. With the exception of the limit from the unfolded atmospheric spectrum, which can be seen as a quasi-differential limit, the limits are on the integrated flux over the energy range which contains 90% of the signal. These limits exclude some models like [21]. Note that both experiments enter new territory, being below the gamma bound [17] where sensitivities are not *a priori* too weak to hope for a discovery (see also section 4.2).

3.3. *Search for point sources*

Searches for neutrino point sources require good pointing resolution and are thus restricted to the ν_μ channel. Both

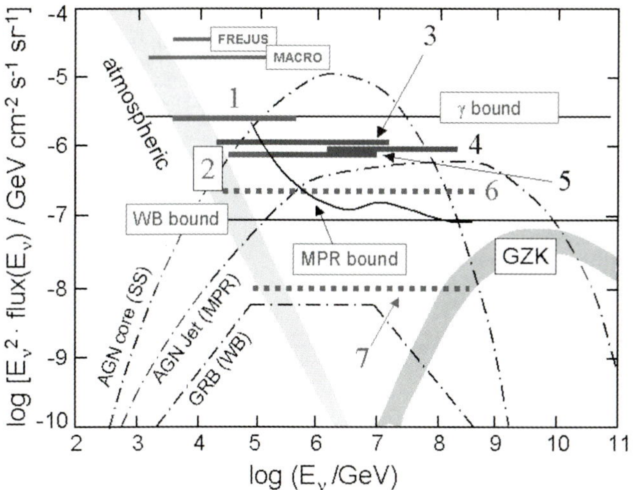

Fig. 5. Theoretical and experimental limits and model predictions to the diffuse flux of extraterrestrial neutrinos, compared to the flux of atmospheric neutrinos. *1)* AMANDA-B10, upward muons [16], *2)* AMANDA-II, upward muons (prelim.), *3)* NT-200, cascades (prelim. [19]), *4)* AMANDA-II, cascades [18], *5)* AMANDA-B10, UHE events [20], *6)* AMANDA-II and NT200+: expectation for 4 years, *7)* IceCube expectation for 3 years. Shown are the limits to the flux of *all* flavors, with the assumption that $v_e : v_\mu : v_\tau = 1 : 1 : 1$ at Earth. Experimental limits obtained for muons alone have been multiplied by a factor of 3. Theoretical bounds and predictions which have been originally given for muons and assuming $v_e : v_\mu : v_\tau = 1 : 2 : 0$ (i.e.without consideration of oscillations) have been multiplied by a factor 1.5. For the MPR and WB bound see [22, 23].

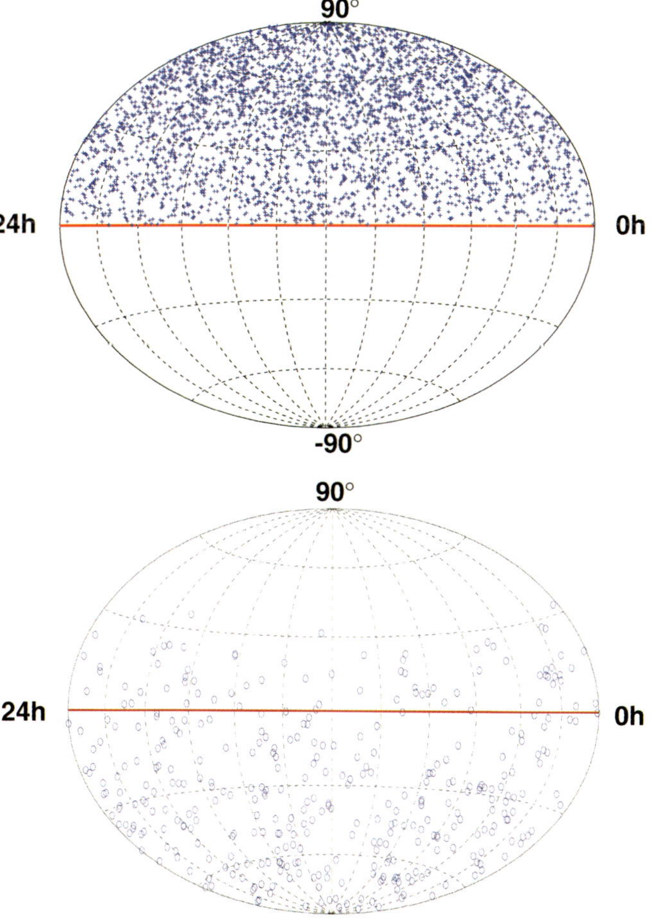

Fig. 6. Sky map in equatorial coordinates. *Top:* 3329 neutrino candidates recorded with AMANDA-II in 2000–2003. *Bottom:* 372 neutrino candidates recorded with NT-200 in 1998–2003 (preliminary).

experiments have produced sky-plots based on data taken over several years. Figure 6 shows 3329 upward moving muons recorded by AMANDA during four years (2000–2003), and 372 Baikal events recorded during 5 years (1998–2003). Since good pointing resolution is mandatory, and since this can be achieved only for tracks crossing the array, the smaller NT-200 cannot compete with AMANDA. We note, however, that for hard neutrino spectra (e.g. E^{-2}) NT-200 is yet the largest telescope at the Northern hemisphere and nicely complements AMANDA in the South. In the following, only AMANDA data will be discussed in more detail.

AMANDA events were selected to maximize the model rejection potential for an E^{-2} neutrino spectrum convoluted with the background spectra due to atmospheric neutrinos and misreconstructed atmospheric muons. The *sensitivity* of the analysis, defined as the average upper limit one would expect to set on a non-atmospheric neutrino flux if no signal is detected, is about $E^2 \Phi_v(E) < 0.6 \times 10^{-7} \,\text{cm}^{-2}\,\text{s}^{-1}\,\text{GeV}$, averaged over the lower hemisphere and for a hypothetical E^{-2} signal spectrum. This is three times lower than the one-year limit for the year 2000 [24]. It corresponds to a neutrino flux above 1 TeV of $F_v(>1\,\text{TeV}) < 6 \cdot 10^{-10}\,\text{cm}^{-2}\,\text{s}^{-1}$ – about a factor five above the Crab gamma flux and close to the gamma flux from Markarian-501 in its flaring phase [25].

The final sample of 3369 neutrino candidates (3329 from below, with 3438 expected atmospheric neutrinos) was searched for point sources with two methods. In the first, the sky is divided into a fine-meshed grid of overlapping bins which are tested for a statistically significant excess over the background expectation (estimated from all other bins in the same declination band). This search yielded no evidence for extraterrestrial point sources. The second method is an unbinned search, in which the sky locations of the events and their uncertainties from reconstruction are used to construct a sky map of significance in terms of fluctuation over background. The hot spots on this map are well within the expectation from a random event distribution. In a search for 33 preselected sources, the strongest excess was observed from the direction of the Crab nebula, with 10 events where 5 are expected – again no significant effect given the number of trials. One thus sees no evidence for point sources with an E^{-2} energy spectrum based on the first four years of AMANDA-II data, and it seems unlikely that another few years of data will change this result. Obviously, a much larger array like IceCube is neccessary to detect *steady* sources. However, the picture may change principally for *transient* sources, where, for searches during known gamma flares (like e.g. the blazars Mkr-421, Mkr-501, ES1959+650), the signal-to-noise ratio may improve dramatically.

3.4. *Search for neutrinos from GRBs*

A special case of point source analysis is the search for neutrinos coincident with gamma ray bursts (GRBs) detected by satellite-borne detectors. Here, the timing of the neutrino event serves as an additional selection handle which significantly reduces background.

Both collaborations have used the GRB sample collected by the BATSE satellite detector which was decommissioned in 2000. The AMANDA (Baikal) and BATSE data taking periods were overlapping in 1997–2000 (1998–2000). Samples containing 312 (368) bursts triggered by BATSE from this period have been analyzed by AMANDA (Baikal). Data were searched for an excess of events in a 10 min (100 second) window around the GRB

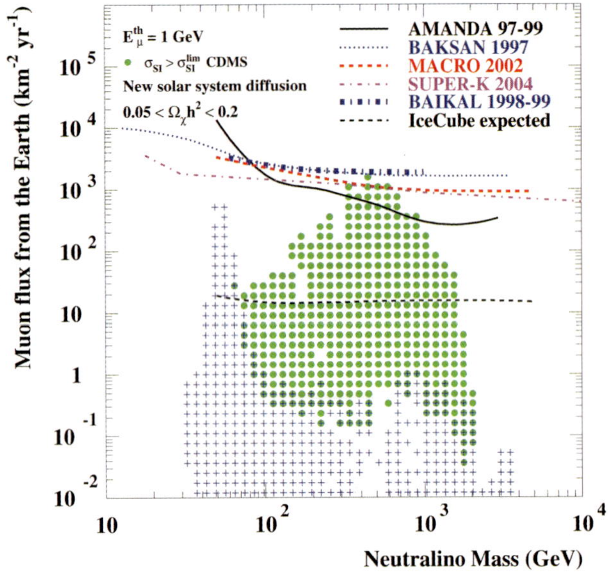

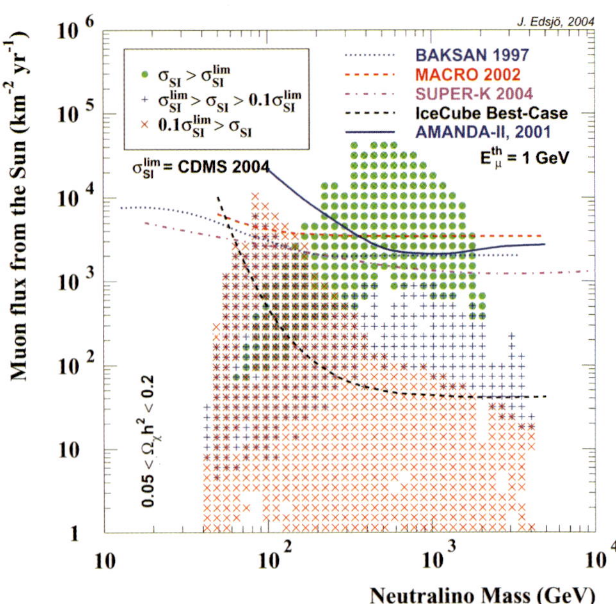

Fig. 7. Limits on the muon flux due to neutrinos from neutralino annihilations in the center of the Earth (left) and the Sun (right). The symbols correspond to model predictions [3] within the allowed parameter space of the MSSM. The dots represent models disfavored by direct searches with CDMS II [27]. Crosses (+) cover models which could be tested by the direct searches ten times more sensitive than the present ones. Amanda and Baikal limits are preliminary.

time. No coincident neutrino event was observed in the case of AMANDA, and one (over a background of 0.46) for Baikal. Assuming a broken power-law energy spectrum as proposed by Waxmann and Bahcall [23], the 90% C.L. upper limit on the expected neutrino flux at the Earth derived by AMANDA is $E^2\Phi_\nu(E) < 4 \times 10^{-8}\,\text{cm}^{-2}\,\text{s}^{-1}\,\text{sr}^{-1}\,\text{GeV}$. This is approximately a factor 15 above the Waxmann-Bahcall flux prediction.

Other classes of bursts are being included in the analysis, like the so-called non-triggered BATSE bursts and triggers from the Third Interplanetary Network (IPN3), since 2000 the major source of GRB detection.

3.5. Dark matter search

The Minimal Supersymmetric extension of the Standard Model (MSSM) provides a promising dark matter candidate in the neutralino, which could be the lightest supersymmetric particle. Neutralinos can be gravitationally trapped in massive bodies, and can then via annihilations and the decay of the resulting particles produce neutrinos. Dark matter can therefore indirectly searched for by looking for fluxes of neutrinos from the center of the Earth or the Sun.

Both collaborations have searched for vertically up-going tracks from the center of the Earth, for AMANDA-B10 using data from 1997–1999, for NT-200 using data from 1998/1999. No indication of an excess over atmospheric neutrinos was found. The 90% C.L. upper limit on the muon flux from the center of the Earth are compared in Fig. 7 (left) to limits obtained from underground experiments and to MSSM model predictions excluded by direct search.

With its larger mass and a higher capture rate due to additional spin-dependent processes, the Sun is more effective than the Earth in catching WIMPs. AMANDA-II data from 2001 data yield no indication of a WIMP signature. The preliminary upper limit on the muon flux from the Sun is compared to MSSM predictions in Fig. 7 (right). For heavier neutralino masses, the limit obtained with less than one year of AMANDA-II data is already competitive with limits from indirect searches with detectors that have several years of integrated livetime. It should be noted that the two

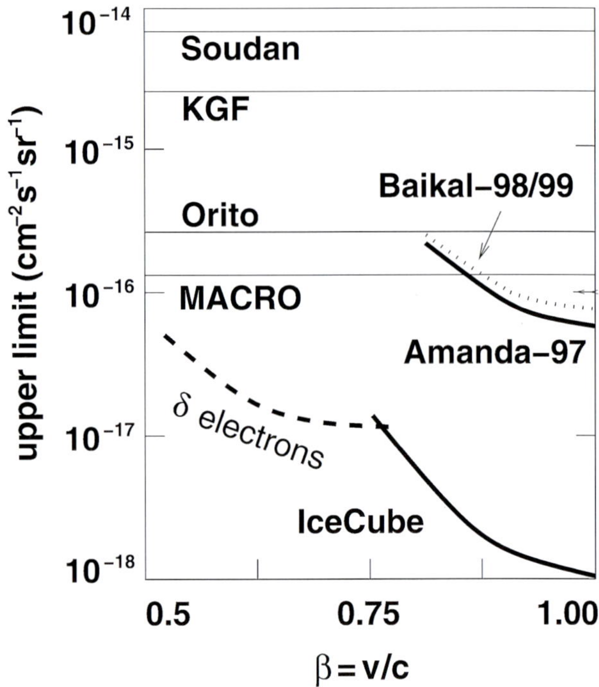

Fig. 8. Upper limits on the flux of relativistic monopoles obtained in different experiments.

methods are complementary since they (a) probe the WIMP distribution in the solar system at different epochs and (b) are sensitive to different parts of the velocity distribution. Direct searches are sensitive to high-energy recoils and therefore to the high-velocity tail of the WIMP flux, indirect searches are more sensitive to low-velocity WIMPs since those are easier captured by celestial bodies.

3.6. Magnetic monopoles

A magnetic monopole with unit magnetic Dirac charge $g = 137/2 \cdot e$ and velocities above the Cherenkov threshold in water ($\beta > 0.75$) would emit Cherenkov radiation smoothly along its path, exceeding that of a bare relativistic muon by a factor of 8300. This is a rather unique signature. Figure 8 summarizes the limits obtained until now. A cube kilometer detector could improve the sensitivity of this search by nearly two orders of magnitude. The search could be extended to even lower velocities by detection of the δ electrons generated along the monopole path.

Limits on the flux of particles moving with with less than $10^{-3}c$, like GUT magnetic monopoles catalyzing baryon decay, Q-balls or nuclearites [4] have been obtained with early stages of the Baikal detector [10, 28]. The trigger system of the future Icecube is flexible enough to search effectively for such particles and to lead to much stronger limits than those of [28].

3.7. *Detection of Supernova bursts*

Due to the low external noise rate, AMANDA is sensitive to the increase of individual counting rates of all PMs resulting from a Supernova burst [29]. AMANDA-II can detect 90% of supernovae within 9.4 kpc with less than 15 fakes per year. This is sufficiently robust for AMANDA to contribute to the SuperNova Early Warning System (SNEWS) with neutrino detectors in the Northern hemisphere. IceCube will monitor the full Galaxy. An alarm would confirm other records within SNEWS but also provide directional information: If several detectors spread around the world would measure the signal front with an accuracy of a few ms, one might determine the supernova direction by triangulation. The high statistics of hits recorded by Icecube would allow a precise measurement of the set-in of the burst [30].

4. IceCube

4.1. *The detector*

With 4800 optical modules on 80 strings, horizontally spaced by 125 m, IceCube [9] covers an area of approximately 1 km^2, with the OMs at depths of 1.4 to 2.4 km below surface. Each string carries 60 OMs, vertically spaced by 17 m. The strings are arranged in a triangular pattern. The configuration of IceCube is shown in Fig. 9. At each hole, one station of the IceTop air shower array [32] will be positioned. An IceTop station consists of two ice tanks of total area 7 m^2.

A first IceCube string has been successfully deployed on January 27, 2005. For IceCube, a new drilling system has been constructed. The power for heaters and pumps of the EHWD is now $\sim$5 MW, compared to 2 MW for AMANDA. This, and the larger diameter and length of the water transporting hoses, results in only 40 hours needed to drill a 2400 m deep hole (three times faster than with the old AMANDA drill). Mounting, testing and drop of a string with 60 DOMs takes about 20 hours, and deployment of ultimately 18 strings per season seems feasible.

The present AMANDA-II detector will be integrated into IceCube. IceCube will deliver efficient veto information for low energy cascade-like events or short horizontal tracks recorded in AMANDA. Horizontal tracks could be related to neutrinos steming from WIMP annihilations in the Sun.

The IceCube OM contains a 10-inch diameter PM HAMAMATSU R-7081. Different to AMANDA, the PM anode signal is digitized within the OM and sent to the surface via electrical twisted-pair cables. Waveforms of the signals are recorded with 250 MHz over the first 0.5 µs and 40 MHz over 5 µs. The fine sampling is done with the Analog Transient Waveform Recorder (ATWR), an ASIC with four channels, each capable to capture 128 samples with 200–800 Hz. The 40 MHz sampling is performed by a commercial FADC. Each pulse time stamped with 7 ns r.m.s. See [31] for more details.

4.2. *Physics Performance*

Figure 10 shows the IceCube effective area for muons after $\sim$10^{-6} reduction of events from downward muons, as a function of the muon zenith angle [33]. Whereas at TeV energies IceCube is blind towards the upper hemisphere, at PeV and beyond the aperture extends above the horizon and allows observation of the Southern sky.

The IceCube sensitivity to diffuse fluxes after three years of data taking is shown in Fig. 5 (section 3.2). The dashed-dotted lines indicate the Stecker and Salamon model for photo-hadronic interactions in AGN cores [21] and of the model of Mannheim,

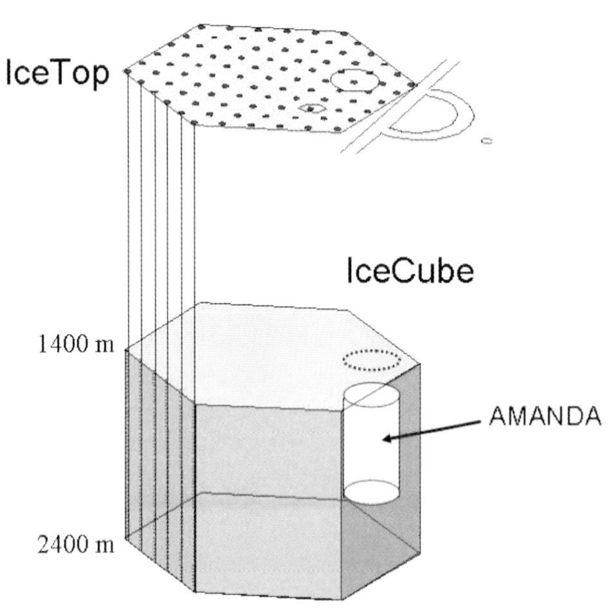

Fig. 9. Schematical View of IceCube, IceTop and AMANDA.

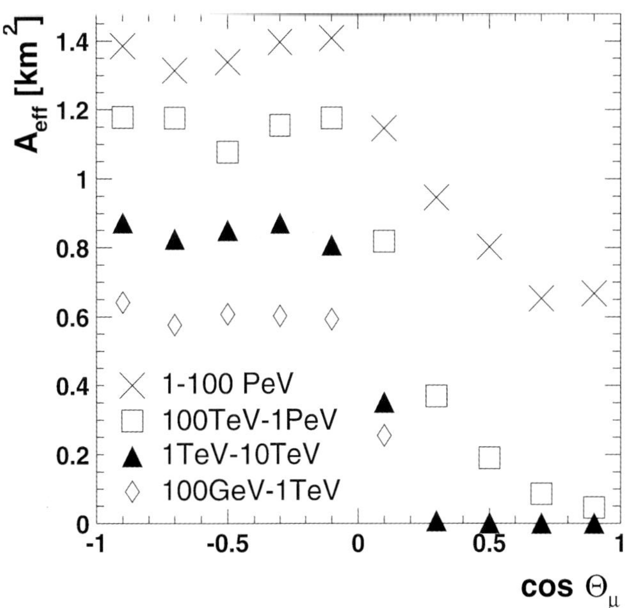

Fig. 10. IceCube effective area for muon as a function of zenith angle. $\cos\theta = -1$ denotes vertically upward moving muons, $\cos\theta = 0$ marks the horizontal direction.

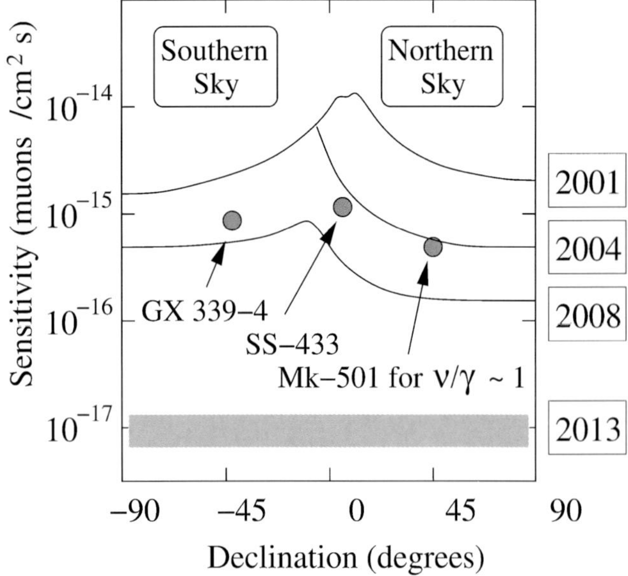

Fig. 11. Scenario for the improvement of experimental sensitivities to TeV point sources. Expected steps for the Northern sky are obtained from Amanda alone and Amanda together with the first 25–30 strings of IceCube, on the Southern sky from the Mediterranean detectors Antares and Nestor (2008). In 2013, both hemispheres will be instrumented with cubic kilometer arrays indicated by the grey band. Shown are also predicted fluxes for two microquasars [35] – one at northern and one on the southern hemisphere - which are just in reach for Amanda and the Mediterranean arrays. As a benchmark, I show also the flux which would be expected if Mk-501, a source spectacular in TeV gamma rays, would produce a flux of TeV muon neutrinos similar to that of gammas emitted in its flaring phase.

Protheroe and Rachen on neutrino emission from photo-hadronic interactions in AGN jets [22]. In case of no signal observed, these models could be rejected with model rejection factor of 10^{-3} and 10^{-2} respectively. Also shown is the GRB estimate by Waxman and Bahcall [23] which would yield of the order of ten events coinciding with a GRB, for 1000 monitored GRBs.

For not too steep angles the IceCube pointing resolution is 0.6 to 0.8 degrees, improving with energy. We expect that evaluation of waveform information will improve these numbers significantly, at least at high energies, and increase the potential for point source identification.

Figure 11 sketches a possible scenario for the point source search over the next decade. Best present limits are from MACRO, Super-Kamiokande (Southern sky) and AMANDA (Northern sky). Baikal limits for the Southern sky will appear soon. This picture will not change until the medium stage Mediterranean detectors come into operation. The ultimate sensitivity for the TeV-PeV range is likely reached by the cubic kilometer arrays. This scale is set by many model predictions for neutrinos from cosmic accelerators or from dark matter decay. A discovery with AMANDA is not yet excluded – be it a flaring blazar or a supernova; for IceCube, hundred times larger than AMANDA

and thousand times larger than underground detectors, it seems extremely likely. However, irrespective of any specific model prediction, IceCube will hopefully also keep the promise for any detector opening a new window to the Universe: to detect *unexpected* phenomena.

Acknowledgements

I thank my colleagues in the collaborations Baikal, AMANDA and IceCube for the long, fruitful cooperation and for helpful discussions.

References

1. Markov, M., Proc. 1960 Int. Conf. High Energy Physics, Rochester (1960), 578.
2. Halzen, F., these proceedings, and Halzen, F. and Hooper, D., Rept. Prog. Phys. **65**, 1065 (2002).
3. Edsjö, J., Proc. Int. Conf. Neutrino-2004, Paris, and references therein.
4. Bakari, D. *et al.*, hep-ex/0004019; Wick, S. *et al.*, Astropart. Phys. **18**, 663 (2003).
5. Aubert, J. J., these proceedings.
6. Ahrens, J. *et al.*, Astropart. Phys. **21**, 565 (2004).
7. Dzilkibaev, Zh.-A., for the Baikal Collaboration, Proc. Int. Conf. Neutrino-2004, Paris.
8. Woschnagg, K., for the Amanda Collaboration, Proc. Int. Conf. Neutrino-2004, Paris (2004), astro-ph/0409423.
9. Botner, O., for the IceCube Collaboration, Proc. Int. Conf. Neutrino-2004, Paris.
10. Belolaptikov, I. *et al.*, Astropart. Phys. **7**, 263 (1997).
11. Belolaptikov, I. *et al.*, Astropart. Phys. **12**, 75 (1999).
12. Spiering, C., for the Baikal Coll., Proc. VLVNT Workshop, Amsterdam (2003), 26, and astro-ph/0404096.
13. Andrés, E. *et al.*, Nature **410**, 441 (2001).
14. Daum, K. *et al.*, Z. Phys. C **66**, 417 (1995).
15. Volkova, L. V., Sov. J. Nucl. Phys. **31**, 784 (1980).
16. Ahrens, J. *et al.*, Phys. Rev. Lett. **90**, 251101 (2003).
17. Berezinsky, V. S. and Smirnov, A. Yu., Astrop. Sp. Sci. **32**, 461 (1975).
18. Ackermann, M. *et al.*, Astropart. Phys. **22**, 127 (2004), astro-ph/0405218.
19. Belolaptikov, I. *et al.*, to be published (see also astro-ph/0404096).
20. Ackermann, M. *et al.*, Astropart. Phys. **22**, 339 (2005).
21. Stecker, F. W. and Salamon, M. H., Space Sci. Rev. **75**, 341 (1996).
22. Mannheim, K., Protheroe, R. and Rachen, J., Phys. Rev. D **63**, 023003 (2001).
23. Waxmann, E. and Bahcall, J., Phys. Rev. D **59**, 023002 (1999).
24. Ahrens, J. *et al.*, Phys. Rev. Lett. **92**, 071102 (2004).
25. See e.g. Horan, D. and Weekes, D. C., New Astron. Rev. **48**, 527 (2004), and astro-ph/0310391.
26. Ahrens, J. *et al.*, Phys. Rev. D **66**, 032006 (2002).
27. Akerib, D. S. *et al.*, astro-ph/0405033.
28. Bezrukov, L. B. *et al.*, Sov. J. Nucl. Phys. **52**, 54 (1990).
29. Ahrens, J. *et al.*, Astropart. Phys. **16**, 345 (2002).
30. Neunhöffer, T., for the Amanda Collaboration, Proc. 27th ICRC, Hamburg (2001), 1125.
31. Karle, A., for the IceCube Coll., Proc. Int. Conf. Neutrino-2002, Munich 388, (2002), and Spiering, C., for the IceCube Coll., Proc. VLVNT Workshop, Amsterdam 21, (2003), and astro-ph/0404090.
32. Stanev, T. and Ulrich, R., for the IceCube Collaboration, astro-ph/0501046.
33. Ahrens, J. *et al.*, Astropart. Phys **20**, 507 (2004), astro-ph/0305196.
34. Spiering, C., J. Phys. G, Nucl. Part. Phys. **29**, (2003), astro-ph/030368.
35. DiStefano, C. *et al.*, Phys. Rev. D **66**, 063004 (2002).

Physica Scripta. Vol. T121, 119–125, 2005

Status of Radio and Acoustic Detection of Ultra-High Energy Cosmic Neutrinos and a Proposal on Reporting Results

David Saltzberg[1,*]

[1] Dept. of Physics and Astronomy, University of California, Los Angeles (UCLA), 475 Portola Plaza, Los Angeles, California, 90095-1547, USA

Received January 8, 2005; accepted February 18, 2005

PACS numbers: 98.70.Sa, 95.55.Vj, 96.40.Tv

Abstract

Neutrino astronomy offers the possibility to perform extra-galactic observations well beyond the photon absorption cutoff above 5×10^{13} eV. Based on observations of cosmic rays, we already know that astrophysical sources produce particles with at least a million times more energy than this photon cutoff. Once discovered, either the nature of the sources themselves or the cross sections of ultra-high energy neutrinos with terrestrial matter may reveal exotic physical processes that are inaccessible to modern accelerators. Some of these processes may be due to as-yet unknown physics at the grand unification scale or beyond. Neutrino telescopes based on optical techniques currently operating and under construction have apertures measured in several km^3-sr. Radio and acoustic detection techniques have been demonstrated in laboratory experiments and are currently used for instrumentation of apertures 10 to 10,000 times larger than optical techniques for neutrinos above 10^{16} eV. I discuss the status of current and proposed neutrino telescope projects based on these techniques. These telescopes have already ruled out some of the more exotic predictions for neutrino intensity. The upcoming generation of radio-based and acoustic-based detectors should be sensitive to cosmic neutrinos above 10^{18} eV originating through the so-called GZK process. A comparison of different neutrino telescopes using a common aperture variable shows how they are complementary in the trade-offs of volume versus threshold. I include a proposal for how neutrino telescopes should report their sensitivities to facilitate direct comparisons among them and to allow testing of neutrino brightness models that appear even after publication of the experimental results.

1. Cosmic neutrinos for astrophysicists and particle physicists

Both astrophysical and particle-physics questions drive the quest to detect high-energy ($> 10^{12}$ eV) cosmic neutrinos. To date, the only two detected neutrino sources, albeit at lower energies (5–50 MeV), have illustrated how such observations can significantly impact both fields. First, the observation of the expected flux of neutrinos from the Sun provided astrophysicists with the first direct view of the central engine, not just surface, of the Sun and confirmed the predictions of the standard solar model. For the particle physicists, measurements of the flavor content of this neutrino flux yielded two compelling facts about neutrinos: neutrinos have mass and the weak neutrino eigenstates are not their mass eigenstates. Second, observations of neutrinos emitted from SN1987a in the Large Magellanic Cloud found the predicted prompt neutrino production from the cataclysmic astrophysical event itself, estimated to be of order 99% of the energy release. For particle physicists, the lack of dispersion of the arrival times of these neutrinos over such a long baseline yielded direct limits on the neutrino mass complementary to the contemporary laboratory measurements, and subject to entirely different systematic uncertainties. To date, the observation of every cosmic neutrino source has resulted in a Nobel Prize.

Since neutrino cross sections increase with energy, while backgrounds due to atmospheric neutrino production decrease, most searches for other sources of cosmic neutrinos operate at high energies. Potential sources for these neutrinos include black holes accreting matter in active galactic nuclei (AGN) [1] and in fireballs associated with gamma-ray bursts (GRB). Other considered sources include so-called topological defects (TD) [2], remnants of the early universe which might produce particles with masses comparable to various unification scales such as 10^{25-28} eV, that in turn decay to particles producing neutrinos. Observations of AGN and GRB neutrino sources would likely yield insight into astrophysical processes responsible for accelerating the highest energy cosmic rays. Alternatively, observation of TD neutrinos could provide insight into elementary particle physics unobtainable through direct production at terrestrial accelerators.

An additional, well constrained theoretically, source of cosmic neutrinos is produced by the GZK process, the inelastic collision of cosmic ray protons with energies above $10^{19.5}$ eV with the 2.7 K microwave background photons. The collisions occur with a mean free path short compared to cosmological distance scales [3] (of order 4 Mpc for protons with energy 6×10^{20} eV). The energy threshold for protons to interact this way, $10^{19.5}$ eV, corresponds to the center-of-mass threshold energy for pion production, about 1200 MeV. The pions' decay chain produces the so-called GZK neutrinos with energies above 10^{17} eV [4], sometimes called the "guaranteed" neutrinos. Assuming our local intensity of such protons is no different than the rest of the universe, the intensity of neutrinos is predicted with only an order-of-magnitude uncertainty [5]. The astrophysical interest in these neutrinos is not that they allow astronomers to look deep into the core of an astrophysical object, but rather that significant deviation from the predicted level would point to a lack of understanding of the relevant energy flows. Note that these predictions are based on proton intensities at 10^{18} and 10^{19} eV, and do not use measurements at or above the GZK cutoff at $10^{19.5}$ eV. The only significant loophole is that the sources themselves could have an independent internal cutoff at $10^{19.5}$ eV, which would be a cruel coincidence.

The observation of any source of cosmic neutrinos would also provide a unique laboratory for particle physicists. Neutrino interactions above 10^{18} eV would probe the electroweak and other interactions at center-of-mass energies beyond those obtainable with modern particle accelerators. Because the standard-model neutrino cross sections are so small, deviations from the standard model due to micro-black hole production or other non-perturbative effects [6] could make them dramatically larger, even by several orders of magnitude. The value of the cross section could be studied by the absorption of neutrinos from a cosmic source through the Earth's crust and perhaps atmosphere [7].

---

*Email address: saltzberg@physics.ucla.edu

By contrast, even though cosmic-ray proton or ion interactions probe similarly large center-of-mass energies, their interactions are hadronic, so these additional channels would produce only a small change to their already large cross sections.

2. Quantifying cosmic neutrino detection

In neutrino astronomy, as in other types of astronomy, it is important to distinguish the concepts of *flux*, *intensity*, and *brightness*. I will use these terms as they appear in radiative transfer theory [8], with the modification that I will consider the transfer of number of neutrinos rather than heat. *Energy* here will refer to the individual energy of each neutrino, E, in analogy to the photon frequency, v, in radiative transfer theory. This convention is consistent with the comprehensive review by Learned and Mannheim [9].

In Table I, I give the names and units used for the relevant variables. Detectors measure a number of neutrinos, N_v, through an area, A, for a given solid angle, $\Delta\Omega$, during a livetime, Δt, perhaps with some event-by-event energy information. For these proceedings I assume that sources produce a neutrino output that is constant in time; however, based on high energy gamma-ray observations of AGN, observers should be prepared to see flaring objects. For diffuse sources, the relevant variables are intensity, I, and brightness, I_E. Often these are denoted as $\mathrm{d}^3N/\mathrm{d}A\,\mathrm{d}\Omega\,\mathrm{d}t$ and $\mathrm{d}^4N/\mathrm{d}A\,\mathrm{d}\Omega\,\mathrm{d}t\,\mathrm{d}E$, respectively. For point sources, the relevant variables are the flux variables given in the table, which I will not consider further here. Sometimes "intensity" is referred to as "diffuse flux". Brightness from isotropic models is commonly abbreviated as $\mathrm{d}N/\mathrm{d}E$, leaving the differentials with respect to area and solid angle as implied, but typically indicated in the units.

The reader should be careful to note the distinction between intensity and brightness. Neutrino intensity is calculated or measured as an integral over energy, so it is not itself a function of energy. The reader will be reminded of the extent of the integral by denoting the intensity as "$I|_{E_1}^{E_2}$". In practice, theorists typically provide predictions on neutrino brightness, I_E, while experimentalists provide sensitivity to neutrino intensities, $I|_{E_1}^{E_2}$.

For quantifying detection and sensitivities I consider that there exists an unknown brightness of ultra-high energy neutrinos which might depend on E, θ, and ϕ. Assuming the brightness is isotropic, it depends only on E. The brightness I_E corresponds to no physical observable and for the experimentalist exists only under

an integral sign:

$$I|_{E_1}^{E_2} = \int_{E_1}^{E_2} I_E(E)\,\mathrm{d}E, \tag{1}$$

where $I|_{E_1}^{E_2}$ is the number of neutrinos per area per steradian per time with energy between E_1 and E_2. (For this reason, brightness may be thought of as an "intensity density", in analogy with the use of "flux density" for F_E, but this is not the common term.)

The properties of a neutrino telescope determine the number of events it will detect. First, note that a simple flat particle counter with 100% efficiency is typically described by the area, A, and solid angle coverage, $\Delta\Omega$. In general, one only considers the product $\alpha \equiv [A\Delta\Omega]$ vs. energy which usually cannot be uniquely factored (For example, even for a simple flat paddle that is sensitive to particles crossing in either direction, $\alpha = [A\Delta\Omega]$ does not unambiguously factor. It has $\alpha = A \times 2\pi$ not $A \times 4\pi$ due to projection effects. This relative factor of one-half can be absorbed either into an effective area or effective solid angle.) A neutrino telescope typically has a very low efficiency for neutrinos but still can be thought of as having an equivalent area with 100% efficiency, which is typically a strong function of neutrino energy. Thus during a time interval Δt, a telescope observe N_v events in the interval $[E_1, E_2]$:

$$N_v = \left[\sum_{i=1}^{3} \int_{E_1}^{E_2} \alpha^i(E)\, I_E^i(E)\,\mathrm{d}E \right] \Delta t, \tag{2}$$

where i is a sum over the three neutrino species. This may be considered the defining equation of $\alpha(E)$ for each neutrino species.

If the number of expected background events is much less than one, one may compare the sensitivities of neutrino telescopes directly using $[A\Delta\Omega](E)\Delta t$. Note that the comparison must be a function of neutrino energy. However, in comparing experiments, one sometimes needs to take into account the expected backgrounds to properly compare the sensitivities. Here I define a quantity [11] $[A\Delta\Omega](E)\Delta t/N_{3\sigma}$, where $N_{3\sigma}$ is the number of events the experiment would need to see a 3σ excess over background. (Note that although high-energy gamma ray detectors often use 5σ as their standard, they are usually looking for point sources in many places on the sky. Here one is concerned with detection of an isotropic intensity for which there are many times fewer trials.) For a background-free experiment, take $N_{3\sigma} = 1$.

Experiments commonly report an effective water-equivalent volume times steradians, $[V^{w.e.}\Delta\Omega]$ rather than $[A\Delta\Omega]$. The variable $[V^{w.e.}\Delta\Omega]$ is defined so that the two descriptions become equivalent in the thin-target approximation via

$$[V^{w.e.}\Delta\Omega](E) \equiv [A\Delta\Omega](E) \times \ell_{int}^{H_2O}, \tag{3}$$

where $\ell_{int}^{H_2O}$ is the mean interaction length for the neutrino in liquid water at that energy. Specifically, $\ell_{int}^{H_2O} = m_{amu}/(\sigma\rho^{H_2O})$, where ρ^{H_2O} is the density of water, m_{amu} is the atomic mass unit and σ is the neutrino cross section per nucleon [12]. In the rare case, typically at very high energies, that the active detector medium itself is large enough to produce some neutrino shielding, it would be less ambiguous to report just $[A\Delta\Omega](E)$.

There are two reasons why the variable $[V^{w.e.}\Delta\Omega]$ is sometimes used. First, the detectors physically occupy volumes; so one can attempt to more sensibly factor this into effective $V_{eff}^{w.e.}$ and $\Delta\Omega_{eff}$. Second, this value does not appear to depend on the neutrino cross section. However, for some high energy telescopes a large portion of their aperture comes from events outside the nominally instrumented volume. Also since neutrino telescopes

Table I. *Definitions of relevant quantities used here. Note that the term "flux density" finds more than one use in the literature, but the more common is for F_E.*

Term	Symbol	With arguments	Units for neutrinos
neutrinos	N_v	N_v	(unitless)
neutrino intensity	I	$I(\theta, \phi)$	$[\text{length}]^{-2}\,[\text{sr}]^{-1}[\text{time}]^{-1}$
neutrino brightness, *or* neutrino specific intensity	I_E,	$I_E(E, \theta, \phi)$	$[\text{length}]^{-2}\,[\text{sr}]^{-1}[\text{time}]^{-1}[\text{energy}]^{-1}$
net flux *flux density*	F_E	$F_E(E)$	$[\text{length}]^{-2}[\text{time}]^{-1}[\text{energy}]^{-1}$
total integrated flux, *or* surface flux *or* flux density	F	F	$[\text{length}]^{-2}[\text{time}]^{-1}$

above 10^{14} eV need to include upstream shielding of neutrinos by the Earth, the independence is far from complete. In practice, each telescope may have a very different dependence on the value of neutrino cross section. Since one usually needs the experiments' own Monte Carlo simulations to know the variance of the sensitivity with cross section, neutrino telescope collaborations would do better to publish either $[A\Delta\Omega](E)$ or $[V^{w.e.}\Delta\Omega](E)$ for values of 75%, 100% and 125% of the nominal cross sections used.

Looking at the typical neutrino brightness predictions [1], [2], [5] in the context of equations 2 and 3, it becomes apparent that for neutrino energies 10^{13} eV to 10^{15} eV, one needs detectors with $[V^{w.e.}\Delta\Omega]$ of order 1–10 km^3-sr. Already these sizes demand instrumenting large natural volumes since they are larger than anything that can be reasonably manufactured. Several telescopes are currently running or under construction that employ optical techniques in the clearest media known: Antarctic ice (Amanda, IceCube) and deep water (Baikal, Antares, Nestor, and Nemo). The limiting size of such detectors is determined by the effective attenuation length of light (including scattering), 50–100 m in these media. This length determines spacing and the practicality of deployment, yielding detectors that may achieve apertures up to 10 km^3-sr.

For energies above 10^{18} eV, one can attempt to detect the GZK neutrinos whose brightness is fairly well understood. The desired size increases to at least 1,000 km^3-sr. So while the standard optical detection techniques are well suited to energies up to 10^{15} eV, larger volumes are needed at higher energies. As will be described in Sections 3 and 4, neutrino-induced showers also produce radio emission up to tens of GHz and acoustic emission to hundreds of kHz. Since in some materials the attenuation length for these emissions may exceed a kilometer, several groups are exploiting these techniques to achieve the extremely large apertures. (Additional information may also be found in another recent review [13].) Two other techniques also offer the possibility of such larger apertures: atmospheric observations from space and exploiting mountain ranges and the finite lifetime of the τ lepton; these will be discussed in Section 5. Based on some lessons learned from this exercise, I present a proposal on reporting results in Section 6.

3. Radio detection

When a neutrino interacts in material, it produces a shower of relativistic particles. For energies above 10^{15} eV, about 20% of the neutrino energy appears as a relativistic hadronic shower through the "inelasticity" of the neutrino interaction with matter. The hadronic shower eventually becomes electromagnetic (electrons, positrons, and photons) through production of π^0 mesons which quickly decay to photons. For charged-current interactions, which are 70% of the interactions at these energies, a charged lepton, e, μ, or τ is also produced with the remaining energy, each of which eventually deposits some electromagnetic energy as well. These particles are typically moving faster than the speed of light in the material, c/n_{refr}, and produce the optical Cherenkov radiation that is the basis for many neutrino telescopes.

In the early 1960's, Gurgen Askaryan realized [14] that a strong radio component (up to $\sim$10 GHz, lasting of order 1 ns) of Cherenkov emission would be produced as well. For interactions in matter, a (10–30)% charge excess of electrons over positrons will develop since the target material contains electrons at rest and no positrons. For wavelengths longer than the lateral size of the shower (a few centimeters in radius) the emission becomes coherent. Unlike the optical radiation, which is incoherent, the power of radio emission grows as the square of the energy of the shower. At 10^{18} eV, about 10^8 electrons radiate in phase, producing a large electromagnetic pulse (of order 500 V/m at 1 m) Askaryan's calculations were subsequently confirmed by independent groups using modern shower simulations based on EGS [15] and with GEANT [16], which are now in basic agreement. The predicted characteristics and intensity of emission were further demonstrated in a series of experiments at SLAC [17]. The square-law dependence of the power in incident neutrino energy, which helps for high neutrino energies, unfortunately also implies a relatively high threshold due to the presence of irreducible thermal noise in any detection medium. Thresholds as low as 10^{17} eV are achieved with this technique, perhaps extendible as low as 10^{16} eV. In his original work, Askaryan presciently pointed out that detectors could be based on large ice sheets, natural salt formations and the lunar surface material which to date still form the basis of this technique.

The RICE [18] telescope consists of 20 dipole antennas sensitive after filtering from 200 to 500 MHz placed on the same strings as the Amanda neutrino telescope at the South Pole. The array comprises a $200 \times 200 \times 200$ m array. The ice on the Antarctic plateau has exceptionally long attenuation lengths, in excess of 1000 m, as extracted from airborne ground-penetrating radar and surface measurements in Greenland, Iceland, laboratory measurements and a new South Pole measurement [19]. Hence, the sensitive volume is much larger than the array itself, achieving above 10 km^3-sr above 10^{18} eV with nearly full-time duty cycle. RICE has placed limits on cosmic neutrinos with energies above 10^{16} eV based on a 3-year dataset and data taking is ongoing. The RICE aperture is compared to current and expected techniques in Figure 1 and Table II. Studies of an extension of this technique, X-RICE [20], to holes spread over an area up to 10^4 km^2 in Antarctica could be sensitive to tens to hundreds of GZK neutrinos per year.

Based on an idea of Zheleznyk and Dagkesamanskii [21], the emission from neutrino interactions in the outer 10–20 m of the Moon's surface is detectable above its thermal

Table II. *Estimated number of events from various models that would be observed in some characteristic telescopes which are being constructed or considered. The WB model is a E^{-2} spectrum with the Waxman-Bahcall [47] coefficient. Disclaimer: these numbers include my own estimates of the detector apertures as described in the text. A more accurate comparison could be achieved if more of the experiments follow the proposal described in the text. Here the Amanda numbers correspond to ν_μ only; they have also recently reported sensitivity to other flavors.*

| | | N_{events} | | | |
| | | | GZK | | |
Telescope	Duration	Top. Def. (PS)	(min)	(max)	WB
Amanda-II	3 live years	0.6	0.02	0.1	125
ANITA	45 live days	44	4.8	18	6.5
Auger	3 live years	0.7	1.0	3.0	1.1
EUSO	2.7 live years	18	0.9	3.6	1.9
IceCube	3 live years	1.1	0.5	1.3	281
RICE	3 live years	3.3	0.9	2.8	1.2
SALSA-1000 Rx	3 live years	50	58	194	56

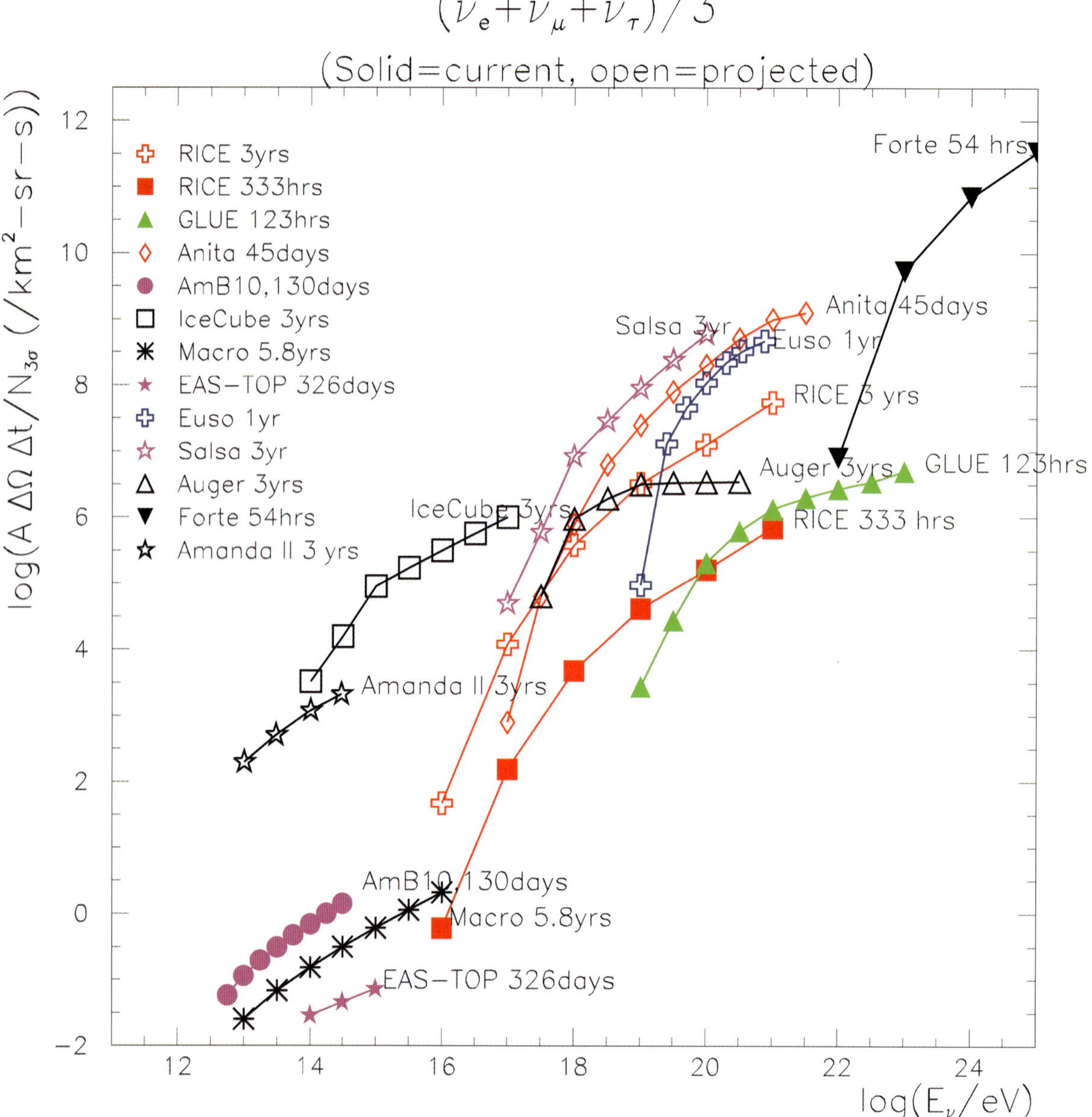

Fig. 1. My attempt to convert several telescopes' sensitivity to a common $[A\Delta\Omega]\Delta t/N_{3\sigma}$, as defined in the text. The quantity presented here is the average over all three neutrinos species, for illustrative purposes, but an analysis breaking it down by flavor, such as in Ref. [40], may also be done. The values presented here should be taken as representative, awaiting official numbers from the collaborations themselves, so comparisons should be taken to within a $\log_{10}(2) = 0.3$ "grain-of-salt" on the vertical and horizontal scales. Current results are presented with solid symbols, projected results are with open symbols. Note that some projected telescopes are funded and under construction, while others are in the conceptual stage. Times are live time. CAVEAT: Because this plot presents the average over three neutrino flavors, telescopes that obtain most of their sensitivity from one flavor (Amanda, Auger) would do better relative to other experiments in pure-flavor plots. When space allows, three separate plots for the individual flavors would be better.

black-body emission for energies above 10^{20} eV by terrestrial radio telescopes. A 12-hour search using the Parkes 64 m dish in Australia [22] did not see any events but had significant radio-frequency interference. A search using radio telescopes at the Goldstone tracking station (GLUE) [23] used two antennas in coincidence to eliminate terrestrial background. GLUE did not see any events after 123 hours of observations with a threshold of 10^{20} eV and achieving $[V^{w.e.}\Delta\Omega]$ of order 500 km^3-sr, although unlike RICE, the duty cycle due to available telescope time corresponded to only 5×10^{-3}. The GLUE aperture is compared to current and expected techniques in Figure 1 and Table II. Future attempts using this technique are underway at the Kalyazin radiotelescope [24].

The FORTE experiment [25] used a space-based platform with a log-periodic antenna array designed for VHF lightning observations to achieve a total of 2.3 days integrated observation of Greenland ice. (Its orbit did not provide a view of Antarctic ice.) Since the radio emission travels through the ionosphere, there is a characteristic frequency-dependent delay which allowed suppression of backgrounds. Because FORTE operated at a lower frequency and bandwidth than other experiments (22 MHz BW from 30 to 300 MHz) it suffered from a higher threshold, 10^{22} eV. However at these frequencies the Cherenkov radio emission diffracts due to the finite length of the shower and thereby fills a large solid angle. FORTE thereby achieves an impressive $[V^{w.e.}\Delta\Omega] \sim 100,000$ km^3-sr. Its duty cycle was $\sim 3 \times 10^{-3}$. The FORTE aperture is compared to current and expected techniques in Figure 1.

The ANITA [26] detector combines many of the most attractive features of the above experiments into one project designed to be sensitive to the GZK neutrinos. ANITA will be a balloon-borne payload of dual-ridged gain horn antennas (200–1200 MHz after filtering) flying for a total of 60 days beginning in the 2006–07 season. At an altitude of 37 km above the Antarctic ice, ANITA will have a threshold of approximately 10^{17} eV and with an instantaneous view of 1.5 million km^3 of ice, yielding an aperture of $\sim$20,000 km^3-sr. Assuming an 18-day flight per year, the duty cycle is an annualized 0.05. A test flight in the 2003–04 season, ANITA-lite [26], indicated a sufficiently radio-quiet environment and successfully demonstrated the essential in-flight systems. This engineering flight worked well enough that a new limit may be extracted from these data. The ANITA aperture is compared to other expected results in Figure 1 and Table II.

A complementary approach that is still in the conceptual stage is to instrument large natural salt formations with antennas much in the style of the RICE detector. Salt domes which are large and dry offer attenuation lengths in excess of a few hundred meters which, accounting for the increased density, correspond to nearly a kilometer water-equivalent attenuation length. Example detectors include the SND [27], SALSA [28] and ZESANA [29] concepts which could instrument many tens of cubic kilometers of salt. Because the antennas are closer to the events than ANITA, the threshold could be an order-of-magnitude lower. Even though $[V^{w.e.}\Delta\Omega]$ is comparable or even lower than ANITA, because the duty cycle is much larger, the ultimate sensitivity of salt-based detectors is the largest of any considered and has the lowest threshold. In addition, there would be closer to full-sky coverage and a larger duty cycle for flaring sources. Because salt-domes are generally covered by an overburden of rock, they are naturally shielded from man-made electromagnetic sources on or above the Earth's surface. The SALSA aperture is compared to other expected results in Figure 1 and Table II.

4. Acoustic detection

In addition to the radio detection idea, Askaryan described how ultra-high energy neutrino interactions could be detected by underwater acoustic techniques [30]. A 10^{21} eV neutrino shower under water deposits 150 J of heat through ionization in a highly localized region. As a result, a hydrothermal pressure impulse lasting of order 10 μs propagates outward as a thin pancake. Verification of the production of this pulse was performed at accelerators at Brookhaven, Uppsala, and ITEP [31].

Noise sources in the ocean include turbulence, surface waves, wind, oceanic traffic, precipitation, biological systems and thermal noise. The noise spectral density goes through a minimum in the 10 kHz range which is also where the power spectral density of the neutrino-induced pressure pulse would peak. Work is underway to characterize the noise environment in solids such as Antarctic ice and salt domes.

Over the years, several detectors have been proposed, generally using hydrophone arrays in the deep ocean. Considerable interest has been renewed lately. R&D for acoustic sensors and site selection was discussed recently at a workshop dedicated to the acoustic technique [32] and can be found among the slides. Hydrophone arrays near Kamchatka (SADCO) [33], Rona (U.K.) [34], and TREMAIL near the Antares site have been used for preliminary environmental measurements and testing of reconstruction techniques. At the acoustic workshop, a large number of groups showed significant and broad progress in development of hydrophones and pre-amplifiers for the Nemo, Antares, and Lake Baikal sites. The existing Lake Baikal sensors

may even show a hint of excess consistent with an acoustical neutrino signal [35] and their data-taking continues. Detector development for solid media such as Antarctic ice and salt domes is also underway [36]. Well-understood calibrators for solid and liquid acoustic detectors are critical; methods under development include submerged implosions and noise from airplanes. Until recently, sensitivity estimates have been largely analytic, but Monte Carlo simulation tools for acoustic detection are being developed [37] that will aid future design and sensitivity estimates.

Here I discuss one acoustic-based telescope in detail, the SAUND detector [38], since it has been carried through to completion, including a publication. Located in the deep sea off of the Bahamas, the SAUND collaboration instrumented 7 hydrophones arranged in a star pattern using a subset of the U.S. Navy's AUTEC array in the Bahamas to detect the impulse in frequency bands 7–50 kHz. The hydrophones are deep (1.6 km) and sheltered by several islands and shoals with little shipping traffic. The observed noise floor, consistent with $v^{-1.7}$ Knudsen noise set the threshold at $\sim$$10^{21}$ eV but could be reduced by finding or building a hydrophone array with closer spacing. Imploding light bulbs at various distances and depths tested the event reconstruction and were consistent with attenuation lengths of 500 to 1000 m. Refraction effects become significant at 1000 m and beyond. The collaboration ran a physics run for 195 days. Although the $[V^{w.e.}\Delta\Omega]$ of about 100 km^3-sr is still too small for these energies, it should be noted that SAUND's geometry was not optimal for the pancake-like shape of the acoustic pulse. There is hope that arrays with optimized geometry and also solids such as large salt domes could provide one or two orders of magnitude more aperture through larger signal and lower backgrounds. The SAUND group is preparing to install SAUND-2 to instrument 1500 km^3 of sea water [39].

A hybrid detector combining both radio and acoustic techniques holds some promise. Due to the extreme difference in the velocities of propagation of radio versus sound waves, and their different polarization properties, a simultaneous detection of the same event using the two techniques might yield interesting information. However, if the thresholds and sizes of the two arrays are significantly different, they would be unlikely to make any simultaneous observations.

Considerable new activity has begun in the last couple of years in the field of acoustic detection techniques for neutrino astronomy after a relatively quiet two decades. The radio and acoustic fields are now so rich that future summaries ought to be divided among two speakers to do the fields justice.

5. Other neutrino telescope techniques

We wish to compare the radio and acoustic techniques to other neutrino telescopes that have run or will turn on that are sensitive to neutrinos above 10^{12} eV. Often it is difficult based on the published materials to convert the reported limits into the sensitivity curve $[A\Delta\Omega]\Delta t/N_{3\sigma}$ versus energy. I have converted their limits in a hopefully reasonable way [40] and show the comparisons in Figure 1. Numbers of expected events based on these estimates and equation (2) are shown in Table II.

In the 10^{12} to 10^{15} eV regime, several detectors have looked for muons produced by v_μ charged-current neutrino interactions or pion decay in the material below them. MACRO was a small detector but ran for nearly 6 live years and set limits on the neutrino intensity between 10^{13} and 10^{16} eV. Recently Amanda-B10, which is a much larger detector, set a stronger limit on the ν_μ intensity between 5×10^{13} and 5×10^{14} eV in less time. The Lake Baikal neutrino telescope [41] reports limits with a threshold around 10^{13} eV for ν_μ interactions. Using a different technique EAS-TOP [42] ran for 326 days and reported limits for E between 10^{14} and 10^{15} eV based on the rate of horizontal extensive air showers. The Amanda-II detector [43], consisting of 19 strings will set even strong bounds. As shown in the figure, from 10^{14} to 10^{17} eV the IceCube [44] detector will greatly increase the current sensitivity. Telescopes based in the Mediterranean (Antares, Nemo and Nestor) [45], using sea water rather than Antarctic ice will both overlap and extend the sky coverage of IceCube. The sensitivities of a few representative optical-based telescopes are compared with the radio and acoustic techniques in Figure 1.

Above about 10^{16} eV, a few other techniques besides radio and acoustic offer the possibility of extending beyond $10 \, \mathrm{km}^3$-sr. The Auger collaboration would be sensitive to ν_τ interactions in nearby mountains; due to the finite lifetime of the τ lepton, they would be able to detect its decay with $[V^{w.e.} \Delta\Omega]$ as large as $10 \, \mathrm{km}^3$-sr above 10^{18} eV. This has the advantage that it can be done parasitically with an existing experiment. The EUSO collaboration [46] showed how observing large volumes of the atmosphere from an orbiting platform could be sensitive to $[V^{w.e.} \Delta\Omega]$ of order $100 \, \mathrm{km}^3$-sr with a high duty cycle, subject to the detector being launched. Still, as shown in the figure, the apertures achievable with radio detection are larger. Radio-based detectors in embedded in salt or ice, with their relatively low threshold, large duty cycle and large sky coverage may eventually be the most versatile ultra-high energy radio-based neutrino detector.

6. Proposal for how neutrino telescopes should report results

Because I do not have access to each collaboration's internal Monte Carlo simulations and/or calculations I took some liberties of interpretation in converting the reported numbers into a common framework. The most common problem was that some limits already included the neutrino cross section and even particular models for neutrino brightness that were hard to untangle. Sometimes "effective area" or "effective volume" was quoted without stating for which solid angle it was defined. Sometimes the "effective volume" did not state if it was for water-equivalent or for the actual medium's density. Sometimes it was not even clear if deadtimes were already included in the reported livetime or absorbed into the limit as an efficiency. As a result of this effort, I propose here how neutrino telescopes should report their sensitivity in the future so that telescopes can be meaningfully compared. Another advantage of following this proposal is that if a new model arrives long after the collaboration is still able and willing to run its simulation tools, meaningful limits can still be extracted via Equation (2). (Dramatically different interaction models, for example changing neutrino cross sections by orders of magnitude, would still require separate study.)

- Quote $[A\Delta\Omega]$ and/or $[V^{w.e.}\Delta\Omega]$ as a function of energy *for each neutrino species*, which I will call here the aperture. Note that this includes upstream attenuation of neutrinos by the Earth.
- Specify which cross sections were used.
- Specify the livetime Δt the observations correspond to. Deadtime corrections belong here, not in the quoted aperture.
- Give the expected background events during the livetime Δt, if any. If this depends on the threshold applied, these steps should be repeated for a few thresholds.
- Quote the aperture for cross sections 75% and 125% the nominal value used. If the effect is non-linear, more variations would be appropriate.
- Quote the aperture for some variation of the inelasticity, y, as well.
- Number of events observed, if any. If significant, this list should be repeated for a range of detector thresholds and observed events.

With the above information, comparisons could be directly made without subsequent interpretation. Also limits on new models could be set using old data.

7. Conclusions

Detection of cosmic neutrinos offers a fertile ground for addressing interesting questions in astrophysics as well as elementary particle physics. The generation of neutrino telescopes coming online within the next 3–5 years will greatly increase the existing apertures and probe theoretically interesting regions. Below 10^{16} eV, neutrino astronomy is the domain of the optical Cherenkov detectors. Above 10^{17} eV, radio-based detectors offer the largest apertures and are well matched to the expected spectrum of GZK neutrinos. The acoustic technique also shows promise if ongoing work to demonstrate lower energy thresholds is successful. The variety of techniques applied to the challenge of neutrino astronomy is impressive. I have presented a proposal for describing the sensitivity of these widely varying techniques in a common framework.

Acknowledgments

The author thanks the Nobel Symposium Committee for its hospitality and organization of an excellent workshop. This work was partially supported by the U.S. Department of Energy's Office of Science and the National Aeronautics and Space Administration. I thank the many representatives of the collaborations mentioned here as well as Amy Connolly, Bob Cousins, and Jay Hauser for helpful discussions.

References

1. Mannheim, K., Protheroe, R. and Rachen, J., Phys Rev. D **63**, 023003 (2001).
2. Yoshida, S. *et al.*, Phys. Rev. Lett. **81**, 5505 (1998); Protheroe, R. and Stanev, T., Phys. Rev. Lett. **77**, 3708 (1996).
3. Greisen, K., Phys. Rev. Lett. **16**, 748 (1966); Zatsepin, G. and Kuzmin, V., J. Exp. Theor. Phys. Lett. **4**, 78 (1966).
4. Berezinsky, V. and Zatsepin, G., Phys. Lett. B **28**, 423 (1969); Sov. J. Nucl. Phys. **11**, 111 (1970).
5. The estimates quoted here are from Engel, R., Seckel, D. and Stanev, T., Phys. Rev. D **64**, 093010 (2001). See references therein for other calculations.
6. See one compilation in Han, T. and Hooper, D., New J. Phys. **6**, 150 (2004). See also Alvarez-Muniz, J. *et al.*, Phys. Rev. D **65**, 124015 (2002); Fiore, R. *et al.*, Phys. Rev. D **68**, 093010 (2003).
7. Kusenko, A. and Weiler, T., Phys. Rev. Lett. **88**, 161101 (2002); Alvarez-Muniz, J. *et al.*, Phys. Rev. Lett. **88**, 021301 (2002).
8. Rybicki, G. and Lightman, A., "Radiative Processes in Astrophysics," (Wiley-Interscience, 1979).
9. Learned, J. and Mannheim, K., Ann. Rev. Nucl. Part. Sci. **50**, 679 (2000).
10. Williams, D., Ph.D. Dissertation, University of California Los Angeles, (2004).
11. Note that the parameter here is $N_{3\sigma}$ divided by what Super-K calls a "green's function" in their fluence analysis, but adapted here to a diffuse intensity case: Fukuda, F. *et al.* (Super-K collaboration), Astro. Phys. J. **578**, 317 (2002).

12. The most common values of cross section used are Gandhi, R. *et al.*, Phys. Rev. D. **58**, 093009 (1998).

13. Nahnhauer, R., Proc. XXI Conf. on Neutrino Physics and Astrophysics, astro-ph/0411715 (2004).

14. Askaryan, G., Soviet Physics JETP **14**, 441 (1962); Askaryan, G., Soviet Physics JETP **21**, 658 (1965).

15. Zas, E., Halzen, F. and Stanev, T., Phys. Rev. D **45**, 362 (1992); Alvarez-Muniz, J. and Zas, E., Phys. Lett. B **434**, 396 (1998) and references therein.

16. Frichter, G., Ralston, J. and McKay, D., Phys. Rev. D **53**, 1684 (1996); Razzaque, S. *et al.*, in Proc. RADHEP-2000, AIP#579 (2001).

17. Saltzberg, D. *et al.*, Phys. Rev. Lett. **86**, 2802 (2001); Gorham, P. *et al.*, astro-ph/0412128 (2004).

18. Kravchenko, I. *et al.*, Astropart. Phys. **19**, 15 (2001); Kravchenko, I. *et al.* Astropart. Phys. **20**, 195 (2003); RICE = Radio in Ice Cherenkov Experiment.

19. For example, Bogorodsky, V. and Gavrilo, V., "Ice: Physical Properties," (Modern Methods of Glaciology, Leningrad, 1980); Bogorodsky, V., Bentley, C. and Gundmandsen, P., "Radioglaciology," (Reidel Publishing, Dordrecht, 1985). A comprehensive list of sources is available in the second entry of Ref. [18]. A direct measurement at the South Pole was recently completed, see Barwick, S. *et al.*, J. Glac., in press with preprint at http://www.lns.cornell.edu/~dzb/tem/RF-eps-im.pdf.

20. Seckel, D. and Frichter, G., Proc. 26th ICRC, AIP#516 (1999); Seckel, D., in Proc. RADHEP-2000, AIP#579 (2001).

21. Zheleznykh, I., Proc. Neutrino **88**, 528 (1988); Dagkesamanskii, R. and Zheleznykh, I., Soviet Physics JETP **50**, 233 (1989).

22. Hankins, T., Ekers, R. and O'Sullivan, J., Mon. Not. R. Astron. Soc. **283**, 1027 (1996).

23. Gorham, P. *et al.*, Phys. Rev. Lett. **93**, 041101 (2004); GLUE = Goldstone Lunar Ultra-high energy neutrino Experiment.

24. Zheleznykh, I. and Dagkesamanskii, R., private communication (2004). See also Beresnyak, A. astro-ph/0310295.

25. Lehtinen, N. *et al.*, Phys. Rev. D **69**, 013008 (2004); FORTE = Fast On-orbit Recording of Transient Events.

26. Silvestri, A., Proc. Int'l School of Cosmic Ray Astrophysics, astro-ph/0411007 (2004); ANITA = ANtarctic Impulsive Transient Antenna.

27. Chiba, M. *et al.*, in Proc. RADHEP-2000, AIP#579; An updated manuscript is in preparation, Chiba, M., private communication (2004); SND = Salt Neutrino Detector.

28. Gorham, P. *et al.*, Nucl. Instr. Meth **A490**, 476 (2002); Saltzberg, D., in Proc. SPIE, vol. 4858, p. 191 (2003); Williams, D. and Connolly, A., in D. Williams UCLA Ph.D dissertation, chapter 5 (2004); Gorham, P. *et al.* astro-ph/0412128 (2004); SALSA = SALt dome Shower Array.

29. van den Berg, A. *et al.*, in KVI Annual Report (2004); van den Berg, A., in Symposium on "Core Business: heat production of the Earth" (2004); ZESANA = ZEchstein SAlt Neutrino Array.

30. Askaryan, G., Sov. J. At. Energy **3**, 921 (1957); Askaryan, G. and Dolgoshein, B., J. Exp. Theor. Phys. Lett. **25**, 213 (1977). For a more complete history see John Learned's slides in Ref. [32] and from RADHEP-2000.

31. Sulak, L. *et al.*, Nucl. Instr. Meth. **161**, 203 (1979); See Nahnhauer, R. and Rostovtsev, A., talks in Ref. [32]. Albul, V. *et al.*, Instrum. Exp. Tech. **44**, 327 (2001).

32. See http://hep.stanford.edu/neutrino/SAUND/workshop/ and links therein; SAUND = Study of Acoustic Ultra-high energy Neutrino Detection.

33. Dedenko, L. *et al.*, in Proc. RADHEP-2000 (2001); See Zheleznykh, I., talk in Ref. [32]; SADCO = Sea Acoustic Detection of Cosmic Objects.

34. See talks by Rhodes, C., Thompson, L. and Waters, D., in Ref. [32].

35. Budnev, N. *et al.*, in manuscript presented at Ref. [32].

36. See talks by Nahnhauer, R. and Fink, M., in Ref. [32].

37. See talks by Waters, D. and Niess, V., in Ref. [32].

38. Vandenbrouke, J., Gratta, G. and Lehtinen, N., to appear in Astrophys. J., astro-ph/0406105; Lehtinen, N. *et al.*, Astropart. Phys. **17**, 279 (2002).

39. Gratta, G., private communication (2005).

40. Details of the calculation can be found at http://www.physics.ucla.edu/~saltzbrg/uhenu.ps (2003).

41. Spiering, C., these proceedings; Spiering, C., in Proc VLVνT Workshop, astro-ph/0404096.

42. Aglietta, M. *et al.*, Phys. Lett. B **333**, 555 (1994).

43. Barwick, S., Proc. 27th ICRC (2001); http://area51.berkeley.edu/manuscripts/20010601xx-ICRC_AmIIv3.pdf

44. Karle, A. *et al.*, astro-ph/0209556 (2002).

45. Sokalski, I. *et al.*, (Antares Collab.) hep-ex/0501003; Tsirigotis, A. *et al.*, (Nestor Collab) Eur. Phys. J. C33, **S956** (2004); Migneco, E. *et al.* (Nemo Collab.), Proc. VLVνT Workshop, 5. (2004).

46. Bottai, S. and Giurgola, S., Proc. 28th ICRC, 1113 (2003); EUSO = Extreme Universe Space Observatory.

47. Waxman, E. and Bahcall, J., Phys. Rev. **D59**, 023002 (1999); Waxman, E. and Bahcall, J., Phys. Rev. **D64**, 023002 (2001).

Physica Scripta. Vol. T121, 126–130, 2005

Detection of Neutrino-Induced Air Showers

A. A. Watson[1]

School of Physics and Astronomy, University of Leeds, Leeds LS2 9JT, UK

Received December 11, 2004; accepted in revised form February 17, 2005

PACS numbers: 96.40−z; 96.40.Tv; 96.40.Pq; 98.70.Sa

Abstract

High energy neutrinos can arise from a variety of processes. Interactions of ultra high energy cosmic rays with radiation and matter lead to secondary particles, some of which create neutrinos. Interactions of the 2.7 K CMB radiation with protons gives rise to pions and neutrons, and neutrons produced in the photodisintegration of heavy nuclei are also a source of neutrinos. There is speculation that super-heavy relic particles with masses of $\sim 10^{21}$ eV are created in the early Universe: if such particles exist then their decay channels are expected to contain neutrinos. The different sources of neutrinos are summarised. High energy neutrinos can be detected through the extensive air showers that they create in the atmosphere and the potential of the Pierre Auger Observatory, now nearing completion, and of the planned EUSO and ASHRA instruments, to detect neutrino-induced air showers, will be described.

1. Introduction

It is now known that cosmic rays of energy above 10^{19} eV exist. These are commonly called Ultra High Energy Cosmic Rays (UHECRs): the highest energy detected thus far is 3×10^{20} eV [1]. The UHECR beam is widely thought to be dominated by baryonic primaries (from protons to iron nuclei) and consequently, as explained below, is expected to contain neutrinos with characteristic energies of $\sim 10^{16}$ eV and $\sim 10^{18}$ eV. The energy and number of these neutrinos depend upon the mode of production and on the distribution of the sources. There is now a realistic possibility of detecting neutrinos of $\sim 10^{18}$ eV using instruments designed to detect giant extensive air showers whilst neutrinos of 10^{16} eV may be detectable by the ICECUBE instrument as set out in an accompanying paper by C Spiering. In this paper I will describe where neutrinos with energies above 10^{16} eV might come from and explain the principles behind the detection of the more energetic ones. I will use the Pierre Auger Observatory, presently under construction and taking data in Argentina, as an example, but also briefly describe the potential of the EUSO instrument planned for the International Space Station and a ground-based device, ASHRA, planned for the observation of neutrinos that interact in mountains in Hawaii.

2. Sources of very high energy neutrinos

2.1. *Proton primaries as a neutrino source*

If UHECRs are protons, produced as a result of some electromagnetic acceleration process in a distant source (Active Galactic Nuclei or AGNs may be such sources), then these protons can interact with matter or radiation close to the AGN, or with radiation in the inter-galactic or interstellar space that they cross *en route* to earth, giving rise to neutrinos through the following reactions:

$$p_{cr} + p \rightarrow p + p + N(\pi^+ + \pi^- + \pi^0) \tag{1a}$$

and

$$p_{cr} + \gamma \rightarrow \Delta^+ \rightarrow p + \pi^0 \quad \text{or} \quad n + \pi^+. \tag{1b}$$

In these equations p_{cr} represents a cosmic ray proton. In reaction (1a) N denotes the multiplicity of pion production. Charged pions and muons are sources of electron and muon neutrinos and near to the sources the ratio of muon to electron neutrinos will be $4 : 2$. However, because of neutrino mass and the associated oscillations, and the immense distances from the sources, the neutrino flux at the earth is expected to consist of nearly equal numbers of electron, muon and tau neutrinos. Reaction (1a) can take place with any protons or nuclei that are encountered as targets but the matter density of intergalactic and interstellar space is so low that material close to sources is the most probable target for what is a sort of cosmic beam dump. The photon field in reaction (1b) can either be close to the source or be one of the radiation fields that the proton traverses while travelling to earth. In particular, the 2.7 K radiation field becomes very important when the proton energy is in the region of 10^{20} eV. Neutrinos can arise from the decay chains of the charged pions of reaction (1a) and from the second decay mode of (1b), from both the π^+ decay chain or from the decay of the neutron. Note that the mean travel distance before decay of a neutron of 10^{20} eV is only about 1 Mpc.

The number of neutrinos created can be predicted with reasonable accuracy if it is assumed that the primary particles are protons produced throughout the Universe. Of course, it is not known whether this is the case as there remain uncertainties about the origin of the very highest energy cosmic rays and the existence or otherwise of the Greisen-Zatsepin-Kuzmin (GZK) cut-off. The muon and electron neutrinos coming from the charged pion decay chain carry about 5% of the pion energy and so are in the range $3 \times 10^{18} < E < 2 \times 10^{20}$ eV. The electron neutrinos from neutron decay carry only about 4×10^{-4} of the neutron energy so that these neutrinos are created in the range $\sim 2 \times 10^{16} < E < 10^{18}$ eV.

2.2. *Iron nuclei as the neutrino source*

The reaction 1(b) occurs when the Lorentz factor, Γ, of the cosmic ray proton is $\sim 10^{11}$, with the threshold being about 6×10^{19} eV. Thus the same reaction, photopion production, will take place when a nucleus of mass A has an energy above the threshold energy of $(A \times 6 \times 10^{19}$ eV$)$ or 3×10^{21} eV for a Fe nucleus. The remarks made above in the context of protons concerning the energy of the daughter neutrinos apply. However, for nuclei of mass A, a further reaction is important for neutrino production: this is photodisintegration. When Γ of a nucleus is $\sim 10^{10}$, single nucleons can be removed from the nucleus with the nucleon that is removed retaining the Lorentz factor of the parent

[1] a.a.watson@leeds.ac.uk

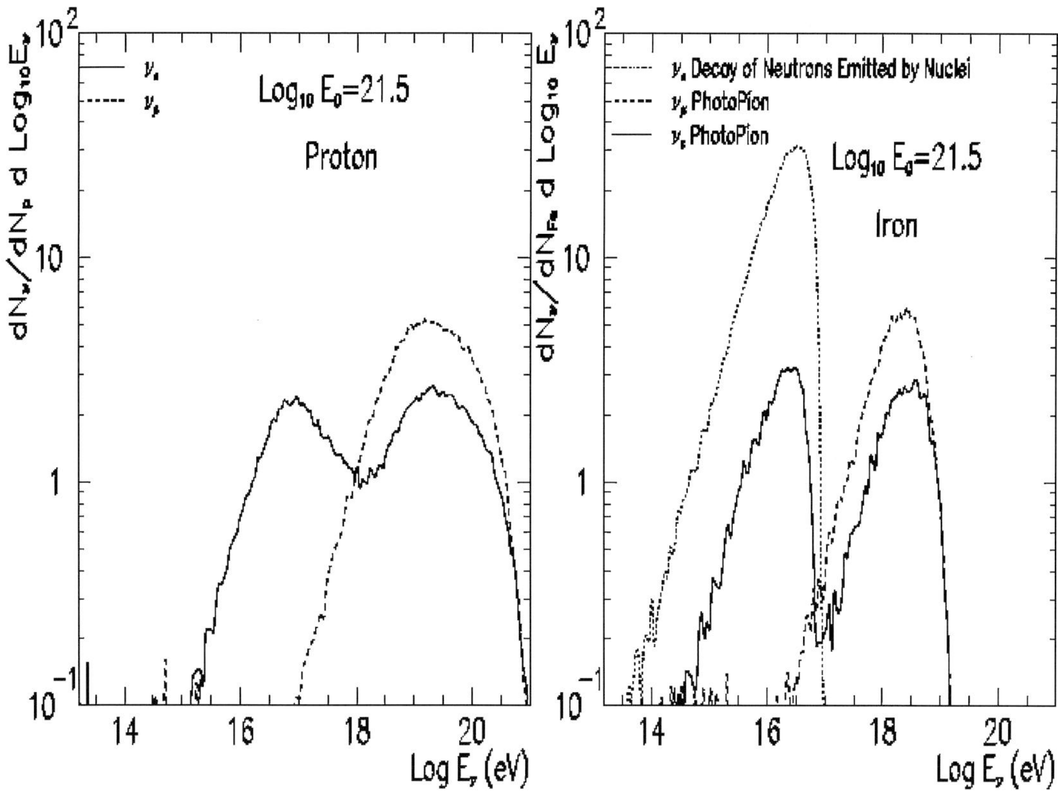

Fig. 1. The neutrino yield for a proton primary (left) and an iron primary (right). In each case the initial energy was chosen as $10^{21.5}$ eV and the propagation distance was 300 Mpc. The different origins of the neutrinos are shown. The dotted line in the right-hand diagram shows the neutrino flux that arises from the decay of neutrons from photodisintegration processes. The figure is from [3].

particle: there is a large cross-section at the giant dipole resonance, near ~ 10 MeV in the laboratory frame for many nuclei. Thus photodisintegration is an important source of neutrinos of energy $\sim 10^{16}$ eV. Multiple nucleon production occurs at higher energies.

2.3. *Summary of some recent calculations*

The fluxes of cosmogenic neutrinos associated with protons and heavy nuclei have recently been calculated by several authors. An important calculation for the proton beam was described in [2]. Similar calculations, but with extension to the propagation of heavier nuclei, have been reported in [3, 4]. As an example, figure 1 shows the neutrino flux reported in [3] for protons and iron nuclei, supposing each has an initial energy of $10^{21.5}$ eV and started propagation at 300 Mpc from the point of observation. There are features in these plots that quantify the statements made above. For proton primaries, most of the neutrinos are muon neutrinos with energies peaked near 10^{18} eV. For iron primaries, the flux of higher energy neutrinos is reduced and there is a large flux of electron neutrinos, from neutron decay, with their energy peaked at 10^{16} eV. The results of this illustrative calculation have been extended [3] to a distribution of cosmological sources using the parameterisation of Waxman [5]. The outcome is shown in figure 2. Here the good agreement with the original work of [2] for proton primaries is apparent. The solid line in figure 2 corresponds to the Waxman and Bahcall limit [6] that has been derived with the same source evolution model and other parameters [5] together with a maximally thin source where the energy input into neutrinos and protons is equal. It is evident that the numbers of neutrinos produced during propagation are comparable at the higher energies to those from maximal production in the sources.

The neutrino fluxes estimated in [3] and [4] are similar. At the present stage of our understanding of the observational and origin problems, it is not evident that more detailed calculations are particularly useful.

2.4. *Neutrinos from super-heavy relic particles*

An idea to explain the UHECRs that have been reported beyond 100 EeV is that super-heavy relic particles with masses of $\sim 10^{12}$ GeV, produced in the early Universe, may decay to produce what is observed [7]. While details of the fragmentation of these particles remains a matter of debate, it is generally accepted that the resulting UHECR beam would contain copious fluxes of neutrinos and photons at $\sim 10^{20}$ eV. Indeed it has been suggested that most of the UHECR are photons, although experimental evidence suggests that the photon flux is less than 40% of the baryonic flux above 10^{19} eV [8]. The neutrinos arising from the decay would be additional to any cosmogenic neutrinos from more conventional sources and have energies above 10^{19} eV. While the flux of neutrinos is uncertain it is generally agreed that the super-heavy relic particles would cluster in the halo of galaxies so that an anisotropy in the general region of our galactic centre might be expected.

3. Methods of detecting high energy neutrinos using extensive air showers

Neutrino primaries may be detectable by studying very inclined air showers. This idea was first proposed by Berezinsky and Smirnov [9] and was re-examined in the context of the Auger Observatory by Capelle *et al.* [10]. The trick is to study the properties of showers that arise at very large angles ($>70°$) from the vertical.

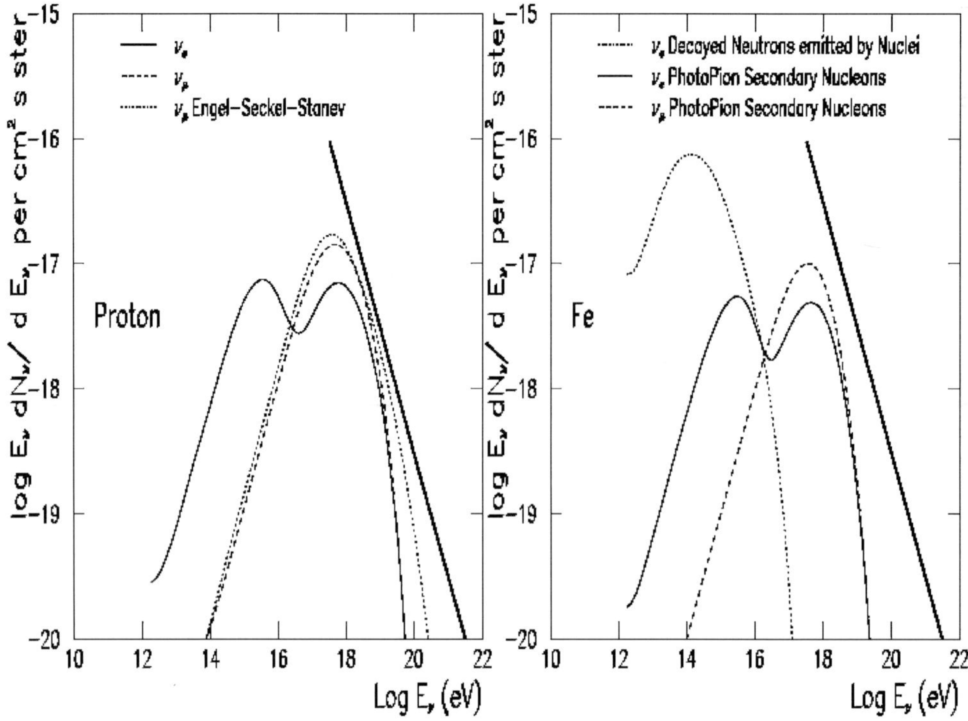

Fig. 2. Electron and muon neutrino fluxes obtained from the nominal choice of astrophysical and cosmological parameters used in [3] and taken [5]. The protons (left) and iron (right) primaries were assumed to have a maximum energy at production of $4Z \times 10^{20}$ eV. The proton flux from the Waxman and Bahcall model [6] is represented by a solid line. The figure is from [3].

A neutrino can interact anywhere in the atmosphere with equal probability. However, if one restricts a search to large zenith angles then it should be possible to identify occasions when the neutrino has interacted deep in the atmosphere. The mode of identification depends on the detection technique.

3.1. *Prospects for detection with the Pierre Auger Observatory*

A neutrino-induced shower arriving at a large zenith angle has distinctive characteristics that make it possible to envisage detecting it with a conventional, ground-based, air shower array as discussed in [9, 10]. Most showers detected at large zenith angles will have been produced by baryonic primaries. The vast majority of the particles detected in such events will be high energy muons as at $>70°$ the large atmospheric thickness of more than $2500\,\mathrm{g\,cm}^{-2}$ (for the depth of the Auger Observatory) filters out the electromagnetic radiation that arises from neutral pion decay. The muons are accompanied by a small fraction of electromagnetic component (around 20%) that is in time and spatial equilibrium with the muons. This electromagnetic component has its origin in muon bremsstrahlung, pair production, knock-on electrons and muon decay. These showers have a large radius of curvature as the source of the muons is far from the shower detector. The particles in the shower disc arrive tightly bunched in time and the distribution of the signal size is rather flat across the array. Such events have been known about for some time and studied in some detail [11, 12] using data from the Haverah Park array. By contrast, a shower produced by a neutrino, if it interacts in the volume of air over the detector, will have a curved shower front, a steep fall-off of particle signal with distance from the shower core and a distinctively broad time spread of the particles at the detectors. The only instrument which is currently large enough to have any prospect of detecting neutrinos, and with the ability to exploit these characteristics, is the Pierre Auger Observatory.

The Pierre Auger Observatory is planned as an instrument with sites in the Northern and Southern Hemisphere. Each site will contain 1600 Cherenkov detectors holding 12 tonnes of water spread out over $3000\,\mathrm{km}^2$. The water tanks will be overlooked by a set of 4 fluorescence detectors each capable of detecting the faint light produced from N_2 molecules excited by the shower particles as they traverse the atmosphere. This hybrid method is very powerful for understanding the shower development and for making a model-independent estimate of the primary energy. The southern part of the Observatory, sited in Pampa Amarilla near Malargüe, Argentina, is nearing completion. As at December 2004, 2 of the 4 fluorescence detectors have been completed and are overlooking 557 water tanks. All these devices are taking data. The Pierre Auger Observatory is now the largest shower detector ever constructed and by March 2005 will have an achieved an exposure that exceeds the famous AGASA array. A description of the prototype instrument has been given in [13]. Distinctive and novel features of the water detectors of the Auger Observatory are the FADC records that are obtained from each of the three 9" photomultipliers that view the individual water volumes. In figure 3 the FADC records in a near vertical shower (13°) are compared with those in one at 76° from the zenith where the shower penetrated $\sim$4 times as much matter. These events were registered with the prototype array. The broad FADC traces, the curved shower front (4 km as compared with 27 km) and the steep fall-off of signal size with distance from the shower core evident in the near vertical event are the signatures that will be sought in inclined events and, if any are found, used to assess the events as neutrino candidates [14].

Although the air volume up to 1 km above the $3000\,\mathrm{km}^2$ contains about $3\,\mathrm{km}^3$ water equivalent, it is not clear that any neutrinos will be detectable. The ν_μ and ν_e from the GZK processes probably have too low a flux though some ν_τ, expected in the primary beam because of neutrino oscillations, may be discovered

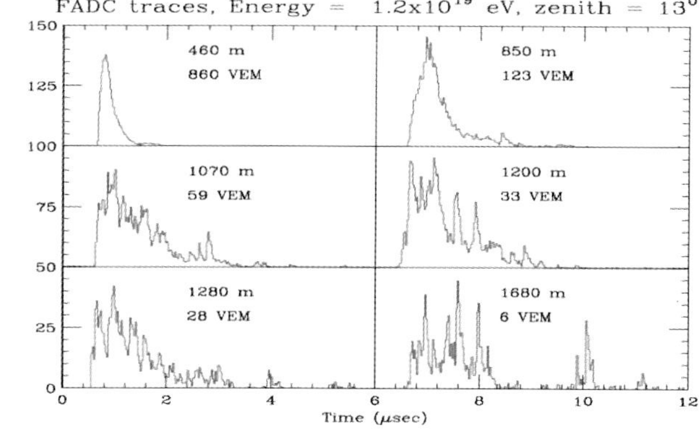

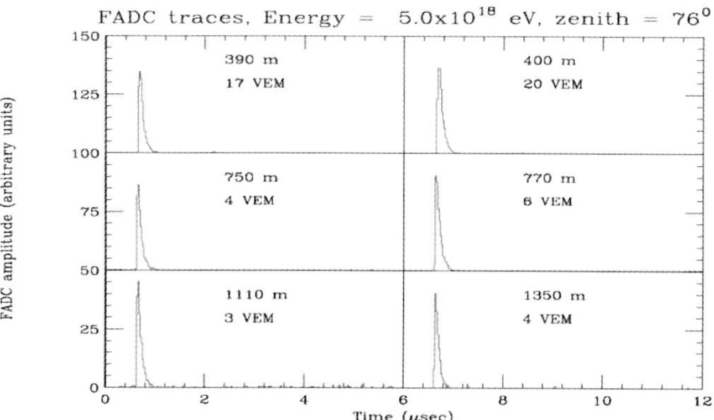

Fig. 3. The upper plot shows FADC traces from a 13° shower: the radius of curvature was 4 km, and the density ratio 134 over a distance ratio of 3.7. By contrast the lower figure shows a shower of 76° which has a radius of curvature of 27 km and a density ratio of 7.5 for a distance ratio of 3.5. Picture from Auger Collaboration data: from J. W. Cronin [14].

[15] as might be neutrinos from some models for AGNs [16] and GRBs [17].

The sensitivity of the Auger Observatory to tau neutrinos of $> 10^{18}$ eV is particularly interesting as here one can take advantage of ν_τ interactions in the rock or the Andes Mountains in the vicinity of the detector. At 10^{18} eV the mean free path in rock is about 300 km and taus will travel about 50 km, escaping from the rock to create showers over the surface detector. At 10^{18} eV at least 5 water-tanks will be struck. Only a few such events are expected each year but these will be readily detectable because of their distinctive signature and direction [15].

3.2. *Prospects for detection with EUSO, the extreme Universe Observatory*

Attaining the ability to monitor an air mass greater than is possible with the Auger Observatory has led to the concept of observing the fluorescence light produced by showers from a detector in space. A promising line is the development of an idea due to Linsley [18] in the form of the EUSO instrument that has been conceived to fly on the International Space Station. This instrument would have the capability to monitor $\sim 10^5$ km^2 sr (after allowing for an estimated on-time of 8%). The neutrino events will be identified as developing deep in the atmosphere at large angles from the zenith and the threshold will correspond to an energy of 5×10^{19} eV. An Italian-led collaboration has driven the design [19]. Unfortunately, at the time of writing (December 2004), the future of this imaginative project is unclear because of uncertainties about funding for the Space Station. This is

disappointing as the technique offers one of the only ways to push to energies beyond whatever limits are found with the Auger instruments.

3.3. *Prospects for detection with the ASHRA instrument*

A ground based experiment, ASHRA, is planned for Hawaii, where interactions of tau neutrinos in mountains will be sought using fluorescence telescopes with very high angular resolution. This is a novel device which will consist of 12 light collection detectors covering 2π steradians visible from one site with an 80 mega pixel array of CMOS sensors. The current candidate station sites are locations near the summits of the three mountains of Mauna Loa, Hualalai, and Mauna Kea on the Hawaii Big Island. These have been chosen after taking into account the need for redundant observation with stereo aperture for UHECRs, the atmospheric purity, the rate of fine weather, low light pollution and the accessibility. Details of the project, which is Japanese led, with support from scientists in Taiwan and the USA, can be found in [20]. In addition to searching for neutrino interactions in the nearby mountains, the ASHRA detector will be used for more conventional studies of UHECRs and for ground based gamma ray astronomy about 100 GeV. A particular feature of the instrument is the very fine pointing accuracy, claimed to be 1 arc min of resolution. If a neutrino signal is detected this angular resolution will be of great value in locating the sources of the neutrinos and of some of the charged cosmic rays, which are expected to be spread over more than a degree because of intergalactic and galactic magnetic fields.

4. Conclusions

The detection of high energy neutrinos by the extensive air shower technique has still to be demonstrated but with the Pierre Auger Observatory there seems some prospect of discovering, at the very least, tau neutrinos from cosmogenic processes. What will be seen depends on such unknowns as the neutrino cross-section, the nature of UHECR sources and the nature of the baryonic primaries. Indeed, as is clear from figures 1 and 2, if the neutrino energy spectrum could be measured then there would be important insight into the nature of the primary particles as iron primaries will produce fewer high energy neutrinos above 10^{18} eV than will protons. However, the present challenge is to detect any neutrinos at all: the measurement of the spectrum with the detail required to aid mass resolution will take some decades.

Acknowledgements

I would like to thank the organisers for the invitation to participate in the Nobel Symposium on Neutrino Physics. Many enlightening discussions with my co-authors of reference [3] and with Enrique Zas are gratefully acknowledged. Work on ultra high energy cosmic rays at the University of Leeds is supported by PPARC.

References

1. Bird, D. *et al.*, Phys. Rev. Lett. **71**, 3401 (1993).
2. Engel, R., Steckel, D. and Stanev, T., Phys. Rev. **D64**, 09301 (2001).
3. Ave, M., Busca, N., Olinto, A., Watson, A. A. and Yamamoto, T., Proc. CRIS04, Catania, Nucl. Phys. B Proc. Suppl. vol. 136 (2004); Ave, M., Busca, N., Olinto, A., Watson, A. A. and Yamamoto, T., Astropart. Phys. **23**, 19 (2005).

4.   Hooper, D., Taylor, A. and Sarkar, S., Astropart. Phy. **23**, 11 (2005), astro-ph/0407618.
5.   Waxman, E., Astrophys. J. **452**, L1 (1995).
6.   Waxman, E. and Bahcall, J., Phys. Rev. D **59**, 023002 (1999).
7.   Berezinsky, V., Kachelreiss, M. and Vilenkin, A., Phys. Rev. Lett. **22**, 4302 (1997); Benakli, K., Ellis, J. and Nanopolous, D. V., Phys. Rev. D **59**, 047301 (1999); Birkel, M. and Sarkar, S., Astropart. Phys. **9**, 297 (1998); Chisholm, J. R. and Kolb, E. W., astro-ph/0306288, submitted to Phys. Rev. D; Chung, D. J. H., Kolb, E. W. and Riotto, A., Phys. Rev. Lett. **81**, 4048 (1998); Rubin, N. A., M. Phil. Thesis, University of Cambridge, (1999); Sarkar, S. and Toldra, R., Nucl. Phys. B **621**, 495 (2002).
8.   Ave, M. *et al.*, Phys. Rev. Lett. **85**, 2244 (2000).
9.   Berezinsky, V. S. and Smirnov, A. Yu., Astrophys. Space Sci. **32**, 461 (1975).
10. Capelle, K. S. *et al.*, Astropart. Phys. **8**, 321 (1998).
11. Andrews, D. *et al.*, Proc. 11th ICRC (Budapest) Acta Phys. Acad. Sci. Hung. **29** Suppl 3, 337 (1970); Hillas, A. M. *et al.*, Proc. 11th ICRC (Budapest) Acta Phys. Acad. Sci. Hung. **29** Suppl 3, 533 (1970).
12. Ave, M. *et al.*, Astropart. Phys. **14**, 109 (2000).
13. Auger Collaboration: Abraham, J. *et al.*, Nucl. Instr. Meth. A **523**, 50 (2004).
14. Cronin, J. W., Proc. TAUP Conference, Seattle (2003), astroph/0402487.
15. Bertou, X. *et al.*, Astropart. Phys. **17**, 183 (2002).
16. Mannheim, K., Astropart. Phys. **3**, 295 (1995).
     Stecker, F. W. and Salamon, M. H., Space Sci. Rev. **75**, 341 (1996).
17. Waxman, E., Phys. Rev. Lett. **75**, 386 (1995).
18. Linsley, J., USA Astronomy Survey Committee (Field Committee) Documents (1979).
     Linsley, J., Proc. 19th Int. Cos. Ray Conf. (La Jolla) **3**, 438 (1995).
19. www.euso-mission.org.
20. www.icrr.u-tokyo.ac.jp/~ashra.

Physica Scripta. Vol. T121, 131–136, 2005

Prospect for Relic Neutrino Searches

Graciela B. Gelmini[1,*]

[1]Physics and Astronomy Department, University of California Los Angeles (UCLA), 475 Portola Plaza, Los Angeles, CA 90095-1547, USA

Received December 21, 2004; accepted January 11, 2005

PACS numbers: 14.60.Lm; 14.60.Pq; 95.35.+d; 95.85.Ry

Abstract

Neutrinos from the Big Bang are theoretically expected to be the most abundant particles in the Universe after the photons of the Cosmic Microwave Background (CMB). Unlike the relic photons, relic neutrinos have not so far been observed. The Cosmic Neutrino Background (CνB) is the oldest relic from the Big Bang, produced a few seconds after the Bang itself. Due to their impact in cosmology, relic neutrinos may be revealed indirectly in the near future through cosmological observations. In this talk we concentrate on other proposals, made in the last 30 years, to try to detect the CνB directly, either in laboratory searches (through tiny accelerations they produce on macroscopic targets) or through astrophysical observations (looking for absorption dips in the flux of Ultra-High Energy (UHE) neutrinos, due to the annihilation of these neutrinos with relic neutrinos at the Z-resonance).

We concentrate mainly on the first possibility. We show that, given present bounds on neutrino masses, lepton number in the Universe and gravitational clustering of neutrinos, all expected laboratory effects of relic neutrinos are far from observability, awaiting future technological advances to reach the necessary sensitivity. The problem for astrophysical searches is that sources of UHE neutrinos at the extreme energies required may not exist. If they do exist, we could reveal the existence, and possibly the mass spectrum, of relic neutrinos, with detectors of UHE neutrinos (such as ANITA, Auger, EUSO, OWL, RICE and SalSA).

1. Introduction

Neutrinos from the Big Bang are theoretically expected to be the most abundant particles in the Universe after the photons of the Cosmic Microwave Background (CMB). The Cosmic Neutrino Background CνB can contain the three active neutrinos of the Standard Model of Elementary Particles (SM), which are Dirac or Majorana particles with masses between about 0.01 and 1 eV, one or more sterile neutrinos (present in trivial extensions of the SM) and possibly light bosons coupled to neutrinos. Unlike the CMB, the CνB has not been yet observed. Its detection would provide insight into early moments of our Universe, from before Big-Bang the Nucleosynthesis (BBN) until now. In fact, the CνB is the oldest relic from the Big Bang, produced a few seconds after the Bang itself. Thus, it impacts cosmology from the BBN (which finished about 20 minutes later), to the emission the CMB (380 kyr later), to the formation the Large Scale Structure of the Universe (a Gyr later). Due to this impact, relic neutrinos may be revealed indireclty in the near future through cosmological observations [1].

In this talk we concentrate on other proposals, made in the last 30 years, to try to detect the CνB, either in laboratory searches or through astrophysical observations. We will concentrate in the first possibility and mention the second briefly.

In laboratory experiments cosmic neutrinos could be revealed through the tiny accelerations they produce on macroscopic targets, accelerations which are quadratic or linear in the Fermi coupling constant. Forces quadratic in the Fermi constant are for

---

*Email address: gelmini@physics.ucla.edu

sure present and are largest for Dirac neutrinos. Torques linear in the Fermi coupling constant could be present only if there is a net Lepton number in the background, i.e. a difference between relic neutrinos and antineutrinos. If present, this effect has a comparable magnitude for both Majorana and Dirac neutrinos. We show that, given present bounds on neutrino masses, Lepton number in the Universe and gravitational clustering of neutrinos, all expected laboratory effects of relic neutrinos are far from observability, awaiting future technological advances to reach the necessary sensitivity.

Astrophysical searches would look for absorption dips in the flux of Ultra-High Energy neutrinos, due to the annihilation of these neutrinos with relic neutrinos at the Z-resonance. The problem with this idea is that sources of UHE neutrinos at the extreme energies required (10^{22} eV) may not exist. If they do exist, we could reveal the existence, and possibly the mass spectrum, of relic neutrinos, with detectors of UHE neutrinos, such as ANITA, Auger, EUSO, OWL, Rice and SalSA.

2. The standard relic neutrino background

The standard relic neutrino background is assumed to consist of the three active neutrinos of the SM, with relic abundances dictated purely by the interactions present in the SM, and a negligible Lepton number in the Universe. However we know that there must be neutrino physics beyond the SM, because we know now experimentally that neutrinos have masses, which they do not in the SM. In fact the solar mass-square difference $\Delta m_{12}^2 \simeq 8.1 \times 10^{-5}$ eV^2 and the atmospheric mass difference $\Delta m_{23}^2 \simeq 2.2 \times 10^{-3}$ eV^2 [2] imply that there are at least three neutrino mass eigenstates. Since the larger mass entering into a mass-square difference must be larger than or equal to $\sqrt{\Delta m^2}$, we know that two of the three masses must be larger than 0.9×10^{-2} eV. Cosmological bounds dictate that all active neutrino masses are smaller than about 1 eV [3].

The thermal history of neutrinos starts at temperatures $T \gg$ MeV, at which the three active neutrinos ν_α of the SM (α stands for e, μ or τ) were in equilibrium, its reaction rate being larger than the expansion rate of the Universe, $\Gamma_\nu \gg H$. Thus neutrinos had an equilibrium distribution

$$f_{\nu_\alpha}(p) = \left[\exp\left(\frac{E - \mu_\alpha}{T} \right) + 1 \right]^{-1} \tag{1}$$

where $E \simeq p$, since $m_\alpha \ll$ MeV and μ_α's are the chemical potentials. The standard assumption is that $\mu_\alpha = 0$. At $T \simeq$ MeV, the neutrino interaction rate, $\Gamma_\nu = \langle \sigma_\nu n_\nu \rangle$, falls below the expansion rate, $H = \sqrt{8\pi\rho/3M_P^2}$, and neutrinos decouple at a temperature between 2 and 3 MeV, depending on the flavour. But even after neutrinos are decoupled, while they are relativistic they maintain the equilibrium distributions $f_{\nu_\alpha}(p)$. Just after neutrinos decouple, at $T \simeq m_e = 0.5\,\text{MeV}$, $e^{\pm}$ pairs annihilate and transfer their entropy into photons, increasing their temperature T with respect to the temperature of neutrinos, which becomes $T_\nu = (4/11)^{1/3}\,T$. Therefore now $T_\nu = 1.9\,^\circ\text{K} = 1.7\ 10^{-4}\,\text{eV}$, which means that at least two of the active neutrinos in the CνB (with masses above $10^{-2}\,\text{eV}$) are non-relativistic. The Dirac or Majorana nature of neutrinos becomes important for non-relativistic neutrinos.

From this history we obtain the usual expressions for the number density

$$n_\nu = n_{\nu^c} = \tfrac{3}{22}n_\gamma = \tfrac{3\zeta(3)}{11\pi^2}\,T^3 = 56/\text{cm}^3, \tag{2}$$

the relativistic energy density (for $T_\nu > m_\nu$)

$$\rho_\nu + \rho_{\bar\nu} = \tfrac{7}{8}\left(\tfrac{T_\nu}{T}\right)^4 \rho_\gamma = \tfrac{7\pi^2}{120}\left(\tfrac{4}{11}\right)^{4/3} T^4, \tag{3}$$

and the non-relativistic energy density (for $T_\nu < m_\nu$) $\rho_\nu = m_\nu n_\nu$ (using Ω, the density in units of the critical density, and h, the reduced Hubble constant), i.e.

$$\Omega_\nu h^2 = \sum_\alpha m_\alpha/94\,\text{eV}, \tag{4}$$

which, with only the interactions present in the SM, are the same for Majorana or Dirac neutrinos. The reason is that although Dirac neutrinos have four possible states and Majorana only two, the two additional states of Dirac neutrinos are not populated in the early Universe, because of the smallness of neutrino masses.

The number of relativistic neutrino species, N_ν, is used to parametrize ($\rho_{\text{relativistic}} - \rho_\gamma$) in terms of the present density of one relativistic standard neutrino species (computed in the limit of instantaneous decoupling), i.e.

$$\rho_{\text{relativistic}} = \rho_\gamma + \rho_\nu + \rho_x = [1 + \tfrac{7}{8}(\tfrac{4}{11})^{4/3}\,N_\nu]\rho_\gamma. \tag{5}$$

Lower bounds on N_ν larger than zero have been obtained, both during BBN as well as from CMB measurements. The BBN two-σ bound $1.6 \le N_\nu \le 3.2$ [4] constitutes, in fact, a detection of relic neutrinos, since during BBN at least ν_e, $\bar\nu_e$ are needed (for the weak interactions of p, n). However, the two-σ bound obtained by WMAP from measurements of the CMB anisotropy, $0.9 \le N_\nu \le 8.3$ [4], measures only the relativistic energy density in the Universe at the time of CMB emission (380 kyr after the Bang), which may not necessarily consist of relic neutrinos.

3. Non-Standard neutrino backgrounds

A non-standard relic neutrino background can have less or more neutrinos that the standard background. In inflationary models, the beginning of the radiation dominated era of the Universe results from the decay of coherent oscillations of a scalar field, and the subsequent thermalization of the decay products into a thermal bath with the so called "reheating temperature", T_R. This temperature may have been as low as $0.7\,\text{MeV}$ [5] (a very recent analysis strengthens this bound to $\sim 4\,\text{MeV}$ [6]). It is well known that a low reheating temperature inhibits the production of particles which would have become non-relativistic or decoupled at T above or close to T_R [7, 8]. The final number density of active neutrinos starts departing from the standard number for $T_R \simeq 8\,\text{MeV}$, stays within 10% of it for $T_R > 5\,\text{MeV}$, and for $T_R = 1\,\text{MeV}$ the number of $\nu_{\mu,\tau}$ is about 2.7% of the standard number. We may even have no relic neutrinos left in extreme

models in which neutrinos annihilate into light bosons at late times [9].

A non-standard relic neutrino background with a neutrino asymmetry would have more neutrinos than a standard background (in which the neutrino chemical potentials are assumed to be zero). While charge neutrality requires the asymmetry in charged leptons to be the same as that in protons, for which $(n_B - n_{\bar B})/n_\gamma \simeq O(10^{-10})$, no such requirement limits the asymmetry in neutrinos. With a relic neutrino asymmetry, the number of neutrinos and antineutrinos of the same flavour is different, $n_{\nu_\alpha} \neq n_{\bar\nu_\alpha}$. The relic neutrino energy density always increases with a neutrino asymmetry,

$$N_\nu = 3 + \frac{15}{7}\sum_\alpha\left[2\left(\frac{\xi_\alpha}{\pi}\right)^2 + \left(\frac{\xi_\alpha}{\pi}\right)^4\right]$$
$$= 3 + \sum_\alpha 0.22(2\,\xi_\alpha^2 + 0.10\,\xi_\alpha^4). \tag{6}$$

We see here that for any non-zero value of the dimensionless chemical potential $\xi_\alpha \equiv \mu_{\nu_\alpha}/T$ (chosen as parameter because it is constant while the expansion of the Universe is adiabatic), N_ν is larger than for a zero value, even if for any value of ξ smaller than 1 the increase is very small. For example, $\Delta N_\nu = 4 \times 10^{-3}$ for $|\xi| = 0.1$ and $\Delta N_\nu \simeq 1$ for $|\xi| = 1.5$.

The net lepton number,

$$L_\alpha \equiv \frac{n_{\nu_\alpha} - n_{\bar\nu_\alpha}}{n_\gamma} = \frac{\pi^2}{12\zeta(3)}\left(\xi_\alpha + \frac{\xi_\alpha^3}{\pi^2}\right)\left(\frac{T_\nu}{T_\gamma}\right)^3$$
$$= 0.25(\xi_\alpha + 0.10\xi_\alpha^3) \tag{7}$$

can be sizeable even with values of ξ somewhat smaller than 1. For example $(n_{\nu_\alpha} - n_{\bar\nu_\alpha}) \simeq 10/\text{cm}^3$ for $|\xi| = 0.1$ and $(n_{\nu_\alpha} - n_{\bar\nu_\alpha}) \simeq 190/\text{cm}^3$ for $|\xi| = 1.5$.

So $|\xi| \geq 0.1$ produce small ν-density increases but significant ν-asymmetries. This is important for what follows, because the most conservative upper bound on ξ is about 0.1.

In the absence of significant extra contributions to the radiation density of the Universe during BBN (except for that implied by the neutrino asymmetries), bounds from BBN alone allow for larger chemical potentials for ν_μ and ν_τ than for ν_e: $|\xi_e| < 0.2$, $|\xi_{\mu,\tau}| < 2.6$. The reason is that the effects of ξ_e and of $\xi_{\mu,\tau}$ compensate each other, but while the neutron to proton ratio increases with $\sqrt{\rho_{rad}}$ (which increases with $\xi_{\mu,\tau}$ of any sign), it is proportional to $\exp(-\xi_e)$, thus it decreases faster with positive ξ_e [10]. However, due to the large mixings between neutrinos, fast neutrino oscillations equilibrate all ξ, so the bound on all neutrino asymmetries must be equal to the smallest upper bound imposed by BBN, $|\xi_{e,\mu,\tau}| < 0.1$ [11]. These oscillations could be suppressed [12] if neutrinos are coupled to light bosons (Majorons [13]), whose interactions produce a large effective potential for neutrinos in the early Universe, and, if so, the previously mentioned bounds would apply. Even without the suppression of oscillations, the smallest upper bound imposed by BBN can be larger than 0.1, if an independent source of radiation is present during BBN, so that $\Delta N_\nu = N_\nu - 3$ and the neutrino chemical potentials are independent parameters. Then, BBN combined with CMB data (provided by WMAP) require $-0.1 \le \xi_e \le 0.3$ for $-2 \le \Delta N_\nu \le 5$ [14].

Thus, in the following we will take as conservative upper bounds $\xi_{e,\mu,\tau} \simeq 0.1$ which implies $(n_\nu - n_{\bar\nu}) \simeq 10/\text{cm}^3$ and $(n_\nu + n_{\bar\nu}) \simeq 112/\text{cm}^3$, or, as an extreme upper bound $\xi_{\mu,\tau} \simeq 3$, i.e. $L \simeq 2.5$, $(n_\nu - n_{\bar\nu}) \simeq n_\nu \simeq 1,050/\text{cm}^3$.

4. Gravitational clustering of relic neutrinos

Gravitational clustering of neutrinos in our galaxy or galaxy cluster may enhance the relic neutrino density making it easier to detect the CνB on Earth. Already in 1979, Tremaine and Gunn [15] produced a kinematical constraint, which shows that neutrinos as light as we now know they are, would not significantly cluster. Light neutrinos with masses $m_v <$ eV can be gravitationally bound only to the largest structures, large clusters of galaxies. We can see this using simple velocity arguments. Only cosmic neutrinos with velocities smaller than the escape velocity of a given structure can be bound to it. The escape velocity from a large galaxy like ours is about $600\,\mathrm{km/s}$ and from a large cluster of galaxies is about $2,000\,\mathrm{km/s}$. Considering that the average velocity modulus of non-relativistic neutrinos of mass m and temperature T_v is (using Maxwell-Bolztman distribution) $\langle|\vec{\beta}_v|\rangle = \sqrt{8kT_v/\pi m} = \sqrt{4.3\ 10^{-4}\ \mathrm{eV}/m}$ (namely $\langle|\vec{v}_v|\rangle = 6,200\,\mathrm{km/s}$ for $m = 1\,\mathrm{eV}$, and $\langle|\vec{v}_v|\rangle = 19,600\,\mathrm{km/s}$ for $m = 0.1\,\mathrm{eV}$), it is obvious that only about a third of 1eV mass neutrinos, and a very small fraction of lighter neutrinos, could be gravitationally bound to large clusters at present. Fermi degenerate neutrinos (those with $\xi > 1$) may have even larger average velocities depending on their chemical potential, but the conclusions remain the same. For $\xi \gg 1$, we have $\langle|\vec{\beta}_v|\rangle = \sqrt{6\xi T_v/5\,m} \simeq \sqrt{\xi 1.68\ 10^{-4}\ \mathrm{eV}/m}$ (namely $\langle|\vec{v}_v|\rangle = \sqrt{\xi}$ $12,300\,\mathrm{km/s}$ for $m = 0.1\,\mathrm{eV}$), and both expressions coincide for $\xi = 2.5$. In all cases the amount of neutrinos in the tail of the velocity distribution with velocities smaller that $600\,\mathrm{km/s}$, which would be gravitationally bound to galaxies, is much smaller.

More recently, in 2002, Singh and Ma [16] studied the clustering of neutrinos in cold dark matter halos. They found that neutrino overdensities decrease with cluster halo mass and distance to the center (and we are in the periphery of the Virgo supercluster), so overdensities close to Earth could be at most of $O(1)$ for neutrino masses close to $1\,\mathrm{eV}$, and for $m_v \leq 0.1\,\mathrm{eV}$ neutrino clustering is insignificant.

5. Prospects for laboratory searches: effects of $O(G_F^2)$

Given the characteristic relic neutrino energy $\langle E_{v_\alpha}\rangle \simeq T_v \simeq 10^{-4}\,\mathrm{eV}$, the relic v-nucleon cross sections are very small. For Dirac neutrinos,

$$\sigma_{v-N} \approx \begin{cases} G_F^2 m_v^2/\pi \simeq 10^{-56}\,(m_v/\mathrm{eV})^2\,\mathrm{cm}^2 & \text{for (NR-Dirac)} \\ G_F^2 E_v^2/\pi \simeq 5 \times 10^{-63}\,\mathrm{cm}^2 & \text{for (R)} \end{cases}, \quad (8)$$

where R stands for relativistic, NR for non relativistic and G_F is the Fermi coupling constant. Majorana and Dirac neutrinos are indistinguishable while relativistic. For NR Majorana neutrinos the cross sections are even smaller: since only the $\gamma_\mu\gamma_5$ weak interaction coupling remains, a factor β_v^2 appears, where β is the neutrino velocity. With $(n_v + n_{\bar{v}}) \simeq 100\,\mathrm{cm}^{-3}$, incoherent scattering off nucleons leads to rates smaller than $10^{-6}\,\mathrm{yr}^{-1}$ per kiloton, for the most favourable case NR Dirac neutrinos with eV mass.

Nuclear coherence enhancement factors, of order $A^2 \simeq 10^4$ (with A the atomic number), do not help much. But coherence over the relic v wavelength, $\lambda_v = 2\pi\hbar/4T_v \approx 2.4\,\mathrm{mm}$ (or $1.2\,\mathrm{mm} \times \mathrm{eV}/m_v$ for clustered neutrinos), makes an enormous difference since a volume λ_v^3 contains more than 10^{20} nuclei. Since destructive interference occurs if target size is larger than λ_v the largest enhancement is obtained with a material less than half filled with grains of size λ_v [17, 18].

Even with this sizeable cross section, the net momentum imparted by relic neutrinos on a target on Earth would be zero on average, if the CνB would be on average at rest with respect to the Earth. But this is clearly not so. A reasonable guess is that the CνB is at rest with the CMB, and the Sun's motion with respect to the CMB (derived from COBE-DMR dipole anisotropy) is $v_{sun} = 369.0 \pm 2.5$ km/sec, i.e. the speed of the Earth with respect to the CνB is $\beta_{earth} = 1.231 \times 10^{-3}$.

A momentum of the order of the neutrino momentum, $\Delta\vec{p} \approx \vec{p}_v$, is imparted in each neutrino collision. Due to the bulk velocity of the "neutrino wind" on Earth, $-\vec{\beta}_{earth}$, there is a preferred direction for $\Delta\vec{p}$, thus $\langle\Delta\vec{p}\rangle \approx -\vec{\beta}_{earth}(3T_v/c)$ (or $\approx -\vec{\beta}_{earth}\,cm_v$ for clustered neutrinos).

The resulting accelerations for relativistic (Dirac or Majorana) neutrinos are [19]. (ρ is the target density in units of $\mathrm{gr\,cm}^{-3}$)

$$a_R^D \simeq 2 \times 10^{-33}\mathrm{cm/s}^2 \quad f(\rho/10). \quad (9)$$

For non-clustered non-relativistic Dirac neutrinos (i.e. for most relic neutrinos), the cross-sections are [19]

$$a_{NC-NR}^D \simeq 3 \times 10^{-27}\mathrm{cm/s}^2 f\,(m_v/\mathrm{eV})^2 \quad (\rho/10), \quad (10)$$

and for clustered non-relativistic Dirac neutrinos, the cross sections are [19]

$$a_{C-NR}^D \simeq 10^{-26}\mathrm{cm/s}^2 \quad f(\rho/10). \quad (11)$$

In these equations the factor f accounts for the possible enhancement due to clustering (as well as, possibly, a large lepton asymmetry), $1 \leq f \equiv (n_v + n_{v^c})/100\,\mathrm{cm}^{-3} < 10$.

As mentioned above, non-relativistic Majorana neutrinos have only a $\gamma_\mu\gamma_5$ coupling. The coherent interactions due to the static limit of this coupling are, however, suppressed by β_v, the ratio of the "small" and "large" components of the spinor. Recall that for non-relativistic spinors the lower components are "smaller" than the upper components by a factor of β. Thus the analog of Eqs. (10) and (11) for non-relativistic Majorana neutrinos are suppressed by an extra factor of $\beta_v^2 \approx 10^{-6}$.

6. Prospects for laboratory searches: effects linear in G_F

Coherent interations of a low energy neutral particle, with a medium in which the interatomic spacing is much smaller than the deBroglie particle wavelength (recall $\lambda_v \simeq 2.4\,\mathrm{mm}$), change the particle momentum from p to p'. Then, one can define an index of refraction $n = p'/p$, and $n - 1 \sim G_F$. However, early proposals to use "neutrino optics", either total reflection [20] or refraction (or refraction in a superconducting surface which would induce a current) [21] were incorrect. Cabibbo and Maiani [22] and Langacker, Leveille and Sheiman [23] in 1982, proved that the force due to linear momentum or energy exchange on a target immersed in a uniform neutrino field cancel to order G_F, in fact [22]

$$\vec{F} = -\frac{\Delta\vec{p}_v}{\Delta t} \simeq G_F \int d^3x \rho_A(x)\vec{\nabla}n_v(x), \quad (12)$$

where ρ_A is the atomic number density of the target, and $\vec{\nabla}n_v(x)$ is the gradient of the local neutrino number density. This gradient is zero (since n_v due to gravitational effects is uniform at the scale of possible detectors), except for scattered waves due to weak interactions, which are of order G_F, and thus lead to forces $F \simeq G_F^2$. In fact, Smith and Lewin [18] proposed in 1983 to generate large artificial neutrino density distortions to induce a neutrino density gradient and thus a force, but there is no known way to do this.

There is only one possible mechanical effect of order $G_F G_F$, proposed by L. Stodolsky [24] in 1974: a torque of order G_F can arise if both the target (e.g. consisting of magnetized iron) and the ν-background have a polarization. The ν-background must have a non-zero net flux of weak-interactions-charge (i.e. of neutrinos minus antineutrinos) to produce a net torque on polarized electrons. Since the Earth is moving with respect to the CvB, there is a net flux of particles reaching us, with $\langle \vec{\beta}_\nu \rangle = \vec{\beta}_{earth}$. Thus we only need a lepton asymmetry in those particles to have a net flux of weak-charge $\sim -\vec{\beta}_{earth}(n_\nu - n_{\bar{\nu}})$.

The Stodolsky effect consists of an energy split of the two spin states of non-relativistic electrons in the CvB. This energy split is proportional to the difference between the densities of neutrinos and antineutrinos in the neutrino background for Dirac neutrinos and relativistic Majorana, and proportional to the net helicity of the background for non-relativistic Majorana neutrinos, as we will now see. The distinction of Dirac or Majorana neutrinos is important only for non-relativistic neutrinos.

The Hamiltonian density of the neutrino-electron interaction is

$$\mathcal{H}(x) = \frac{G_F}{\sqrt{2}}(\bar{e}\gamma^\mu(g_V - g_A\gamma_5)e)[\bar{\nu}\gamma_\mu(1 - \gamma_5)\nu]. \tag{13}$$

In the non-relativistic limit the electron current $(\bar{e}\gamma^\mu\gamma_5 e)$ yields a factor $\vec{\sigma}_e \cdot \vec{\beta}_e, \vec{\sigma}_e$ (from the time and space components of the current). For Dirac neutrinos (for which $\nu \neq \bar{\nu}$), the γ_μ term dominates, and $[\bar{\nu}\gamma_\mu\nu]$ yields $\bar{\nu}1\nu, \ldots \sim (n_\nu - n_{\bar{\nu}})$, which is non zero if there is a net Lepton number in the CvB. The effect, originally derived by Stodolski [24] only for Dirac neutrinos, is proportional to the product $\vec{\sigma}_e \cdot \vec{\beta}_e(n_\nu - n_{\bar{\nu}})$. For Majorana neutrinos (for which $\nu = \bar{\nu}$), only the $\gamma_\mu\gamma_5$ term remains, and $[\bar{\nu}\gamma_\mu\gamma_5\nu]$ yield in the non-relativistic limit $\bar{\nu}(\vec{\sigma}_\nu \cdot \vec{\beta}_\nu, \vec{\sigma}_\nu)\nu \sim (n_{\nu_\ell} - n_{\nu_r})$, which is non-zero only if there is a net helicity in the CvB [19].

Here we call left (right) chirality eigenstates $\nu_L(\nu_R)$, and left (right) helicity eigenstates ν_ℓ (ν_r). The general expression for the energy of one electron in the CvB, in the electron-rest frame, to first order in β_{earth}, both for Dirac and Majorana neutrinos were first derived in Ref. [19]. Let us see the most relevant particular cases, starting from early times, before the decoupling of neutrinos.

Since neutrinos are lighter than about 1 eV, they were relativistic at decoupling ($T_{\text{dec}} \geq O(\text{MeV})$). Relativistic neutrinos are only in left-handed chirality states (and anti-neutrinos only in right-handed chirality states). These are the only states produced by weak interactions. For relativistic neutrinos chirality and helicity coincide (up to mixing terms of order $m_\nu/E_\nu \simeq m_\nu/T_\nu$). Thus, at decoupling, neutrinos ν_L were only in left-handed helicity states, and antineutrinos ν_R^c (or ν_R in the case of Majorana neutrinos) in right-handed ones. In this case, the term in the Hamiltonian of one electron in the CvB linear in the spin of the electron $\vec{s}_e$, is

$$H_R^D = H_R^M = -\sqrt{2}G_F g_A 2\vec{s}_e \cdot \vec{\beta}_{earth}(n_{\nu_L} - n_{\nu_R^c}). \tag{14}$$

Helicity is an eigenstate of propagation and, therefore, it does not change while neutrinos propagate freely, even if they become non-relativistic. Recall that two of the neutrinos mass eigenstates are non-relativistic at present. For Majorana neutrinos chirality acts as lepton number, so we are calling "neutrinos" those particles produced at $T > T_{dec}$ as ν_L, and "anti-neutrinos" those produced as ν_R. Thus, neglecting intervening interactions, non-relativistic background neutrinos are in left-handed helicity

eigenstates (which consist of equal admixtures of left- and right-handed chiralities) and anti-neutrinos are in right-handed helicity eigenstates (which also consist of equal admixtures of left- and right-handed chiralities). If the non-relativistic neutrinos are Dirac particles, only the left-handed chirality states (right for anti-neutrinos) interact, since the other chirality state is sterile, while if the neutrinos are Majorana, both chirality states interact (the right-handed "neutrino" state is the right-handed anti-neutrino). In the most favourable case of a large $\xi = \mu/T$, for $m_\nu \leq 0.1$ eV, the term in the Hamiltonian linear in the electron spin, for non-clustered non-relativistic relic neutrinos (which are most of them), in the electron rest frame is

$$H_{NC-NR}^D \simeq \tfrac{1}{2}H_{NC-NR}^M \simeq 0.85\langle|\vec{\beta}_\nu|\rangle^{-1}H_R^D \leq \frac{7}{\sqrt{\xi}}H_R^D, \tag{15}$$

where $\langle|\vec{\beta}_\nu|\rangle = \sqrt{6\xi T_\nu/5\,m} \simeq \sqrt{\xi 1.7\,10^{-4}\,\text{eV}/m}$ is the characteristic velocity (the average of the velocity modulus) of relic neutrinos in the CvB rest frame. The dominant term shown here, comes from the space part of the neutrino current ($\vec{\sigma}_\nu$ in the helicity base is $h\hat{\beta}_\nu \sim \langle|\vec{\beta}_\nu|\rangle^{-1}$, where $\hat{\beta}_\nu$ is the unit vector of the neutrino velocity in the relic neutrino rest frame). Notice that there is no β factor penalty for Majorana neutrinos.

Slow enough non-relativistic neutrinos eventually fall into gravitational potential wells, become bound and, after a characteristic orbital time, their helicities become well mixed up, since momenta are reversed and spins are not. Thus, gravitationally bound relic neutrinos have well mixed helicities, no net helicity remains. For these clustered non-relativistic neutrinos, the Stodolsky effect cancels for Majorana neutrinos, but only changes by a factor for Dirac neutrinos.

$$H_{C-NR}^D = \tfrac{1}{2}H_R^D, \qquad H_{C-NR}^M = 0. \tag{16}$$

As we argued above, most relic neutrinos however are not gravitationally bound at present, because they are too light.

The Stodolsky effect requires a Lepton asymmetry $n_\nu - n_{\nu^c} = f\,100\,\text{cm}^{-3}$ (where the maximum possible overdensity factor is $f(\xi)_{max} = 0.1\text{--}10$) to obtain an energy difference, ΔE, between the two helicity states of an electron in the direction of the bulk velocity of the neutrino background, $\langle \vec{\beta}_\nu \rangle = \vec{\beta}_{earth}$. In the case of Dirac or relativistic Majorana neutrinos of density n_ν, with a very large lepton asymmetry favouring, say, neutrinos ν_L (so that $n_{\nu_L} = n_{\nu_\ell} = n_\nu$) we have

$$\Delta E \simeq f2\sqrt{2}G_F g_A |\vec{\beta}_{earth}|n_\nu. \tag{17}$$

This is the equation we will use to estimate the maximum possible accelerations due to this effect. We should recall however that $\Delta E_{NC-NR}^M \leq 8\Delta E$ and that $\Delta E_{C-NR}^M = 0$.

The energy difference ΔE implies a torque of magnitude $\Delta E/\pi$ applied on the spin of the electron. Since the spin is "frozen" in a magnetized macroscopic piece of material with N polarized electrons, the total torque applied to the piece has a magnitude $\tau = N\Delta E/\pi$. Given a linear dimension R and mass M of the macroscopic object, its moment of inertia is parametrized as $I = MR^2/\gamma$, where γ is a geometrical factor. In the typical case of one polarized electron per atom in a material of atomic number A, the number N above is $N = (M/gr)N_{AV}/A$ (using cgs units), where N_{AV} is Avogadro's number. Thus, the effect we are considering would produce an angular acceleration of order $\alpha = \tau/I$ and a linear acceleration of order $a_{G_F} = R\alpha$ in the magnet given by

$$a_{G_F} \simeq 10^{-27}\frac{\text{cm}}{\text{s}^2}f \cdot \left(\frac{\gamma}{10}\right)\left(\frac{100}{A}\right)\left(\frac{\text{cm}}{R}\right)\left(\frac{\beta_{earth}}{10^{-3}}\right). \tag{18}$$

The local density enhancement factor due to a net lepton number or clustering is $0 \leq f \equiv (n_v - n_{v^c})/100\,\mathrm{cm}^{-3} < 100$.

We should compare the accelerations mentioned in Eqs. (10), (11) and (18) above with the smallest measurable acceleration at present, with "Cavendish" type torsion balances, which is about $10^{-12}\,\mathrm{cm/s}^2$ [25]. This is about 15 orders of magnitude larger. As an example of an attempt to produce a relic neutrino detector, we can mention a particular design of a torsion oscillator proposed by C. Hagmann [26] in 1999. In this proposed detector, the uncertainty principle gives a minimum measurable acceleration (QL stands for quantum limit),

$$a_{\mathrm{QL}} = 5 \times 10^{-24}\,\mathrm{cm/s}^2 (10\,\mathrm{kg}/m)^{1/2} (1\mathrm{day}/\tau_0)^{1/2} (10^6\mathrm{s}/\tau) \quad (19)$$

where τ_0 is the oscillation period, and τ measurement time. This acceleration is larger by a factor $(10^3\text{--}10^4)f^{-1}$ than the largest possible accelerations mentioned above. We would need a detector with a lower quantum limit, and that can operate at that limit. To beat the quantum limit P. Smith proposed to go to tons of mass target [18, 27] or to the opposite, to sub-micron granules whose single displacements can be measured (which involves nanotechnology not yet in place) [28].

Still there are other difficulties. Seismic and gravitational variations are a problem. For example, the gravity gradient due to the Moon produces a varying torque about 10^{10} larger than the relic neutrino force. A possible solution would be a concentric balance [28]. Solar neutrinos would produce equal or larger accelerations (and possibly dark matter particles could too) [19]. So directionality would be needed to separate a signal from relic neutrinos (Smith and Lewin [27] proposed using laminated materials).

The evident conclusion is that laboratory effects of the CvB are still far from observability, therefore awaiting future technology.

7. Prospects for astrophysical searches

Only at the Z-resonance the cross section of astrophysical neutrinos with the CvB is large enough to reveal its existence [29]. A simple argument is that the cross section at the Z-resonance, $\sigma_{annih}(E^{\mathrm{res}}) \simeq 4 \times 10^{-32}\,\mathrm{cm}^2$, yields a mean free path for Ultra-High Energy (UHE) neutrinos larger but not by much than the Hubble distance, ℓ_{Hubble}, the size of the visible Universe, $\ell_{m.f.p.} \simeq 35 \times \ell_{\mathrm{Hubble}}$. Thus the probability of interaction is non-negligible, about 0.03. Otherwise, for interactions outside the Z-resonance, the Universe is transparent to UHEv.

If an intense enough flux of UHEv would exist the resonant annihilation with the CvB would leave an absorption dip at

$$E^{\mathrm{res}}_{v_i} = \frac{m_Z^2}{2\,m_{v_i}} = 4.2 \times 10^{22}\,\mathrm{eV} \left(\frac{0.1\,\mathrm{eV}}{m_{v_i}}\right). \quad (20)$$

A recent examination [30] of the possibility of relic neutrino absorption spectroscopy through this mechanism shows that, if intense enough sources of UHEv at energies $E_v \geq 10^{22}\,\mathrm{eV}$ and above would exist, the different masses of relic neutrinos could produce separated absorption dips. Experiment such as ANITA, Auger, EUSO, OWL, RICE, and SalSA able to detect such UHE neutrinos are expected to be available in the near future [31]. The problem with this idea, is that the only sources proposed to produce such UHEv fluxes are topological deffects. Even Active Galactic Nuclei are not thought to be able to produce neutrinos with energies above $10^{21}\,\mathrm{eV}$.

Another signature of the annihilation of UHEv with relic neutrinos at the Z-resonance, is the emission of UHE p, n, γ and v, in what T. Weiler called "Z-bursts". In fact, Z-bursts were proposed [32] as the possible origin of UHE Cosmic Rays above the "Greisen-Zatsepin Kuzmin cutoff" at $E_{GZK} \simeq 5 \times 10^{19}\,\mathrm{eV}$ observed, for example, by AGASA [33]. This mechanism is now considered unlikely to be correct (and if it is correct, it would imply a relic neutrino mass $m_v \simeq 0.3\,\mathrm{eV}$ [34]). If they do not produce the UHECR, Z-bursts are subdominant, and thus not a good signal to detect the CvB.

8. Conclusions

In the past, neutrinos were thought to be sufficiently massive ($m_v \geq 20\,\mathrm{eV}$) to cluster in our galaxy and make up the local dark matter halo. If so, they could have had large local overdensities, even $f \simeq 10^7$. Also chemical potentials had a much larger upper bound until a few years ago, $\xi \leq 6.9$ [35]. All this made for much more optimistic predictions of the laboratory effects of relic neutrinos.

Now, we know that relic neutrinos have sub-eV masses, the enhancement factors f due to clustering could only be at most of $O(1)$, and the maximum allowed chemical potential is likely $\xi \leq 0.1$. We know also, that at least two of the three active neutrino mass eigenstates have masses $m_{v_{2,3}} > 10^{-2}\,\mathrm{eV}$, thus most relic neutrinos are non-relativistic and non-clustered, and have an average velocity with respect to the Earth similar to that of the CMB, $\beta_v \simeq 10^{-3}$. These values lead to estimations of the effect of the "cosmic neutrino wind" on macroscopic targets that are smaller than those made even a few years ago.

There are for sure forces on macroscopic targets of order G_F^2, and possibly torques of order G_F. $O(G_F^2)$ forces are largest when due to coherent elastic scattering out of target grains of the size of the neutrino deBroglie length $\lambda_v \simeq 0.2\,\mathrm{cm}$. The largest effect is for Dirac neutrinos with mass as large as possible (i.e. close to eV), as can be seen in Eqs. (10) and (11). The only $O(G_F)$ effect is a torque on a polarized target, if there is a net flux of weak charge due to the CvB. It requires a lepton asymmetry in the relic neutrinos. The effect is largest for non-relativistic non-clustered Majorana or Dirac neutrinos, for which the acceleration could be up to one order of magnitude larger than in Eq. (18). These accelerations are tiny, and there are difficult backgrounds. New experimental ideas are needed to reach the required sensitivity.

With respect to the prospects for astrophysical searches, Z-pole absorption dips in UHEv flux could reveal relic neutrinos and their masses. The problem with this idea is the existence of sources of UHEv with $E_v \geq 10^{22}\,\mathrm{eV}$. If these sources exist, this would be the most promising search mechanism in the foreseeable future, using detectors such as ANITA, Auger, EUSO, OWL, RICE, and SalSA.

Acknowledgements

This work is supported in part by the DOE Grant DE-FG03-91ER40662, Task C and NASA Grant NAG5-13399.

References

1. See the talk of M. Tegmark in this Symposium.
2. Maltoni, M. *et al.*, hep-ph/0405172.
3. Spergel D. N. *et al.*, Astrophys. J. Suppl. **148**, 175 (2003); Tegmark, M. *et al.*, [SDSS Collaboration], astro-ph/0310723; Hannestad, S., J. Cosmol. Astropar. Phys. **0305**, 004 (2003).
4. Barger, V. *et al.*, Phys. Lett. B **566**, 8 (2003).

5. Kawasaki, M., Kohri, K. and Sugiyama, N., Phys. Rev. Lett. **82**, 4168 (1999) and Phys. Rev. D **62**, 023506 (2000).

6. Hannestad, S., Phys. Rev. D **70**, 043506 (2004).

7. Giudice, G. F., Kolb, E. W. and Riotto, A., Phys. Rev. D **64**, 023508 (2001).

8. Giudice, G. F. *et al.*, Phys. Rev. D **64**, 043512 (2001).

9. Kolb, E. and Turner, M., Phys. Lett. B **159**, 102 (1985); Beacom, J. F., Bell, N. F. and Dodelson, S., Phys. Rev. Lett. **93**, 121302 (2004).

10. Kang, H. and Steigman, S., Nucl. Phys. B**372**, 494 (1992); Hansen, S. H. *et al.*, Phys. Rev. D **65**, 023511 (2002).

11. Lunardini, C. and Smirnov, A. Y., Phys. Rev. D **64**, 073006 (2001); Dolgov, A. D. *et al.*, Nucl. Phys. B **632**, 363 (2002); Abazajian, K. N., Beacom, J. F. and Bell, N. F., Phys. Rev. D **66**, 013008 (2002).

12. Babu, K. S. and Rothstein, I. Z., Phys. Lett. B **275**, 112 (1992); Bento, L. and Berezhiani, Z., Phys. Rev. D **64**, 115015 (2001); Dolgov, A. D. and Takahashi, F., Nucl. Phys. B **688**, 189 (2004).

13. Chikashige, Y., Mohapatra, R. N. and Peccei, R. D., Phys. Rev. Lett. **45**, 1926 (1980); Gelmini, G. B. and Roncadelli, M., Phys. Lett. B **99**, 411 (1981); Georgi, H. M., Glashow, S. L. and Nussinov, S., Nucl. Phys. B **193**, 297 (1981).

14. Barger, V. *et al.*, Phys. Lett. B **569**, 123 (2003).

15. Tremaine, S. and Gunn, J., Phys. Rev. Lett **42**, 407 (1979).

16. Singh, S. and Ma, C. P., Phys. Rev. D **67**, 023506 (2003).

17. Zeldovich, Y. and Khlopov, M., Sov. Phys. Usp. **24**, 755 (1981); Shvartsman, V. *et al.*, J. Exp. Theo. Phys. Lett. **36**, 277 (1982).

18. Smith, P. F. and Lewin, J. D., Phys. Lett. B **127**, 185 (1983).

19. Duda, G., Gelmini, G. and Nussinov, S., Phys. Rev. D **64**, 122001 (2001).

20. Opher, R., Astron. Astrophys. **37**, 135 (1974).

21. Lewis, R. R., Phys. Rev. D **21**, 663 (1980).

22. Cabibbo, N. and Maiani, L., Phys. Lett. B **114**, 115 (1982).

23. Langacker, P., Leveille, J. and Sheiman, J., Phys. Rev. D **27**, 1228 (1983).

24. Stodolsky, L., Phys. Rev. Lett. **34**, 110 (1974).

25. Adelberger, E. *et al.*, Ann. Rev. Nucl. Part. Sci. **41**, 269 (1991).

26. Hagmann, C., astro-ph/9902102, pub. in "Asilomar 1998, Particle phys. and the early universe", 460.

27. Smith, P. F. and Lewin, J. D., Acta Phys. Polon. B **15**, 1201 (1984) and Astroph. J. **318**, 738 (1987); See also Smith, P. F., Proc. "4th Int. Wokshop on the Identification of Dark Matter" York, UK, 610 (2002) and references therein.

28. Smith, P. F., Proc. of Texas/ESO-CERN4 (Brighton 1990), 425 (1990) and RAL-91-017, published in Santa Monica Astropart. (1990): 311–325 (QB981:I35:1990).

29. Weiler, T., Phys. Rev. Lett. **49**, 234 (1982) and Astrophys. J. **285**, 495 (1984); Roulet, E., Phys. Rev. **D47**, 5247 (1993).

30. Eberle, B., Ringwald, A., Song, L. and Weiler, T. J., Phys. Rev. D **70**, 023007 (2004).

31. ANtarctic Impulse Transient Array, http://www.ps.uci.edu/~anita/; Pierre Auger Observatory, http://www.auger.org/; Extreme Universe Space Observatory, http://www.euso-mission.org/; Orbiting Wide-angle Light-collectors, http://owl.gsfc.nasa.gov/; Radio Ice Cerenkov Experiment, http://www.bartol.udel.edu/~spiczak/rice/rice.html; Saltdome Shower Array, Gorham, P. *et al.*, Nucl. Instrum. Meth. A **490**, 476 (2002).

32. Weiler, T., Astropart. Phys. **11**, 303 (1999); Fargion, D., Mele, B. and Salis, A., Astrophys. J. **517**, 725 (1999).

33. Yoshida, S. *et al.*, [AGASA Collaboration], in: Proc. 27th International Cosmic Ray Conference, Hamburg, Germany, 2001, vol. 3, pp. 1142–1145.

34. See, for example, Gelmini, G., Varieschi, G. and Weiler, T. J., Phys. Rev. D **70**, 113005 (2004).

35. Kang, H. S. and Steigman, G., Nucl. Phys. B **372**, 494 (1992).

Physica Scripta. Vol. T121, 137–141, 2005

Leptogenesis in the Early Universe[1]

Tsutomu Yanagida[1,2]

[1]Department of Physics, University of Tokyo, Tokyo 113-0033, Japan
[2]Research Center for the Early Universe, University of Tokyo, Tokyo 113-0033, Japan

Received December 14, 2004; accepted January 11, 2005

Abstract

The presence of superheavy Majorana neutrinos is an important prediction of unified theories beyond the standard model. An integration of the heavy Majorana neutrinos induces very small masses for neutrinos via the seesaw mechanism. Thus, small neutrino masses is a reflection of a unification at very high energy scale.

It is quite natural to consider that the heavy Majorana neutrinos are produced thermally in the early universe. Then, the heavy neutrinos begin to decay into Higgs and lepton, or Higgs and anti-lepton when the temperature of the universe cools down to the masses of heavy Majorana neutrinos. The decays of the heavy neutrinos may produce the lepton-number asymmetry if CP invariance is violated in the decay processes. This lepton-number asymmetry is converted into the baryon-number asymmetry in the present universe (leptogenesis) through nonperturbative effects of the electroweak gauge theory. We show, in this talk, that the neutrino masses suggested from atmospheric and solar neutrino-oscillation experiments are just in the range favorable for the thermal leptogenesis. With the aid of the observed neutrino masses we find that the thermal leptogenesis takes place at temperatures $T \gtrsim 2 \times 10^9 \, \text{GeV}$.

However, the above scenario suffers from the gravitino problem if one extends the standard model to the supergravity framework. This is because too many gravitinos are produced at the temperature required for the thermal leptogenesis and their decays destroy light elements created by the big-bang nucleosynthesis. We show that a leptogenesis via inflaton decay is an interesting alternative to the thermal leptogenesis. This scenario is free from the gravitino problem if the gravitino has a relatively large mass as $m_{3/2} \simeq 3$–$6 \, \text{TeV}$.

The importance of neutrinoless double β decay experiments and measurement of CP violation in neutrino oscillation experiments is also emphasized to explain the present universe's baryon asymmetry.

1. Introduction

The baryon-number asymmetry in the universe is one of the most fundamental parameters for describing the nature of the present universe. Galaxies would not be formed and hence we would not even exist if the universe's baryons were too many or too few (beyond two orders of magnitude of the present value either way). Therefore, it is a big challenge in particle physics and cosmology to account for the observed value of baryon-number asymmetry from the basic laws of Nature. In 1967, it was pointed out by Sakharov [1] that the baryon-number asymmetry may be explained if the following three conditions operate: (1) the baryon-number conservation is violated, (2) C and CP invariance is broken, and (3) there is a deviation from thermal equilibrium in the baryon-number violating process. The CP violation was already found in the neutral K meson decays and we know that the C and CP invariance is no intrinsic law of Nature.

It was observed, in 1976, by 't Hooft [2] that the baryon number is no longer conserved at the quantum level in the standard electroweak gauge theory. This is due to the presence of infinitely many vacua of nonabelian gauge theory, which are classified by topological charges. The quantum tunneling between topologically distinct vacua corresponds to aninstanton and this

tunneling causes a change of baryon number B. However, this baryon-number violating effect is strongly suppressed by the factor $\exp(-16\pi/g^2)$ and hence it was not considered as a relevant and realistic process for a long time. Ten years later than the 't Hooft observation, Kuzmin, Rubakov and Shaposhnikov noticed that the suppression does not work at a temperature close to the electroweak phase transition [3]. The barrier between the distinct vacua becomes smaller or vanishing at high temperatures and the vacuum transition easily takes place. This so-called "sphaleron" process violates also lepton number L, whereas a combination B–L is strictly conserved. The B–L conservation is crucial in discussing the origin of the present baryon-number asymmetry as discussed below.

The above KRS mechanism would immediately provide an interesting scenario for creating the baryon-number asymmetry during the electroweak phase transition [4]. This requires the electroweak phase transition to be of first order, which leads to a prediction on the upper bound of the Higgs boson mass as $m_H < 80 \, \text{GeV}$. The experimental lower limit on the Higgs mass is now $m_H > 114 \, \text{GeV}$, which implies that the electroweak phase transition is not first order. There have been proposed various ways to solve this problem, but we have seen no convincing proposal, so far. Furthermore, the effect of CP violation in the standard model is too small to explain the observed value of the baryon-number asymmetry even if the first order phase transition is realized.

It seems, therefore, more successful to consider the baryon-number generation in some baryon-number violating process at high temperatures far above the electroweak phase transition. However, the produced baryon number is erased by the sphaleron process if the baryon number is created with $\Delta B = \Delta L$. This requires B–L violating interactions at high temperatures. Such high-energy B–L violating interactions most likely induce effective operators of B–L violation at low energies. The lowest dimensional operator is $(1/M)HH\ell\ell$ in the standard model, where H and ℓ are SU(2) doublets of Higgs and leptons, respectively. M represents the high energy scale of the B–L violation. In the electroweak symmetry-breaking vacuum this effective operator produces a small neutrino mass. Thus, the small neutrino mass has a logical link with the baryon-number asymmetry in the present universe [5].

The above discussion may suggest an intriguing possibility that a high-energy theory beyond the standard model explains simultaneously two independent observations, that is, the universe's baryon-number asymmetry and the small neutrino masses. In this talk, we show that they are indeed naturally explained by one hypothesis, the so-called seesaw mechanism [6, 7]. We also show that the neutrino masses suggested from neutrino oscillation experiments are well consistent with the observed value of the present baryon-number asymmetry.

The presence of superheavy Majorana neutrinos is an important prediction in a large class of unification models. An integration

---

[1]Extended version of the talk at Nobel Symposium 129 on Neutrino Physics, Sweden, 19–24 August, 2004.

of the heavy Majorana neutrinos induces very small masses for neutrinos via the seesaw mechanism. Thus, very small neutrino masses is a window to ultrahigh-energy physics beyond the standard model [6]. The neutrino masses suggested from neutrino oscillation experiments imply masses of the heavy Majorana neutrinos $N_i(i = 1$–$3)$ to be in a range of $M_i \simeq 10^{10-15}$ GeV. These superheavy neutrinos N_i could be produced through scattering processes of thermal particles in the early universe if the universe's temperature T is higher than their masses M_i. The heavy Majorana neutrinos begin to decay when the temperature of the universe cools down to their masses. They have two distinct decay channels; one is to a Higgs boson + a lepton and the other is to a Higgs boson + an antilepton. If CP is violated in the decay processes, decay rates of the two channels become different from each other, yielding the lepton-number asymmetry in the final state. This lepton-number asymmetry is converted to the baryon-number asymmetry in the early universe through the sphaleron process [3]. This is called leptogenesis [5]. We give, in the next section, a brief review on recent detailed analyses [8] of leptogenesis, which show that the present baryon-number asymmetry can be naturally explained with the aid of observed neutrino masses. It is also shown that thermal leptogenesis requires the temperature $T \gtrsim 2 \times 10^9$ GeV.

The seesaw mechanism is easily incorporated in the framework of supergravity (SUGRA). However, there is a serious problem in the SUGRA, that is the gravitino problem. If the gravitino is unstable, it has a long lifetime and decays after the big-bang nucleosynthesis (BBN) for an interesting range of the gravitino mass, $m_{3/2} \simeq 100$ GeV–10 TeV. The decay products destroy light elements produced by the BBN and hence the relic abundance of the gravitino is constrained from above to keep the success of the BBN. This leads to an upper bound of the reheating temperature T_R after inflation, since the abundance of the gravitino is propotional to the reheating temperature. A recent detailed analysis has derived a strigent upper bound $T_R \lesssim 10^{6-7}$ GeV when the gravitino decay has hadronic modes [9]. This upper bound is much lower than the temperature for leptogenesis, $T_R \gtrsim 2 \times 10^9$ GeV. Therefore, thermal leptogenesis seems hard to work in SUGRA for $m_{3/2} \simeq 100$ GeV–10 TeV, unless some enhancement mechanism operates on the lepton-asymmetry production [10].

There have been proposed various solutions to the above problem [11, 12, 13, 14, 15]. We consider that the nonthermal leptogenesis via inflaton decay [11] is the most interesting, where the superheavy Majorana neutrinos $N_i(i = 1$–$3)$ are produced directly in the inflaton decay.[2] In section 3, we show that a general argument on the inflaton-decay scenario gives us a lower bound on the mass of the heaviest light neutrino as $m_{\nu_3} \gtrsim 0.01$ eV. We also find that this scenario predicts a lower bound of the gravitino mass as $m_{3/2} \gtrsim 3$–4 TeV, together with the gravitino constraint from the BBN [9]. In subsection 3.1, we adopt a chaotic inflation model, since it is free from the initial condition problem [17] and also from the so called η problem. Here, a shift symmetry [18] plays a crucial role on generating a flat potential required for the chaotic inflation in SUGRA. In subsection 3.2, we discuss the nonthermal leptogenesis in the chaotic inflation model and show how naturally the universe's baryon asymmetry is explained in this scenario. Here, the reheating temperature is estimated as

low as 10^{6-7} GeV and hence the model is free from the gravitino problem. The last section is devoted to conclusions.

2. Thermal leptogenesis

The superheavy Majorana neutrinos $N_i(i = 1$–$3)$ are the most important prediction of the seesaw mechanism as stressed in the introduction. Out-of-equilibrium condition is easily satisfied by the delayed decay [19] of the superheavy neutrinos for producing the lepton-number asymmetry. The asymmetry is converted into the baryon-number asymmetry in the universe [5] through the sphaleron process (leptogenesis). In this section we give a brief review on the recent progress [8] of the thermal leptogenesis.

We add three families of heavy Majorana neutrinos $N_i(i = 1$–$3)$ in the standard model. The renormalizable Lagrangian for the Majorana neutrinos is given by

$$\mathcal{L} = \bar{N}_i \,\partial\!\!\!/ N_i + \frac{M_i}{2} N_i N_i + h_{ij} \bar{N}_i \ell_j H + h.c. \tag{1}$$

Then, the heavy Majorana neutrino N has two decay channels,

$$\begin{aligned} N &\to \ell + H, \\ &\to \ell^* + H^*. \end{aligned} \tag{2}$$

If the Yukawa coupling h_{ij} has a CP-violating phase, we have a difference between branching ratios for the two decay channels. This leads to a net lepton-number generation. We assume a mass hierarchy, $M_1 < M_2 < M_3$, for simplicity. We see that the lepton numbers generated in the decays of the heavier Majorana neutrinos $N_{2,3}$ are most likely washed out by the lightest Majorana neutrino N_1. Thus, we consider here lepton-number generation by the decay of the lightest heavy Majorana neutrino N_1.

Interference between tree-level and one-loop diagrams generates lepton-number asymmetry [5, 20] as,

$$\begin{aligned} \epsilon &\equiv \frac{\Gamma(N_1 \to H + \ell) - \Gamma(N_1 \to H^* + \ell^*)}{\Gamma_{N_1}} \\ &\simeq -\frac{3}{16\pi (hh^\dagger)_{11}} \left[\mathrm{Im}(hh^\dagger)_{13}^2 \frac{M_1}{M_3} + \mathrm{Im}(hh^\dagger)_{12}^2 \frac{M_1}{M_2} \right], \end{aligned} \tag{3}$$

where h is a 3×3 matrix of Yukawa coupling defined in Eq. (1). The lepton asymmetry parameter ϵ is written as [21]

$$\epsilon = -\frac{3}{16\pi} \frac{M_1}{\langle H \rangle^2} m_{\nu_3} \delta_{\mathrm{eff}}. \tag{4}$$

Here, the effective CP-violating phase δ_{eff} is given by

$$\delta_{\mathrm{eff}} = \frac{\mathrm{Im}\left[h_{13}^2 + \frac{m_{\nu_2}}{m_{\nu_3}} h_{12}^2 + \frac{m_{\nu_1}}{m_{\nu_3}} h_{11}^2 \right]}{|h_{13}|^2 + |h_{12}|^2 + |h_{11}|^2}, \tag{5}$$

and the seesaw mass formula $m_\nu \simeq m^2/M$ with $m = h\langle H \rangle$ [6, 7] has been used. Then, we obtain the ϵ parameter as

$$\epsilon \simeq -1 \times 10^{-6} \left(\frac{M_1}{10^{10}\,\mathrm{GeV}} \right) \left(\frac{m_{\nu_3}}{0.05\,\mathrm{eV}} \right) \delta_{\mathrm{eff}}, \tag{6}$$

where we have taken $\langle H \rangle \simeq 174$ GeV.

If the out-of-equilibrium condition is well satisfied, we obtain

$$\frac{n_L}{s} \simeq 10^{-9} \times \delta_{\mathrm{eff}} \left(\frac{M_1}{10^{10}\,\mathrm{GeV}} \right), \tag{7}$$

where n_L and s are density of lepton number and entropy, respectively. This lepton-numberasymmetry is converted into the

---

[2] Ref. [16] discusses the production of heavy Majorana neutrinos in inflaton decay. However, the mechanism of generating the lepton-number asymmetry is very different from the leptogenesis we discuss presently.

baryon-number asymmetry through the sphaleron process and we obtain [22]

$$\frac{n_B}{s} \simeq -\frac{28}{79}\frac{n_L}{s}, \tag{8}$$

for the standard model. Then, we find the baryon-number asymmetry in the present universe predicted as

$$\frac{n_B}{s} \simeq 3 \times 10^{-10} \times \delta_{\text{eff}}\left(\frac{M_1}{10^{10}\,\text{GeV}}\right). \tag{9}$$

The observation [23],

$$\frac{n_B}{s} \simeq 0.9 \times 10^{-10}, \tag{10}$$

implies $M_1 \simeq (0.3 - 1) \times 10^{10}\,\text{GeV}$ for $\delta_{\text{eff}} \simeq 1$–0.3.

The above arguement assumes that the out-of-equilibrium condition for the N_1 decay is completely satisfied. Now we discuss when the out-of-equilibrium condition is satisfied. The decay rate of N_1 is given by

$$\Gamma \simeq \frac{1}{8\pi}h_{1k}h_{k1}^{\dagger}M_1. \tag{11}$$

The out-of-equilibrium condition, that is $\Gamma < \Gamma_{\text{exp}}$ at temperature $T \simeq M_1$, gives

$$\bar{m}_{\nu_1} \equiv h_{1k}h_{k1}^{\dagger}\frac{|\langle H \rangle|^2}{M_1} < 2 \times 10^{-3}\,\text{eV}. \tag{12}$$

Here, Γ_{exp} is the expansion rate of the universe. If the above condition is violated, the produced lepton-number asymmetry is reduced. But, the effect is only power suppression and it may be compensated by raising the mass M_1. However, if the above condition is largely violated, there arises another effect to suppress the asymmetry. The N_1-exchange diagrams give lepton-number violating scattering processes acting on lepton and Higgs channels, which may wash out [25] the lepton-number asymmetry produced by the N_1 decay. Since this effect gives an exponential suppression, we have a strong condition on $\bar{m}_{\nu_1}$ and the total sum of the light neutrino masses. Assuming a mass hierarchy for the light neutrinos, $m_{\nu_{1,2}} < m_{\nu_3}$, we obtain $m_{\nu_3} \lesssim 0.2\,\text{eV}$ [8]. On the other hand, if $\bar{m}_{\nu_1}$ is too small compared with $10^{-3}\,\text{eV}$, the production of N_1 is suppressed and we get too small baryon-number asymmetry. This leads to a lower bound of the neutrino mass as $m_{\nu_3} > 10^{-5}\,\text{eV}$ [24]. It is remarkable that the neutrino mass suggested from the atmospheric neutrino oscillation experiments, $\sqrt{\Delta m_{23}^2} \simeq 0.05\,\text{eV}$, satisfies the above conditions very well.

It is clear that the thermal leptogenesis requires the temperature T of the early universe to be higher than the mass of the lightest heavy Majorana neutrino N_1. This implies $T \gtrsim 2 \times 10^9\,\text{GeV}$. As explained in the introduction, this high temperature produces too many gravitinos in SUGRA to keep the success of BBN. Thus, the thermal leptogenesis has a serious problem as long as the gravitino mass is of order 1 TeV when we extend the standard model to the SUGRA framework. We consider that the most interesting alternative is a leptogenesis via inflaton decay [11], which is discussed in the following sections.

3. Nonthermal leptogenesis via inflaton decay

The inflationary early universe is the most attractive hypothesis in modern cosmology, since it not only solves longstanding problems in cosmology, that is the horizon and the flatness problems [26], but also accounts for the origin of density fluctuations [27]. Let us consider, in this section, that the inflaton Φ decays dominantly into a pair of the lightest heavy Majorana neutrinos, that is $\Phi \rightarrow N_1 + N_1$, assuming the other decay modes including N_2 and N_3 are energetically forbidden, for simplicity. Then, the produced N_1's decay subsequently into $H + \ell$ or $H^* + \ell^*$. If the reheating temperature T_R is lower than the mass M_1 of the heavy neutrino N_1, the out-of-equilibrium condition [1] is automatically satisfied.

The chain decays of $\Phi \rightarrow N_1 + N_1$ and $N_1 \rightarrow H + \ell$ or $H^* + \ell^*$ reheat the universe and produce not only the lepton-number asymmetry but also entropy of the thermal bath. Then, the ratio of the lepton number to entropy density after the reheating is estimated as [11]

$$\frac{n_L}{s} \simeq -\frac{3}{2}\epsilon\frac{T_R}{m_\Phi}$$

$$\simeq 3 \times 10^{-10}\left(\frac{T_R}{10^6\,\text{GeV}}\right)\left(\frac{M_1}{m_\Phi}\right)\left(\frac{m_{\nu_3}}{0.05\,\text{eV}}\right), \tag{13}$$

where m_Φ is the inflaton mass and we have taken $\delta_{\text{eff}} = 1$. Notice that the lepton asymmetry parameter ϵ in the N_1 decay is written as in the SUSY theory [21]

$$\epsilon = -\frac{3}{8\pi}\frac{M_1}{\langle H \rangle^2}m_{\nu_3}\delta_{\text{eff}}. \tag{14}$$

This lepton-number asymmetry is converted into the baryon-number asymmetry through the sphaleron effects and we obtain for the SUSY standard model [22]

$$\frac{n_B}{s} \simeq -\frac{8}{23}\frac{n_L}{s}. \tag{15}$$

From the observation [23] in Eq. (10) we derive a constraint

$$m_{\nu_3} \gtrsim 0.01\,\text{eV}. \tag{16}$$

Here, we have assumed $T_R \lesssim 10^7\,\text{GeV}$ to satisfy the cosmological constraint on the gravitino abundance [9] discussed in the introduction. It is very interesting that the neutrino mass suggested from the atmospheric neutrino oscillation experiments, $\sqrt{\Delta m_{23}^2} \simeq 0.05\,\text{eV}$, just satisfies the above constraint. On the other hand, assuming $m_{\nu_3} \simeq 0.05\,\text{eV}$ we obtain $T_R \gtrsim 2 \times 10^6\,\text{GeV}$. This lower limit predicts the gravitino mass as $m_{3/2} \gtrsim 3$–4 TeV together with the BBN constraint in [9].

3.1. A chaotic inflation model in SUGRA

There have been proposed various types of inflation models so far. Among them we consider that the chaotic inflation model is the most attractive since it can realize an inflationary expansion of the universe even at the Planck time and hence it is free from the initial condition problem [17]. However, the chaotic inflation is not easily realized in SUGRA [28]. The reason is that the minimal supergravity potential has an exponential factor, $\exp(\phi^*\phi/M_G^2)$, which prevents any scalar field ϕ from having a value larger than the reduced Planck scale $M_G \simeq 2.4 \times 10^{18}\,\text{GeV}$. However, the inflaton φ should have an initial value much larger than M_G at the Planck time to cause chaotic inflation.

The above problem can be solved by making use of a shift symmetry of the inflaton chiral multiplet $\Phi(x, \theta)$. We assume [18] that theory is invariant under a shift of Φ;

$$\Phi \rightarrow \Phi + iCM_G, \tag{17}$$

where C is a real constant parameter. Hereafter, we take the unit where $M_G = 1$. The Kahler potential is, thus, a function of $\Phi + \Phi^*$, that is $K(\Phi, \Phi^*) = K(\Phi + \Phi^*)$. Now it is clear that the exponential factor $\exp(K(\Phi + \Phi^*))$ discussed above does not contain the imaginary part of the scalar components of Φ, which is called φ. Thus, φ may have an initial value much larger than unity and plays the role of inflaton.

As long as the shift symmetry is exact, however, the inflaton φ never has a potential and inflation is never generated. Therefore, we must introduce a breaking of the shift symmetry. For this purpose we introduce a suprion chiral field Ξ and extend the shift symmetry as [29]

$$\Phi \to \Phi + iC,$$
$$\Xi \to \frac{\Phi}{\Phi + iC}\Xi. \tag{18}$$

Notice that the product $\Xi\Phi$ is invariant of the shift symmetry and the vacuum value $\langle \Xi \rangle = m \ll 1$ represents the breaking of the symmetry. In addition to the shift symmetry we impose R symmetry to suppress the constant term in the superpotential. This compels us to introduce another chiral multiplet $X(x, \theta)$ with the R charge 2 to get a superpotential. Then, we have a superpotential invariant under the shift and R symmetries,[3]

$$W = X(c_1\Xi\Phi + c_2(\Xi\Phi)^2 + c_3(\Xi\Phi)^3 + \cdots), \tag{19}$$

where we have considered that Ξ and Φ carry vanishing R charges and the coefficients c_i are constants of order 1. We see that the first term in the superpotential dominates unless $\langle \Xi \rangle \Phi = m\Phi \geq 1$. Thus, we assume the dominance of the first term in the following analysis, that is,[4]

$$W \simeq mX\Phi. \tag{20}$$

The Kahler potential is given by[5]

$$K = \tfrac{1}{2}(\Phi + \Phi^*)^2 + XX^* + \cdots. \tag{21}$$

Then, the potential for the inflaton φ is given by

$$V(\varphi) \simeq \tfrac{1}{2}m^2\varphi^2. \tag{22}$$

This is nothing but the inflaton potential of the chaotic inflation model and the normalization of the density fluctuations at the COBE scale gives

$$m \simeq 10^{13}\,\text{GeV} \simeq 10^{-5}. \tag{23}$$

We should note here that we have discarded a linear term of X in the superpotential, since it disturbs the inflation dynamics. However, it is easy to forbid such an unwanted term in the superpotential by imposing an extra U(1) symmetry, in which X and Ξ have the U(1) charges $-n$ and $+n$. We assume this U(1) symmetry is broken by $\langle \Theta \rangle = \vartheta$ where the suprion field Θ carries the U(1) charge -1. Then, we see that the linear term of X in the superpotential is completely forbidden for $n > 0$, while the term $X\Xi\Phi$ is allowed.

3.2. *Leptogenesis in the chaotic inflation model*

Let us consider the inflaton decay. The inflaton chiral multiplet Φ may couple to a pair of the heavy Majorana neutinos N_i, since the Φ is a gauge singlet and has a vanishing R charge.[6] The superpotential contributes to the inflaton φ decay is given by,

$$W = \lambda_i\langle\Theta\rangle^{n+2\gamma_i}\langle\Xi\rangle\Phi N_i N_i,$$
$$= \lambda_i\vartheta^{n+2\gamma_i}m\Phi N_i N_i, \tag{24}$$

where λ_i are constants of order 1 and we have assumed the U(1) charge for N_i to be $+\gamma_i$. If $M_i < m_\Phi/2$ the inflaton φ can decay into the N_i pair and the decay rate is given by

$$\Gamma_\varphi \simeq (\lambda\vartheta^{n+2\gamma_i})^2\frac{m^3}{32\pi} \simeq 10\vartheta^{2(n+2\gamma_i)}\,\text{GeV}, \tag{25}$$

for $m \simeq 10^{13}$ GeV. Then, the reheating temperature is estimated as

$$T_R \simeq \lambda_i\vartheta^{n+2\gamma_i} \times 10^9\,\text{GeV}. \tag{26}$$

For a given reheating temperature we estimate the baryon-number asymmetry produced via the inflaton decay as

$$\frac{n_B}{s} \simeq 1 \times 10^{-10}\left(\frac{T_R}{10^6\,\text{GeV}}\right)\left(\frac{M_1}{m_\Phi}\right)\left(\frac{m_{v_3}}{0.05\,\text{eV}}\right),$$
$$\simeq 1 \times 10^{-7}\vartheta^{n+2\gamma_i}\left(\frac{M_i}{m_\Phi}\right)\left(\frac{m_{v_3}}{0.05\,\text{eV}}\right), \tag{27}$$

for $\lambda_i = 1$.

It is an intriguing possibility to identify the above extra U(1) symmetry with the Froggatt and Nielsen U(1) [30]. If it is the case, the mass of the heavy Majorana neutrinos, M_i, is given by

$$M_i \simeq \vartheta^{2\gamma_i}M_0, \tag{28}$$

where[7]

$$M_0 \simeq \vartheta^{2\beta_i} \times 10^{15}\,\text{GeV}. \tag{29}$$

Here, β_i is the U(1) charge for lepton-doublet chiral multiplets ℓ_i, and we have assumed an hierarchy $m_{v_3} > m_{v_{1,2}}$ and used $m_{v_3} \simeq 0.05$ eV. It is clear that the dominant contribution to n_B/s comes from the decay of N_i whose mass is near the inflaton mass $\lesssim m_\Phi/2 \simeq 10^{13}$ GeV. A phenomenological analysis gives us $\vartheta \simeq 1/17$ [31]. Thus, we consider the case of $\gamma_i + \beta_i = 1$ which provides $M_i \simeq \vartheta^2 \times 10^{15}$ GeV $\simeq 3 \times 10^{12}$ GeV. Then, we find

$$\frac{n_B}{s} \simeq \tfrac{1}{3} \times 10^{-7}\vartheta^{n+2-2\beta_i}. \tag{30}$$

The observation $n_B/s \simeq 0.9 \times 10^{-10}$ [23] suggests the exponent $n + 2 - 2\beta_i = 2$. For $n = 1$ we get $\beta_i = 1/2$ leading to a prediction, $\tan\beta \equiv \langle H\rangle/\langle\bar{H}\rangle \simeq 10$–15. Notice, however, that the prediction on $\tan\beta$ depends on the unkown Froggatt and Nielsen charge n for the suprion field Ξ.

4. Conclusions

The seesaw mechanism relates small neutrino masses to ultrahigh-energy new physics [6]. Thus, the small neutrino masses and mixing angles are a window to the new physics beyond the standard model. The superheavy Majorana neutrinos is a prediction of the seesaw mechanism. In this talk, we have shown that the decays of the superheavy Majorana neutrinos N_i produce the lepton-number asymmetry, which is converted into the baryon-number symmetry in our universe (leptogenesis).

The thermal leptogenesis is very attractive, since we do not need any extra assumption except for the presence of the superheavy Majorana neutrinos and the reheating temperature

---

[3] Absence of other possible terms such as $W \propto X$ is explained below.

[4] We take $c_1 = 1$, for simplicity.

[5] The term $(\Phi + \Phi^*)$ in the Kahler potential does not affect the inflation dynamics [29].

[6] The coupling to a pair of Higgs multiplets is forbidden by a discrete B–L symmetry [29].

---

[7] M_0 represents the B–L breaking scale.

$T_R \gtrsim 2 \times 10^9$ GeV. This reheating temperature, however, causes a serious cosmological problem in SUGRA, namely too many gravitinos are produced to keep the success of the BBN with such high reheating temperatures, if the gravitino mass $m_{3/2} \simeq$ 100 GeV–10 TeV and the gravitino is unstable. This problem forces us to consider nonthermal leptogenesis. We have found that a nonthermal leptogenesis via inflaton decay is an interesting alternative to the thermal leptogenesis and it works very well to explain the baryon-number asymmetry $n_B/s \simeq 0.9 \times 10^{-10}$ in the present universe.

However, in this last section, we briefly review attempts to solve the gravitino problem in the thermal leptogenesis.

The first one is the proposal by Pilaftsis who considers quasi-degenerate heavy Majorana neutrinos ($M_1 \simeq M_2$) [10]. In this model the lepton-asymmetry parameter ϵ is enhanced by a factor of $M_1/(M_1 - M_2)$ and hence the decays of both N_1 and N_2 may produce enough asymmetry even for $T_R \lesssim 10^{6-7}$ GeV.

The second is the proposal by Bolz, Buchmuller and Plumacher [32] who consider the case where the gravitino is the stable lightest SUSY particle (LSP). In this case the next to LSP is the subject of the cosmological constraint, since its decay products may destroy the light elements created by the BBN as for an unstable gravitino. Detailed analyses show that this scenario survives marginally only for a small parameter region of $m_{3/2} \simeq$ 10–100 GeV [33].

The third solution is given by a gauge-mediation model in which the gravitino is a stable LSP of mass $m_{3/2} \lesssim 1$ GeV. If the gravitino mass is $m_{3/2} \simeq$ 1–30 eV, we have no gravitino problem. Dark matter may be the axion. For $m_{3/2} \simeq$ 100 keV–1 GeV, there is an interesting possibility that a late-time entropy production in a class of gauge mediation models renders naturally the gravitino to be a dominant component of the dark matter [34]. In this scenario the reheating temperature may be as high as $T_R \simeq 10^{13}$ GeV.

The last solution is to assume an anomaly mediation with a gravitino mass $\gtrsim 100$ TeV. In this case the gravitino decays before the BBN and hence there is no cosmological problem. However, the gravitino decay mode always contains one LSP and hence the relic abundance of the gravitino must be constrained from above so that the density of the nonthermal LSP produced by the gravitino decay do not exceed the dark-matter density. This condition leads to $T_R \lesssim 10^{11}$ GeV [35] which is consistent with the thermal leptogenesis.

We stress here that the above solutions predict distinct particle spectra at the TeV scale, which may be testable in future collider experiments such as LHC. If all of them are excluded experimentally, we should consider the nonthermal leptogenesis seriously. On the other hand, if one of them is confirmed the thermal leptogenesis will become more convincing.

Acknowledgements

The author is grateful to the organizers of Nobel Symposium 129 on Neutrino Physics in Sweden for the hospitality during the stay. This work is partially supported by Grant-in-Aid Scientific Research (s) 14102004.

References

1. Sakharov, A. D., ZhETF Pis'ma **5**, 32 (1967).
2. 't Hooft, G., Phys. Rev. Lett. **37**, 8 (1976).
3. Kuzmin, V. A., Rubakov, V. A. and Shaposhnikov, M. E., Phys. Lett. **155**, 36 (1985).
4. For a review see, Cohen, A. G., Kaplan, D. B. and Nelson, A. E., Ann. Rev. Nucl. Part. Sci. **43**, 27 (1983).
5. Fukugita, M. and Yanagida, T., Phys. Lett. B **174**, 45 (1986).
6. Yanagida, T., in Proc. of the Workshop on "The Unified Theory and the Baryon Number in the Universe", Tsukuba, Japan, Feb. 13–14, (1979), (eds. O. Sawada and S. Sugamoto), (KEK Report KEK-79-18, 1979, Tsukuba) p. 95; Progr. Theor. Phys. **64**, 1103 (1980); Ramond, P. in a Talk given at Sanibel Symposium, Palm Coast, Fla., Feb. 25–Mar. 2, (1979), preprint CALT-68-709. See also Glashow, S., in Proc. of the Cargése Summer Institute on "Quarks and Leptons", Cargése, July 9–29, (1979), (eds. M. Lévy *et al.*, (Plenum 1980, New York), p. 707.
7. Minkowski, P., Phys. Lett. **67**B, 421 (1977). This early paper discusses the seesaw matrix in a calculation of $\mu \to e + \gamma$ decay amplitude. Since the context of this paper is limited, it does not establish the link to unification-scale physics, a crucial merit of the seesaw mechanism.
8. Buchmuller, W., Di Bari, P. and Plumacher, M., hep-ph/0302092; hep-ph/0406014; Giudice, G. F., Notari, A., Raidal, M., Riotto, A. and Strumia, A., hep-ph/0310123.
9. Kawasaki, M., Kohri, K. and Moroi, T., astro-ph/0408426.
10. Pilafsis, A., Phys. Rev. D**56**, 5431 (1997); Ellis, J., Raidal, M. and Yanagida, T., Phys. Lett. B**546**, 228 (2002).
11. Kumekawa, K., Moroi, T. and Yanagida, T., Progr. Theor. Phys. **92**, 437 (1994); Asaka, T., Hamaguchi, K., Kawasaki, M. and Yanagida, T., Phys. Lett. B**464**, 12 (1999); Phys. Rev. D**61**, 083512 (2000); Giudice, G. F., Peloso, M., Riotto, A. and Tkachev, I., J. High Energy Phys. **08**, 014 (1999).
12. Murayama, H., Suzuki, H., Yanagida, T. and Yokoyama, J., Phys. Rev. Lett. **70**, 1912 (1993); Ellis, J. R., Raidal, M. and Yanagida, T., Phys. Lett. B**581**, 9 (2004).
13. Murayama, H. and Yanagida, T., Phys. Lett. B**322**, 349 (1994); Hamaguchi, K., Murayama, H. and Yanagida, T., Phys. Rev. D **65**, 043512 (2002).
14. Murayama, H. and Yanagida, T., in ref. [13]; Dine, M., Randall, L. and Thomas, S., Nucl. Phys. B**458**, 291 (1996).
15. Grossman, Y., Kashti, T., Nir, Y. and Roulet, E., Phys. Rev. Lett. **91**, 251801 (2003); D'Ambrosio, G., Giudice, G. F. and Raidal, M., Phys. Lett. B**575**, 75 (2003).
16. Lazarides, G. and Shafi, Q., Phys. Lett. **258**B, 305 (1991).
17. Linde, A., Phys. Lett. **129**B, 177 (1983); See also Linde, A., in a Talk given at Nobel Symposium 2003 "Cosmology and String Theory", Sigtunastiftelsen, Sweden; hep-th/0402051.
18. Kawasaki, M., Yamaguchi, M. and Yanagida, T., Phys. Rev. Lett. **85**, 3572 (2000).
19. Weinberg, S., Phys. Rev. Lett. **42**, 850 (1979); Yoshimura, M., Phys. Lett. **88**B, 294 (1979).
20. Flanz, M., Paschos, E. A. and Sakar, U., Phys. Lett. B **345**, 248 (1995); Covi, L., Roulet, E. and Vissani, F., Phys. Lett. B **384**, 169 (1996); Buchmuller, W. and Plumacher, M., Phys. Lett. B **431**, 354 (1998).
21. Hamaguchi, K., Murayama, H. and Yanagida, T., in ref. [13]; Davidson, S. and Ibarra, A., Phys. Lett. B**535**, 25 (2002).
22. Harvey, J. A. and Turner, M. S., Phys. Rev. D**42**, 3344 (1990).
23. Spergel, D. N. *et al.*, Astrophy. J. Suppl. **148**, 175 (2003).
24. Buchmuller, W., Peccei, R. and Yanagida, T., unpublished in (2004).
25. Fukugita, M. and Yanagida, T., Phys. Rev. D**42**, 1285 (1990).
26. Guth, A. H., Phys. Rev. D**23**, 347 (1981).
27. Guth, A. H. and Pi, So-Y., Phys. Rev. Lett. **49**, 1110 (1982); Hawking, S., Phys. Lett. **115**B, 295 (1982); Starobinsky, A. A., Phys. Lett. **117**B, 175 (1982).
28. Goncharov, A. S. and Linde, A. D., Phys. Lett. **139**B, 27 (1984); Murayama, H., Suzuki, H., Yanagida, T. and Yokoyama, J., Phys. Rev. D**50**, R2356 (1994).
29. Kawasaki, M., Yamaguchi, M. and Yanagida, T., Phys. Rev. D**63**, 103514 (2001).
30. Froggatt, C. D. and Nielsen, H. B., Nucl. Phys. B**147**, 277 (1979).
31. Sato, J. and Yanagida, T., Phys. Lett. B**430**, 127 (1998); Buchmuller, W. and Yanagida, T., Phys. Lett. B**445**, 399 (1999).
32. Bolz, M., Buchmuller, W. and Plumacher, M., Phys. Lett. B**443**, 209 (1998).
33. Fujii, M., Ibe, M. and Yanagida, T., Phys. Lett. B**579**, 6 (2004) [hep-ph/0310142]; Ellis, J. R., Olive, K. A., Santoso, Y. and Spanos, V. C., Phys. Lett. B**588**, 7 (2004) [hep-ph/0312262]; Feng, J. L., Su, S. and Takayama, F., hep-ph/0404198; Roszkowski, L. and de Austri, R. R., hep-ph/0408227.
34. Fujii, M. and Yanagida, T., Phys. Lett. B**549**, 273 (2002) [hep-ph/0208191]; Fujii, M., Ibe, M. and Yanagida, T., Phys. Lett. B**579**, 6 (2004) [hep-ph/0310142].
35. See, for an explicit model, Ibe, M., Kitano, R., Murayama, H. and Yanagida, T., hep-ph/0403198.

Physica Scripta. Vol. T121, 142–146, 2005

Neutrinos and Big Bang Nucleosynthesis

Gary Steigman[*]

Departments of Physics and Astronomy, The Ohio State University, Columbus, OH 43210, USA

Received December 17, 2004; accepted January 11, 2005

PACS numbers: 26.35.+c, 95.30.Cq, 98.80.Es, 98.80.Ft

Abstract

The early universe provides a unique laboratory for probing the frontiers of particle physics in general and neutrino physics in particular. The primordial abundances of the relic nuclei produced during the first few minutes of the evolution of the Universe depend on the electron neutrinos through the charged-current weak interactions among neutrons and protons (and electrons and positrons and neutrinos), and on all flavors of neutrinos through their contributions to the total energy density which regulates the universal expansion rate. The latter contribution also plays a role in determining the spectrum of the temperature fluctuations imprinted on the Cosmic Background Radiation (CBR) some 400 thousand years later. Using deuterium as a baryometer and helium-4 as a chronometer, the predictions of BBN and the CBR are compared to observations. The successes of, as well as challenges to the standard models of particle physics and cosmology are identified. While systematic uncertainties may be the source of some of the current tensions, it could be that the data are pointing the way to new physics. In particular, BBN and the CBR are used to address the questions of whether or not the relic neutrinos were fully populated in the early universe and, to limit the magnitude of any lepton asymmetry which may be concealed in the neutrinos.

1. Introduction

During its early evolution the universe was hot and dense, passing brief epochs as a universal particle accelerator and as a cosmic nuclear reactor. As a consequence, through its evolution the entire universe provides a valuable alternative to terrestrial accelerators and reactors as probes of fundamental physics at the highest energies and densities. Several decades of progress have validated this Particle Astrophysics and Particle Cosmology approach to testing and constraining models of High Energy Physics and Cosmology; for early work see, e.g., [1, 2, 3]. This strategy has proven especially useful in connection with the physics of neutrinos (e.g., masses, mixing, number of flavors, etc.).

Neutrinos play two different, but equally important roles in Big Bang Nucleosynthesis (BBN). On the one hand, electron-type neutrinos (and antineutrinos), through their charged current, weak interactions help to regulate the neutron-proton ratio, which plays a key role in determining the abundance of helium (^{4}He) emerging from BBN when the universe is $\sim$20 minutes old. For example, since the ^{4}He yield is largely fixed by the supply of neutrons available at BBN, an asymmetry between ν_e and $\bar{\nu}_e$ (neutrino "degeneracy") will drive the relative abundance of neutrons up or down, thereby increasing or decreasing the relic abundance of ^{4}He. On the other hand, **all** flavors of neutrinos were relativistic at BBN ($\sim$few MeV $\gtrsim T \gtrsim 30$ keV), contributing significantly to the total density, which determines the expansion rate of the universe at that epoch. The competition between the universal expansion rate (the Hubble parameter, H) and the nuclear and weak interaction rates is key to regulating the relic abundances of the light nuclides (D, ^{3}He, ^{4}He, ^{7}Li) synthesized during BBN.

This latter effect of (light, relativistic) neutrinos on the expansion rate also plays a role some 400 kyr later, at "recombination" (protons and electrons combine to form neutral hydrogen) when the Cosmic Background Radiation (CBR) is set free from the tyranny of electron scattering to propagate throughout the Universe. By influencing the age of the Universe and the size of the sound horizon at recombination, the neutrinos help to fix the scales of the CBR temperature anisotropies observed by WMAP and other CBR detectors; see, e.g., [4] and references therein. Here, however, neutrino degeneracy plays no role except, perhaps, by increasing the neutrino energy density and, thereby, affecting the expansion rate. This latter effect is, generally, subdominant.

Since neutrinos influence the early evolution of the Universe at these two, widely separated epochs ($\sim$20 minutes and $\sim$400 kyr later), the relics from BBN (light nuclides) and the temperature anisotropies imprinted on the CBR provide two, largely independent windows on neutrino physics. These connections and what we have learned from them are reviewed here. For further details and references, see [5, 6]; this review is largely based on these two papers. After introducing some notation in the next section, the constraints from the CBR are reviewed in §3. §4 provides an overview of BBN and of the current status of the comparisons between the observational data and the predictions of the standard model (SBBN) as well as of some general extensions of the the standard model (non-standard BBN). In §5 the constraints from the CBR and from BBN are combined to identify the allowed ranges of the baryon and neutrino parameters. We conclude in §6 with a summary and with an identification of the successes of the standard models of particle physics and cosmology and of some of the challenges confronting them.

2. Notation

To set the scene for the discussion to follow, it is useful to first introduce some notation. We are interested in three, key parameters: the baryon density, the number of "equivalent" neutrinos, and a measure of a neutrino-antineutrino asymmetry.

As the universe expands, the baryon density decreases. A dimensionless measure of the baryon density is provided by the ratio of baryons to photons (in the CBR). Following $e^{\pm}$ annihilation, this ratio is preserved during the subsequent evolution of the universe. The parameter η is defined by the present (i.e., post-BBN, post-recombination) value of this ratio: $\eta \equiv (n_B/n_\gamma)_0$; $\eta_{10} \equiv 10^{10}\eta$. An equivalent measure of the baryon density is provided by the baryon density parameter, Ω_B, the ratio (at present) of the baryon density to the critical density. In terms of the present value of the Hubble parameter, $H_0 \equiv 100h$ km s^{-1} Mpc^{-1}, these two measures are related by

$$\eta_{10} \equiv 10^{10}(n_B/n_\gamma)_0 = 274\Omega_B h^2. \tag{1}$$

[*]Email address: steigman@mps.ohio-state.edu

In the standard model of particle physics there are three families of light neutrinos ($N_v = 3$) which, in the standard model of cosmology, are relativistic at BBN and also at recombination. The early evolution of the universe is "radiation dominated", *i.e.*, the energy density is dominated by the contributions from relativistic particles, including the neutrinos. The universal expansion rate, as measured by the Hubble parameter, depends on the density: $H \propto \rho^{1/2}$. Any additional (non-standard) contributions to the energy density (such as, *e.g.*, from additional flavors of neutrinos) will result in a speed-up of the expansion rate,

$$S \equiv H'/H = (\rho'/\rho)^{1/2} > 1. \tag{2}$$

Any non-standard contribution to the density may be written in terms of what would be the energy density due to an equivalent number of "extra" neutrinos ΔN_v ($N_v \equiv 3 + \Delta N_v$). Prior to $e^\pm$ annihilation, this may be written as

$$\frac{\rho'}{\rho} \equiv 1 + \frac{7\Delta N_v}{43}. \tag{3}$$

Thus, either S, the expansion rate factor or, ΔN_v, the number of equivalent neutrinos, provide equally good measures of the early universe expansion rate. While it is easy to imagine *extra* contributions to the energy density from new physics beyond the standard model, it must be noted that it is possible for ΔN_v to be negative, leading to a slower than standard, early universe expansion rate ($S < 1$). For example, models where the decay of a massive particle, produced earlier in the evolution of the universe, reheats the universe to a temperature which is not high enough to (re)populate a thermal spectrum of the standard neutrinos ($T_{RH} \lesssim 7\,\text{MeV}$), will result in $\Delta N_v < 0$ and $S < 1$ [7].

For any neutrino flavor i, an asymmetry ("neutrino degeneracy") between the numbers of v_i and $\bar{v}_i$, relative to the number of CBR photons, can be quantified by the net lepton number L_i, the neutrino chemical potential μ_i or, by the dimensionless degeneracy parameter $\xi_i \equiv \mu_i/T$:

$$L_i \equiv \frac{n_{v_i} - n_{\bar{v}_i}}{n_\gamma} = \frac{\pi^2}{12\zeta(3)}\left(\xi_i + \frac{\xi_i^3}{\pi^2}\right). \tag{4}$$

Although we are interested in lepton asymmetries which are orders of magnitude larger than the baryon asymmetry ($B \sim \eta \lesssim 10^{-9}$), the values of ξ_i ($i = e, \mu, \tau$) considered here are sufficiently small ($|\xi_i| \lesssim 0.1$) so that the "extra" energy density contributed by such degenerate neutrinos is negligible.

$$\Delta N_v(\xi_i) = \frac{30}{7}\left(\frac{\xi_i}{\pi}\right)^2 + \frac{15}{7}\left(\frac{\xi_i}{\pi}\right)^4 \lesssim 0.01. \tag{5}$$

In this case, the results to be presented below for $\xi \neq 0$ will correspond to $N_v = 3$ ($S = 1$). In fact, if the three active neutrinos (v_e, v_μ, v_τ) mix only with each other, all individual neutrino degeneracies will equilibrate via neutrino oscillations to, approximately, the electron neutrino degeneracy before BBN begins [8]. Thus, the magnitude of the **electron** neutrino degeneracy constrained by BBN is of special interest when limiting the total net lepton asymmetry of the universe: $L \approx 3L_e$; $\Delta N_v(\xi) \approx 3\Delta N_v(\xi_e)$.

For the standard models of particle physics and cosmology, $\Delta N_v = 0$ ($S = 1$) and $\xi \equiv \xi_e \approx \xi_\mu \approx \xi_\tau = 0$, and the value (range of values) of η identified by BBN and the CBR should agree, restricting the allowed deviations from zero of ΔN_v and/or ξ_e.

3. CBR

In Figures 1 and 2 are shown the CBR temperature anisotropy angular power spectra for different choices of the baryon density (Fig. 1) and of N_v (Fig. 2). Non-zero values of ΔN_v change the energy density in radiation, which shifts the redshift of the epoch of equal matter (Cold Dark Matter and Baryons) and radiation densities. This results in changes to the angular scales and the amplitudes of the "acoustic" peaks in Figures 1 & 2; see, *e.g.*, [5] and further references therein. WMAP is a much more sensitive baryometer than it is a chronometer. While the best fit values for the baryon density and N_v are $\eta_{10} = 6.3$ and $N_v = 2.75$ ($S = 0.98$) respectively, the 2σ range for the baryon density parameter is $5.6 \leq \eta_{10} \leq 7.3$, whereas for N_v it is $0.9 \leq N_v \leq 8.3$ ($0.81 \leq S \leq 1.36$) [6]. Thus, although the WMAP best fit value of N_v is less than the standard model value of 3, it is clear that this difference is not at all statistically significant. It will be interesting to see if the much tighter CBR constraint on the baryon density parameter (~ 6–8%) is consistent with the value of this parameter identified by SBBN.

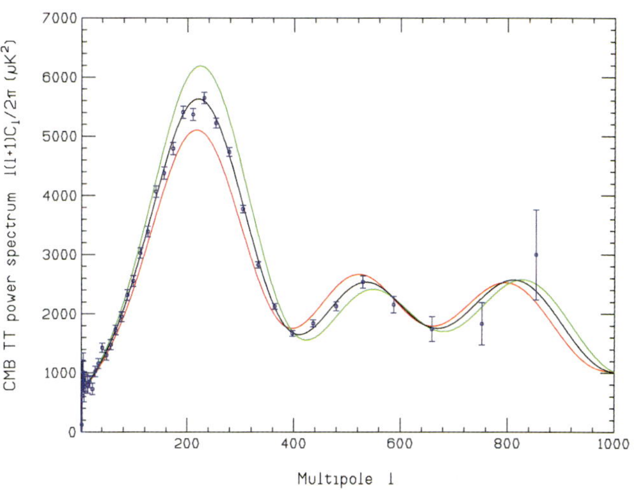

Fig. 1. The CBR temperature anisotropy angular power spectrum for three choices of the baryon density, $\Omega_B h^2 = 0.018$, 0.023 (best fit), 0.028, from bottom to top respectively near $l \approx 200$. The data points with error bars are from Spergel *et al.* [4].

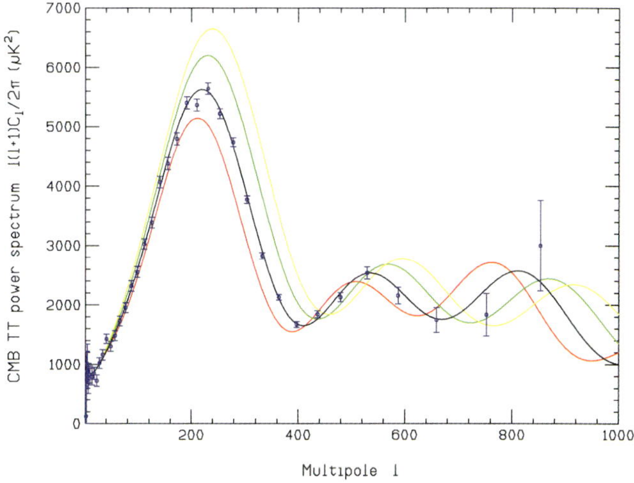

Fig. 2. The CBR temperature anisotropy angular power spectrum as in Fig. 1, now for four choices of the equivalent number of neutrinos $N_v = 1$, 2.75 (best fit), 5, 7, from bottom to top respectively near $l \approx 200$. The data points with error bars are from Spergel *et al.* [4].

4. BBN

Since the relic abundances of D, ^{3}He, and ^{7}Li produced during BBN are *rate* limited (nuclear reaction rates), each of these nuclides is a candidate baryometer. Among these, Deuterium is the baryometer of choice for several reasons. The BBN-predicted abundance of D is sensitive to the baryon density ($y_D \equiv 10^5(D/H)_P \propto \eta^{-1.6}$). The post-BBN evolution of D is simple in that whenever gas is incorporated into stars, D is completely destroyed. Thus, observations of the deuterium abundance anywhere, at any time, provide a *lower* bound to the primordial D abundance (and, therefore, an *upper* bound to η). It is expected that if D can be observed in regions which have experienced minimal stellar processing, the deuterium abundance inferred from such data should be very close to the BBN abundance.

The good news is that there are data from observations of neutral D and neutral H in high redshift, low heavy element ("metallicity") abundance QSO absorption line systems (QSOALS); see Figure 3. As may be seen from Fig. 3, the bad news is that there are only five such systems with good enough data to derive meaningful D abundances [9]. And, even for these, specially selected targets, there is the possibility of confusion between D and H absorption spectra which are identical, save for the wavelength/velocity shift between them. That is, small amounts of hydrogen at the "wrong" redshift (interlopers) can masquerade as deuterium. Further, since the hydrogen absorption in such systems is saturated, it is often difficult to identify the number of systems which contribute to the total absorption and this can lead to errors in the inferred amount of H in determining the D/H ratio. With these caveats in mind, it is important to understand that systematic, rather than statistical uncertainties may dominate the error budget. Indeed, for the data summarized by Kirkman *et al.* [9] and shown in Fig. 3, χ^2 exceeds 16 for 4 degrees of freedom! Following Kirkman *et al.*, the error bars

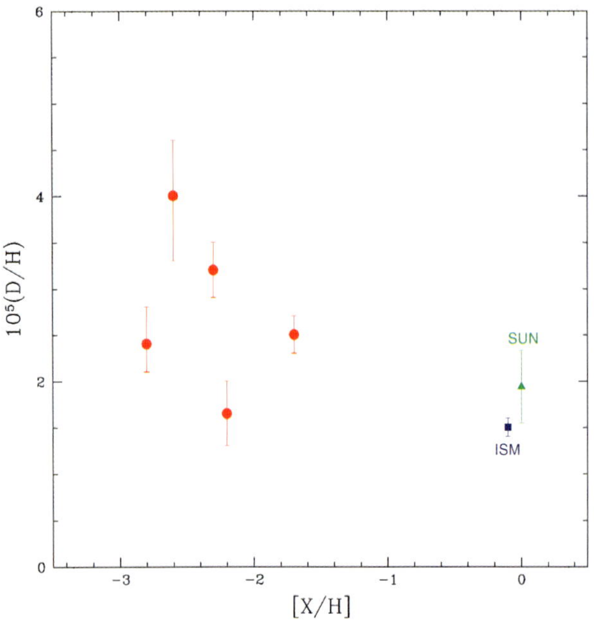

Fig. 3. The deuterium abundance (by number relative to hydrogen), $y_D \equiv 10^5(D/H)$, derived from high redshift, low metallicity QSOALS [9] (filled circles). The metallicity is on a log scale relative to solar; depending on the line-of-sight, X may be oxygen or silicon. Also shown is the solar system abundance (filled triangle) and that from observations of the local ISM (filled square).

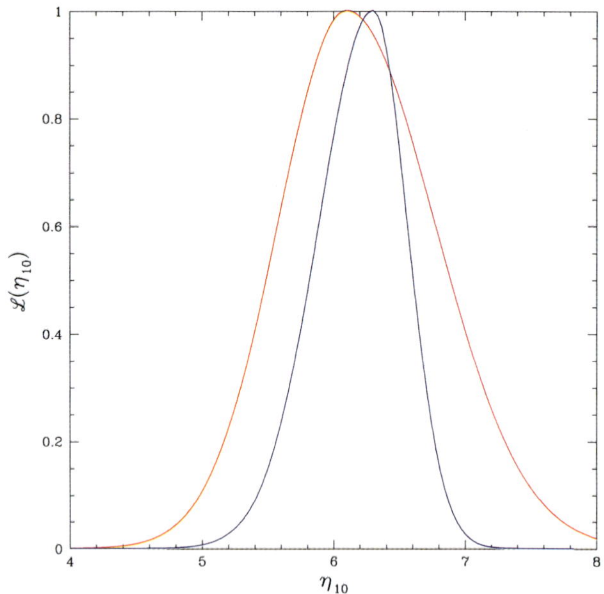

Fig. 4. The likelihood distributions for the baryon density parameter η_{10} inferred from the CBR (WMAP) and from SBBN and deuterium; see the text. The broader distribution, centered at $\eta_{10} \approx 6.1$ is from SBBN and the narrower distribution, centered at $\eta_{10} \approx 6.3$ is from the CBR.

are inflated here to account for this excessive dispersion, and a primordial abundance $y_D = 2.6 \pm 0.4$ is adopted. For SBBN this estimate for the abundance of primordial D corresponds to $\eta_{10} = 6.1^{+0.7}_{-0.5}$. This is in excellent (spectacular!) agreement with the completely independent estimate above in §3 from the CBR. In Figure 4 are shown the likelihood distributions for η derived from the CBR (WMAP) and from SBBN (D).

^{3}He is a much less useful baryometer than is D. In the first place, its BBN-predicted abundance is less sensitive to the baryon density parameter (^{3}He/H $\propto \eta^{-0.6}$). In addition, as gas is incorporated into stars and the stellar-processed material is returned to the interstellar medium when they die, ^{3}He is produced, destroyed and, some survives. This complicated history makes it much more difficult to account for the post-BBN evolution of ^{3}He. Finally, ^{3}He is only observed within the Galaxy [10] and, therefore, its abundance samples only a limited range in metallicity. Nonetheless, while there is a clear oxygen abundance gradient with location within the Galaxy (higher in the center, lower in the suburbs, indicating more stellar processing in the interior), the ^{3}He abundance shows no such gradient (either with position or with metallicity). However, the Bania, Rood, and Balser [10] recommended value of $y_3 \equiv 10^5(^3He/H) = 1.1 \pm 0.2$ corresponds to $\eta_{10} \approx 5.6^{+2.0}_{-1.7}$, in broad agreement with the SBBN-D and CBR-WMAP determinations. So far, so good.

In contrast to deuterium (and to ^{3}He), the abundance of ^{4}He has increased in the post-BBN universe as stars have burned hydrogen to helium (and beyond). Therefore, to avoid model-dependent evolutionary uncertainties, it is best to concentrate on determining the ^{4}He abundance (mass fraction, Y_P) in the most metal-poor sample available and to let the data speak for themselves concerning any correlation between the helium and heavy element abundances. The best such data come from observing the emission lines formed when ionized helium and hydrogen capture electrons in regions of hot, ionized gas (HII regions). There exists at present, thanks largely to the work of Izotov and Thuan [11], a very large sample of helium abundance determinations in low metallicity, extragalactic

H II regions. This newer, more uniform data set complements earlier, more heterogeneous samples [12]. For the WMAP estimate of the baryon density, including its uncertainty, the SBBN-predicted primordial abundance of ^{4}He is $Y_\mathrm{P} = 0.2482 \pm 0.0007$. Unfortunately, *none* of the Y_P estimates [11, 12] agree with the SBBN prediction, all being low by roughly 2σ. Indeed, from their 2004 sample of 82 data points Izotov & Thuan [11] derive such a small uncertainty, that their central value is low by nearly 6σ! For the numerical results presented in §5 below, a primordial mass fraction $Y_\mathrm{P} = 0.238 \pm 0.005$ [13] is adopted.

The ^{4}He abundance determinations are most likely examples of extremely precise, yet inaccurate, determinations of an important cosmological parameter. It has long been known that there are a variety of systematic uncertainties which are likely to interfere with an accurate Y_P determination. In a very recent, detailed study of *some* of these identified systematic uncertainties, Olive & Skillman [14] suggest the true errors likely exceed previous estimates by factors of 2–3 or more ($\Delta Y_\mathrm{P} \approx 0.013$). Given such large uncertainties, it is not surprising that the extant ^{4}He abundance data are, within the errors, consistent with the predictions of SBBN and the CBR (and/or SBBN plus D) determined baryon density. Nonetheless, it might be premature to ignore this challenge to the standard models of particle physics and of cosmology. Perhaps the tension between D and ^{4}He is an early warning of new physics beyond the standard models. Before pursuing this option in the next section, ^{7}Li is considered here.

As with ^{4}He, there is conflict in the comparisons between the SBBN predictions and the ^{7}Li relic abundance estimates derived from observations. Here, too, the potential for systematic errors looms large. ^{7}Li is produced in the Galaxy by cosmic ray spallation/fusion reactions and observations of super-lithium-rich red giants provide evidence that (at least some) stars can be net producers of lithium. Therefore, to infer the BBN yield of ^{7}Li, the data should be limited to those from the most metal-poor halo stars in the Galaxy. For the WMAP baryon density, the SBBN expected ^{7}Li abundance is $[\mathrm{Li}]_\mathrm{P} \equiv 12 + \log(\mathrm{Li/H})_\mathrm{P} = 2.65^{+0.05}_{-0.06}$. In contrast, for a selected data set of the lowest metallicity halo stars, Ryan *et al.* [15] derive a primordial abundance of $[\mathrm{Li}]_\mathrm{P} \approx 2.0$–$2.1$. In deriving the stellar lithium abundances, the adopted stellar temperature plays a key role. When using the infrared flux method effective temperatures, studies of halo and Galactic Globular Cluster stars [16] suggest a higher abundance: $[\mathrm{Li}]_\mathrm{P} = 2.24 \pm 0.01$. Very recently, Melendez & Ramirez [17] have reanalyzed 62 halo stars using an improved infrared flux method effective temperature scale, confirming the higher lithium abundance; they find $[\mathrm{Li}]_\mathrm{P} = 2.37 \pm 0.05$. If this were the true primordial ^{7}Li abundance, then the SBBN value of the baryon density parameter would be $\eta_{10} = 4.5 \pm 0.4$, in conflict with the CBR-WMAP and/or SBBN-D estimates. Indeed, all of the current observational estimates of the abundance of primordial lithium are significantly lower than the SBBN expectation.

As with ^{4}He, the problem may be traced to the astrophysics rather than to the cosmology. Since the low metallicity halo stars used to estimate the primordial abundance of lithium are the oldest stars in the Galaxy, they have had the most time to modify (by dilution and/or destruction) their surface abundances. While mixing of the stellar surface material with the interior would destroy or dilute prestellar lithium, the very small dispersion among the observed values of [Li] derived from the lowest metallicity halo stars suggests this effect may not be large

enough to bridge the ≈ 0.3 dex gap between the observed and CBR/SBBN-predicted abundances; see, *e.g.*, [18] and further references therein.

5. CBR and BBN combined

In contrast to D (and ^{3}He and ^{7}Li), ^{4}He is an insensitive baryometer, but its primordial abundance is a useful, early universe chronometer. If the standard model expansion rate is changed ($S \neq 1$, $\Delta N_v \neq 0$), this will affect the neutron-proton ratio at BBN and change the SBBN-expected value of Y_P. The current conflict between the SBBN-predicted and the observationally inferred values of Y_P requires a slowdown in the early universe expansion rate ($S < 1$; $\Delta N_v < 0$). A joint BBN fit to the observationally inferred D and ^{4}He abundances suggests that $\eta_{10} \approx 5.7$ and $S \approx 0.94$ ($\Delta N_v \approx -0.70$) can relieve the SBBN tension between D and ^{4}He. However, it can be seen in Figure 5 that while these values are entirely consistent with the constraints from the CBR, $N_v = 3$ is consistent with both BBN and the CBR at $\sim 2\sigma$ [5]. Nonetheless, this combination of parameters does not resolve the conflict with ^{7}Li. Although a slowdown in the expansion rate has the effect of increasing ^{7}Li (more time for production), this is compensated by the somewhat lower baryon density (slower reaction rates), which has the opposite effect. The result is that for the choices of S and η which resolve the conflicts between D and ^{4}He (and between WMAP and ^{4}He), the predicted primordial abundance of ^{7}Li is $[\mathrm{Li}]_\mathrm{P} \approx 2.62 \pm 0.10$, still some 0.2–0.3 dex higher than that inferred from the data.

Another example of new physics with the potential to resolve the D $-^4$He conflict while leaving the CBR–D agreement unaffected is a neutrino-antineutrino asymmetry ($\xi \equiv \xi_e \neq 0$); see [6] and references therein. Through its effect on the neutron-proton ratio at BBN, such an asymmetry can change the BBN-predicted ^{4}He abundance. For $\xi_e > 0$, there are more v_e than $\bar{v}_e$ and this drives the n–p ratio down, decreasing the BBN-predicted ^{4}He abundance. For $\xi_e \lesssim 0.1$, the extra energy density contributed by these degenerate neutrinos is small, and $N_v = 3$ remains a good approximation. As a result, such a lepton asymmetry has negligible effect on the CBR temperature anisotropies and the good agreement with the WMAP data is unaffected. The effects

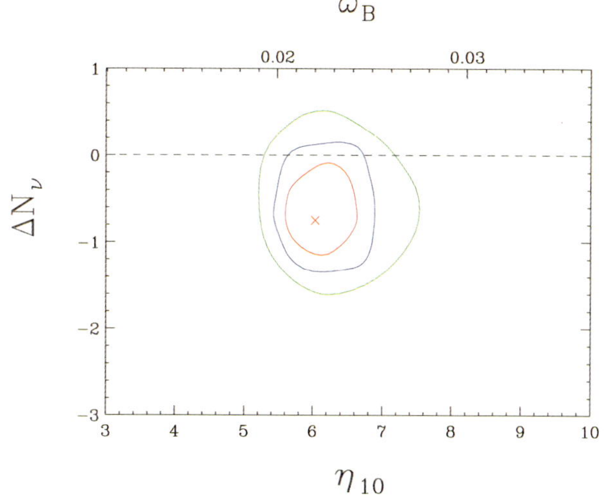

Fig. 5. The 1σ, 2σ, and 3σ contours in the $\Delta N_\nu - \eta_{10}$ plane (from Barger *et al.* [5]) consistent with the WMAP CBR data and the BBN predicted and observed abundances of D and ^{4}He. The cross marks the best fit point; see the text. Note that the upper horizontal axis is for $\omega_\mathrm{B} \equiv \Omega_\mathrm{B} h^2$.

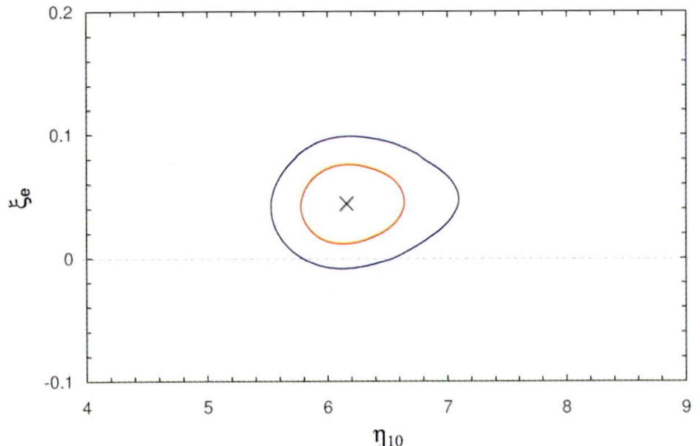

Fig. 6. The 1σ and 2σ contours in the $\xi_e - \eta_{10}$ plane (from Barger *et al.* [6]) consistent with the WMAP CBR data and the BBN predicted and observed abundances of D and ^{4}He. The cross marks the best fit point; see the text.

of such an asymmetry on the predicted abundances of D, ^{3}He, and ^{7}Li are subdominant to that on ^{4}He. The best fit parameter choices which resolve the tension between D and ^{4}He, while preserving the good agreement with the WMAP data are $\eta_{10} = 6.2$ ($\Omega_\text{B} h^2 = 0.23$) and $\xi_e = 0.044$. In Figure 6 are shown the 1σ and 2σ contours consistent with the WMAP data and with BBN (D and ^{4}He).

Finally, we note that if all three parameters (η, S, and ξ) are allowed to be free, much larger ranges in them will remain consistent with BBN (D and ^{4}He) and with the CBR; see, *e.g.*, [6] and the more recent paper of Kneller and Steigman [19]. Fixing the D and ^{4}He abundances, Kneller & Steigman find two approximate, but quite accurate, BBN relations among these three parameters.

$$590(S - 1) \approx 116\eta_{10} - 697, \tag{6}$$

and

$$145\xi_e \approx 106(S - 1) + 6.31. \tag{7}$$

Consistent with the WMAP CBR data, values of ΔN_ν and ξ_e in the ranges $-2 \lesssim \Delta N_\nu \lesssim +5$ and $-0.1 \lesssim \xi_e \lesssim +0.3$ are permitted [6]. However, even with this freedom it is still not possible to reconcile the BBN-predicted and the observed relic lithium abundances [19]. For the values of these parameters which are consistent with BBN and the CBR, $[\text{Li}]_\text{P} \approx 2.6$.

6. Summary and conclusions

BBN and the CBR probe the evolution of the Universe (and its constituents) at two, widely separated epochs in its early evolution. Confronting the predictions of BBN and the CBR with the relic abundance and WMAP data enables independent tests of the standard models of particle physics and cosmology. Qualitatively, the standard models pass these tests with flying colors, permitting BBN and CBR constraints to be put on new neutrinos physics ($N_\nu \neq 3$?; $\xi_e \neq 0$?). When considered in quantitative detail however, there are some challenges to

the standard models at the $\sim 2\sigma$ level. Many would take this as evidence for success and declare victory. However, if these tensions are taken at face value, they might be alerting us to problems with the astronomy, the astrophysics, the cosmology, the particle physics or, combinations of them. It is not unlikely that the apparent conflicts may result from the data, its analysis, and/or the extrapolations to the early universe. While the community awaits the new surprises to be encountered at the LHC, a Linear Collider, or the next generation of terrestrial or space-based telescopes, it should be kept in mind that these challenges could be pointing the way to new physics, especially new neutrino physics, beyond the standard models of particle physics and/or cosmology.

Acknowledgements

I wish to express my sincere appreciation and thanks to the organizers of this symposium and to the Nobel Foundation for its sponsorship. Special thanks are due Per Olof Hulth and Tommy Ohlsson for their tireless efforts to smooth the way and ensure that my participation would be so enjoyable and scientifically successful. The research described here has been supported at The Ohio State University by a grant from the US Department of Energy (DE-FG02-91ER40690).

References

1. Weinberg, S., "Gravitation and Cosmology," (Wiley, New York, 1972).
2. Steigman, G., Ann. Rev. Nucl. Part. Sci. **29**, 313 (1981).
3. Zeldovich, Ya. B. and Novikov, I. D., "Relativistic Astrophysics," vol. 2 (University of Chicago Press, Chicago, 1983).
4. Spergel, D. N. *et al.*, Astrophys J. Suppl. **148**, 175 (2003).
5. Barger, V., Kneller, J. P., Lee, H.-S., Marfatia, D. and Steigman, G., Phys. Lett. B **566**, 8 (2003).
6. Barger, V., Kneller, J. P., Langacker, P., Marfatia, D. and Steigman, G., Phys. Lett. B **569**, 123 (2003).
7. Kawasaki, M., Kohri, K. and Sugiyama, N., Phys. Rev. Lett. **82**, 4168 (1999); Hannestad, S., Phys. Rev. D **70**, 043506 (2004).
8. Lunardini, C. and Smirnov, A. Y., Phys. Rev. D **64**, 073006 (2001); Dolgov, A. D. *et al.*, Nucl. Phys. B **632**, 363 (2002); Abazajian, K. N., Beacom, J. F. and Bell, N. F., Phys. Rev. D **66**, 013008 (2002); Wong, Y. Y., Phys Rev. D **66**, 025015 (2002).
9. Kirkman, D., Tytler, D., Suzuki, N., O'Meara, J. and Lubin, D., Astrophys. J. Suppl. **149**, 1 (2003).
10. Bania, T. M., Rood, R. T. and Balser, D. S., Nature **415**, 54 (2002).
11. Izotov, Y. I. and Thuan, T. X., Astrophys. J. **500**, 188 (1998); ibid, Astrophys. J. **602**, 200 (2004).
12. Lequeux, J., Peimbert, M., Rayo, J. F., Serrano, A. and Torres-Peimbert, S., Astron. Astrophys. **80**, 155 (1979); Pagel, B. E. J., Simonson, E. A., Terlevich, R. J. and Edmunds, M. G., Mon. Not. R. Astron. Soc. **255**, 325 (1992); Olive, K. A., Skillman, E. D. and Steigman, G., Astrophys. J. **489**, 1006 (1997).
13. Olive, K. A., Steigman, G. and Walker, T. P., Phys. Rep. **333**, 389 (2000).
14. Olive, K. A. and Skillman, E. D., astro-ph/0405588 (2004).
15. Ryan, S. G., Norris, J. E. and Beers, T. C., Astrophys. J. **523**, 654 (1999); Ryan, S. G., Beers, T. C., Olive, K. A., Fields, B. D. and Norris, J. E., Astrophys. J. **530**, L57 (2000).
16. Bonifacio, P. and Molaro, P., Mon. Not. R. Astron. Soc. **285**, 847 (1997); Bonifacio, P., Molaro, P. and Pasquini, L., Mon. Not. R. Astron. Soc. **292**, L1 (1997).
17. Melendez, J. and Ramirez, I., Astrophys. J. Lett. **615**, L33 (2004).
18. Pinsonneault, M. H., Steigman, G., Walker, T. P. and Narayanan, V. K., Astrophys. J. **574**, 398 (2002).
19. Kneller, J. P. and Steigman, G., New J. Phys. **6**, 117 (2004).

Physica Scripta. Vol. T121, 147–152, 2005

Extra Galactic Sources of High Energy Neutrinos*

Eli Waxman

Physics faculty, Weizmann Institute of Science, Rehovot 76100, Israel

Received February 8, 2005; accepted February 10, 2005

PACS numbers: 96.40.Tv, 98.70.Rz, 98.70.Sa, 14.60.Pq

Abstract

The main goal of the construction of large volume, high energy neutrino telescopes is the detection of extra-Galactic neutrino sources. The existence of such sources is implied by observations of ultra-high energy, $\geq 10^{19}$ eV, cosmic-rays (UHECRs), the origin of which is a mystery. The observed UHECR flux sets an upper bound to the extra-Galactic high energy neutrino intensity, which implies that the detector size required to detect the signal in the energy range of 1 TeV to 1 PeV is ≥ 1 giga-ton, and much larger at higher energy. Optical Cerenkov neutrino detectors, currently being constructed under ice and water, are expected to achieve 1 giga-ton effective volume for 1 TeV to 1 PeV neutrinos. Coherent radio Cerenkov detectors (and possibly large air-shower detectors) will provide the $\gg 1$ giga-ton effective volume required for detection at $\sim 10^{19}$ eV. Detection of high energy neutrinos associated with electromagnetically identified sources will allow to identify the sources of UHECRs, will provide a unique probe of the sources, which may allow to resolve open questions related to the underlying physics of models describing these powerful accelerators, and will provide information on fundamental neutrino properties.

1. Introduction and summary

The detection of MeV neutrinos from the Sun enabled direct observations of nuclear reactions in the core of the Sun, as well as studies of fundamental neutrino properties [1]. Existing MeV neutrino "telescopes" are also capable of detecting neutrinos from supernova explosions in our local Galactic neighborhood, at distances <100 kpc, such as supernova 1987A. The detection of neutrinos emitted by SN1987A provided a direct observation of the core collapse process and constraints on neutrino properties [2]. The main goal of the construction of high energy, >1 TeV, neutrino telescopes [3] is the extension of the distance accessible to neutrino astronomy to cosmological scales.

The existence of extra-Galactic high-energy neutrino sources is implied by cosmic-ray observations. The cosmic-ray spectrum extends to energies $\sim 10^{20}$ eV, and is likely dominated beyond $\sim 10^{19}$ eV by extra-Galactic sources of protons [6] (see, however, [7]). The origin of the highest energy, $>10^{19}$ eV, cosmic rays (UHECRs) is a mystery. It has been suggested that modifications of the basic laws of physics are required in order to account for the existence of UHECRs. Such "new physics" models commonly postulate the existence of very massive particles, the decay of which produces the observed UHECRs, and generally predict large fluxes of $\sim 10^{20}$ eV neutrinos (see, e.g., [8] for review). I focus here on UHECR production models, which do not invoke modifications to the basic laws of physics. In these models it is assumed that UHECRs are charged particles, most likely protons, accelerated electromagnetically to high energy in astrophysical objects. In this case, some fraction of the protons are expected to produce pions as they escape their source by either hadronic collisions with ambient gas or photo-production with source photons, leading to electron and muon neutrino production through the decay of charged pions.

In §2 a phenomenological, model independent discussion of extra-Galactic high energy neutrino sources is presented. In §2.1 we show that the stringent constraints, which are imposed on the properties of possible UHECR sources by the high energies observed, rule out almost all source candidates, and suggest that γ-ray bursts (GRBs) and active galactic nuclei (AGN) are the most plausible sources. We furthermore demonstrate that these constraints also imply that the UHECR sources may be detectable as point sources at ~ 1 TeV to ~ 1 PeV neutrino energies by km-scale (i.e. giga-ton scale) neutrino telescopes. The required large effective volume will be achieved by optical Cerenkov detectors being constructed under ice and water [3].

In §2.2 we discuss the constraints imposed by cosmic-ray observations on the diffuse extra-Galactic high energy neutrino intensity produced by sources which, like GRBs and AGN jets, are optically thin for high-energy nucleons to $p\gamma$ and $pp(n)$ interactions. The upper bound (fig. 3, eq. (6)), which came to be known as the Waxman-Bahcall (WB) bound [4], implies that km-scale (i.e. giga-ton) neutrino telescopes are needed to detect the expected diffuse extra-Galactic flux in the energy range of ~ 1 TeV to ~ 1 PeV, and much larger effective volume is required at higher energy. Implications of the bound to predictions of neutrino emission from AGN (figure 4) are briefly discussed in §2.3. We note, that the WB bound may be evaded by postulating the existence of sources which are optically thick for protons to $p\gamma$ or $pp(n)$ interactions [4, 5]. However, the existence of such "hidden," or "neutrino only," factories is not motivated by measurements of the cosmic-ray flux or by electromagnetic observations.

In §2.4 we discuss "GZK neutrinos". If the sources of UHECRs are extra-Galactic and the particles are indeed protons, then these particles lose energy by interacting with the cosmic microwave background photons [9]. Protons of energy exceeding the threshold for pion production, $\sim 5 \times 10^{19}$ eV, lose most of their energy over a time short compared to the age of the universe (the "GZK effect"). The decay of the pions generates a background of high energy neutrinos ([10]; for detailed updated discussion see [11] and references therein). The intensity of this background, at neutrino energies $\sim 10^{19}$ eV, should be similar to the WB bound. Coherent radio Cerenkov detectors, and possibly large air-shower detectors, will provide the large effective volume required for the detection of the "GZK neutrinos" [12]. Their detection will help to determine the identity of the cosmic-ray particles and will constrain the redshift evolution of UHECR sources.

In §3 we discuss in some detail high energy neutrino emission from GRBs. This discussion illustrates the applicability of the phenomenological, model independent arguments presented in §2 through a discussion of (a model of) a particular, plausible extra-Galactic source of UHECRs and high energy neutrinos. In §3.1

---

*Summary of talk presented at the Nobel Symposium 129: Neutrino Physics, Enköping, Sweden, 2004.

148   *Eli Waxman*

we present a brief discussion of the fireball model of GRBs, and briefly present the arguments suggesting a connection between GRBs and UHECR sources. In §3.2 we show that production of 100 TeV neutrinos in the region where GRB γ-rays are produced is a generic prediction of the fireball model. It is a direct consequence of the *assumptions* that energy is carried from the underlying engine, most likely a (few) solar mass black hole, as kinetic energy of protons and that γ-rays are produced by synchrotron emission of shock accelerated particles. The detection of the predicted neutrino signal will therefore provide strong support for the validity of underlying model assumptions, which is difficult to obtain using photon observations (due to the high optical depth in the vicinity of the GRB "engine"). The predicted neutrino intensity, $\approx 20\%$ of the WB intensity bound, implies a detection of ~ 20 neutrino induced muons events per yr in a km-scale neutrino detector (since these events should be correlated in time and direction with GRB γ-rays, the search for GRB neutrinos is essentially background free). Neutrinos may be produced also in other stages of fireball evolution, at energies different than 100 TeV. The production of these neutrinos is dependent on additional model assumptions. As an example, we discuss in §3.3 the production of TeV neutrinos expected in the "collapsar" scenario, where GRB progenitors are associated with the collapse of massive stars.

The discussion of GRB neutrino emission demonstrates that in addition to identifying the sources of UHECRs, high-energy neutrino telescopes can also provide a unique probe of the physics of these sources. Moreover, detection of high energy neutrinos from extra-Galactic sources may also provide information on fundamental neutrino properties [13]. High energy neutrinos are expected to be produced in astrophysical sources by the decay of charged pions, which lead to the production of muon and electron neutrinos. However, oscillation to ν_τ's [14] imply that neutrino telescopes should detect equal numbers of ν_μ's and ν_τ's. Up-going τ's, rather than μ's, would be a distinctive signature of such oscillations. Detection of neutrinos from GRBs could be used to test the simultaneity of neutrino and photon arrival to an accuracy of ~ 1 s, checking the assumption of special relativity that photons and neutrinos have the same limiting speed. These observations would also test the weak equivalence principle, according to which photons and neutrinos should suffer the same time delay as they pass through a gravitational potential. With 1 s accuracy, a burst at 1 Gpc would reveal a fractional difference in limiting speed of 10^{-17}, and a fractional difference in gravitational time delay of order 10^{-6} (considering the Galactic potential alone). Previous applications of these ideas to supernova 1987A (see [15] for review), yielded much weaker upper limits: of order 10^{-8} and 10^{-2} respectively.

2. Phenomenological considerations

2.1. *The luminosity constraint for UHECR sources and its implications for high energy neutrino telescopes*

The essence of the challenge of accelerating particles to $>10^{19}$ eV can be understood using the following simple arguments. Consider an astrophysical source driving a flow of magnetized plasma, with characteristic magnetic field strength B and velocity v. Imagine now a conducting wire encircling the source at radius R, as illustrated in fig. 1. The potential generated by the moving plasma is given by the time derivative of the magnetic flux Φ and is therefore given by $V = \beta BR$ where $\beta = v/c$. A proton which

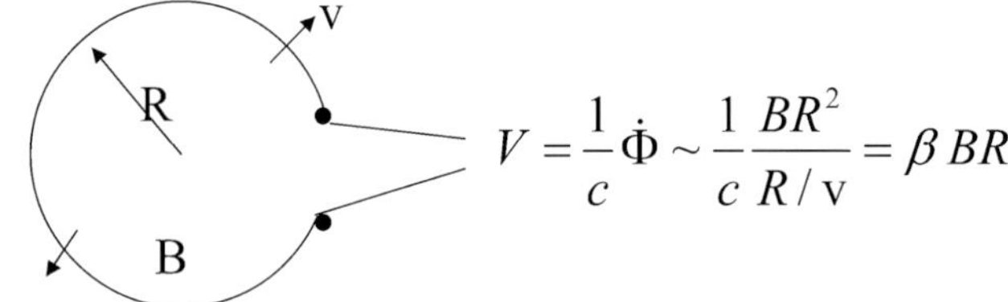

$$V = \frac{1}{c}\dot{\Phi} \sim \frac{1}{c}\frac{BR^2}{R/v} = \beta BR$$

Fig. 1. Potential drop generated by an outflow of magnetized plasma.

is allowed to be accelerated by this potential drop would reach energy $E_p \sim \beta e BR$. The situation is somewhat more complicated in the case of a relativistic outflow, where $\Gamma \equiv (1 - \beta^2)^{-1/2} \gg 1$. In this case, the proton is allowed to be accelerated only over a fraction of the radius R, comparable to R/Γ. To see this, one must realize that as the plasma expands, its magnetic field decreases, so the time available for acceleration corresponds to the time of expansion from R to, say, $2R$. In the observer frame this time is R/c, while in the plasma rest frame it is $R/\Gamma c$. Thus, a proton moving with the magnetized plasma can be accelerated over a transverse distance $\sim R/\Gamma$. This sets a lower limit to the product of the magnetic field and source size, which is required to allow acceleration to E_p, $BR > \Gamma E_p/e\beta$. This constraint also sets a lower limit to the rate L at which energy should be generated by the source. The magnetic field carries with it an energy density $B^2/8\pi$, and the flow therefore carries with it an energy flux $> vB^2/8\pi$ (some energy is carried also as plasma kinetic energy), which implies $L > vR^2B^2$ and therefore

$$L > \frac{\Gamma^2}{\beta}\left(\frac{E_p}{e}\right)^2 c = 10^{45.5}\frac{\Gamma^2}{\beta}\left(\frac{E_p}{10^{20}\,\mathrm{eV}}\right)^2\,\mathrm{erg/s}. \tag{1}$$

Only two types of sources are known to satisfy this requirement. The brightest steady sources are active galactic nuclei (AGN). For them Γ is typically between 3 and 10, implying $L > 10^{47}$ erg/s, which may be satisfied by the brightest AGN [16]. The brightest transient sources are GRBs. For these sources $\Gamma \simeq 10^{2.5}$ implying $L > 10^{50.5}$ erg/s, which is generally satisfied since the typical observed MeV-photon luminosity of these sources is $L_\gamma \sim 10^{52}$ erg/s (It was recognized early on, e.g. [17] and references therein, that while highly magnetized neutron stars may also satisfy the constraint $BR > \Gamma E_p/e\beta$, it is hard to utilize the potential drop in their electro-magnetic winds for acceleration to ultra-high energy; see, e.g., ref. [18] for a recent discussion).

It is instructive to compare eq. (1) to the constraints imposed on a neutrino source by requiring it to be detectable by a km-scale neutrino telescope. Consider the minimum flux of a source that can be detected by a neutrino telescope with effective area A (in the plane perpendicular to the source direction) and exposure time T. The probability that a muon produced by the interaction of a muon neutrino with a nucleon will cross the detector is given by the ratio of the muon and neutrino mean free paths (Note, that since the mean free path of muons of energy >0.3 TeV is of order 1 km, a detector with effective cross-sectional area A corresponds to a detector with an effective volume $\sim A \times 1$ km.) For water and ice, this probability is $P_{\nu\mu} \approx 10^{-4}(E_\nu/100\,\mathrm{TeV})^\alpha$, with $\alpha = 1$ for $E_\nu < 100$ TeV and $\alpha = 0.5$ for $E_\nu > 100$ TeV (e.g. [19]). Thus, the neutrino flux required for the detection of N events is

$$f_\nu \approx 5 \times 10^{-12} N \left(\frac{E_\nu}{100\,\mathrm{TeV}}\right)^{1-\alpha}\left(\frac{AT}{\mathrm{km^2 yr}}\right)^{-1}\,\mathrm{erg/cm^2 s}. \tag{2}$$

A lower limit to the source flux is also set by the requirement that the signal would exceed the background produced by atmospheric neutrinos. The flux of atmospheric neutrinos, averaged over zenith angle, is approximately given by $\Phi_\nu^A \approx 5 \times 10^{-8}(E_\nu/100\,\mathrm{TeV})^{-\beta}\mathrm{GeV/cm^2\,s\,sr}$, with $\beta = 1.7$ for $E_\nu < 100\,\mathrm{TeV}$ and $\beta = 2.0$ for $E_\nu > 100\,\mathrm{TeV}$. For a neutrino detector with angular resolution $\Delta\theta$, the source flux for which the signal constitutes a 5σ detection over the atmospheric background flux is

$$f_\nu \approx 3 \times 10^{-12}\left(\frac{E_\nu}{100\,\mathrm{TeV}}\right)^{-\beta/2}\left(\frac{\Delta\theta}{1\,\mathrm{deg}}\right)\left(\frac{AT}{\mathrm{km^2yr}}\right)^{-1/2}\mathrm{erg/cm^2\,s}. \tag{3}$$

Comparing eq. (2) and eq. (3), we find that for km-scale detectors, the atmospheric neutrino background poses a less stringent constraint on the source flux than the requirement of a detectable signal, except at low energies $\ll 100\,\mathrm{TeV}$.

The luminosity of a cosmologically distant source which corresponds to the flux of eq. (2) is

$$L_\nu \approx 10^{46} N d_{L,28}^2 \left(\frac{E_\nu}{100\,\mathrm{TeV}}\right)^{1-\alpha}\left(\frac{AT}{\mathrm{km^2yr}}\right)^{-1}\mathrm{erg/s}. \tag{4}$$

Here, $d_L = 10^{28}d_{L,28}$ cm is the luminosity distance and $d_{L,28} \sim 1$ for sources at redshift $z = 1$. The neutrino luminosity is, of course, a lower limit to the total energy output rate from the source. The lower limit set by eq. (4) to the luminosity of a steady source that may be detectable at $\sim 10^{1\pm1}$ TeV neutrino energies by a km-scale neutrino detector, $L > 10^{46}$ erg/s, is similar to the constraint of eq. (1) imposed by requiring proton acceleration to $\sim 10^{20}$ eV. As mentioned following eq. (1), the only steady sources bright enough to possibly satisfy these constraints are AGN. Eqs. (1) and (4) also imply similar luminosity constraints on GRBs. For a typical GRB duration of $10^{1.5}$ s, eq. (4) implies $L > 10^{52}$ erg/s, similar to the constraint imposed by eq. (1) for the typical Lorentz factor appropriate for these sources, $\Gamma \simeq 10^{2.5}$.

The phenomenological arguments presented above imply that the luminosity that must be produced by a source of UHECRs, eq. (1), may allow such a source to be detectable by a km scale neutrino detector at $\sim 10^{1\pm1}$ TeV neutrino energy, see eq. (4). Eq. (4) also implies that the detection of sources at $\gg 10^2$ TeV neutrino energy would require detectors with effective volume $\gg 1\,\mathrm{km^3}$. The most plausible sources of UHECRs, GRBs and AGN, are therefore also the most likely to be detectable neutrino sources. Such detection would be possible only if the neutrino luminosity of the source constitutes a significant fraction of the total source luminosity. For GRBs, this issue is discussed in some more detail in §3, within the context of current GRB models.

2.2. *The Waxman-Bahcall bound*

Cosmic-ray observations suggest that the cosmic-ray flux is dominated above 10^{19} eV by extra-Galactic sources of protons, and at lower energy by Galactic heavy nuclei sources. Under the assumption that the UHECRs are extra-Galactic protons, the observed flux of UHECRs allows to determine the rate (per unit time and volume) at which high energy protons are produced [20]. Figure 2, adapted from [21], presents a comparison of available UHECR data with the predictions of a model, where extra-galactic protons in the energy range $E_p \le 10^{21}$ eV are produced by cosmologically-distributed sources at a rate and spectrum given by

$$E_p^2\frac{\mathrm{d}\dot{N}_p}{\mathrm{d}E_p} = 0.65 \times 10^{44}\mathrm{erg\,Mpc^{-3}\,yr^{-1}}\,\phi(z). \tag{5}$$

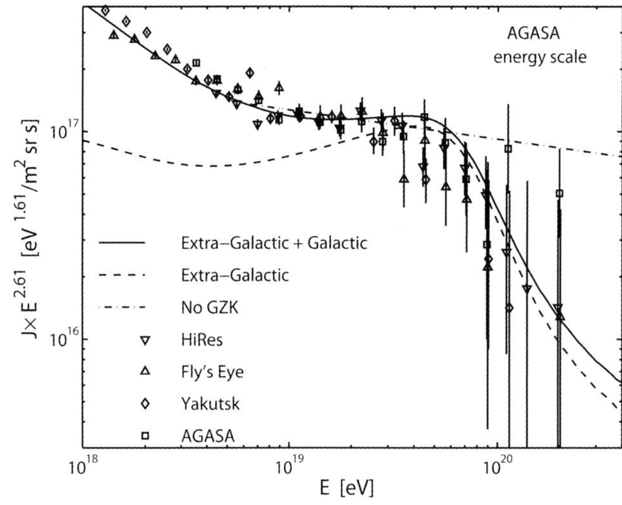

Fig. 2. The solid curve shows the energy spectrum derived from the two-component model discussed in §2.2 (with $\phi(z) \propto (1+z)^3$ up to $z = 2$, following the evolution of star formation rate). The dashed curve shows the extra-Galactic component contribution. The "No GZK" curve is an extrapolation of the $E^{-2.75}$ energy spectrum derived for the energy range of 6×10^{18} eV to 4×10^{19} eV [6]. Data taken from [22, 24–26] (AGASA's energy scale was chosen).

Here, $\phi(z)$ accounts for redshift evolution and $\phi(z=0) = 1$. The spectrum above 10^{19} eV is only weakly dependent on $\phi(z)$ since proton energy loss limits their propagation distance. For the heavy nuclei component dominating at lower, $< 10^{19}$ eV, energy the Fly's Eye experimental fit [22], $\mathrm{d}N/\mathrm{d}E \propto E^{-3.50}$, was used. The power-law spectrum of accelerated particles, $\mathrm{d}N/\mathrm{d}E \propto E^{-2}$, has been observed for both non-relativistic and relativistic shocks, and is believed to be due to Fermi acceleration in collisionless shocks [23] (although a first principles understanding of the process is not yet available).

Figure 2 demonstrates that model predictions are in good agreement with the data of all experiments in the energy range 10^{19} eV to 10^{20} eV. The uncertainty in the derived energy production rate, eq. (5), due to systematic uncertainties in the absolute energy calibration of the experiments is $\approx 20\%$ [21]. As explained in detail in [21], the various experiments are consistent with each other when systematic errors in the absolute energy scale of the events are taken into account. The relative systematic shifts in absolute energy calibration between Fly' Eye and the other experiments, $\{-11\%, +7.5\%, -19\%\}$ for $\{$AGASA, HiRes, Yakutsk$\}$, required to bring into agreement the fluxes measured at 10^{19} eV by the different experiments, are well within the published systematic errors. Above 10^{20} eV the Fly's Eye, HiRes and Yakutsk experiments are in agreement with each other and with the model, while the AGASA experiment reports a flux higher by a factor ~ 3.

The energy production rate, eq. (5), sets an upper bound to the neutrino intensity produced by sources which, like GRBs and AGN jets, are optically thin for high-energy nucleons to $p\gamma$ and $pp(n)$ interactions. For sources of this type, the energy generation rate of neutrinos can not exceed the energy generation rate implied by assuming that all the energy injected as high-energy protons is converted to pions (via $p\gamma$ and $pp(n)$ interactions). The resulting upper bound (for muon and anti-muon neutrinos, neglecting mixing) is [4]

$$E_\nu^2\Phi_\nu < 2 \times 10^{-8}\xi_z\left[\frac{(E_p^2\mathrm{d}\dot{N}_p/\mathrm{d}E_p)_{z=0}}{10^{44}\,\mathrm{erg/Mpc^3yr}}\right]\mathrm{GeV\,cm^{-2}\,s^{-1}\,sr^{-1}}. \tag{6}$$

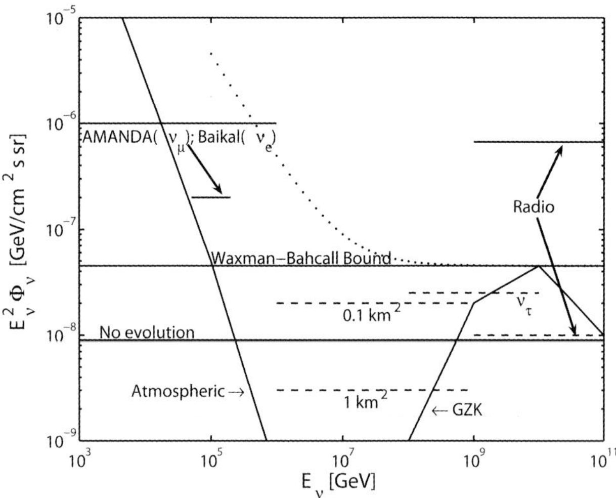

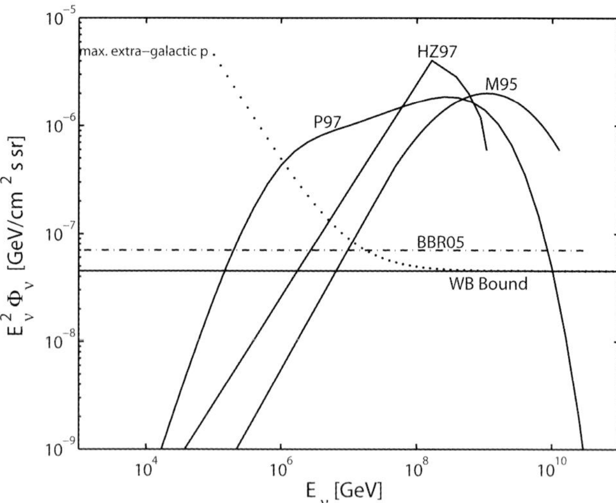

Fig. 3. The upper bound imposed by UHECR observations on the extra-Galactic high energy muon neutrino intensity (lower-curve: no evolution of the energy production rate, upper curve: assuming evolution following star formation rate), compared with the atmospheric neutrino background and with the experimental upper bounds (solid lines) of optical Cerenkov experiments, BAIKAL [27] and AMANDA [3, 28], and of coherent Cerenkov radio experiments (RICE [29], GLUE [30]; see also [31]). The dotted curve is the maximum contribution due to possible extra-Galactic component of lower-energy, $<10^{17}$ eV, protons, that may be "hidden under" the Galactic heavy nuclei flux [4]. The curve labelled "GZK" shows the intensity due to interaction with micro-wave background photons. Dashed curves show the expected sensitivity of 0.1 Gton (AMANDA, ANTARES, NESTOR) and 1 Gton (IceCube, NEMO) optical Cerenkov detectors [3], of the coherent radio Cerenkov (balloon) experiment ANITA [31] and of the Auger air-shower detector (sensitivity to v_τ) [12]. Space air-shower detectors (EUSO) may also achieve the sensitivity required to detect fluxes lower than the WB bound at energies $>10^{18}$ eV [12].

Fig. 4. The WB bound and various model predictions for the diffuse neutrino intensity produced by AGN jets. Early models (M95 = Mannheim 1995, HZ97 = Halzen & Zas 1997, P97 = Protheroe 1997 [32]) predicted an intensity well above the WB bound, and are therefore inconsistent with UHECR observations. More recent models (BBR05 = Becker, Biermann & Rhode [33]) predict fluxes close to the WB bound.

ξ_z is (a dimensionless parameter) of order unity, which depends on the redshift evolution of $E_p^2 d\dot{N}_p/dE_p$ (see eq. (5)). In order to obtain a conservative upper bound, we adopt $(E_p^2 d\dot{N}_p/dE_p)_{z=0} = 10^{44}$ erg/Mpc^{3}yr and a rapid redshift evolution, $\Phi(z) = (1+z)^3$ up to $z = 2$, following the evolution of star formation rate. This evolution yields $\xi_z \approx 3$.

The WB upper bound is compared in fig. 3 with current experimental limits, and with the expected sensitivity of planned neutrino telescopes. The figure indicates that km-scale (i.e. giga-ton) neutrino telescopes are needed to detect the expected extra-Galactic flux in the energy range of ~ 1 TeV to ~ 1 PeV, and that much larger effective volume is required to detect the flux at higher energy.

2.3. *The WB bound and AGN jet model predictions*

Figure 4 presents a comparison of the WB bound with various model predictions for the diffuse neutrino intensity produced by AGN jets. Early models [32] predicted an intensity well above the WB bound, and are therefore inconsistent with UHECR observations. More recent models [33], with more detailed consideration of plasma conditions in the jets, predict fluxes close to the WB bound.

2.4. *GZK neutrinos*

Protons of energy exceeding the threshold for pion production, $\sim 5 \times 10^{19}$ eV, lose most of their energy over a time short compared to the age of the universe. $>10^{20}$ eV protons, for example, lose most of their energy over $<3 \times 10^8$ yr. Assuming that high-energy protons are produced by extra-Galactic sources at the rate implied by UHECR observations (eq. (5)), the proton energy loss to pion production produces a neutrino intensity similar to the WB-bound. The expected GZK neutrino intensity is schematically shown in fig. 3. Since most of the pions are produced in interactions with photons of energy corresponding to the Δ-resonance, each of the resulting neutrinos carry approximately 5% of the proton energy. The neutrino background is therefore close to the WB bound above $\sim 5 \times 10^{18}$ eV, where neutrinos are produced by $\sim 10^{20}$ eV protons. The intensity at lower energies is lower, since protons of lower energy do not lose all their energy over the age of the universe. The GZK intensity in figure 3 decreases at the highest energies since it was assumed that the maximum energy of protons produced by UHECR sources is 10^{21} eV. The results of detailed calculations of the expected GZK neutrino intensity [11] are in agreement with the qualitative analysis presented above.

The detection of GZK neutrinos will be a milestone in neutrino astronomy. Most important, neutrino detectors with sensitivity better than the WB-bound at energies $>10^{18}$ eV will test the hypothesis that the UHECR are protons (possibly somewhat heavier nuclei) of extra-Galactic origin. Measurements of the flux and spectrum would constrain the redshift evolution of the sources. The GZK neutrino intensity presented in fig. 3 is obtained for rapidly evolving sources. For non-evolving sources the intensity is lower by a factor of ≈ 5.

The uncertainty in the experimental determination of the UHECR flux at energies $>10^{20}$ eV lead many authors to speculate that there may be a new source of ultra-high energy cosmic-rays and neutrinos beyond 10^{20} eV, producing a neutrino intensity higher than the WB bound at these neutrino energies. Most of these models involve "new physics": Decay of super-massive dark matter particles and topological defects (e.g. [8]) have been proposed to produce ultra-high energy neutrinos with small associated cosmic ray proton flux, thus possibly producing a flux of neutrinos exceeding the WB-bound. In our view, it is reasonable to extend the WB limit beyond 10^{20} eV by simply extrapolating the horizontal line in fig. 3 to higher energies. The validity of this extrapolation will be tested by future measurements of the spectrum of ultra-high energy cosmic rays and neutrinos.

3. Gamma-ray bursts (GRBs): an illustrative example

3.1. *The fireball model and the association of GRBs and UHECRs*

Gamma-ray bursts (GRBs) are short, typically tens of seconds long, flashes of gamma-rays, carrying most of their energy in $>1\,\mathrm{MeV}$ photons. The detection in the past few years of "afterglows," delayed X-ray, optical and radio emission from GRB sources, proved that the sources lie at cosmological distances, and provided strong support for the following scenario of GRB production [34, 35]. The energy source is believed to be rapid mass accretion on a newly formed solar-mass black hole. Recent observations suggest that the formation of the central compact object is associated with type Ib/c supernovae [36]. The energy release drives an ultra-relativistic, $\Gamma \sim 10^{2.5}$, plasma outflow. The emission of γ-rays is assumed to be due to internal collisionless shocks within the relativistic wind, the "fireball," which occur at a large distance from the central black-hole due to variability in the wind emitted from the central "engine." It is commonly assumed that electrons are accelerated to high energy within the collisionless shocks, and that synchrotron emission from these shock accelerated electrons produces the observed γ-rays. If protons are present in the wind, as assumed in the fireball model, they would also be accelerated to high energy in the region were electrons are accelerated.

GRBs were suggested to be UHECR sources in [37, 38]. The GRB-UHECR association was based in [37] on two major arguments (see [39] for a pedagogical review). First, the constraints imposed on the relativistic wind by requiring that it would produce the observed γ-rays were shown to be remarkably similar to those imposed by the requirement that the wind would allow proton acceleration (in the internal collisionless shocks) to $10^{20}\,\mathrm{eV}$. Independent physical arguments lead, in both cases, to similar lower limits on the expansion Lorentz factor, $\Gamma \geq 300$, and on the ratio of magnetic field and electron energy densities, $u_B/u_e \geq 0.1$. Second, the rate (per unit volume) at which energy is generated by GRBs in γ-rays was shown to be similar to the rate at which energy should be generated in high energy protons in order to account for the observed UHECR flux. Both arguments have been strengthened by more recent GRB and UHECR observations [40].

3.2. *100 TeV neutrinos*

Protons accelerated in the fireball internal shocks lose energy through photo-meson interaction with fireball photons. The decay of charged pions produced in this interaction results in the production of high energy neutrinos. The key relation is between the observed photon energy, E_γ, and the accelerated proton's energy, E_p, at the threshold of the Δ-resonance. In the observer frame,

$$E_\gamma E_p = 0.2\,\mathrm{GeV}^2 \Gamma^2. \tag{7}$$

For $\Gamma \approx 10^{2.5}$ and $E_\gamma = 1\,\mathrm{MeV}$, we see that characteristic proton energies $\sim 10^{16}\,\mathrm{eV}$ are required to produce pions. Since neutrinos produced by pion decay typically carry 5% of the proton energy, production of $\sim 10^{14}\,\mathrm{eV}$ neutrinos is expected [13].

The fraction of energy lost by protons to pions, f_π, is $f_\pi \approx 0.2$ [13, 39]. Assuming that GRBs generate the observed UHECRs, the expected GRB muon-neutrino flux may be estimated using eq. (6) [4, 13],

$$E_\nu^2 \Phi_\nu \approx 0.3 \times 10^{-8} \frac{f_\pi}{0.2}\,\mathrm{GeV}\,\mathrm{cm}^{-2}\,\mathrm{s}^{-1}\,\mathrm{sr}^{-1}. \tag{8}$$

This neutrino spectrum extends to $\sim 10^{16}\,\mathrm{eV}$, and is suppressed at higher energy due to energy loss of pions and muons [4, 13, 41]. Eq. (8) implies a detection rate of ~ 20 neutrino-induced muon events per year (over $4\pi\,\mathrm{sr}$) in a cubic-km detector. Since GRB neutrino events are correlated both in time and in direction with gamma-rays, their detection is practically background free.

3.3. *TeV neutrinos*

The 100 TeV neutrinos discussed in the previous section are produced at the same region where GRB γ-rays are produced. Their production is a generic prediction of the fireball model. It is a direct consequence of the assumptions that energy is carried from the underlying engine as kinetic energy of protons and that γ-rays are produced by synchrotron emission of shock accelerated particles. Neutrinos may be produced also in other stages of fireball evolution, at energies different than 100 TeV. The production of these neutrinos is dependent on additional model assumptions. We discuss below some examples related to the GRB progenitor. For a more detailed discussion see [35, 39] and references therein.

The most widely discussed progenitor scenarios for long-duration GRBs involve core collapse of massive stars. In these "collapsar" models, a relativistic jet breaks through the stellar envelope to produce a GRB. For extended or slowly rotating stars, the jet may be unable to break through the envelope. Both penetrating (GRB producing) and "choked" jets can produce a burst of $\sim 5\,\mathrm{TeV}$ neutrinos by interaction of accelerated protons with jet photons, while the jet propagates in the envelope [42]. The estimated event rates may exceed $\sim 10^2$ events per yr in a km-scale detector, depending on the ratio of non-visible to visible fireballs. A clear detection of non-visible GRBs with neutrinos may be difficult due to the low energy resolution for muon-neutrino events, unless the associated supernova photons are detected. In the two-step "supranova" model, interaction of the GRB blast wave with the supernova shell can lead to detectable neutrino emission, either through nuclear collisions with the dense supernova shell or through interaction with the intense supernova and backscattered radiation field [43].

References

1. Bahcall, J. N., these proc. (hep-ph/0412068), and references therein.
2. Raffelt, G., these proc. (hep-ph/0501049), and references therein.
3. Halzen, F., these proc. (astro-ph/0501593), and references therein.
4. Waxman, E. and Bahcall, J. N., Phys. Rev. **D59**, 023002 (1999); Bahcall, J. N. and Waxman, E., Phys. Rev. **D64**, 023002 (2001).
5. Mannheim, K., Protheroe, R. J. and Rachen, J. P., Phys. Rev. **D63**, 023003 (2001) (astro-ph/9812398, v1, v2, and v3).
6. Nagano, M. and Watson, A. A., Rev. Mod. Phys. **72**, 689 (2000); Abu-Zayyad, T. *et al.*, Astrophys. J. **557**, 686 (2001).
7. Watson, A. A., astro-ph/0408110.
8. Bhattacharjee, P. and Sigl, G., Phys. Rep. **327**, 109 (2000).
9. Greisen, K., Phys. Rev. Lett. **16**, 748 (1966); Zatsepin, G. T. and Kuzmin, V. A., J. Exp. Theor. Phys. **4**, 78 (1966).
10. Berezinsky, V. S. and Zatsepin, G. T., Phys. Lett. **28B**, 423 (1969).
11. Engel, R., Seckel, D. and Stanev, T., Phys. Rev. **D64**, 093010 (2001).
12. Saltzberg, D., these proc. (astro-ph/0501364), and references therein.
13. Waxman, E. and Bahcall, J. N., Phys. Rev. Lett. **78**, 2292 (1997).
14. Bilenky, S. M., these proc. (hep-ph/0410090); Gonzalez-Garcia, M. C., these proc. (hep-ph/0410030), and references therein.
15. Bahcall, J. N., "Neutrino Astrophysics", (Cambridge University Press, NY, 1989), pp. 438–460.
16. Lovelace, R. V. E., Nature **262**, 649 (1976).
17. Hillas, A. M., Ann. Rev. Astron. Astrophys. **22**, 425 (1984).
18. Arons, J., Astrophys. J. **589**, 871 (2003).

19. Gaisser, T. K., Halzen, F. and Stanev, T., Phys. Rep. **258**, 173 (1995).
20. Waxman, E., Astrophys. J. **452**, L1 (1995).
21. Bahcall, J. N. and Waxman, E., Phys. Lett. B **556**, 1 (2003).
22. Bird, D. J. *et al.*, Astrophys. J. **424**, 491 (1994).
23. Blandford, R. and Eichler, D., Phys. Rep. **154**, 1 (1987); Bednarz, J. and Ostrowski, M., Phys. Rev. Lett. **80**, 3911 (1998); Keshet, U. and Waxman, E., Phys. Rev. Lett. **94**, 111102 (2005).
24. Efimov, N. N. *et al.*, in Proc. International Symposium on Astrophysical Aspects of the Most Energetic Cosmic-Rays, (edited by M. Nagano and F. Takahara), (World Scientific, Singapore, 1991), p. 20.
25. Hayashida, N. *et al.*, Astrophys. J. **522**, 225 (1999) and astro-ph/0008102.
26. Abu-Zayyad, T. *et al.*, Phys. Rev. Lett. **92**, 151101 (2004).
27. Balkanov, V. *et al.*, Nucl. Phys. B (Proc. Suppl.) **110**, 504 (2002).
28. Ahrens, J. *et al.*, Phys. Rev. Lett. **90**, 251101 (2003).
29. Kravchenko, I. *et al.*, Astropart. Phys. **19**, 15 (2003).
30. Gorham, P. W. *et al.*, Phys. Rev. Lett. **93**, 041101 (2004).
31. Silvestri, A. *et al.*, astro-ph/0411007.
32. Mannheim, K., Astropart. Phys. **3**, 295 (1995); Halzen, F. and Zas, E., Astrophys. J. **488**, 669 (1997); Protheroe, R. J., IAU Colloq. 163, ASP Conf. **121**, 585 (1997).
33. Atoyan, A. and Dermer, C. D., Phys. Rev. Lett. **87**, 221102 (2001); Alvarez-Muñiz, J. and Meszaros, P., Phys. Rev. **D70**, 123001 (2004); Becker, J. K., Biermann, P. L. and Rhode, W., astro-ph/0502089.
34. Piran, T., Phys. Rep. **333**, 529 (2000); Waxman, E., Lect. Notes Phys. **598**, 393 (2003) (astro-ph/0303517).
35. Mészáros, P., Ann. Rev. Astron. Astrophys. **40**, 137 (2002).
36. Galama, T. J. *et al.*, Nature **395**, 670 (1998); Stanek, K. Z. *et al.*, Astrophys. J. **591**, L17 (2003); Hjorth, J. *et al.*, Nature **423**, 847 (2003).
37. Waxman, E., Phys. Rev. Lett. **75**, 386 (1995).
38. Vietri, M., Astrophys. J. **453**, 883 (1995); Milgrom, M. and Usov, V., Astrophys. J. **449**, L37 (1995).
39. Waxman, E., Lect. Notes Phys. **576**, 122 (2001) (astro-ph/0103186).
40. Waxman, E., Astrophys. J. **606**, 988 (2004).
41. Rachen, J. P. and Mészáros, P., Phys. Rev. **D58**, 123005 (1998).
42. Mészáros, P. and Waxman, E., Phys. Rev. Lett. **87**, 171102 (2001); Razzaque, S., Mészáros, P. and Waxman, E., Phys. Rev. **D69**, 023001 (2004).
43. Guetta, D. and Granot, J., Phys. Rev. Lett. **90**, 201103 (2003); Razzaque, S., Mészáros, P. and Waxman, E., Phys. Rev. Lett. **90**, 241103 (2003); Dermer, C. D. and Atoyan, A., Phys. Rev. Lett. **91**, 071102 (2003).

Physica Scripta. Vol. T121, 153–155, 2005

Cosmological Neutrino Bounds for Non-Cosmologists

Max Tegmark[1,2]

[1] Dept. of Physics, Massachusetts Institute of Technology, Cambridge, MA 02139, USA
[2] Department of Physics, University of Pennsylvania, Philadelphia, PA 19104, USA

Received February 18, 2005; accepted February 21, 2005

PACS numbers: 98.80.−k, 14.60.Pq

Abstract

I briefly review cosmological bounds on neutrino masses and the underlying gravitational physics at a level appropriate for readers outside the field of cosmology. For the case of three massive neutrinos with standard model freezeout, the current 95% upper limit on the sum of their masses is 0.42 eV. I summarize the basic physical mechanism making matter clustering such a sensitive probe of massive neutrinos. I discuss the prospects of doing still better in coming years using tools such as lensing tomography, approaching a sensitivity around 0.03 eV. Since the lower bound from atmospheric neutrino oscillations is around 0.05 eV, upcoming cosmological measurements should detect neutrino mass if the technical and fiscal challenges can be met.

1. Introduction

In the last few years, an avalanche of new cosmological data has revolutionized our ability to measure key cosmological parameters. Measurements of the cosmic microwave background (CMB), galaxy clustering, gravitational lensing, the Lyman alpha forest, cluster abundances and type Ia supernovae paint a consistent picture where the cosmic matter budget is about 5% ordinary matter, 25% dark matter, 70% dark energy, less than 1% curvature and less than 5% massive neutrinos [1–5]. The cosmic initial conditions are consistent with approximately scale-invariant inflation-produced adiabatic fluctuations, with no evidence yet for primordial gravitational waves [1–5]

How precisely do such cosmological observations give us information about neutrino masses, and how should the current limits be interpreted? For detailed discussion of post-WMAP astrophysical neutrino constraints, see [1–3, 6 11] and in particular the excellent and up-to-date reviews [12, 13]. For a review of the theoretical and experimental situation, see [14]. The purpose of this symposium contribution is merely to provide a brief summary of the constraints and the underlying physics at a level appropriate for readers outside of the field of cosmology.

2. The physics underlying cosmological neutrino bounds

Why do cosmological observations place strong bounds on neutrino masses? The short answer is that neutrinos affect the growth of cosmic clustering and this clustering can be accurately measured (Figure 1).

The CMB tells us that the Universe used to be almost perfectly uniform spatially, with density variations from place to place only at the level of 10^{-5}. Gravitational instability caused these tiny density fluctuations to grow in amplitude into the galaxies and the large-scale structure that we observe around us today. The reason for this growth is simply that gravity is an attractive force: if the density at some point exceeds the mean density by some relative amount δ, then mass will be pulled in from surrounding regions and δ increases over time. A classic result is that if all the matter contributing to the cosmic density is able to cluster (like dark matter or ordinary matter with negligible pressure), then fluctuations grow as the cosmic scale factor a [15]:

$$\delta \propto a, \tag{1}$$

i.e., fluctuations double in amplitude every time the Universe doubles it linear size a.

If some fraction of the matter density is gravitationally inert and unable to cluster, the fluctuation growth will clearly be slower. If only a fraction Ω_* can cluster, then equation (1) is generalized to [16]:

$$\delta \propto a^p, \tag{2}$$

where

$$p = \frac{\sqrt{1 + 24\Omega_*} - 1}{4} \approx \Omega_*^{3/5} \tag{3}$$

and the approximation in the last step is surprisingly accurate. Such gravitationally inert components can include dark energy and (on sufficiently small scales) photons and neutrinos. Early on, the cosmic density was completely dominated by photons, so $p \approx 0$ and fluctuations essentially did not start growing until the epoch of matter-domination (MD). At recent times, the cosmic density has become dominated by dark energy Λ, causing fluctuations to gradually stop growing after a net growth factor of about $a_{\Lambda D}/a_{\mathrm{MD}} \approx 4700$ [17].

Massive nonrelativistic neutrinos are unable to cluster on small scales because of their high velocities. Between matter domination and dark energy domination, they constitute a roughly constant fraction $f_\nu = 1 - \Omega_*$ of the matter density. Equation (2) therefore gives a net fluctuation growth factor

$$\left(\frac{a_{\Lambda D}}{a_{\mathrm{MD}}}\right)^p \approx 4700^p \approx 4700^{(1-f_\nu)^{3/5}} \approx 4700 e^{-4f_\nu} \tag{4}$$

from matter-domination (MD) until today, where we have assumed $f_\nu \ll 1$ in the last step[1]. We see that the basic reason that a small neutrino fraction has a large effect is simply that 4700 is a large number, so that a small change in the exponent p makes a noticeable difference.

A key point to remember is that what mattered above was the neutrino *density*, specifically the fractional contribution f_ν of neutrinos to the total density. To translate observational constraints on the neutrino mass density into constraints on neutrino masses, we need to know the neutrino number density. Assuming that this number density is determined by standard model neutrino

---

[1] Taylor expanding gives an exponent $5.07 f_\nu$. One obtains a coefficient closer to 4 when including the fact that raising f_ν makes matter-radiation equality happen earlier, reducing the suppression as detailed in [17].

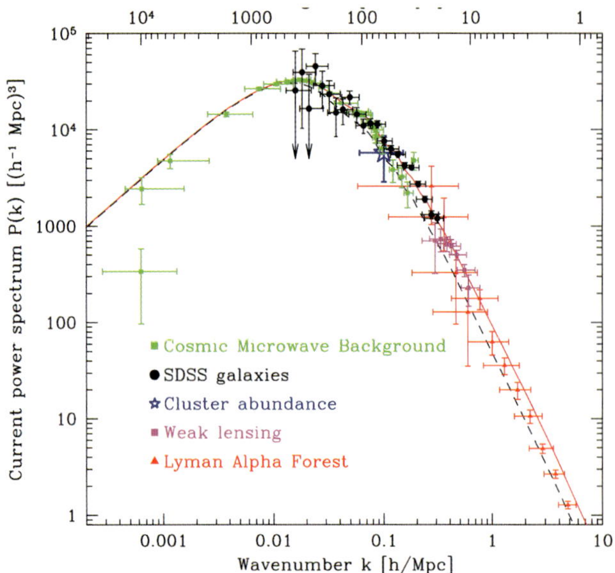

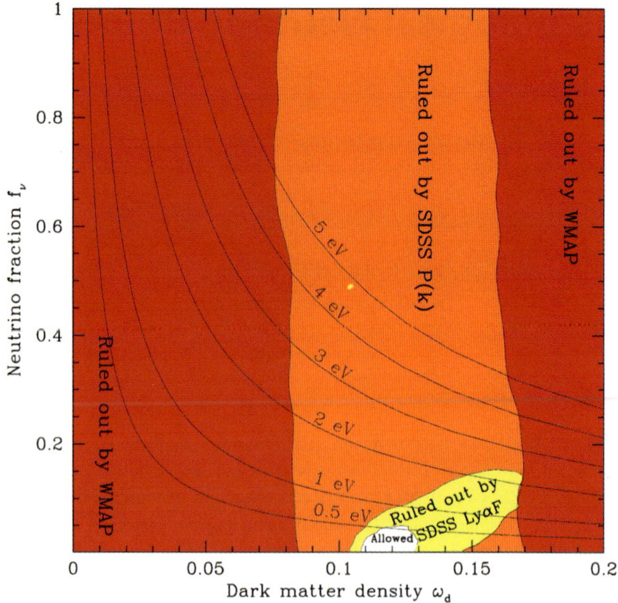

Fig. 1. Cosmological constraints on the current matter power spectrum $P(k)$ reprinted from [26]. See [26] for details about the modeling assumptions underlying this figure. The solid curve shows the theoretical prediction for a "vanilla" flat scalar scale-invariant model with matter density $\Omega_m = 0.28$, Hubble parameter $h = 0.72$ and baryon fraction $\Omega_b/\Omega_m = 0.16$. The dashed curve shows that replacing 7% of the cold dark matter by neutrinos, corresponding to a neutrino mass sum $M_\nu = 1$ eV, suppresses small-scale power by about a factor of two.

Fig. 2. 95% constraints in the (ω_d, f_ν) plane, reprinted from [2] and [3]. The shaded red/dark grey region is rules out by WMAP CMB observations alone. The shaded orange/grey region is ruled out when adding SDSS galaxy clustering information [2] and the yellow/light grey region is ruled out when including SDSS Lyman α Forest information as well [3]. The six curves correspond to M_ν, the sum of the neutrino masses, equaling 0.5, 1, 2, 3, 4 and 5 eV, respectively – barring sterile neutrinos, no neutrino can have a mass exceeding $\sim M_\nu/3$.

freezeout gives a number density around $112/\text{cm}^3$ and [15][2]

$$f\nu \approx \frac{M_\nu}{\omega_m \times 94.4\,\text{eV}} \approx \frac{M_\nu}{14\,\text{eV}}, \tag{5}$$

where

$$M_\nu \equiv \sum_{i=1}^{3} m_\nu^i \tag{6}$$

is the sum of the three neutrino masses and $\omega_m = h^2 \Omega_m \approx 0.15$ is the measured matter density in units of $1.8788 \times 10^{-26}\,\text{kg/m}^3$[2].

The power spectrum $P(k)$ shown in Figure 1 is the variance of the fluctuations δ in Fourier space, so massive neutrinos suppress it by the same factor as they suppress δ^2, *i.e.*, by a factor [23]:

$$\frac{P(k; f_\nu)}{P(k; 0)} \approx e^{-8f_\nu}. \tag{7}$$

This means that a neutrino mass sum $M_\nu = 1$ eV cuts the power roughly in half on the small scales where neutrinos cannot cluster.

The length scale below which neutrino clustering is strongly suppressed is called the neutrino free-streaming scale, and roughly corresponds to the distance neutrinos have time to travel while the Universe expands by a factor of two. Intuitively, neutrinos clearly will not cluster in an overdense clump so small that its escape velocity is much smaller than the typical neutrino velocity. On scales much larger than the free-streaming scale, on the other hand, neutrinos cluster just as cold dark matter and give $\Omega_* = 1$ and $p = 1$ above. This explains the effect of neutrinos on the power spectrum seen in Figure 1: suppression only on small scales, thereby changing the overall shape of $P(k)$ in a characteristic

way. This shape change distinguishes the neutrino effect from that of dark energy, which lowers Ω_* and hence the power spectrum amplitude on *all* scales by the same factor, preserving the shape of $P(k)$.

3. What are the constraints?

Cosmological observations have thus far produced no convincing detection of neutrino mass, but strong upper limits as illustrated in Figure 2. This figure shows that the WMAP CMB-measurements alone [24] tell us almost nothing about neutrino masses and are consistent with neutrinos making up 100% of the dark matter. Rather, the power of WMAP is that it constrains other cosmological parameters so strongly that it enables large-scale structure data to measure the small-scale $P(k)$-suppression that massive neutrinos cause. Combining WMAP with Sloan Digital Sky Survey (SDSS) [25] galaxy clustering measurements [26] gives the most favored value $M_\nu = 0$ and the 95% upper limit $M_\nu < 1.7\,\text{eV}$ [2]. Including information about SDSS or 2dFGRS [27] galaxy bias tightens this bound to $M_\nu < 0.6$–$0.7\,\text{eV}$ [1, 28]. Including SDSS measurements of the so-called Lyman α forest (intergalactic gas backlit by quasars) further tightens the bound to $M_\nu < 0.42\,\text{eV}$ (95%).

These upper limits are complemented by the lower limit from neutrino oscillation experiments. Atmospheric neutrino oscillations show that there is at least one neutrino (presumably mostly a linear combination of ν_μ and ν_τ) whose mass exceeds a lower limit around $0.05\,\text{eV}$ [14, 21]. Thus the atmospheric neutrino data corresponds to a lower limit $\omega_\nu \gtrsim 0.0005$, or $f_\nu \gtrsim 0.004$. The solar neutrino oscillations occur at a still smaller mass scale, perhaps around $0.008\,\text{eV}$ [14, 22, 29]. These mass-splittings are substantially smaller than $0.42\,\text{eV}$, suggesting that all three mass eigenstates would need to be almost degenerate for neutrinos to weigh in near our upper limit. Since sterile neutrinos are disfavored from being thermalized in the early

---

[2]The neutrino energy density must be very close to the standard freezeout density [18–20], given the large mixing angle solution to the solar neutrino problem and near maximal mixing from atmospheric results – see [21, 22] for up-to-date reviews. Any substantial asymmetries in neutrino density from the standard value would be "equilibrated" and produce a primordial ^{4}He abundance inconsistent with that observed.

universe [30, 31], it can be assumed that only three neutrino flavors are present in the neutrino background; this means that none of the three neutrinos can weigh more than about $0.42/3 = 0.14\,\text{eV}$. The mass of the heaviest neutrino is thus in the range 0.05–$0.14\,\text{eV}$.

A caveat about non-standard neutrinos is in order. As mentioned above, the cosmological constraints to first order probe only the *mass density* of neutrinos, ρ_ν, which determines the small-scale power suppression factor, and the *velocity dispersion*, which determines the scale below which the suppression occurs. For the low mass range we have discussed, the neutrino velocities are high and the suppression occurs on all scales where SDSS is highly sensitive. We thus measure only the neutrino mass density, and our conversion of this into a limit on the mass sum assumes that the neutrino number density is known and given by the standard model freezeout calculation. In more general scenarios with sterile or otherwise non-standard neutrinos where the freezeout abundance is different, the conclusion to take away is an upper limit on the total light neutrino mass density of $\rho_\nu < 4.8 \times 10^{-28}\,\text{kg/m}^3$ (95%). To test arbitrary nonstandard models, a future challenge will be to independently measure both the mass density and the velocity dispersion, and check whether they are both consistent with the same value of M_ν.

4. Outlook

Although cosmological neutrino bounds have recently improved dramatically, there is ample room for further improvement in the near and intermediate future. The basic reason for this is that the weakest link in current constraints is their dependence on other cosmological parameters. For instance, the galaxy clustering constraints cannot directly exploit the dramatic effect of neutrinos on the *amplitude* of the small-scale power spectrum shown in Figure 1, merely the slight change in its *shape*, and this shape change can be partially mimicked by changing other cosmological parameters such as the spectral index from inflation which effectively tilts the $P(k)$. The reason for this shortcoming is that galaxy surveys do not measure the clustering amplitude of matter directly, merely the clustering amplitude of luminous matter (galaxies), which is known to differ by a constant factor that is measured empirically.

Weak gravitational lensing bypasses this shortcoming. Light from distant galaxies or CMB patterns is deflected in a measurable way by the gravitational pull of *all* intervening matter, regardless whether it is luminous or dark, baryonic or non-baryonic, allowing the matter power spectrum $P(k)$ to be measured in a clean assumption-free way. This booming field has the potential to attain

a neutrino mass sensitivity of $0.03\,\text{eV}$ or better [32–34]. Since the lower bound from atmospheric neutrino oscillations is around $0.05\,\text{eV}$, upcoming cosmological measurements should detect neutrino mass if the technical and fiscal challenges can be met.

Acknowledgements

The author wishes to thank Angélica de Oliveira-Costa for helpful comments and Uroš Seljak and Matias Zaldarriaga for publishing the CMBFast software [35], which was used for the above figures.

This work was supported by NASA grant NAG5-11099, NSF CAREER grant AST-0134999, and fellowships from the David and Lucile Packard Foundation and the Research Corporation.

References

1. Spergel, D. N. *et al.*, Astrophys. J. Suppl. **148**, 175 (2003).
2. Tegmark, M. *et al.*, Phys. Rev. D **69**, 103501 (2004).
3. Seljak, U. *et al.*, astro-ph/0407372 (2004).
4. Cole, S., astro-ph/0501174 (2005).
5. Eisenstein, D. J., astro-ph/0501171 (2005).
6. Hannestad, S., J. Cosmol. Astropart. Phys. **05**, 004 (2003).
7. Elgaroy, O. and Lahav, O., J. Cosmol. Astropart. Phys. **0304**, 004 (2003).
8. Bashinsky, S. and Seljak, U., astro-ph/0310198 (2003).
9. Hannestad, S., astro-ph/0310133 (2003).
10. Hannestad, S., astro-ph/0411475 (2004).
11. Hannestad, S., hep-ph/0412181 (2003).
12. Hannestad, S., New J. Phys. **6**, 108 (2004).
13. Elgaroy, O. and Lahav, O. hep-ph/0412075 (2004).
14. King, S., hep-ph/0310204 (2003).
15. Kolb, E. W. and Turner, M. S., "The Early Universe" (Addison Wesley: New York, 1990).
16. Bond, J. R., Efstathiou, G. and Silk, J., Phys. Rev. Lett. **45**, 1980 (1980).
17. Tegmark, M., Vilenkin, A. and Pogosian, L., Phys. Rev. D **71**, 103523 (2005).
18. Dolgov, A. D. *et al.*, Nucl. Phys. B **632**, 363 (2002).
19. Abazajian, K. N., Beacom, J. F. and Bell, N. F., Phys. Rev. D **66**, 013008 (2002).
20. Wong, Y. Y., Phys. Rev. D **66**, 025015 (2002).
21. Kearns, E. T., hep-ex/0210019 (2002).
22. Bahcall, J. N., J. High En. Phys. **0311**, 004 (2003).
23. Hu, W., Eisenstein, D. J. and Tegmark, M., Phys. Rev. Lett. **79**, 3806 (1998).
24. Bennett, C. L. *et al.*, Astrophys. J. Suppl. **148**, 1 (2003).
25. York, D. G. *et al.*, Astron. J. **120**, 1579 (2000).
26. Tegmark, M. *et al.*, Astrophys. J. **606**, 702 (2004).
27. Percival, W. J. *et al.*, Mon. Not. R. Astron. Soc. **327**, 1297 (2001).
28. Seljak, U. *et al.*, astro-ph/0406594 (2004).
29. Ahmed, S. N. *et al.*, nucl ex/0309004 (2003).
30. Di Bari, P., Phys. Rev. D **65**, 043509 [Addendum-ibid. D **67**, 127301 (2003)] (2002).
31. Abazajian, K. N., Astropart. Phys. **19**, 303 (2003).
32. Hu, W., Astrophys. J. **522**, L21 (1999).
33. Kaplinghat, M., Knox, L. and Song, Y. S., astro-ph/0303344 (2003).
34. Abazajian, K. N. and Dodelson, S., Phys. Rev. Lett. **91**, 041301 (2003).
35. Seljak, U. and Zaldarriaga, M., Astrophys. J. **469**, 437 (1996).

Physica Scripta. Vol. T121, 156–160, 2005

Neutrino Intrinsic Properties: The Neutrino-Antineutrino Relation

Boris Kayser[*]

Fermilab, MS 106, P.O. Box 500, Batavia IL 60510, USA

Received February 20, 2005; accepted March 3, 2005

PACS number: 14.60.Pq

Abstract

Are neutrinos their own antiparticles? We explain why they very well might be. Then, after highlighting the fact that, to determine experimentally whether they are or not, one must overcome the smallness of neutrino masses, we discuss the one approach that nevertheless shows great promise. Finally, we turn to the consequences of neutrinos being their own antiparticles. These consequences include unusual electromagnetic properties, and manifestly CP-violating effects from "Majorana" phases that have no quark analogues.

1. Introduction

The recent discovery that neutrinos have nonzero masses and mix implies that these particles must have very interesting intrinsic properties. At the Symposium, we discussed what is known, and what we would like to learn, about a number of these properties. In this written version of the talk, we would like to focus on one property: the relation between a neutrino and its antineutrino.

2. The neutrino-antineutrino relation

One of the most interesting questions about the intrinsic nature of neutrinos, raised by the discovery of neutrino mass, is the question of whether neutrinos are their own antiparticles. Is each neutrino mass eigenstate v_i identical to its antiparticle $\overline{v}_i$, or distinct from it? If $\overline{v}_i = v_i$, we call the neutrinos Majorana particles, while if $\overline{v}_i \neq v_i$, we call them Dirac particles.

Of course, we know that the electron is distinct from its antiparticle, the positron, because these two particles carry opposite electric charge. However, a neutrino carries no electric charge, and may not carry any other conserved charge-like quantum number. It might be thought that there is a conserved lepton number L, defined by

$$L(v) = L(\ell^-) = -L(\overline{v}) = -L(\ell^+) = 1, \tag{1}$$

that distinguishes neutrinos v and charged leptons ℓ^- on the one hand from antineutrinos $\overline{v}$ and anti-charged leptons ℓ^+ on the other hand. However, there is no evidence that such a conserved quantum number exists. If it does not exist, then nothing distinguishes $\overline{v}_i$ from v_i. The neutrinos are then Majorana particles, identical to their antiparticles.

Many theorists believe that, indeed, the lepton number L defined by Eq. (1) is not conserved. One reason for this belief is the nature of the very successful Standard Model (SM). As originally proposed, the SM conserves L. However, it contains no neutrino masses. Nor does it contain any chirally right-handed neutrino fields, v_R, but only left-handed ones, v_L.

Now that we know neutrinos have masses, we must extend the SM to accommodate them. Suppose that we try to do this in a manner that will preserve the conservation of L. Then, for a neutrino v, we add to the SM Lagrangian a "Dirac mass term" of the form

$$\mathcal{L} = -m_D \overline{v_L} v_R + h.c. \tag{2}$$

Here, m_D is a constant, and v_R is a right-handed neutrino field that we were obliged to add to the SM in order to construct the Dirac mass term. A Dirac mass term does not mix neutrinos and antineutrinos, so it conserves L.

In the SM, left-handed fermion fields belong to electroweak-isospin doublets, but right-handed ones are isospin singlets. In particular, v_R will carry no electroweak isospin. Thus, once v_R is present, all the SM principles, including electroweak-isospin conservation, allow the occurrence of the "Majorana mass term"

$$\mathcal{L}_M = -m_M \overline{v_R^c} v_R + h.c. \tag{3}$$

Here, m_M is a constant, and v_R^c is the charge conjugate of v_R. Like v_R, v_R^c carries no electroweak isospin. Thus, $\mathcal{L}_M$ is electroweak-isospin conserving, as required. However, any Majorana mass term of this form converts a v into a $\overline{v}$, or a $\overline{v}$ into a v. Thus, it does not conserve L.

If we insist that the SM, extended to accommodate neutrino masses, remain L conserving, then, of course, Majorana mass terms are forbidden. However, if we do not impose L conservation by hand, but require only the general SM principles, such as electroweak-isospin conservation and renormalizability, then Majorana mass terms such as the one in Eq. (3) are allowed. It is then very natural to expect that they are present in nature, so that L is not conserved and the neutrinos are Majorana particles.

The most popular explanation of why neutrinos are so light is the see-saw mechanism, discussed at this Symposium by P. Ramond. This mechanism includes Majorana mass terms. Hence, it predicts that L is not conserved and that neutrinos are Majorana particles.

How can we test whether neutrinos are their own antiparticles? Let us first discuss an approach [1] that would not work, but that clearly illustrates why most approaches would not work. Suppose that a neutrino mass eigenstate v_i is produced in π^+ decay, as depicted in Fig. 1(a). We assume, as shown there, that the v_i is emitted to the left in the π^+ rest frame. Owing to the left-handed

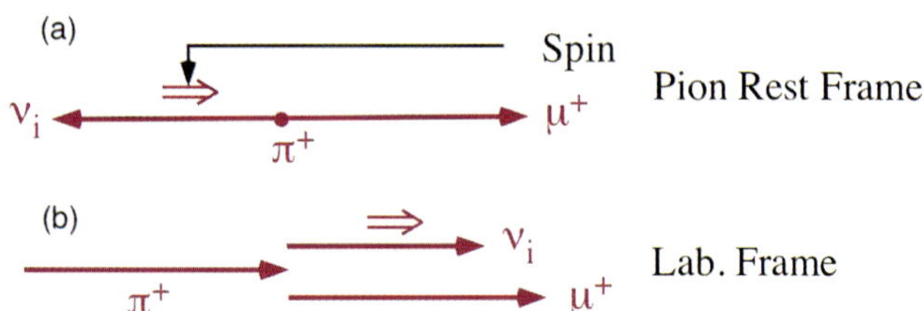

Fig. 1. A pion decay into a muon and a neutrino mass eigenstate. (a) The decay as seen in the pion rest frame. (b) The same decay as seen in the laboratory frame when the pion is moving faster in that frame than the neutrino is moving in the pion rest frame.

---

[*]Email address: boris@fnal.gov

character of the SM weak interactions, the v_i will have left-handed (LH) helicity, so that its spin vector will point to the right, as shown. Now imagine that the parent π^+ is moving to the right in the laboratory frame at a speed greater that that of the v_i in the π^+ rest frame. Then, as indicated in Fig. 1(b), the v_i will be moving to the right in the laboratory frame. Moreover, its spin vector will be pointing to the right in this frame, just as it was in the π^+ rest frame. Thus, in the laboratory frame, the v_i will have *right-handed* (RH) helicity.

Now, the statement that a neutrino v_i is its own antiparticle means that, *for a given helicity h,*

$$\overline{v}_i(h) = v_i(h). \tag{4}$$

It is very difficult to test whether this equality holds in nature because, thanks to the left-handed character of the weak interactions, weak processes produce neutrinos that are almost exclusively of LH helicity, but antineutrinos that are almost exclusively of RH helicity. Thus, the comparison called for by Eq. (4) cannot be made. However, in our hypothetical experiment with π^+ decays, we have created a sample of v_i that are tagged as neutrinos, rather than antineutrinos, because they are produced together with μ^+ particles, but which have RH helicity in the laboratory frame because of the very rapid motion of their parent pions. Are these RH v_i identical to RH $\overline{v}_i$, in conformity with Eq. (4)? That is, do these RH v_i interact as RH $\overline{v}_i$ do? A RH $\overline{v}_i$ enjoys the SM charged-current interaction. When this particle strikes a target at rest, this interaction allows it to produce a charged antilepton, ℓ^+. Can the RH v_i in our hypothetical π^+ decay experiment do the same?

Unfortunately, we cannot turn our hypothetical experiment into a real one to answer this question. The reason is that neutrino masses are much too small. Suppose, for example, that v_i has a mass m_i around 0.05 eV, a value suggested by atmospheric neutrino oscillation data. A π of mass m_π, moving faster in the laboratory frame than its daughter v_i moves in the π rest frame, must have a laboratory energy $E_\pi(\text{Lab})$ obeying

$$\frac{E_\pi(\text{Lab})}{m_\pi} > \frac{E_{v_i}(\pi \text{ Rest Frame})}{m_i}. \tag{5}$$

For $m_i \sim 0.05$ eV, this requires that $E_\pi(\text{Lab}) \gtrsim 10^5$ TeV. Furthermore, a v_i from π decay will not get its helicity reversed by the forward motion of the π unless, in the π rest frame, its angle of emission is within $\sim m_i/E_{v_i}(\pi \text{ Rest Frame})$ of dead backward with respect to the π beam direction [1]. Thus, the fraction of all π-decay v_i that get their helicity reversed is only $\sim [m_i/E_{v_i}(\pi \text{ Rest Frame})]^2$. For $m_i \sim 0.05$ eV, this fraction is 10^{-18}. Obviously, our hypothetical experiment is completely impossible in practice.

This practical impossibility illustrates a very general point: We are trying to find a way to demonstrate that $\overline{v}_i = v_i$, or, equivalently, that lepton number L is not conserved. In this effort, we are assuming that the interactions of neutrinos are correctly described by the SM. Now, the SM interactions conserve L, so the L nonconservation that we seek can only come from Majorana neutrino mass terms. Thus, it must vanish when the neutrino masses vanish. Hence, any attempt to demonstrate that $\overline{v}_i = v_i$ or that L is not conserved will be challenged by the smallness of neutrino masses. Our experiment based on π decay cannot meet this challenge. Indeed, the only approach that shows considerable promise of being able to meet it is the search for neutrinoless double beta decay.

Neutrinoless double beta decay ($0v\beta\beta$) is the reaction Nucl $\rightarrow$ Nucl$' + e^- + e^-$, in which one nucleus decays into another plus

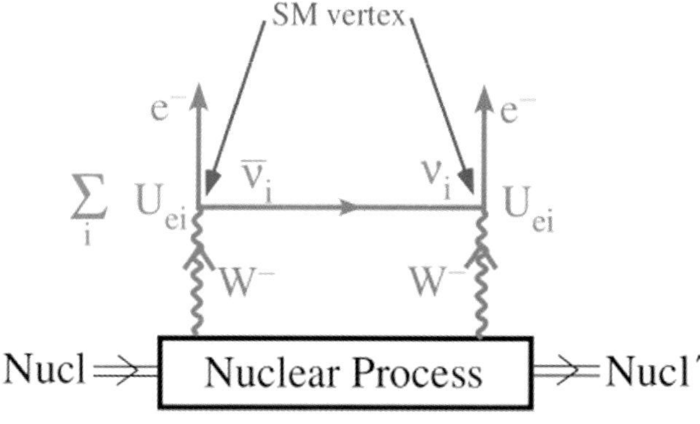

Fig. 2. The neutrino-exchange mechanism for $0v\beta\beta$.

two electrons and nothing else. Manifestly, this reaction would not conserve L. Thus, observing it at any nonzero level [2] would establish that neutrinos are identical to their antiparticles. Like any L-nonconserving process, $0v\beta\beta$ is suppressed by the smallness of neutrino masses. However, if we choose as our parent nucleus one that cannot decay by α or single β emission, and wait long enough, we might see it decay by $0v\beta\beta$ emission. To be sure, any nucleus that can decay in this L-nonconserving way can also decay via the L-conserving process Nucl $\rightarrow$ Nucl$' + e^- + e^- + \overline{v} + \overline{v}$. However, this two-neutrino double beta decay is phase-space suppressed, giving the neutrinoless mode a chance to be observed.

The dominant mechanism for $0v\beta\beta$ is expected to be the neutrino-exchange diagram in Fig. 2, in which one or another of the neutrino mass eigenstates v_i is exchanged. The neutrino-electron-W-boson vertices in this diagram are assumed to be SM weak vertices, which conserve L. Thus, if $\overline{v}_i$ is distinct from v_i, the exchanged particle emitted by the leptonic weak vertex on the left side of the diagram must be a $\overline{v}_i$. But when this same exchanged particle is absorbed by the leptonic weak vertex on the right side of the diagram, it must be a v_i. Thus, this diagram does not exist if $\overline{v}_i \neq v_i$, but only if $\overline{v}_i = v_i$.

Apart from an overall coupling strength, the amplitude for a v_i to create a charged lepton of "flavor" α at a SM weak vertex is $U_{\alpha i}$, where U is the unitary leptonic mixing matrix. Hence, there is a factor of U_{ei} at each of the leptonic weak vertices in Fig. 2. As indicated in that figure, the amplitude for $0v\beta\beta$, Amp $[0v\beta\beta]$, is a coherent sum over the contributions of the different v_i. Just as if it had been born in an e^--producing β decay, the exchanged v_i in Fig. 2 is emitted in a state which is almost totally of right-handed helicity, but which contains a small piece, of order m_i/E_{v_i}, having left-handed helicity. Here, as before, m_i is the mass of v_i, and E_{v_i} is its energy. When the exchanged v_i is absorbed, the absorbing SM left-handed current can only absorb its left-handed component without further suppression. Since this component is $\mathcal{O}[m_i/E_{v_i}]$, the contribution of v_i exchange to $0v\beta\beta$ is proportional to m_i. Hence, recalling the two factors of U_{ei} in Fig. 2, and summing over all the v_i contributions,

$$\text{Amp}[0v\beta\beta] \propto \left| \sum_i m_i U_{ei}^2 \right| \equiv m_{\beta\beta}. \tag{6}$$

The quantity $m_{\beta\beta}$ is known as the effective Majorana neutrino mass for neutrinoless double beta decay.

As we have stressed, if neutrino interactions are governed by the SM, then any L nonconservation in nature must vanish with the neutrino masses. Eq. (6) makes this vanishing explicit for the case of $0v\beta\beta$.

The fact that Amp $[0\nu\beta\beta]$ depends on neutrino masses means that a measurement of the rate for $0\nu\beta\beta$ would provide information on these masses. This was discussed at this Symposium by S. Petcov.

3. Consequences of Majorana character

Suppose that, as many theorists suspect, neutrinos are indeed their own antiparticles. What would be the physical consequences?

3.1. *Electromagnetic properties*

If neutrinos are their own antiparticles, then they are not only electrically neutral, but also devoid of any internal charge distribution. To see why, suppose, for example, that a neutrino has a charge distribution consisting of a positively-charged core surrounded by a compensating negatively-charged shell. The CPT-mirror image of this neutrino would have a negatively-charged core surrounded by a positively-charged shell. But if the neutrino is its own antiparticle, then (apart from a reversal of its spin), it must be identical to its CPT-mirror image. Thus a neutrino that is its own antiparticle cannot have a charge distribution.

A Majorana neutrino cannot have a magnetic or electric dipole moment either. From CPT invariance it follows that a fermion and its antiparticle have equal and opposite dipole moments. But a Majorana neutrino is its own antiparticle. Hence, it can have no dipole moments.

Both Majorana and Dirac neutrinos can have *transition* dipole moments connecting one neutrino mass eigenstate to another. Such moments make possible radiative decays such as $\nu_2 \to \nu_1 + \gamma$. However, in the SM, extended to included neutrino mass, such decays are exceedingly slow [3]. For example, if the mass of ν_2 is ~ 0.05 eV and ν_1 is massless, the lifetime for $\nu_2 \to \nu_1 + \gamma$ is $\sim 10^{49}$ yr. Visibly fast neutrino decays, electromagnetic or non-electromagnetic, would entail physics beyond even the extended SM. It is interesting that for some years, the Super-Kamiokande atmospheric ν_μ disappearance data could be explained either in terms of neutrino oscillation or in terms of neutrino decay [4]. However, the decay hypothesis is now excluded by Super-Kamiokande results, including their L/E (Distance/Energy) analysis [5].

3.2. *CP violation in neutrino oscillation*

Neutrino oscillation is completely insensitive to whether neutrinos are of Dirac or Majorana character. In particular, CP violation in neutrino oscillation is completely insensitive to this question. At first sight, this might seem strange. After all, CP violation in neutrino oscillation would manifest itself as a difference between the probabilities for the CP-mirror-image oscillations $\nu_\alpha \to \nu_\beta$ and "$\overline{\nu}_\alpha \to \overline{\nu}_\beta$", where a, $\beta = e$, μ, or τ. If neutrinos are Majorana particles, so that each mass eigenstate ν_i is identical to its antiparticle $\overline{\nu}_i$, is "$\overline{\nu}_\alpha \to \overline{\nu}_\beta$" a different process from $\nu_\alpha \to \nu_\beta$? Indeed it is, as illustrated in Fig. 3 for the example of $\nu_e \to \nu_\mu$ and "$\overline{\nu}_e \to \overline{\nu}_\mu$". As shown in Fig. 3, each of these two processes is defined in practice by the flavors and *the signs of the charges* of the charged leptons at the source and detection ends of the neutrino beamline. When one goes from $\nu_e \to \nu_\mu$ to "$\overline{\nu}_e \to \overline{\nu}_\mu$", the signs of these charges reverse, so that $\nu_e \to \nu_\mu$ and "$\overline{\nu}_e \to \overline{\nu}_\mu$" are different processes, and they can have different probabilities, even when $\overline{\nu}_i = \nu_i$. The amplitude for each process is a coherent sum over the contributions of the various ν_i, as shown in Fig. 3. Irrespective of whether $\overline{\nu}_i = \nu_i$, as assumed in Fig. 3, or not, the contribution of

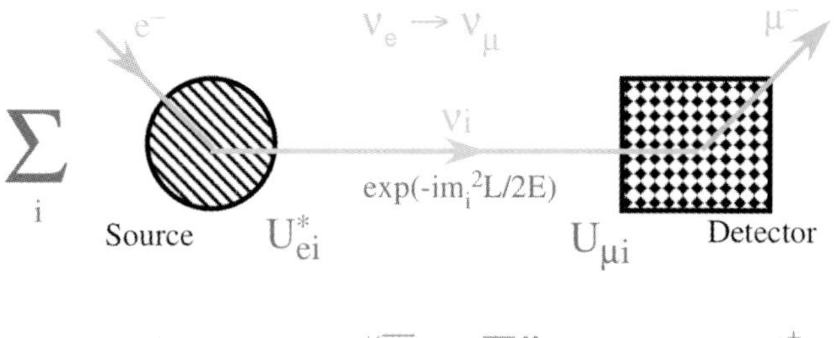

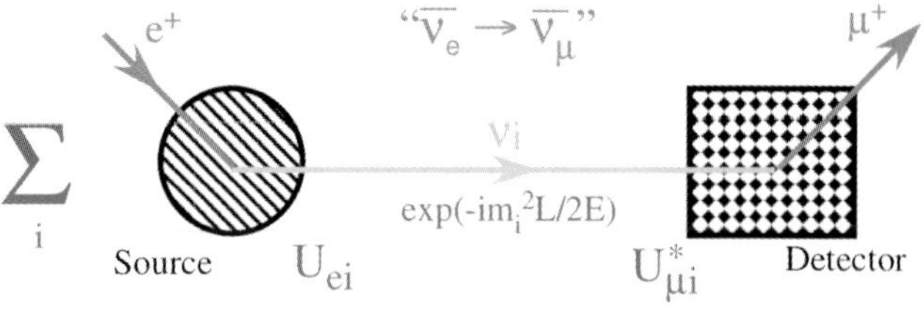

Fig. 3. The amplitudes for the oscillations $\nu_e \to \nu_\mu$ and "$\overline{\nu}_e \to \overline{\nu}_\mu$", showing the charged leptons that define these processes. For each process, we show the elements of the mixing matrix U that appear in the amplitude where the neutrino is produced and where it is detected. We also show the amplitude $\exp(-m_i^2 L/2E)$ for each mass eigenstate ν_i to propagate for a distance L with energy E. The figure assumes that $\overline{\nu}_i = \nu_i$, but nothing would change if in fact $\overline{\nu}_i \neq \nu_i$, so that the propagating neutrino in "$\overline{\nu}_e \to \overline{\nu}_\mu$" is a $\overline{\nu}_i$ rather than a ν_i.

each ν_i to each process involves the factors indicated in this figure. Clearly, if the mixing matrix U contains a CP-violating phase, the probabilities for $\nu_e \to \nu_\mu$ and "$\overline{\nu}_e \to \overline{\nu}_\mu$" can differ, even when $\overline{\nu}_i = \nu_i$.

3.3. *Majorana CP-violating phases*

The 3×3 quark mixing matrix contains just one CP-violating phase. However, if neutrinos are Majorana particles, then the 3×3 leptonic mixing matrix may contain three CP-violating phases [6]. The two extra phases, ξ_1 and ξ_2, are called Majorana phases. The phase ξ_i is associated with the neutrino mass eigenstate ν_i. If, in the absence of the Majorana phases, the leptonic mixing matrix is U^0, then in their presence it is U, where

$$U_{\alpha i} = U^0_{\alpha i} e^{i\frac{\xi_i}{2}}; \quad \text{all } \alpha. \tag{7}$$

Majorana phases have physical consequences only in processes that involve a violation of lepton number L [7]. They do not affect neutrino oscillation, but do influence $0\nu\beta\beta$. Indeed, their influence on $0\nu\beta\beta$ is obvious from Eq. (6) for $m_{\beta\beta}$, the effective Majorana neutrino mass for $0\nu\beta\beta$. Clearly, $m_{\beta\beta}$ depends on the relative phase of U_{e1}^2 and U_{e2}^2; that is, on $\xi_1 - \xi_2$.

The "Dirac" CP-violating phase in the quark mixing matrix can lead to manifest CP violation, by which we mean a manifestly CP-violating inequality between the rate for some process, and the rate for its CP-mirror image. Can Majorana phases cause such a manifestly CP-violating inequality too? In theory, they can cause it in heavy neutrino decay in the early universe, leading to the excess of matter over antimatter that we see in the universe today (see Sec. 4). This phenomenon, known as leptogenesis, was discussed at this Symposium by T. Yanagida. However, we wish to ask whether Majorana phases can cause CP-violating rate inequalities in present-day processes. The answer is yes [8], although these inequalities occur only in processes that would be extremely difficult to observe. An example is $e^+ W^- \to \nu \to \mu^- W^+$, and its CP-mirror image, $e^- W^+ \to \nu \to \mu^+ W^-$. Here, the intermediate Majorana neutrino propagates down a beamline, as in a neutrino oscillation experiment, and the W bosons are attached to nucleons at the neutrino source and at the detector.

The amplitude for $e^+ W^- \to v \to \mu^- W^+$ is given by

$$\text{Amp}\,[e^+ W^- \to v \to \mu^- W^+] = S \sum_i U_{ei} U_{\mu i} \frac{m_i}{E} e^{-im_i^2 \frac{L}{2E}}. \quad (8)$$

Here, the i'th term is the contribution of an intermediate v_i. S is a kinematical factor, m_i is the mass of v_i, L is the distance traveled by the neutrino, and E is its energy. The factor $\exp[-im_i^2 L/2E]$ is the same neutrino propagator that occurs in ordinary neutrino oscillation. The factor m_i/E is a suppression factor reflecting the fact that the incoming e^+ leads to a neutrino in a dominantly right-handed helicity state, whereas the left-handed current that produces the outgoing μ^- would prefer that the neutrino be in a left-handed state. It is this suppression factor that makes this process so hard to observe.

The amplitude for $e^- W^+ \to v \to \mu^+ W^-$ is given by

$$\text{Amp}\,[e^- W^+ \to v \to \mu^+ W^-] = S \sum_i U_{ei}^* U_{\mu i}^* \frac{m_i}{E} e^{-im_i^2 \frac{L}{2E}}. \quad (9)$$

Comparing this with Amp $[e^+ W^- \to v \to \mu^- W^+]$, and consulting Eq. (7), we see that Majorana phases in the U matrix will indeed lead to a CP-violating rate inequality. In fact, this would even be true if there were only two generations of leptons. To see this, let us suppose that there are only two generations: two charged leptons e and μ, and two neutrinos, v_1 and v_2. Then the most general unitary mixing matrix is

$$
U = \begin{array}{cc} v_1 & v_2 \end{array} \\
\begin{bmatrix} c\,e^{i\frac{\xi}{2}} & s \\ -s\,e^{i\frac{\xi}{2}} & c \end{bmatrix} \begin{array}{c} e \\ \mu \end{array}, \quad (10)
$$

where c and s are, respectively, the cosine and sine of a mixing angle θ, ξ is a Majorana phase, and the symbols outside of the matrix label the columns and rows. Using this mixing matrix in Eq. (8) and squaring, we find that the rate Γ for $e^+ W^- \to v \to \mu^- W^+$ is given by [9]

$$\Gamma[e^+ W^- \to v \to \mu^- W^+]$$
$$= K \frac{\sin^2 2\theta}{4E^2}\left[m_1^2 + m_2^2 - 2m_1 m_2 \cos\left(\Delta m^2 \frac{L}{2E} - \xi \right) \right]. \quad (11)$$

Here, $K = |S|^2$, and $\Delta m^2 = m_2^2 - m_1^2$. Using the same mixing matrix in Eq. (9), we find that

$$\Gamma[e^- W^+ \to v \to \mu^+ W^-]$$
$$= K \frac{\sin^2 2\theta}{4E^2}\left[m_1^2 + m_2^2 - 2m_1 m_2 \cos\left(\Delta m^2 \frac{L}{2E} + \xi \right) \right]. \quad (12)$$

We see that if the Majorana phase ξ is present, the rates for the CP-mirror-image processes $e^+ W^- \to v \to \mu^- W^+$ and $e^- W^+ \to v \to \mu^+ W^-$ differ.

When compared with behavior familiar from the quark sector, the rates of Eqs. (11) and (12) have a surprising feature. One may think of the quark mixing matrix as indicating the combination of d, s, and b quarks that couples to a particular up-type quark (such as the top quark), or, equivalently, the combination of u, c, and t quarks that couples to a particular down-type quark (such as the strange quark). Thus, if the masses of u, c, and t were equal, the quark mixing matrix would lose its meaning and all of its physical consequences, because we could no longer distinguish u, c, and t from one another. Similarly, the mixing matrix would lose its meaning and consequences if the masses of d, s, and b were

equal. But what happens to the rates of Eqs. (11) and (12) when the masses of v_1 and v_2, m_1 and m_2, are equal? With $m_1 = m_2 \equiv m$, these rates become

$$\Gamma[e^+ W^- \to v \to \mu^- W^+] = \Gamma[e^- W^+ \to v \to \mu^+ W^-]$$
$$= K \sin^2 2\theta \frac{m^2}{E^2} \sin^2 \frac{\xi}{2}. \quad (13)$$

Evidently, if the Majorana phase ξ is present, so that these rates do not vanish, the mixing angle θ in the leptonic mixing matrix still has physical consequences. To understand why this can happen, we note that through a change of phase conventions, the phase ξ that is associated with neutrino v_1 in the mixing matrix U of Eq. (10) can be removed from the U matrix and attached to the mass of v_1 [8]. The mass of v_1 then becomes a complex quantity, $m_1 e^{i\xi}$. But the mass of v_2, a neutrino with no associated Majorana phase, is real. Thus, there is still a distinction between v_1 and v_2, even when their masses are of equal size. Hence, the leptonic mixing matrix can still have meaning and physical consequences.

4. A concluding question

The exploration of the possible effects of Majorana phases raises a question: Why are there three generations of quarks and leptons?

It is well known that the presently observed preponderance of matter over antimatter in the universe could not have developed without a violation of CP sometime in the early universe. Once, it had been thought that this violation of CP came from the CP-violating phase in the quark mixing matrix. Had this been the case, we could have argued that—

a) The phase in the quark mixing matrix cannot produce CP violation unless there are at least three generations.
b) Without this CP violation, there would not be a preponderance of matter over antimatter in the universe.
c) Without this preponderance, we would not be here to ask why there are three generations.

However, it is now well established that the phase in the quark mixing matrix could not have produced nearly enough CP violation to explain the present excess of matter over antimatter. As a result, there is a lot of interest in the appealing possibility that this excess resulted from leptogenesis. In the see-saw mechanism, each light neutrino v is accompanied by a very heavy neutrino N. Both the light neutrino and its heavy see-saw partner are Majorana particles. In leptogenesis [10], discussed at this Symposium by T. Yanagida, there is a CP violation, coming from Majorana phases, in the decays of the heavy neutrinos N in the early universe. This CP violation leads to unequal rates for the leptonic decays $N \to \ell^+ + \text{Higgs}^-$ and $N \to \ell^- + \text{Higgs}^+$. These unequal rates lead, in turn, to a universe with unequal amounts of matter and antimatter, as observed.

In the two-generation example described in Sec. 3.3, we saw that Majorana phases can already produce manifestly CP-violating rate inequalities, such as the one required for leptogenesis, when there are only *two* generations. So why are there *three*?

Acknowledgements

This work was supported by the U.S. Department of Energy.

References

1. This approach, suggested by B. K. and L. Stodolsky, was discussed in Kayser, B., in "Massive Neutrinos in Astrophysics and in Particle Physics" (Proceedings of the Moriond Workshop), (ed. J. Tran Thanh Van) (Editions Frontieres, Gif-sur-Yvette, France, 1984) p. 11.
2. Some evidence has been reported that perhaps $0\nu\beta\beta$ has already been seen. See Klapdor-Kleingrothaus, H.V. in *Valencia 2003, Astroparticle and High Energy Physics*, available on the JHEP Conference PDF Server.
3. Kim, C. and Pevsner, A., *Neutrinos in Physics and Astrophysics* (Harwood, Chur, Switzerland, 1993).
4. Barger, V., Learned, J., Pakvasa, S. and Weiler, T., Phys. Rev. Lett. **82**, 2640 (1999).
5. Ashie, Y. *et al.*, The Super-Kamiokande Collaboration. Phys. Rev. Lett. **93**, 10180 (2004).
6. Bilenky, S., Hosek, J. and Petcov, S., Phys. Lett. **B94**, 495 (1980); Schechter, J. and Valle, J., Phys. Rev. **D22**, 2227 (1980); Doi, M., Kotani, T., Nishiura, H., Okuda, K. and Takasugi, E., Phys. Lett. **B102**, 323 (1981); Kayser, B. in "CP Violation," (ed. C. Jarlskog), (World Scientific, Singapore, 1989), p. 344.
7. Nieves, J. and Pal, P., Phys. Rev. **D64**, 076005 (2001).
8. de Gouvêa, A., Kayser, B. and Mohapatra, R., Phys. Rev. **D67**, 053004 (2003).
9. Schechter, J. and Valle, J., Phys. Rev. **D23**, 1666 (1981).
10. Fukugita, M. and Yanagida, T., Phys. Lett. **B174**, 45 (1986).

Physica Scripta. Vol. T121, 161–165, 2005

NuTeV and Neutrino Properties

Michael H. Shaevitz

Columbia University, New York, NY 10027, USA

Received January 13, 2005; accepted February 8, 2005

PACS numbers: 13.15.+g, 14.60.Lm, 14.60.Pq, 14.60.St

Abstract

This report explores the results and implications of the weak mixing angle measurement made by the NuTeV neutrino experiment at Fermilab. The NuTeV experiment, using a technique that exploits muon neutrino and antineutrino data to determine the neutral current to charged current ratios, R^ν and $R^{\bar\nu}$, has made the most precise measurement of the weak mixing angle using neutrinos as probes. The result gives a value of $\sin^2\theta_W(on-shell) = 0.2277 \pm 0.0016$ which is about three standard deviations larger than the standard model prediction of 0.2227. Various interpretations for the source of the anomaly are considered including changes to the inputs to the standard model predictions, unexpected symmetry violations, or new physics interpretations involving unanticipated neutrino properties or new particle contributions. Speculations on new precison measurements to further explore this region are also presented, including, for example, a future reactor neutrino-electron elastic scattering measurement. At present the discrepancy is unexplained, but could point to some as yet undiscovered broken quark symmetry, or towards new physics associated with neutrino interactions or mixings.

1. Introduction

For more than thirty years, starting with the discovery of neutral currents using neutrino beams at CERN in 1973, electroweak measurements have provided the most important and precise tests of the standard model (SM) of particle physics. A combination of theoretical and experimental progress has now allowed a new era of precision studies. On the theoretical side, the phenomenology is well understood and high order corrections to many processes have been calculated. The experimental measurement accuracy has become very precise and now allows comparisons to expectation that are sensitive to higher order and new physics corrections. The current measurements of electroweak processes at LEP/SLD, the top and W boson mass at CDF/D0, and precision charged lepton and neutrino experiments have led to strong constraints on the standard model including the prediction of a light Higgs boson and possibly supersymmetry.

The weak neutral current couplings are parameterized in the SM by a single parameter, the weak mixing angle, $\sin^2\theta_W$. This mixing angle sets the scale of the left and right-handed couplings of the quarks and leptons to the Z boson. In the SM, the neutrinos are special being that they are neutral particles with only left handed couplings. In addition, neutrinos are now known to have mixing among flavors and also possibly to sterile partners that have no standard model couplings.

Current measurements of many processes show spectacular agreement with the SM. [1] On the other hand, there a few areas that indicate some inconsistencies. The quark and lepton measurements at LEP/SLD show a disagreement at about the three standard deviation level between the b quark forward-back asymmetry and the electron left-right asymmetry. Leptonic measurements tend to predict a light Higgs mass whereas the quark asymmetries would indicate a much heavier Higgs. In addition, the measurement of the invisible width at LEP gives

a number of active neutrinos which is too small by about two standard deviations, $N_\nu = 2.985 \pm 0.008$. This may indicate a reduced coupling of neutrinos to the Z boson similar to what would be extracted from the NuTeV measurement described in this manuscript.

Neutrino electroweak measurements are complementary to the collider measurements in that they give better constraints on the neutrino couplings and in probing the theory at low Q^2. Neutrino-electron scattering ($\nu + e^- \rightarrow \nu + e^-$) provides a very clean probe, but the small cross sections make precision measurements difficult. The Charm II measurements [2] of $\overset{(-)}{\nu}_\mu e^-$ elastic scattering yielded a measurement of $\sin^2\theta_W = 0.2324 \pm 0.0083$ and $g_V = -0.035 \pm 0.017$ ($SM: -0.0398$) and $g_A = 0.503 \pm 0.017 (SM:-0.5065)$. Elastic $\overset{(-)}{\nu}_e e^-$ scattering measurements have also been used to probe for a neutrino magnetic moment. Current limits from reactor (accelerator) measurements are $\mu_\nu < 1 \times 10^{-10}\mu_B$ ($6.8 \times 10^{-10}\mu_B$) at 90% CL for electron (muon) neutrinos respectively. SM predictions for magnetic moments give $\mu_\nu = 3.2 \times 10^{-10}\mu_B \times m_\nu/\text{eV}$.

Neutrino-quark (nucleon) scattering, on the other hand, provides a process with high statistics at moderate Q^2 but with complications due to the quark distribution modeling. Examples here include elastic ($\nu_\mu + p \rightarrow \nu_\mu + p$) [3] and deep-inelastic scattering ($\nu_\mu + N \rightarrow \nu_\mu + X$). Even though they are not sensitive to the neutrino couplings, parity-violating measurements using charged leptons also probe the electroweak model at low Q^2 and have provided several precision results [4] for comparison, as shown in Figure 1.

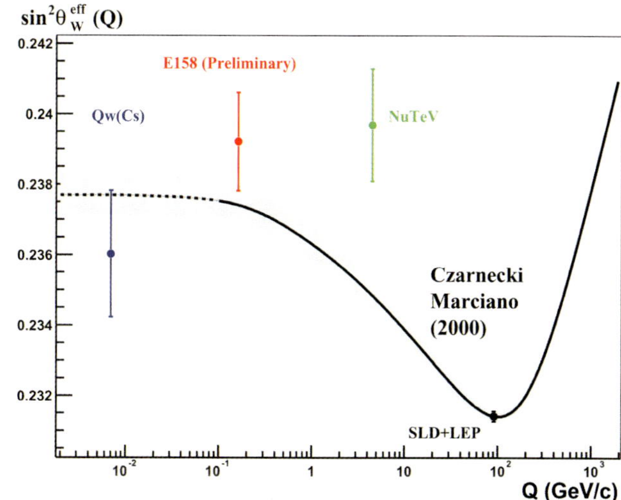

Fig. 1. Various electroweak measurements converted to an effective $\sin^2\theta_W^{eff}$ appropriate for the Moller scattering. The curve is fixed by the precision LEP/SLD measurements at the Z-pole. (From reference [4].)

2. Neutrino electroweak measurements using DIS scattering

The NuTeV experiment uses neutrino deep-inelastic scattering studies of both muon neutrinos and antineutrinos to make a precision determination of the weak mixing angle. The measurement is the most precise in the neutrino sector and is complementary to others in probing different radiative corrections and physics off the Z pole as well as investigating the coupling of neutrinos and light (u, d) quarks.

In determining $\sin^2\theta_W$, past neutrino measurements have used the neutral to charged-current ratio in order to cancel many of the systematics associated with neutrino flux and quark distribution modeling. The neutral-current (NC) to charged-current (CC) cross section ratios for neutrinos and antineutrinos in terms of the $\sin^2\theta_W$ and ρ is given by

$$R^{\nu(\bar{\nu})} = \frac{\sigma_{NC}^{\nu(\bar{\nu})}}{\sigma_{CC}^{\nu(\bar{\nu})}} = \rho^2\left(\frac{1}{2} - \sin^2\theta_W + \frac{5}{9}\sin^4\theta_W\left(1 + \frac{\sigma_{CC}^{\bar{\nu}(\nu)}}{\sigma_{CC}^{\nu(\bar{\nu})}}\right)\right). \quad (1)$$

Before NuTeV, this technique was limited in precision due to the systematic uncertainties associated with modeling the corrections to Eq. (1) due mainly to CC charm production from the strange-sea quarks which contributes to the denominator and not the numerator.

For NuTeV, which uses a combination of neutrino and antineutrino measurements, this systematic limitation has been dramatically reduced. This technique was first proposed by Paschos and Wolfenstein [5] and exploits NC and CC cross section differences to reduce the dependence on sea quark contributions.

$$R^- = \frac{\sigma_{NC}^{\nu} - \sigma_{NC}^{\bar{\nu}}}{\sigma_{CC}^{\nu} - \sigma_{CC}^{\bar{\nu}}} = \rho^2(\tfrac{1}{2} + \sin^2\theta_W) = g_L^2 - g_R^2 \quad (2)$$

where $g_{L,R}^2 = u_{L,R}^2 + d_{L,R}^2$.

For sea quarks one expects that the quark and antiquark distributions will be identical. Since $\sigma(\nu q) = \sigma(\bar{\nu}\bar{q})$, this means that all sea quark contributions to R^- will be zero at leading order. Specifically under these assumptions, the charm and strange sea error will become negligible and the problematic CC charm production will only have contributions from the fairly well-known valence d quark distributions. The significant improvement in sensitivity is shown in Figure 2, where the NuTeV measurement is compared to other previous neutrino determinations.

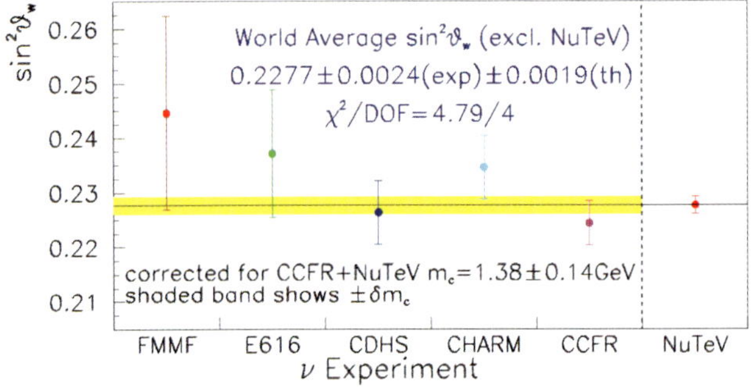

Fig. 2. Comparison of the NuTeV $\sin^2\theta_W$ measurement with those from previous neutrino experiments. All the values have been corrected assuming a charm quark mass of $m_c = 1.38 \pm 0.014\,\text{GeV}$; the shaded band represents the NuTeV value with the m_c uncertainty.

3. The NuTeV neutrino experiment

In order to exploit the technique of Eq. (2), an experiment must be able to collect separate data sets for neutrino and antineutrino beams. Before NuTeV, separated beams at high energy and high intensity were not available. To accomplish this separation, the NuTeV collaboration developed a new type of sign-selected, quadrupole focused beam. The beam used the 800 GeV protons from the Fermilab Tevatron impinging on a BeO target to produce secondary pions and kaons. Dipole magnets downstream of the target directed mesons of appropriate sign into a quadrupole focusing channel that sent the secondary beam into a 300 m evacuated decay region. The sign selection allowed separate ν and $\bar{\nu}$ running, and also removed K^0 mesons that in the past had been a prime source of uncertainty in the electron neutrino contamination. With this design, NuTeV was able to reduce the wrong-sign contamination in the beam to the level of 3×10^{-4} (4×10^{-4}) for neutrino (antineutrino) running while keeping the electron neutrino background to about 1.6%.

Neutrino interactions were observed in the NuTeV detector [6] located approximately 1.5 km downstream of the proton target. The detector consisted of an 18 m long, 690 ton steel-scintillator target followed by an instrumented iron-toroid spectrometer. The target calorimeter was composed of 1683 m $\times$ 3 m $\times$ 5.1 cm steel plates interspersed with liquid scintillation counters and drift chambers. The scintillation counters provided triggering information as well as a determination of the longitudinal event vertex, event length, and visible energy deposition. The mean position of the hits in the drift chambers gave the position of the event vertex. The muon spectrometer measured the momentum and charge of final state muons, which was needed in order to determine the neutrino flux as a function of energy. A tagged muon, hadron, and electron calibration beam was used to continuously monitor and study the detector.

Neutrino events in the detector have a hadronic shower component from the struck quark, and for CC events, an outgoing muon. For inclusion in the analysis, events were required to have 20 GeV of visible energy in the calorimeter and to satisfy fiducial volume requirements cutting events near detector edges. The observed events were separated into NC and CC statistically based on the "event length," which was defined in terms of the number of scintillation counters associated with the event. CC events tend to be much longer than NC events due to the outgoing penetrating muon. From test beam and Monte Carlo studies, the separation length was set at 16 to 18 counters depending on the visible calorimeter energy. Using this procedure yielded 457,000 (101,000) short events and 1,166,000 (250,000) long events for the neutrino (antineutrino) data and gave uncorrected ratios of short to long events with statistical errors of

$$R_{\text{exp}}^{\nu} = 0.3916 \pm 0.0007 \text{ (stat.)}$$

and

$$R_{\text{exp}}^{\bar{\nu}} = 0.4050 \pm 0.0016 \text{ (stat.).}$$

These experimental ratios can then be related to the physics parameters, $R^{\nu(\bar{\nu})}$ and $\sin^2\theta_W$, using a detailed Monte Carlo (MC) that includes the physics and detector model. Example effects that need to be included in the MC are: a model of the quark distributions; neutrino fluxes as a function of position and energy; hadronic shower modeling; and detector response versus energy, position, and time. The five largest corrections associated with

Table I. *Five largest corrections in the determination of $R^{\nu(\bar{\nu})}$ from $R^{\nu(\bar{\nu})}_{exp}$.*

Source	$\delta R^{\nu}_{\exp}$	$\delta R^{\bar{\nu}}_{\exp}$
Short CC Background	−0.068	−0.026
Electron Neutrinos	−0.021	−0.024
EM Radiative Corrections	+0.007	+0.011
Charm Mass	−0.005	−0.012
Cosmic-ray Background	−0.004	−0.019
Statistical Error	±0.0013	±0.0027

Table II. *Uncertainties for both the single parameter $\sin^2 \theta_W$ fit and for the comparison of R^{ν} and $R^{\bar{\nu}}$ with model predictions.*

Source of uncertainty	$\delta \sin^2 \theta_W$	δR^{ν}	$\delta R^{\bar{\nu}}$
Data Statistics	0.00135	0.00069	0.00159
Monte Carlo Statistics	0.00010	0.00006	0.00010
TOTAL STATISTICS	0.00135	0.00069	0.00159
$\nu_e, \bar{\nu}_e$ Flux	0.00039	0.00025	0.00044
Energy Measurement	0.00018	0.00015	0.00024
Shower Length Model	0.00027	0.00021	0.00020
Counter Efficiency, Noise, Size	0.00023	0.00014	0.00006
Interaction Vertex	0.00030	0.00022	0.00017
TOTAL EXPERIMENTAL	0.00063	0.00044	0.00057
Charm Production, Strange Sea	0.00047	0.00089	0.00184
Charm Sea	0.00010	0.00005	0.00004
$\sigma^{\bar{\nu}}/\sigma^{\nu}$	0.00022	0.00007	0.00026
Radiative Corrections	0.00011	0.00005	0.00006
Non-Isoscalar Target	0.00005	0.00004	0.00004
Higher Twist	0.00014	0.00012	0.00013
R_L	0.00032	0.00045	0.00101
TOTAL MODEL	0.00064	0.00101	0.00212
TOTAL UNCERTAINTY	0.00162	0.00130	0.00272

determining the corrected $R^{\nu(\bar{\nu})}$ values from the measured $R^{\nu(\bar{\nu})}_{\exp}$ ratios are given in Table I.

With the Monte Carlo tuned to incorporate all the corrections and dependences, one can then do a simultaneous fit to $R^{\nu}_{\exp}$ and $R^{\bar{\nu}}_{\exp}$ with an additional constraint on the slow-rescaling mass for charm production (m_c) from the NuTeV $\mu^+\mu^-$ data [7]. This procedure is similar to using an explicit R^- calculation in reducing the uncertainties related to sea quark scattering as well as many experimental systematics common to the ν and $\bar{\nu}$ samples. Statistical and systematic uncertainties in the $\sin^2 \theta_W$ as well as $R^{\nu(\bar{\nu})}$ are shown in Table II. The final result [8] for the weak mixing angle is

$$\sin^2 \theta_W^{(on-shell)} = 0.2277 \pm 0.0013(stat.) \pm 0.0009(syst.),$$

with the error being dominated by statistics. This result is more than a factor of two more precise than the previous world average for νN scattering of $\sin^2 \theta_W = 0.2277 \pm 0.0036$. Comparing to the expectation predicted from the combination of the other electroweak results [1], $\sin^2 \theta_W = 0.2227 \pm 0.0004$, the NuTeV differs by about three standard deviations.

This discrepancy is mainly associated with a 2.6 σ difference in the R^{ν} value since the $R^{\bar{\nu}}$ value is in agreement when systematic errors are also included. The agreement of the $R^{\bar{\nu}}$ value is important since this quantity is most sensitive to systematic

uncertainties such as m_c and has reduced sensitivity to $\sin^2 \theta_W$:

$$R^{\nu}_{\exp} = 0.3916 \pm 0.0013 \quad (SM: 0.3950),$$

$$R^{\bar{\nu}}_{\exp} = 0.4050 \pm 0.0027 \quad (SM: 0.4066).$$

In terms of chiral couplings, the difference with the SM is mainly in the left-handed $g_L^2 = u_L^2 + d_L^2$ coupling:

$$g_L^2 = u_L^2 + d_L^2 = 0.3000 \pm 0.0014 \quad (SM: 0.3042),$$

$$g_R^2 = u_R^2 + d_R^2 = 0.0308 \pm 0.0011 \quad (SM: 0.0301).$$

Global electroweak fits to all measurements including the NuTeV result give a $\chi^2/dof = 25/15$ which has a probability of about 4%. This high χ^2 value has a substantial contribution due to NuTeV but also reflects the tension between the heavy quark and lepton asymmetries.

4. Interpretations of the NuTeV result

Many interpretations of the NuTeV result have been put forward, spanning areas associated with changes in the standard model fits, changes in standard physics assumptions about quark asymmetries and nuclear effects, and inclusion of physics beyond the standard model. Each of these ideas has been investigated by NuTeV and others with the goal of determining how big a change to the measurement could be accommodated. These investigations are non-trivial, because the NuTeV analysis is complicated, and many of the model assumptions are directly tied to fits and constraints from data. In addition, the measured short-to-long ratio needs to be tied to NC and CC cross sections. Therefore, accurately estimating how much a given change will modify the NuTeV result requires implementing the change into an analysis using these data constraints.

4.1. *Changes to assumptions in the standard model fits*

The NuTeV analysis uses an enhanced leading-order formalism that implements constraints from both the observed CC single muon and dimuon data. External measurements are used to parameterize the longitudinal structure function, R_{long}, the d to u quark ratio, the charm quark sea, and higher twist effects (see Ref. [8] for details). Next-to-leading order (NLO) estimates from idealized analyses give small changes in the extracted $\sin^2 \theta_W$ value, $\delta \sin^2 \theta_W = 0.0004$ to $+0.0015$, and quark distribution variations are not sizeable for these idealized analyses [10]. To test possible NLO effects, NuTeV and others are developing full NLO ν event generators that include full NLO evolution with gluon terms and heavy charm effects. Initial indications are that these full calculations will not lead to significant changes [12].

Questions have also been raised with respect to the radiative corrections applied to the data. Although one cannot rigorously break out separate contributions, the largest corrections to the neutral to charged-current ratio come from electromagnetic radiative corrections. Those associated with bremsstrahlung from the final state muon in charged current events are especially large but are straightforward to calculate. NuTeV uses the only currently available code from Bardin and Dokuchaeva [13] which when applied shifts the result by $\delta \sin^2 \theta_W = -0.008$. New calculations are now becoming available that have improved treatment of the initial state mass singularities and allow investigations of input parameters and scheme dependence [14]. NuTeV plans to implement these new codes into the analysis to quantitatively determine their impact.

4.2. Changes in standard physics assumptions

The Paschos-Wolfenstein relation in Eq. (2) assumes isospin symmetry, $u_p(x) = d_n(x)$ and $d_p(x) = u_n(x)$, assumes momentum symmetry for the sea quarks, $xs(x) = x\bar{s}(x)$ and $xc(x) = x\bar{c}(x)$, and assumes that any nuclear environment effects are common for both W and Z boson exchange. Violations of these symmetries are possible but subject to constraints from various measurements. The NuTeV collaboration has investigated many of these effects and has provided a weighting function that can be convoluted with a given model to provide the shift that would result for the extracted $\sin^2\theta_W$ value [9].

Isospin symmetry violations could come from quark mass effects ($m_u \neq m_d$) or wave function differences due to the different quark charges. To explain the NuTeV discrepancy would require d_V quarks in the proton to carry about 5% more momentum than the u_V quarks in the neutron. Models of these effects are not very predictive but typically predict values less than about 1.5% [15], [9]. As stated by several authors, making conclusions from phenomenological models is difficult and new data will be needed in order to constrain possible isospin violation effects to the $\sin^2\theta_W$ measurement.

A strange-antistrange quark momentum asymmetry ($xs(x) = x\bar{s}(x)$) in the nucleon has also been put forward as a possible explanation for the NuTeV discrepancy [16]. The NuTeV dimuon results provide a direct measure of this asymmetry through the processes

$$v + s \rightarrow \mu^- + c \rightarrow \mu^- \mu^+ + X,$$

$$\bar{v} + \bar{s} \rightarrow \mu^+ + \bar{c} \rightarrow \mu^+ \mu^- + X.$$

NuTeV has performed a full NLO analysis of the dimuon data [11] and has restricted the strange sea asymmetry to

$$\int xs^-(x)\,\mathrm{d}x = \int \tfrac{1}{2}(xs(x) - x\bar{s}(x))\,\mathrm{d}x = -0.0009 \pm 0.0014.$$

A value for this integral of $+0.0060$ would be needed to explain the NuTeV discrepancy which would be inconsistent with this measurement.

Nuclear effects that are different for W and Z boson exchange could be important for the NuTeV analysis. Most nuclear effects are only large at small Q^2 and would affect W and Z exchange the same. NuTeV is at relatively high Q^2 with mean values of 25 (16) GeV^2 for v ($\bar{v}$) events, and the NuTeV $\sin^2\theta_W$ extraction shows no effect with increasing the visible hadron energy cut. In addition, the quark distribution measurements show no $1/Q^2$ dependence in the NuTeV kinematic region. Enhanced vector dominance models have been proposed but mainly affect sea quarks at low x and would cancel in R^- [9]. Furthermore, these nuclear corrections would change R^v and $R^{\bar{v}}$ more than R^- making their deviation with the SM even larger.

4.3. Physics beyond the standard model

The NuTeV discrepancy could be related to anomalous neutrino properties. Interpreting the NuTeV data in terms of a reduced neutral current coupling ρ_0, by fixing $\sin^2\theta_W$ to its SM value, yields a value of

$$\rho_0^2 = 0.9884 \pm 0.0026(stat.) \pm 0.0032(syst.)$$

which is a $2.8\,\sigma$ deviation from expectation. This is consistent with the LEP determination of the invisible width ("neutrino counting") result which is also $1.9\,\sigma$ low. Thus, these results could be explained by invoking a lower effective coupling of

neutrinos due to some new phenomena (*i.e.* mixing with heavy or sterile neutrinos). Giunti [17] has also speculated that neutrino oscillations of electron to sterile neutrinos could make the intrinsic electron neutrino background in NuTeV smaller, leading to a smaller measured value for $\sin^2\theta_W$. This simple model would require a level of 20% mixing and a high Δm^2 which is inconsistent with current limits. J. S. Ma *et al.* have recently made further investigations, and shown that models with several sterile neutrinos ($3 + 2$ models) when constrained by existing limits can at most explain $0.3\,\sigma$ of the NuTeV discrepancy with the SM [18].

Other possible new physics interpretations have also been considered. These interpretations need to respect the constraints from the precision Z-pole measurements from LEP and SLD. But, to a large extent, these constraints are insensitive to effects not directly involving the Z and are not too constraining with respect to neutrino couplings. In explaining the NuTeV measurements, these new physics models have to provide a change in R^v (or g_L) without changing $R^{\bar{v}}$ (or g_R). One possibility is SUSY particle contributions to loop corrections or R-parity violating SUSY corrections at tree level. Studies of these possibilities show that the effects are generally small and in the wrong direction. In addition, it is almost universal that these corrections change both R^v and $R^{\bar{v}}$ at a similar level. It may be possible in extended SUSY models to produce an v to $\bar{v}$ asymmetry in the corrections and several authors have pursued such models [19].

Considering other types of models, a finely tuned contact interaction where an additional left-handed, quark-lepton vertex interaction is added with a strength of 1% of the weak interaction (giving a scale of about 5 TeV) could explain the NuTeV result. Imposing leptoquarks to explain the effect is generally difficult, since these corrections would increase both the NC and CC rate making the g_L discrepancy larger and tend to violate constraints from π-decay. Finally, one can introduce extra U(1) vector bosons that are tuned to have special reduced couplings to 2nd generation leptons. In all of these cases, one will have to wait for the results from searches at the Tevatron and LHC to see if any of these possibilities might be viable.

Several authors have considered combination models to explain the NuTeV result. Typically, one invokes special neutrino properties combined with changes to standard model parameters and possibly new particles. Loinza *et al.* [20] have suggested a model which has neutrino mixing, a heavy Higgs-boson mass, and new heavy bound states. The Zvv coupling is suppressed by having flavor dependent mixing of the standard neutrinos with new heavy, sterile neutrinos. This suppression leads to many inconsistencies with the Z-pole measurements that are corrected by invoking a Higgs mass greater than 200 GeV. The heavy Higgs mass is inconsistent with the measurements of the W boson mass; these are then accommodated by including new physics associated with new heavy bound states (U-parameter type new physics). For fits where the neutrino coupling has a flavor independent suppression, the models yield a reduction in the Zvv coupling of $(1 - \varepsilon) = 0.003 \pm 0.001$. Allowing for flavor dependent corrections gives better overall fits with, for example, $(1 - \varepsilon) = 0.005 \pm 0.002, 0.003 \pm 0.001$, and 0.001 ± 0.003 for e, μ, and τ suppressions respectively.

5. Future measurements

As outlined above, there are uncertainties in both the experimental and theoretical situation that will require new measurements for clarification. In particular measurements are needed at low Q^2 as well as for processes involving neutrinos. New information on charged lepton scattering at low Q^2 is becoming available both now and in the near future. The SLAC E158 experiment has put out preliminary results [4] ($\sin^2\theta_W(M_Z) = 0.2330 \pm 0.0011$ (stat.) ± 0.0010 (syst.) ± 0.0006 (theory)) for their full data set which shows a $+1.2\ \sigma$ discrepancy from the SM prediction as shown in Fig. 1. Electroweak measurements at JLab include the QWEAK experiment (http://www.jlab.org/qweak/) which will measure polarization asymmetries for electron-proton elastic scattering and the DIS-Parity experiment (http://www.jlab.org/~xiaochao/pac24/) which will study polarized electron-deuteron scattering. Both experiments should have good precision with similar or better uncertainties than the E158 measurements.

In the neutrino sector, the Nomad experiment is attempting to measure the weak mixing angle using their large statistical sample of neutrino only data. The idea is to apply next-to-next-to-leading-order and $1/Q^2$ corrections to account for quark model effects, as well as use dimuon data to constrain the strange and the charm mass. Initial sensitivity estimates have been presented [21] giving errors of $\delta\sin^2\theta_W = 0.002$ (stat.) ± 0.003 (syst.).

Other possible precision electroweak measurements include neutrino-electron scattering, which would eliminate any QCD and quark model corrections. The CHARM II result [2] is presently the most precise for $\nu_\mu e$ scattering but suffers from poor statistics due to the very small cross section. An improved experiment with $\times 25$ better statistics could reach the level of $\delta\sin^2\theta_W = 0.002$ but would require a very large (~ 5000 tons), fine-grained detector such as a liquid argon TPC combined with a high energy neutrino beam that has a rate an order of magnitude greater than NuTeV.

Recently, Conrad *et al.* [22] have suggested that electron anti-neutrinos from a reactor might be used to measure $\bar{\nu}_e e^-$ elastic scattering. The process is a combination of W and Z exchange with the total rate being sensitive to $\sin^2\theta_W$. The reactor provides a very high rate of neutrinos, and when combined with a modest sized (~ 50 ton) detector can give sensitivities at the $\delta\sin^2\theta_W = 0.002$ level. Such an experiment could be realized as part of a two detector $\bar{\nu}_e$ disappearance search for neutrino oscillation. The near detector would be used for the weak mixing angle measurement as well as providing the event normalization for the oscillation search in the far detector. In the near detector, the CC inverse beta-decay process ($\bar{\nu}_e + p \rightarrow e^- + n$) would provide the event normalization for the elastic scattering measurement and the far detector would monitor backgrounds. Using these techniques, the measurement would provide a clean, purely leptonic electroweak measurement using neutrinos and would be complementary in addressing the NuTeV discrepancy [23].

In summary, the NuTeV measurement has the precision to be important for tests of the standard model and the experimental technique and cross checks are robust with respect to systematic uncertainties. In comparison with prediction, the extracted $\sin^2\theta_W$ is about three sigma high indicating a lower effective coupling to left-handed quarks or a neutral current coupling that is $\sim 1.1\%$ smaller than expected. Interpretations have spanned the range from uncertainties in the quark and other model corrections to possible new physics associated with neutrinos and their interactions or mixings. Future measurements will be needed to explore the NuTeV measurement and its possible explanations.

Acknowledgments

The author thanks the NuTeV collaboration for allowing the presentation of their data and Geralyn Zeller and Paul Nienaber for useful discussions on this manuscript. This work was supported by the National Science Foundation.

References

1. The LEP Electroweak Working Group, arXiv:hep-ex/0412015; arXiv:hep-ex/0112021.
2. Geiregat, D. *et al.* [CHARM-II Collaboration], Phys. Lett. B **232**, 539 (1989).
3. Ahrens, L. A. *et al.*, Phys. Rev. D **35**, 785 (1987).
4. Anthony, P. L. *et al.* [SLAC E158 Collaboration], Phys. Rev. Lett. **92**, 181602 (2004) [arXiv:hep-ex/0312035]; plus updates available at the E-158 web site http://www.slac.stanford.edu/exp/e158/.
5. Paschos, E. A. and Wolfenstein, L., Phys. Rev. D **7**, 91 (1973).
6. Harris, D. A. *et al.*, Nucl. Instr. Meth. **A447**, 373 (2000).
7. Goncharov, M. *et al.*, Phys. Rev. **D64**, 112006 (2001).
8. Zeller, G. P. *et al.* [NuTeV Collaboration], Phys. Rev. Lett. **88**, 091802 (2002) [Erratum-ibid. **90**, 239902 (2003)] [arXiv:hep-ex/0110059].
9. Zeller, G. P. *et al.* [NuTeV Collaboration], "On the effect of asymmetric strange seas and isospin-violating parton. Phys. Rev. D **65**, 111103 (2002) [Erratum-ibid. D **67**, 119902 (2003)] [arXiv:hep-ex/0203004].
10. Davidson, S., Forte, S., Gambino, P., Rius, N. and Strumia, A., J. High En. Phys. **0202**, 037 (2002) [arXiv:hep-ph/0112302].
11. Mason, D. [NuTeV Collaboration], arXiv:hep-ex/0405037.
12. Dobrescu, B. A. and Ellis, R. K., Phys. Rev. D **69**, 114014 (2004) [arXiv:hep-ph/0310154].
13. Bardin, D. and Dokuchaeva, V. A., JINR-E2-86-260 (1986); Arbuzov, A. B., Bardin, D. Y. and Kalinovskaya, L. V., arXiv:hep ph/0407203.
14. Diener, K. P. O., Dittmaier, S. and Hollik, W., Phys. Rev. D **69**, 073005 (2004) [arXiv:hep-ph/0310364].
15. Londergan, J. T. and Thomas, A. W., arXiv:hep-ph/0407247; Londergan, J. T. and Thomas, A. W., Phys. Rev. D **67**, 111901 (2003) [arXiv:hep-ph/0303155].
16. Kretzer, S. *et al.*, Phys. Rev. Lett. **93**, 041802 (2004) [arXiv:hep-ph/0312322].
17. Giunti, C. and Laveder, M., arXiv:hep-ph/0202152.
18. Ma, J. S., Conrad, J. M., Sorel, M. and Zeller, G. P., arXiv:hep-ex/0501011.
19. Babu, K. S. and Pati, J. C., arXiv:hep-ph/0203029.
20. Loinaz, W., Okamura, N., Rayyan, S., Takeuchi, T. and Wijewardhana, L. C. R., Phys. Rev. D **70**, 113004 (2004) [arXiv:hep-ph/0403306].
21. Petti, R. [NOMAD Collaboration], arXiv:hep-ex/0411032.
22. Conrad, J. M., Link, J. M. and Shaevitz, M. H., submitted to Phys. Rev. D, arXiv:hep-ex/0403048.
23. Rosner, J. L., Phys. Rev. D **70**, 037301 (2004) [arXiv:hep-ph/0404264].

Physica Scripta. Vol. T121, 166–171, 2005

Absolute Masses of Neutrinos – Experimental Results and Future Possibilities

Christian Weinheimer[1,2]*

[1]Helmholtz-Institut für Strahlen- und Kernphysik, Nussallee 14-16, Rheinische Friedrich-Wilhelms-Universität Bonn, D-53359 Bonn, Germany
[2]*new address:* Institut für Kernphysik, Wilhelm-Klemm-Str. 9, Westfälische Wilhelms-Universität Münster, D-48149 Münster Germany

Received February 10, 2005; accepted February 18, 2005

PACS numbers: 14.60.Pq, 23.40.−s, 29.30.Dn

Abstract

The discovery of neutrino oscillation proved recently that neutrinos have non-vanishing masses in contrast to their description within the Standard Model of particle physics. However, the absolute neutrino mass scale, which is very important for particle physics as well as for cosmology and astrophysics, cannot be revealed by oscillation experiments. Although there are a few ways to determine the neutrino mass scale, the only model-independent method is the investigation of the electron energy spectrum of a β decay near its endpoint. The tritium β decay experiments at Mainz and Troitsk using tritium have recently been finished and have given upper limits on the neutrino mass scale of about $2\,\mathrm{eV/c^2}$. The bolometric experiments using ^{187}Re have finished the first experiments yielding a sensitivity on the neutrino mass of $15\,\mathrm{eV/c^2}$. The new Karlsruhe Tritium Neutrino Experiment (KATRIN) will enhance the sensitivity on the neutrino mass by another order of magnitude down to $0.2\,\mathrm{eV/c^2}$ by an ultra-precise measurement of the tritium β decay spectrum using a very strong windowless gaseous molecular tritium source and a huge ultra-high resolution electrostatic spectrometer of MAC-E-Filter type.

1. Introduction

The recent discovery of neutrino oscillation by experiments with atmospheric, solar, reactor and accelerator neutrinos [1] proved that neutrino mix and that they have non-zero masses in contrast to their current description in the Standard Model of particle physics. Unfortunately, these oscillation experiments are sensitive to the differences of squared neutrino mass states $\Delta m_{ij}^2 = |m^2(v_i) - m^2(v_j)|$, but not directly to the neutrino masses $m(v_i)$ themselves. On the other hand, if one neutrino mass is measured absolutely the whole neutrino mass spectrum can be obtained from the values Δm_{ij}^2 gained by the oscillation experiments[†].

Theories beyond the Standard Model try to explain the smallness of neutrino masses in comparison with the much heavier charged fermions [2]. One prominent explanation is the Seesaw type I mechanism using heavy Majorana neutrinos yielding a hierarchical pattern of neutrino masses [3]. Alternatively, Seesaw type II models usually produce a scenario of quasi-degenerate neutrino masses with the help of a Higgs triplet [4]. Here, all masses are $0.1\,\mathrm{eV/c^2}$ or heavier exhibiting small mass differences between each other to explain the oscillations. In this quasi-degenerate case – due to the huge abundance of relic neutrinos in the universe left over from the big bang – neutrinos would make up not the major, but a significant contribution to the dark matter. Therefore the open question of the values of the neutrino masses

is not only crucial for particle physics to decide between different theories beyond the Standard Model but it is also very important for astrophysics and cosmology.

There are different ways to determine the neutrino mass scale:

- **Cosmology**
  Information on the absolute scale of the neutrino mass can be obtained from astrophysical observations like the energy and matter distribution in the universe at different scales. Usually these analysis's use the combination of Cosmic Microwave Background data (*e.g.* from the WMAP satellite), the distribution of the galaxies in our universe, the so-called "Large Scale Structure", and information from the so-called "Lyman α-Forest" or X-ray clusters to describe the distribution at large, medium and small scales, respectively. In most cases they give upper limits on the mass of the neutrinos on the order of several $0.1\,\mathrm{eV/c^2}$ [5], in some cases non-zero neutrino masses are found [6] illustrating the dependence on the assumptions and on the data used to obtain the cosmological limits. One should not forget that these models describe 95% of the matter and energy distribution of the universe by yet non-understood quantities like the cosmological constant and the Cold Dark Matter. And, the limits on the neutrino mass rely on the existence of the yet not observed relic neutrinos [7].

- **Neutrinoless double β decay**
  One laboratory way to access the neutrino mass scale is the search for the neutrinoless double β decay. This process is a conversion of two neutrons (protons) into two protons (neutrons) within a nucleus at the same time. Usually two electrons (positrons) and two antineutrinos (neutrinos) are emitted. In the case, that the neutrino is a Majorana particle (particle is equal to its antiparticle) the double β decay could occur without emission of any neutrinos. This transition is directly proportional to the neutrino mass (in the absence of right-handed weak charged currents or the exchange of other new particles). The observable of double β decay is the so-called effective neutrino mass

$$m_{\mathrm{ee}} = \sum_i |U_{\mathrm{ei}}^2 \cdot m(v_\mathrm{i})| \qquad (1)$$

which is a coherent sum over all neutrino mass eigenstates $m(v_i)$ contributing to the electron neutrino with their (complex) mixing matrix elements U_{ei}. A subgroup of the most sensitive experiment, the Heidelberg-Moscow experiment using 5 low-background, highly enriched and high-resolution ^{76}Ge detectors in the Gran Sasso underground lab, has claimed evidence for having observed neutrinoless double β decay. Recently new data and a re-analysis of the old data have

---

*Email address: weinheimer@uni-muenster.de
[†]The ambiguity of the sign of the differences of squared masses in $\Delta m_{ij}^2 = |m^2(v_i) - m^2(v_j)| = |m^2(v_j) - m^2(v_i)|$ can in principle be resolved in the presence of matter effects.

been presented [8] showing a line at the position expected for neutrinoless double β decay with 4σ significance. Due to the uncertainties of the nuclear matrix element this signal translates into $0.1\,\mathrm{eV/c^2} \leq m_{ee} \leq 0.9\,\mathrm{eV/c^2}$. Clearly, this yet unconfirmed result requires further checks.

- **Direct neutrino mass determination**

In contrast to the other methods, the direct method does not require further assumptions. The neutrino mass is determined using the relativistic energy-momentum relationship. Therefore $m^2(\nu)$ is the observable in most cases.

The non-observation of a dependence of the arrival time on energy of supernova neutrinos from SN1987a gave an upper limit on the neutrino mass of $5.7\,\mathrm{eV/c^2}$ [9]. Unfortunately nearby supernova explosions are too rare and too less understood to allow a further improvement to sub-eV neutrino masses.

Therefore, the investigation of the electron energy spectrum of a β decay is still the most sensitive model-independent and direct method to determine the neutrino mass. The β spectrum exhibits the value of a non-zero neutrino mass, when the neutrino is emitted non-relativistically. This is the case in the vicinity of the endpoint E_0 of the β spectrum where nearly all decay energy is given to the β electron. Therefore, the mass of the electron neutrino has to be determined by investigating the shape of the β spectrum near its endpoint E_0 precisely (see fig. 1). From fig. 1 it is clearly visible that the main requirement for such an experiment is to cope with the vanishing count rate near the endpoint by providing the strongest possible signal rate at lowest background rate. Additionally, to become sensitive to the shape of the β spectrum a high energy resolution on the order of eV is required.

Tritium is the standard isotope for this study due to its low endpoint of 18.6 keV, its rather short half-life of 12.3 y, its super-allowed shape of the β spectrum, and its simple electronic structure. Tritium β decay experiments have been performed in search for the neutrino mass for more than 50 years. ^{187}Re is a second isotope suited to determine the neutrino mass. The disadvantage of its primordial half-life can be compensated by using large arrays of cryobolometers.

For each neutrino mass state $m(\nu_i)$ contributing to the electron neutrino a kink at $E_0 - m(\nu_i)c^2$ with a size proportional to

$|U_{ei}^2|$ will occur. However, due to the smallness of differences of squared neutrino masses Δm_{ij}^2 observed in oscillation experiments and as a consequence of the limited sensitivity of present and upcoming direct neutrino mass experiments only an incoherent sum or an average neutrino mass can be obtained [10], which can be defined as the *electron neutrino mass* $m(\nu_e)$ by

$$m^2(\nu_e) = \sum_i |U_{ei}|^2 \cdot m^2(\nu_i). \tag{2}$$

Comparing equations (1) and (2) it is obvious that the neutrinoless double β decay and the investigation of the β decay spectrum yield complementary information. In the former case complex phases of the neutrino Majorana mixing matrix U can lead to a partial cancellation. Also the large uncertainties of the nuclear matrix element and the possibility, that the exchange of other more exotic particles adds to the neutrinoless double β decay signal, disfavors double β decay for a precise neutrino mass determination. On the other hand the high sensitivity of the next generation of double β decay experiments and their unique possibility to prove the Majorana nature of neutrinos underline the very high importance of neutrinoless double β decay experiments.

This paper is organized as following: In section 2 the recent direct neutrino mass experiments investigating the β decay spectra of tritium and ^{187}Re are shortly reviewed. Section 3 describes the new KATRIN experiment aiming for a $0.2\,\mathrm{eV/c^2}$ neutrino mass sensitivity. The conclusions are given in section 4.

2. Direct neutrino mass experiments

A major break-through in tritium β decay experiments was achieved in the nineties by a new type of spectrometer, the so-called MAC-E-Filter (Magnetic Adiabatic Collimation followed by an Electrostatic Filter), which was developed independently at Mainz, Germany and at Troitsk, Russia [11, 12]. This integrating spectrometer provides high luminosity and low background combined with a large energy resolution.

The two recent tritium β decay experiments at Mainz and at Troitsk use similar MAC-E-Filters with an energy resolution of $4.8\,\mathrm{eV}$ ($3.5\,\mathrm{eV}$) at Mainz (Troitsk). The spectrometers differ slightly in size: The diameter and length of the Mainz (Troitsk) spectrometer are 1 m (1.5 m) and 4 m (7 m). The major differences between the two setups are the tritium sources: Mainz uses as tritium source a thin film of molecular tritium quench-condensed on a cold graphite substrate, whereas Troitsk has chosen a windowless gaseous molecular tritium source. After the upgrade of the Mainz experiment in 1995–1997 both experiments run with similar signal and similar background rates.

2.1. *The Mainz neutrino mass experiment*

Figure 2 shows the Mainz setup after its upgrade in 1995–1997, which included the installation of a new tilted pair of superconducting solenoids between tritium source and spectrometer and the use of a new cryostat providing tritium film temperatures of below 2 K. The first measure eliminated source correlated background and allowed the source strength to be increased significantly. The second measure avoids the roughening transition of the homogeneously condensed tritium films with time [13]. The upgrade of the Mainz setup was completed by the application of HF pulses on one of the electrodes in between measurements every 20 s, and a full automation of the

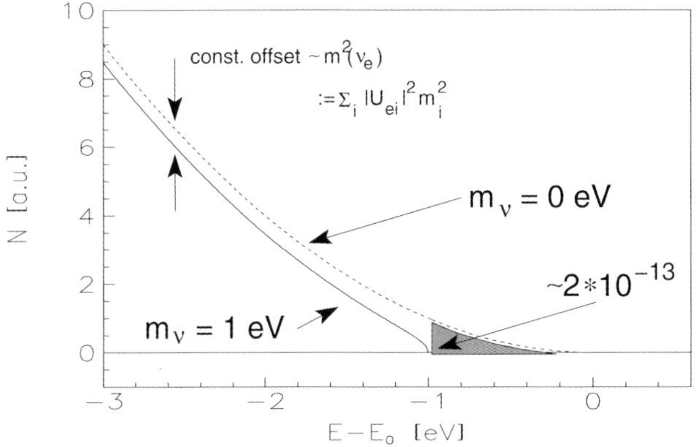

Fig. 1. Expanded tritium β spectrum around tritium endpoint E_0 for $m(\nu_e) = 0$ (dashed line) and for an arbitrarily chosen neutrino mass of $1\,\mathrm{eV/c^2}$ (solid line). The gray shaded area corresponds to a fraction of $2 \cdot 10^{-13}$ of all tritium β decays. The offset between the two curves explains how "$m(\nu_e)$" is defined: as the average over all neutrino mass states with their contribution according to the neutrino mixing matrix U.

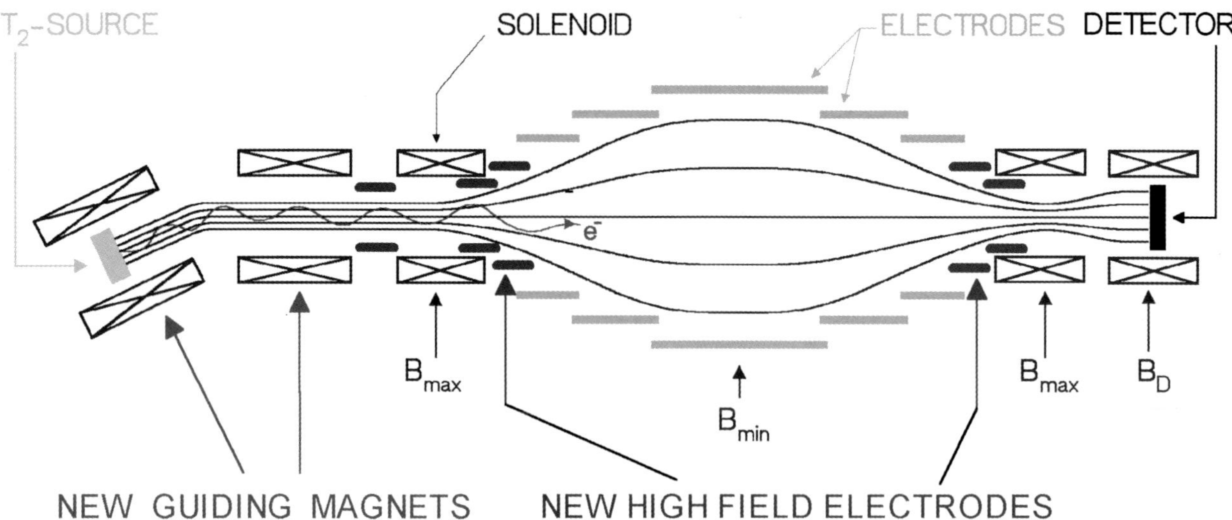

Fig. 2. The upgraded Mainz setup shown schematically. The outer diameter amounts to 1 m, the distance from source to detector is 6 m.

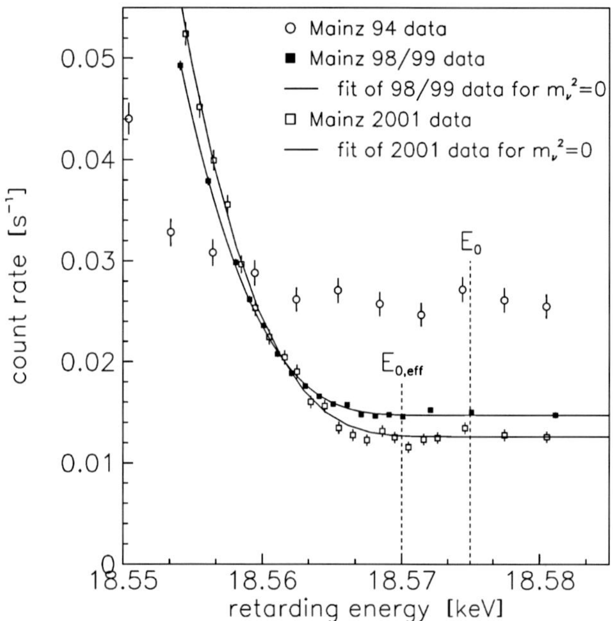

Fig. 3. Averaged count rate of the 1998/1999 data (filled squares) with fit (line) and of the 2001 data (open squares) with fit (line) in comparison with previous Mainz data from 1994 (open circles) as a function of the retarding energy near the endpoint E_0 and effective endpoint $E_{0,eff}$ (accounting for the width of response function of the setup and the mean rotation-vibration excitation energy of the electronic ground state of the $^3\mathrm{HeT}^+$ daughter molecule).

apparatus and remote control. This former improvement lowers and stabilizes the background, the latter one allows long-term measurements.

Figure 3 shows the endpoint region of the Mainz 1998, 1999 and 2001 data in comparison with the former Mainz 1994 data. An improvement of the signal-to-background ratio by a factor 10 by the upgrade of the Mainz experiment as well as a significant enhancement of the statistical quality of the data by longterm measurements are clearly visible. The main systematic uncertainties of the Mainz experiment are the inelastic scattering of β electrons within the tritium film, the excitation of neighbor molecules due to the β decay, and the self-charging of the tritium film by its radioactivity. As a result of detailed investigations in Mainz [14–16] – mostly by dedicated experiments – the systematic corrections became much better understood and their uncertainties were reduced significantly.

The high-statistics Mainz data from 1998–2001 allowed the first determination of the probability of the so-called neighbor excitation[‡] to occur in $(5.0 \pm 1.6 \pm 2.2)\%$ of all β decays [16] in good agreement with the theoretical expectation [17].

The most sensitive analysis on the neutrino mass, in which only the last 70 eV of the β spectrum below the endpoint are used, resulted in the following for Mainz 1998, 1999 and 2001 data [16]

$$m^2(\nu_e) = (-0.6 \pm 2.2 \pm 2.1)\,\mathrm{eV}^2/\mathrm{c}^4 \tag{3}$$

which corresponds to an upper limit of

$$m(\nu_e) < 2.3\,\mathrm{eV}/\mathrm{c}^2 \quad (95\%\,\mathrm{C.L.}) \tag{4}$$

This is the lowest model-independent upper limit of the neutrino mass obtained thus far.

2.2. *The Troitsk neutrino mass experiment*

The windowless gaseous tritium source of the Troitsk experiment [18] is essentially a tube of 5 cm diameter filled with T_2 resulting in a column density of 10^{17} molecules/cm^2. The source is connected to the ultrahigh vacuum of the spectrometer by a series a differential pumping stations.

From its first data taking in 1994 the Troitsk experiment reports an anomalous excess in the experimental β spectrum as a sharp step of the count rate at a varying position of a few eV below the endpoint of the β spectrum E_0 [18], which seems to be an experimental artefact appearing with varying intensity at the Troitsk setup. The Troitsk experiment is correcting for this anomaly by fitting an additional line to the β spectrum run-by-run.

Combining the 2001 results with the previous ones since 1994 gives [19]

$$m^2(\nu_e) = (-2.3 \pm 2.5 \pm 2.0)\,\mathrm{eV}^2/\mathrm{c}^4 \tag{5}$$

from which the Troitsk group deduce an upper limit of

$$m(\nu_e) < 2.05\,\mathrm{eV}/\mathrm{c}^2 \quad (95\%\,\mathrm{C.L.}). \tag{6}$$

The values of eqs. (5) and (6) do not include the systematic uncertainty which is needed to account for, when the timely-varying anomalous excess count rate at Troitsk is described run-by-run by an additional line.

[‡]The sudden change of the nuclear charge during β decay results in the excitation of neighboring molecules with a certain probability.

2.3. *Rhenium β decay experiments*

Due to the complicated electronic structure of ^{187}Re and its primordial half-life the advantage of the 7 times lower endpoint energy E_0 of ^{187}Re with respect to tritium can only be exploited if the β spectrometer measures the entire released energy, except that of the neutrino. This situation can be realized by using a cryogenic bolometer as the β spectrometer, which at the same time contains the β emitter ^{187}Re.

One disadvantage connected to this method is the fact that one measures always the entire β spectrum. Even for the case of the very low endpoint energy of ^{187}Re, the relative fraction of events in the last eV below E_0 is of order 10^{-10} only (compare to figure 1). Considering the long time constant of the signal of a cryogenic bolometer (typically several hundred µs) only large arrays of cryogenic bolometers could deliver the signal rate needed.

Two groups are working on ^{187}Re β decay experiments at Milan (MiBeta) and at Genoa (MANU2) using AgReO$_4$ and metallic rhenium, respectively. Although cryogenic bolometers with an energy resolution of 5 eV have been produced with other absorbers, this resolution has yet not been achieved with rhenium. The lowest neutrino mass limit of $m(\nu_e) < 15\,\text{eV}/\text{c}^2$ comes from MiBeta [20]. Further improvements in the energy resolution and the number of crystals are envisaged aiming for a sensitivity of a few eV/c^2.

3. The KATRIN experiment

The very important tasks presented in the introduction – to distinguish hierarchical from quasi-degenerate neutrino mass scenarios and to check the cosmological relevance of neutrino dark matter for the evolution of the universe – require the improvement of the direct neutrino mass search by one order of magnitude at least.

The KATRIN collaboration has taken this challenge and has proposed to build an ultra-sensitive tritium β decay experiment based on the successful MAC-E- Filter spectrometer technique and a very strong Windowless Gaseous Tritium Source (WGTS) [21] at the Forschungszentrum Karlsruhe, Germany. The international KATRIN collaboration consists of many groups from Czech Republic, Germany, Russia, UK and US, combining the worldwide expertise in tritium β decay, groups providing special knowledge, with the strength and the possibilities of a big national laboratory including Europeans biggest tritium laboratory. Figure 4 shows a schematic view of the proposed experimental configuration.

The windowless gaseous tritium source (WGTS, see fig. 5) allows for the measurement of the endpoint region of the tritium β decay and consequently the determination of the neutrino mass with a maximum of signal strength combined with a minimum of systematic uncertainties from the tritium source. The WGTS consists of a 10 m long cylindrical tube of 90 mm diameter filled with molecular tritium gas of high isotopic purity ($>95\%$). The tritium gas will be continuously injected by a capillary at the middle and pumped out by a series of differential pump stations at the end giving rise to a density profile over the source length of nearly triangular shape with a total column density of $5 \cdot 10^{17}/\text{cm}^2$ providing a count rate about a factor 100 larger than in Mainz and Troitsk. The β electrons are leaving the WGTS directly to both ends following the magnetic field lines at a magnetic field of 3.6 T whereas the pumped-out tritium gas is then purified and re-circulated. Of special importance is the control of the column density on the 1 per mill level by regulating the pressure in the tritium supply buffer vessel and the temperature of the WGTS tube. To allow a very stable and low WGTS temperature the WGTS tube is placed inside a pressure-stabilized LNe cryostat.

The electron transport system (see fig. 5) adiabatically guides β decay electrons from the tritium source to the spectrometer by a system of superconducting solenoids at a magnetic field of 5.6 T. At the same time it is eliminating any tritium flow towards the spectrometer by a differential pumping system consisting of 1 m long tubes inside the magnets alternated by pump ports with turbo molecular pumps yielding a tritium reduction factor of about a factor of 10^7. In the second part the surfaces of the liquid helium

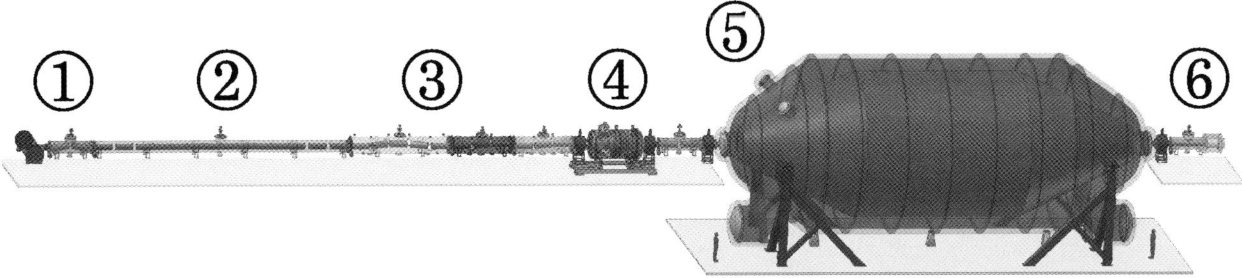

Fig. 4. Schematic view of the KATRIN experiment with the rear monitoring and calibration system (1), the windowless gaseous tritium source (WGTS) (2), the differential and cryopumping electron transport section (3), the pre spectrometer (4), the main spectrometer (5) and the electron detector array (6). The main spectrometer has a length of 24 m and a diameter of 10 m, the overall length over the experimental setup amounts to about 70 m. Not shown is the monitor spectrometer.

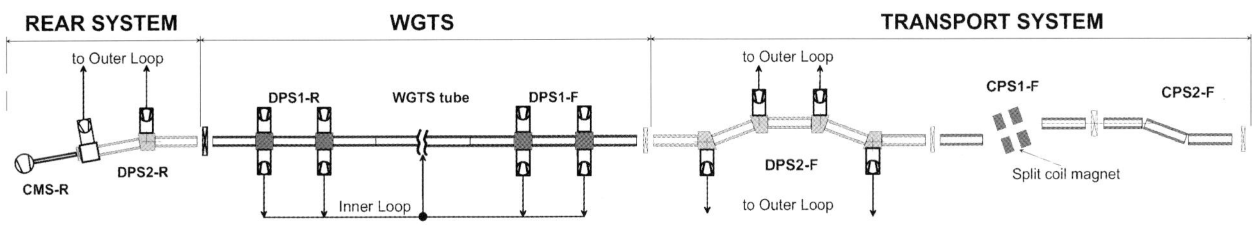

Fig. 5. Schematic view of the KATRIN windowless gaseous tritium source (WGTS) and the differential and cryopumping electron transport section with the optional quench-condensed tritium source (QCTS).

Fig. 6. Schematic view of the KATRIN pre spectrometer consisting of spectrometer vessel and two superconduction solenoids. The spectrometer vessel is set on high voltage and acts as retarding electrode. The inner electrode system is built in the central part as a nearly massless wire structure put on slightly more negative potential than the vacuum vessel to reject secondary electrons from the walls (see text).

cold vacuum tube act as a cryo-trapping section to suppress the tritium partial pressure further to an insignificant level. To reduce the molecular beaming effect, the direct line-of-sight is prohibited by 20 degree bents between each pair of superconducting magnets of 1 m length.

Between the tritium source and the main spectrometer a pre-spectrometer of MAC-E- Filter type will be installed (see fig. 6). It acts as an electron pre-filter running at a retarding energy about 200 eV below the endpoint of the β spectrum to reject all β electrons except the very high energetic ones in the region of interest close to the endpoint E_0. This minimizes the chances that β electrons cause background in the main spectrometer by ionization of residual gas.

A key component of the new experiment will be the large electrostatic main spectrometer with a diameter of 10 m and an overall length of about 23 m. This high-resolution MAC-E-Filter will allow to scan the tritium β decay endpoint at a resolution of $\Delta E = 0.93$ eV, which is – at a much higher luminosity – a factor of 4–5 better than for the MAC-E-Filters in Mainz and Troitsk.

Although limiting the electron input rate by the pre-spectrometer, stringent vacuum conditions have to be fulfilled to suppress background from the main spectrometer. Special selection of materials as well as surface cleaning and out-baking at 350 °C will allow to reach a residual gas pressure of better than 10^{-11} mbar. To reduce the size of surfaces inside the vacuum chamber, not a complex electrode system will be installed, but the vacuum vessel itself will be put on high voltage and thus will create the electric retarding potential.

A new idea of strong background suppression has been developed and successfully tested at the Mainz spectrometer, where it resulted in a factor 10 reduced background rate. It will be applied also to the KATRIN spectrometers: The vessel walls at high potential will be covered by a system of nearly massless wire electrodes, which are put to a slightly more negative potential. Secondary electrons ejected by cosmic rays or environmental radioactivity from the vessel wall, will thus be repelled and prohibited from entering the magnetic flux tube which is connected to the detector.

These new ideas and these new technical solutions will be applied also to the KATRIN pre-spectrometer, which has already being set up at the Forschungszentrum Karlsruhe. The vacuum

tests with the pre-spectrometer have been successfully finished yielding at a temperature of -20 °C a final pressure of less than 10^{-11} mbar and a outgasing rate of less than 10^{-13} mbar l/s cm^2. Both values are better than the KATRIN requirements showing that the material and pump selections as well as the surface treatments are well under control. The wire electrode system for the KATRIN pre-spectrometer has been constructed and awaits installation. When instrumented with a scanning electron gun and a 64-pixel silicon PIN-detector the electromagnetic and background properties of the KATRIN pre-spectrometer will be investigated from spring 2005 on.

The final KATRIN detector requires high efficiency for electrons at $E_0 = 18.6$ keV and low γ background. A high energy resolution of $\Delta E < 600$ eV for 18.6 keV electrons should suppress background events at different energies. The present concept of the detector is based on a large array of about 400 PIN photodiodes surrounded by low-level activity passive shielding and an active veto counter to reduce background.

After publishing the Letter of Intent [21] the KATRIN collaboration has done significant work to increase the sensitivity of the experiment. The major improvements of the setup to increase the statistics are the design of a tritium re-circulating and purification system providing a near to maximum tritium purity of $>95\%$, the increase of the diameter of the windowless tritium source from 75 mm to 90 mm and, correspondingly, of the diameter of the main spectrometer from 7 m to 10 m. Additionally an optimization of the measurement point distribution around the endpoint has been performed.

Instrumental improvements have been developed as well as plans for dedicated experiments and their analysis have been worked out in order to determine systematic corrections and to reduce their uncertainties. The main systematic uncertainties comprise the inelastic scattering within the tritium source and the stability of the retarding voltage of the main spectrometer. The former will be determined and repeatedly monitored with the help of a high-precision electron gun injecting electrons from the rear system. For the latter, a dedicated high-precision high voltage divider is being developed with the support of the Physikalisch Technische Bundesanstalt at Braunschweig, Germany. For redundancy, the retarding high voltage of the main spectrometer is applied in parallel to a third spectrometer, the monitor spectrometer[§], which continuously measures a sharp electron line. Different energetically well-defined sources are in preparation, *e.g.* a condensed ^{83m}Kr conversion electron source or a cobalt photoelectron source irradiated by γs from ^{241}Am. Another systematic uncertainty is the electrical potential distribution within the WGTS, which will be checked by running the WGTS in a second "high temperature regime" of 120–150 K with the conversion electron emitter ^{83m}Kr added to the gaseous molecular tritium. The source contamination by other hydrogen isotopes than tritium will be monitored with the help of laser Raman spectroscopy.

The detailed simulations of the KATRIN experiment yield the following (see fig. 7): A sensitivity of 0.20 eV/c^2 will be achieved with the KATRIN experiment after 3 years of pure data taking. Statistical and systematic uncertainties contribute about equally. This value of 0.20 eV/c^2 corresponds to an upper limit with 90% C.L. in the case that no neutrino mass will be observed.

[§]The Mainz spectrometer will be modified for this purpose into a high-resolution spectrometer with $\Delta E \approx 1$ eV.

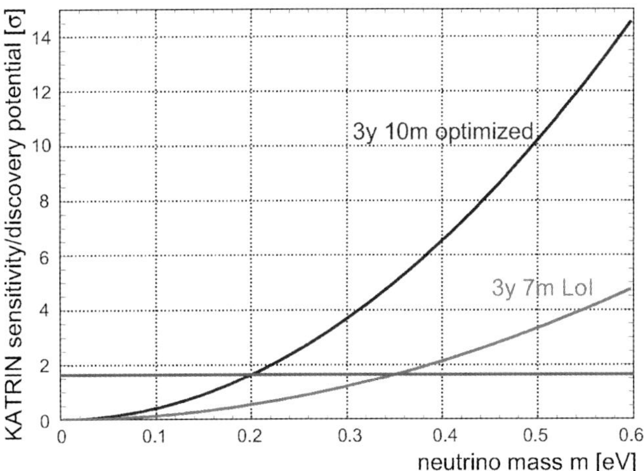

Fig. 7. KATRIN's discovery potential or sensitivity in units of the total uncertainty σ for a 3 years measurement as function of the neutrino mass for the improved KATRIN setup with the 10 m spectrometer (black) and the previous KATRIN design with the 7 spectrometer from the KATRIN Letter of Intent [21] (gray). For a given neutrino mass the y-axis shows the difference of the measured neutrino mass to zero in units of the total standard deviation σ. The horizontal line shows the upper limit with 90% C.L. in case that no neutrino mass is found.

To the contrary, a non-zero neutrino mass of $0.30\,\mathrm{eV}/\mathrm{c}^2$ would be detected with 3σ significance, a mass of $0.35\,\mathrm{eV}/\mathrm{c}^2$ even with 5σ.

The design of the experiment is nearly finished and a detailed description was documented [22]. Some parts (*e.g.* the pre-spectrometer) have been set up already. Four major components have been ordered, the windowless gaseous tritium source WGTS, the differential pumping system, the main spectrometer vessel and the helium liquefier. Many dedicated test experiments are being performed at different places to investigate the inner tritium loop, the cryo-trapping, methods to improve the vacuum conditions, new background suppression methods, calibration sources, detector and data acquisition, etc. The full setup of the KATRIN experiment will be finished and data taking will start in 2008.

4. Conclusions

Neutrino oscillation experiments have pointed to new physics beyond the Standard Model by proving that neutrinos mix and that they have non-zero neutrino masses. The next goal is to determine the absolute scale of the neutrino mass due to its high importance for particle physics, astrophysics and cosmology.

Among various ways to address the absolute neutrino mass scale the investigation of the shape of a β decay spectrum near its endpoint is the only model-independent method. This direct method is complementary to the search for the neutrinoless double β decay and to the information from astrophysics and cosmology. Possible isotopes are ^{187}Re and tritium. The first experiments with the former isotope resulted in a limit on the neutrino mass of $15\,\mathrm{eV}/\mathrm{c}^2$. The tritium β decay experiments at Mainz and Troitsk have been finished yielding upper limits of about $2\,\mathrm{eV}/\mathrm{c}^2$. The new KATRIN experiment – also investigating the tritium β spectrum – will enhance the sensitivity further by one order of magnitude down to $0.2\,\mathrm{eV}/\mathrm{c}^2$ to be able to probe the cosmological relevant neutrino mass region and to distinguish quasi-degererate from hierarchical neutrino mass scenarios.

Acknowledgements

The work of the author for the KATRIN experiment is supported by the German Bundesministerium für Bildung und Forschung and within the virtual institute VIDMAN by the Helmholtz Gemeinschaft.

References

1. Suzuki, Y., McDonald, A., Suzuki, A. and Conrad, J., these proceedings.
2. Mohapatra, R., these proceedings.
3. Ramond, P., these proceedings.
4. King, S., these proceedings.
5. Tegmark, M., these proceedings.
6. Allen, S. W. *et al.*, arXiv:astro-ph/0303076.
7. Beacom, J. F. *et al.*, arXiv:astro-ph/0404585.
8. Klapdor-Kleingrothaus, K. V. *et al.*, Phys. Lett. B **586**, 198 (2004).
9. Eidelman, S. *et al.*, Phys. Lett. B **592**, 1 (2004).
10. Weinheimer, Ch., ch. 2 of "Massive Neutrinos", (ed. G. Altarelli and K. Winter), "Springer Tracts in Modern Physics", (pringer, 2003), p. 25–52.
11. Picard, A. *et al.*, Nucl. Instr. Meth. B **63**, 345 (1992).
12. Lobashev, V. M., Nucl. Inst. Meth. A **240**, 305 (1985).
13. Fleischmann, L. *et al.*, Eur. Phys. J. B **16**, 521 (2000).
14. Aseev, V. N. *et al.*, Eur. Phys. J. D **10**, 39 (2000).
15. Bornschein, B. *et al.*, J. Low Temp. Phys. **131**, 69 (2003).
16. Kraus, C. *et al.*, hep-ex/0412056, Eur. Phys. J. C **40**, 447 (2005).
17. Kolos, W. *et al.*, Phys. Rev. A **37**, 2297 (1988).
18. Lobashev, V. M. *et al.*, Phys. Lett. B **460**, 227 (1999).
19. Lobashev, V. M., Nucl. Phys. A **719**, 153c (2003).
20. Sisti, M. *et al.*, Nucl. Instr. Meth. A **520**, 125 (2004).
21. Osipowicz, A. *et al.* (KATRIN Letter of Intent), arXiv:hep-ex/0109033.
22. Angrik, J. *et al.* (KATRIN Design Report 2004), FZK Scientific Report 7090.

Physica Scripta. Vol. T121, 172–177, 2005

Flavor Theories and Neutrino Masses

Pierre Ramond[1]

Institute for Fundamental Theory, Department of Physics, University of Florida, Gainesville FL 32611, USA

Received January 24, 2005; accepted February 18, 2005

PACS numbers: 12.15Ff, 14.60Kt, 14.60Pq

Abstract

The wealth of new data on neutrinos is easily incorporated in the Electroweak Theory. In this "ν-Standard Model", lepton mixings are distinguished from quark mixings by a unitary matrix coming from the Seesaw. We catalog models in terms of the number of large angles (one or two) in that matrix. Pati-Salam unification implies Cabibbo effects in lepton mixings (MNS). Without such small mixings, the Solar and Atmospheric angles may well be the same, and the CHOOZ angle could vanish: in a wide class of flavor-symmetric models, it is of $\mathcal{O}(\lambda/\sqrt{2})$. We discuss a new approximate chiral family symmetry that preserves Froggat-Nielsen, and relates the $\bar{5}$ of the second and third families, but not the particles in the **10**'s. If exact, the electron and down quark are both massless and the atmospheric angle is maximal. We conclude by discussing possible Wolfenstein parametrizations of the MNS matrix, assuming various types of "Cabibbo Flops".

1. Introduction

The triumphs of the Standard Model [1] can all be traced to its underlying chirality-preserving gauge symmetries. Chiral symmetry-breaking Yukawa couplings are chosen *à la carte*, to reproduce the masses and decay channels of its fermions. Their coupling strengths are adjusted to reproduce the data set, via the Higgs mechanism. Sadly, the recent bounty of neutrino data has not yet brought clarity to this baffling problem, and today, we are no closer to an understanding of the breaking of chirality at fundamental scales.

In the absence of any underlying principle, no satisfactory and credible theory of these Yukawa couplings exists; only glimpses of theory can be found in the literature, carefully crafted to relate a subset of measured parameters. Pati-Salam unification [2], as suggested by the gauge sector, can be applied to Yukawa couplings, yielding some understanding of the relation between quark and lepton masses and mixings. In spite of some qualitative successes, it translates the complications of the data into a plethora of Higgs particles.

In desperation, Froggatt and Nielsen [3] viewed some small Yukawa couplings as *effective couplings* generated by some unknown physics at a larger scale. Mass and mixing hierarchies stem from the dimensions of these higher-order operators. For book-keeping they introduced new Abelian charges: a Yukawa coupling with higher FN charges would be more suppressed than one with lesser FN charges. It could even be that there are *no* Yukawa (chiral-symmetry breaking) couplings at tree level [4], and *all* are generated by the high-scale Froggatt-Nielsen physics.

In the following, we first review the new data and its implications. We then describe the Seesaw modifications [5, 6] to the Standard Model needed to accomodate these data. We emphasize the appearance of a new unitary matrix in the lepton

mixing sector, in terms of which we can catalog models of flavor mixings. Then, we introduce basic notions from Grand-Unification to narrow down the relations between the quark and lepton mixing matrices. When viewed in the context of Froggat and Nielsen, a new flavor symmetry suggests itself, designed to determine some of the prefactors in those schemes. Finally, we examine possible *Cabibbo Flops*, intrusions of Cabibbo-size effects in the lepton mixings and masses [7].

2. The data

There is now "Physics Beyond the Standard Model", as evidenced by the observation of oscillations among neutrino species. Thirty five years of experiments on solar neutrinos, HOMESTAKE [8], GALLEX [9], SAGE [10], SUPERK [11] and SNO [12], yield

$$\Delta m_\odot^2 = |m_{\nu_1}^2 - m_{\nu_2}^2| \sim 7. \times 10^{-5}\,\mathrm{eV}^2,$$

with corroborating evidence on antineutrinos [13]. Neutrinos born in Cosmic ray collisions [14], and on Earth [15] give

$$\Delta m_\oplus^2 = |m_{\nu_2}^2 - m_{\nu_3}^2| \sim 3. \times 10^{-3}\,\mathrm{eV}^2.$$

The best bound to their absolute value of the masses comes from WMAP [16]

$$\sum_i m_{\nu_i} < .71\,\mathrm{eV}.$$

These experimental findings are not sufficient to determine fully the mass patterns. One oscillates between three patterns, *hierarchy*,

$$|m_{\nu_1}| < |m_{\nu_2}| \ll |m_{\nu_3}|,$$

inverse hierarchy

$$|m_{\nu_1}| \simeq |m_{\nu_2}| \gg |m_{\nu_3}|,$$

and *hyperfine*

$$|m_{\nu_1}| \simeq |m_{\nu_2}| \simeq |m_{\nu_3}|.$$

The mixing patterns provide some surprises, since it contains one small angle and two large angles. In terms of the MNS lepton mixing matrix,

$$\begin{pmatrix} \cos\theta_\odot & \sin\theta_\odot & \epsilon \\ -\cos\theta_\oplus \sin\theta_\odot & \cos\theta_\oplus \cos\theta_\odot & \sin\theta_\oplus \\ \sin\theta_\oplus \sin\theta_\odot & -\sin\theta_\oplus \cos\theta_\odot & \cos\theta_\oplus \end{pmatrix},$$

the various experiments yield

$$\sin^2 2\theta_\oplus > 0.85\,, \qquad 0.30 < \tan^2\theta_\odot < 0.65,$$

while there is only a limit [17] on the third so-called CHOOZ angle

$$|\epsilon|^2 < 0.05.$$

[1] Supported in part by the US Department of Energy under grant DE-FG02-97ER41029.

Thus neutrinos have different, albeit tiny, masses. Of course, more experiments are need, as theorists have not yet proposed convincing arguments for their origin. In particular, one would like to know their nature (Majorana or Dirac), the sign of Δm^2, the value of the CHOOZ angle, and if CP-violation is observable in the lepton sector. Answers to these questions have far reaching consequences, in particular for leptogenesis [18], and even for the fundamental nature of neutrino couplings.

3. The ν-standard model

Although quarks and leptons partake equally in electroweak interactions, there are subtle differences in how their mixings and masses are generated. Quark masses and mixings arise solely from the diagonalization of Yukawa matrices generated by the $\Delta I_W = \frac{1}{2}$ breaking of electroweak symmetry. For charge $2/3$

$$\mathcal{U}_{2/3} \begin{pmatrix} m_u & 0 & 0 \\ 0 & m_c & 0 \\ 0 & 0 & m_t \end{pmatrix} \mathcal{V}_{2/3}^\dagger,$$

and charge $-1/3$

$$\mathcal{U}_{-1/3} \begin{pmatrix} m_d & 0 & 0 \\ 0 & m_s & 0 \\ 0 & 0 & m_b \end{pmatrix} \mathcal{V}_{-1/3}^\dagger,$$

resulting in the observable CKM matrix

$$\mathcal{U}_{CKM} \equiv \mathcal{U}_{2/3}^\dagger \, \mathcal{U}_{-1/3}.$$

Phenomenologically it is approximately a unit matrix with small Cabibbo ($\theta_c \sim 13°$) corrections. This implies similar family mixings for up-like and down-like quarks. Quark masses are highly hierarchical, and may be viewed as powers of the Cabibbo angle. The charged lepton Yukawa matrix

$$\mathcal{U}_{-1} \begin{pmatrix} m_e & 0 & 0 \\ 0 & m_\mu & 0 \\ 0 & 0 & m_\tau \end{pmatrix} \mathcal{V}_{-1}^\dagger$$

also stems from $\Delta I_W = \frac{1}{2}$ electroweak breaking, and displays a similar hierarchical structure.

To obtain neutrino masses in the Standard Model, it is simplest to add one right-handed neutrino for each family. This yields another $\Delta I_W = \frac{1}{2}$ Yukawa matrix

$$\mathcal{U}_0 \begin{pmatrix} m_1 & 0 & 0 \\ 0 & m_2 & 0 \\ 0 & 0 & m_3 \end{pmatrix} \mathcal{V}_0^\dagger.$$

If neutrino masses arise solely from these electroweak symmetry breaking couplings, it is difficult to understand the extraordinary mass gap between charged and neutral leptons. However, we have not discussed the masses of the right-handed neutrinos. Their masses are of Majorana type, since they carry no gauge quantum numbers to forbid its disgonal elements (unlike charged leptons, say), and necessarily violate total lepton number.

In the context of effective field theories, one expects their masses to be of the order of lepton number breaking. Total lepton number-violating processes have never been seen resulting in a bound from neutrinoless double β decay experiments. Either they are very large or zero.

If they are zero, the analysis proceeds as in the quark sector, and the observable MNS lepton mixing matrix is just

$$\mathcal{U}_{MNS} \equiv \mathcal{U}_{-1}^\dagger \, \mathcal{U}_0.$$

As for the quarks, it would be generated solely from the isospinor breaking of electroweak symmetry, but this is not likely since the quark and lepton mixing patterns are so different.

In the belief that global symmetries are not fundamental (black hole fodder), we expect their masses to be as large as the Standard Model's cut-off, since they are unprotected by gauge symmetries. This yields the Seesaw where large right-handed neutrino masses engender tiny electroweak neutrino masses, the latter being suppressed by the ratio of the two scales

$$\frac{\Delta I_W = \frac{1}{2}}{\Delta I_W = 0}.$$

Hence the detection of small neutrino masses suggest a large cut-off scale for the ν- Standard Model. The electroweak neutrino mass matrix is now

$$\mathcal{M}_{Seesaw}^{(0)} = \mathcal{M}_{Dirac}^{(0)} \frac{1}{\mathcal{M}_{Majorana}^{(0)}} \mathcal{M}_{Dirac}^{(0)\,T},$$

which we can rewrite as

$$\mathcal{M}_{Seesaw}^{(0)} = \mathcal{U}_0 \, \mathcal{C} \, \mathcal{U}_0^T,$$

in terms of the central matrix [19]

$$\mathcal{C} = \mathcal{D}_0 \, \mathcal{V}_0^\dagger \frac{1}{\mathcal{M}_{Majorana}^{(0)}} \mathcal{V}_0^* \, \mathcal{D}_0.$$

It is diagonalized by the unitary matrix $\mathcal{F}$

$$\mathcal{C} = \mathcal{F} \, \mathcal{D}_\nu \, \mathcal{F}^T,$$

where the mass eigenstates produced in β-decay are (unimaginatively labelled "1", "2", "3")

$$\mathcal{D}_\nu = \begin{pmatrix} m_{\nu_1} & 0 & 0 \\ 0 & m_{\nu_2} & 0 \\ 0 & 0 & m_{\nu_3} \end{pmatrix}.$$

The effect of the Seesaw is to add the unitary $\mathcal{F}$ matrix to the MNS lepton matrix

$$\mathcal{U}_{MNS} = \mathcal{U}_{-1}^\dagger \, \mathcal{U}_0 \, \mathcal{F}.$$

This framework enables us to recast theoretical questions in terms of $\mathcal{F}$. In particular, where do the large angles come from? We can then distinguish models in terms of the number of large angles $(0, 1, 2)$ contained in $\mathcal{F}$.

4. A pinch of Grand Unification

It would be irresponsible to consider extensions of the Standard Model outside the framework of Grand Unification [2, 20, 21, 22]. In its simplest form it generates relations between the $\Delta I_W = \frac{1}{2}$ quark and lepton Yukawa matrices.

In $SU(5)$, the charge $-1/3$ and charge -1 Yukawa matrices are family-transposes of one another.

$$\mathcal{M}^{(-1/3)} \sim \mathcal{M}^{(-1)\,T}.$$

In $SO(10)$, it is the charge $2/3$ Yukawa matrix that is related to the neutral Dirac matrix

$$\mathcal{M}^{(2/3)} \sim \mathcal{M}_{Dirac}^{(0)}.$$

These grand unified patterns result in naive expectations for the unitary matrices

$$\mathcal{U}_{-1/3} \sim \mathcal{V}_{-1}^*; \qquad \mathcal{U}_{2/3} \sim \mathcal{U}_0.$$

They impact on the observable mixings, since just this modicum of grand-unification implies

$$\mathcal{U}_{MNS} = \mathcal{U}_{-1}^\dagger \, \mathcal{U}_0 \, \mathcal{F}$$
$$\sim \mathcal{U}_{-1}^\dagger \, \mathcal{U}_{-1/3} \, \mathcal{U}_{CKM}^\dagger \, \mathcal{F}$$
$$\sim \left(\mathcal{V}_{-1/3}^T \, \mathcal{U}_{-1/3} \right) \, \mathcal{U}_{CKM}^\dagger \, \mathcal{F}.$$

These simple ideas generate suggestive relation between the CKM and MNS mixing matrices. Also, the $\mathcal{F}$-matrix notation delineates the expected departures in mixing patterns between the two sectors. Hence two wide classes of models:

I-) Family-Symmetric models where the $\mathcal{M}_{-1/3}$ Yukawa matrix is symmetric (at least in terms of Cabibbo mixings), that is

$$\mathcal{U}_{-1/3} = \mathcal{V}_{-1/3}^*.$$

This results in the particularly simple relation between the two mixing matrices

$$\mathcal{U}_{MNS} = \mathcal{U}_{CKM}^\dagger \, \mathcal{F}.$$

In these models, the two large angles are necessarily contained in $\mathcal{F}$. A (3×3) unitary matrix with one small and two large angles suggests non-trivial structures for the matrix it diagonalizes, hinting at a Non-Abelian family symmetry in the neutral sector.

Interestingly, these models provide a testable prediction for the size of the CHOOZ angle. With a family-symmetric charge $-1/3$ matrix, the MNS matrix reads

$$\mathcal{U}_{MNS} = \mathcal{U}_{CKM}^\dagger$$
$$\times \begin{pmatrix} \cos\theta_\odot & \sin\theta_\odot & \lambda^\gamma \\ -\cos\theta_\oplus \sin\theta_\odot & \cos\theta_\oplus \cos\theta_\odot & \sin\theta_\oplus \\ \sin\theta_\oplus \sin\theta_\odot & -\sin\theta_\oplus \cos\theta_\odot & \cos\theta_\oplus \end{pmatrix},$$

where we have chosen to fill the zero in the $\mathcal{F}$ matrix by a Cabibbo effect, with γ equal or greater than one. It follows that

$$\theta_{13} \sim \lambda \sin\theta_\oplus \sim \frac{1}{\sqrt{2}} \lambda.$$

It will be interesting to see if this definite prediction of type I models, $\theta_{13} \sim 7 - 9°$, is borne out in future experiments.

II-) Family-Skewed models where the $\mathcal{M}_{-1/3}$ Yukawa matrix is different from its transpose. One can make a compelling arguments for at least one large angle to reside in $\mathcal{U}_{-1}$. If we extend the Wolfenstein [23] expansion of the CKM matrix in powers of the Cabibbo angle λ to include quark mass ratios

$$\frac{m_s}{m_b} \sim \lambda^2 \qquad \frac{m_d}{m_b} \sim \lambda^4,$$

we find the charge $-1/3$ Yukawa matrix

$$\mathcal{M}^{(-1/3)} = \begin{pmatrix} \lambda^4 & \lambda^3 & \lambda^3 \\ \lambda^? & \lambda^2 & \lambda^2 \\ \lambda^? & \lambda^? & 1 \end{pmatrix}.$$

The exponents on the lower diagonal elements are left undetermined. They are determined only by further theoretical assumptions. They can be generated from Froggatt-Nielsen [3] charges, producing a highly skewed Yukawa matrix

$$\mathcal{M}^{(-1/3)} = \begin{pmatrix} \lambda^4 & \lambda^3 & \lambda^3 \\ \lambda^3 & \lambda^2 & \lambda^2 \\ \lambda^1 & 1 & 1 \end{pmatrix}.$$

In the limit of no Cabibbo mixing,

$$\mathcal{M}^{(-1/3)} \approx \begin{pmatrix} 0 & 0 & 0 \\ 0 & 0 & 0 \\ 0 & a & b \end{pmatrix} + \mathcal{O}(\lambda),$$

and

$$\mathcal{U}_{MNS} = \begin{pmatrix} 1 & 0 & 0 \\ 0 & \cos\theta_\oplus & \sin\theta_\oplus \\ 0 & -\sin\theta_\oplus & \cos\theta_\oplus \end{pmatrix} \mathcal{F},$$

where

$$\tan\theta_\oplus = \frac{a}{b},$$

yields an unsuppressed "atmospheric angle". Significantly, such a "large" (not necessarily maximal) atmospheric angle had been anticipated long ago on the basis of $SO(10)$ [24] and more recently within the Froggatt-Nielsen framework [25]. In these models, $\mathcal{F}$ need contain only one large angle, which is very natural, but they yield no generic prediction for the CHOOZ angle.

5. Right-handed hierarchy

A puzzling feature of the data is that $\mathcal{F}$ contains at least one large angle. After all, $\mathcal{F}$ diagonalizes a matrix which includes the presumably hierarchical neutral Dirac Yukawa matrix, generated by isospinor electroweak breaking. This suggests special restrictions put upon the Majorana mass matrix of the right-handed neutrinos. We want to illustrate this point by looking at a 2×2 two-families case [19], and write

$$\mathcal{D}_0 = m \begin{pmatrix} a\,\lambda^\beta & 0 \\ 0 & 1 \end{pmatrix},$$

and define M_1, M_2 to be the eigenvalues of the right-handed neutrino's Majorana mass matrix. This matrix can be diagonalized by a large mixing angle in one of two cases:

– Its matrix elements have similar orders of magnitude $\mathcal{C}_{11} \sim \mathcal{C}_{22} \sim \mathcal{C}_{12}$, in which case we find that

$$\frac{M_1}{M_2} \sim \lambda^{2\beta},$$

suggesting a doubly *correlated hierarchy* between the $\Delta I_W = 0$ and $\Delta I_W = \frac{1}{2}$ Sectors. This agrees well with grand-unified models such as $SO(10)$ and E_6, where each right-handed neutrinos is part of a family.

– A large mixing angle can occur if the diagonal elements are much smaller than the diagonal ones, that is $\mathcal{C}_{11}, \mathcal{C}_{22} \ll \mathcal{C}_{12}$. Then we find

$$\frac{\lambda^\alpha m^2}{\sqrt{-M_1 M_2}} \begin{pmatrix} 0 & a \\ a & 0 \end{pmatrix}.$$

Hence maximal mixing may infer that some of the right-handed neutrinos are Dirac partners of one another, leading to conservation in the right-handed mass matrix, of a relative lepton number $L_1 - L_2$.

6. Beyond Froggatt-Nielsen

The Froggat-Nielsen approach relates only the level of suppression of a Yukawa operator to its (FN) charge, and it is not a full theory. Without further specification or possibly symmetry it is not possible to determine or even relate the prefactors for the Yukawa matrix elements.

When adding new symmetries to the FN formalism, it is convenient to consider two types:

- Symmetries which alter the Froggat-Nielsen charges. Since they do not commute with the FN, they imply a Non-Abelian flavor algebra with the FN charges as part of their Cartan subalgebras. Although very interesting, we leave their study to future investigation, and rather concentrate on
- Symmetries which do not change the FN charges. In that case, a Non-Abelian structure is not required, although it can be incorporated by increasing the rank of the flavor algebra. Such symmetries can infer relations between matrix elements with the same FN charges.

The Yukawa matrix we have already considered,

$$\mathcal{M}^{(-1/3)} = \begin{pmatrix} \lambda^4 & \lambda^3 & \lambda^3 \\ \lambda^3 & \lambda^2 & \lambda^2 \\ \lambda^1 & 1 & 1 \end{pmatrix},$$

corresponds to FN charges

$$X_{FN} = -\mathcal{B}(2,-1,-1) + \eta(2,0,-2),$$

where $\mathcal{B}$ is Baryon number, and $\eta = 1$ for $\mathbf{Q}$ and u and zero for d. For this assignment, this type of symmetry can relate the right-handed down quarks of the second and third families. This symmetry can either be continuous or discrete.

Consider the phase symmetry

$$\overline{\mathbf{d}}_2 + i\overline{\mathbf{d}}_3 \to e^{i\alpha}(\overline{\mathbf{d}}_2 + i\overline{\mathbf{d}}_3),$$

and its conjugate. It can be realized if only that combination appears in the Yukawa couplings, yielding the couplings

$$y_i\mathbf{Q}_i(\overline{\mathbf{d}}_2 + i\overline{\mathbf{d}}_3)H_d,$$

where the y_i are Yukawa couplings. This symmetry requires the matrix elements of the second and third column to be proportional, yielding a massless down quark. It is not our purpose here to build a complete model with this symmetry, which would take us far afield and require its effect on the Higgs and on the first family down quarks. Rather we note that if we apply it to the lepton sector, in accordance with the usual $SU(5)$ identification, we are led to leptonic couplings of the form

$$z_i L_i(\overline{e}_2 + i\overline{e}_3)H_d,$$

which yield a massless charged lepton. In this case, the atmospheric mixing angle coming from the 2–3 family mixing is maximal. Alternatively, one could have simply required the discrete symmetry between the second and third families of the $\overline{\mathbf{5}}$ of $SU(5)$. However, this symmetry must not be extended to the neutral lepton matrix: it can be destroyed since the angle cancels against itself in the Seesaw formula. An elucidation of these features and more detailed analysis of these problems will appear elsewhere [7].

7. Cabibbo flop

Grand-Unification, even in its simplest form, implies Cabibbo-sized effects in the MNS matrix [26]. In the quark sector, Cabibbo mixing is the strongest between the first and second family. Applied to the lepton sector, the solar angle could also be maximal, with a Cabibbo correction of $13°$ [27, 28].

Recently, we [7] have been exploring possible Wolfenstein parametrizations of the MNS matrix, in the hope that some

regularity might emerge from the data, once Cabibbo effects are taken into account.

We begin by asking if the limit $\theta_c \to 0$ makes any theoretical sense. To simplify matters, assume there is only one small parameter in the flavor sector; then the quark and charged lepton masses of the first two families are zero. In the same limit, $\mathcal{U}_{CKM} = 1$, and there are no neutral flavor changes. Of course the mixing between the first two families is undetermined.

We do not know $\mathcal{U}_{MNS}$ in that limit, the starting point of a Wolfenstein parametrization for the lepton mixing matrix.

The measured values of the lepton mixing angles are

$$\theta_\odot = 32.5°\,^{+2.4°}_{-2.3°}; \qquad \theta_\oplus = 45.00°\,^{+10°}_{-10°}; \qquad \theta_{CHOOZ} < 13°.$$

The solar angle is well measured, but the atmospheric angle is not, and could very well be non-maximal. Furthermore, their values could be affected by Cabibbo flop of $\pm13°$, and the CHOOZ angle could well be a Cabibbo effect. Our starting point is

$$\mathcal{U}_{MNS} = \begin{pmatrix} \cos\eta_\odot & \sin\eta_\odot & 0 \\ -\cos\eta_\oplus\sin\eta_\odot & \cos\eta_\oplus\cos\eta_\odot & \sin\eta_\oplus \\ \sin\eta_\oplus\sin\eta_\odot & -\sin\eta_\oplus\cos\eta_\odot & \cos\eta_\oplus \end{pmatrix} + \cdots,$$

with a range of initial angles

$$15° < \eta_\odot < 45°; \qquad 30° < \eta_\oplus < 60°.$$

We write the Wolfenstein expansion of the MNS Matrix in the form

$$\mathcal{U}_{MNS} \equiv \mathcal{W} + \mathcal{O}(\lambda),$$

where the starting matrix is split in two parts, showing the large angles

$$\mathcal{W} = \mathcal{W}_\oplus\,\mathcal{W}_\odot,$$

with

$$\mathcal{W}_\oplus = \begin{pmatrix} 1 & 0 & 0 \\ 0 & \cos\eta_\oplus & -\sin\eta_\oplus \\ 0 & \sin\eta_\oplus & \cos\eta_\oplus \end{pmatrix},$$

$$\mathcal{W}_\odot = \begin{pmatrix} \cos\eta_\odot & \sin\eta_\odot & 0 \\ -\sin\eta_\odot & \cos\eta_\odot & 0 \\ 0 & 0 & 1 \end{pmatrix}.$$

We introduce Cabibbo flop through the unitary matrix

$$\mathcal{V} = I + \Delta(\lambda),$$

with $\Delta(0) = 1$. Unlike the quark sector it does not commute with the starting matrix

$$[\mathcal{W}, \mathcal{V}(\lambda)] \neq 0.$$

This means that Cabibbo effects from the left and from the right or even in between the two starting matrices are not equivalent. Hence we consider basic flops

- Left $\mathcal{U}_{MNS} = \mathcal{V}(\lambda)\mathcal{W},$
- Right $\mathcal{U}_{MNS} = \mathcal{W}\mathcal{V}(\lambda),$
- Middle $\mathcal{U}_{MNS} = \mathcal{W}_\oplus\mathcal{V}(\lambda)\mathcal{W}_\odot$

and we can have one $\mathcal{O}(\lambda)$ correction (single flop), or two (double flop). The present data is not sufficient to single out a particular Wolfenstein parametrization, but the hope is that by considering possible Cabibbo effects on various starting matrices, generic features suggestive of flavor patterns might become obvious. In particular, they would restrict the size of the CHOOZ angle and of the CP-violation.

To illustrate these points, consider the effect of flop matrices, shown here to $\mathcal{O}(\lambda^3)$,

$$\mathcal{V}_{12}(\lambda) = \begin{pmatrix} 1 - \dfrac{a^2}{2}\lambda^2 & a\lambda & b\lambda^2 \\ -a\lambda & 1 - \dfrac{a^2}{2}\lambda^2 & 0 \\ -b\lambda^2 & 0 & 1 \end{pmatrix},$$

$$\mathcal{V}_{23}(\lambda) = \begin{pmatrix} 1 & 0 & b\lambda^2 \\ 0 & 1 - \dfrac{a^2}{2}\lambda^2 & a\lambda \\ -b\lambda^2 & -a\lambda & 1 - \dfrac{a^2}{2}\lambda^2 \end{pmatrix},$$

$$\mathcal{V}_{\text{double}}(\lambda) = \begin{pmatrix} 1 - \dfrac{a^2}{2}\lambda^2 & a\lambda & (b + \dfrac{aa'}{2})\lambda^2 \\ -a\lambda & 1 - \dfrac{a^2 + a'^2}{2}\lambda^2 & a'\lambda \\ \left(\dfrac{aa'}{2} - b\right)\lambda^2 & -a'\lambda & 1 - \dfrac{a'^2}{2}\lambda^2 \end{pmatrix},$$

where we have limited ourselves to $a = \pm 1$; $a' = \pm 1$; $0.8 < b < 1.2$. For instance, a left single flop $\mathcal{V}_{23}$, yields values for the starting angles that are different from the data,

$\eta_\odot^\circ$	$\eta_\oplus^\circ$	$\theta_\odot^\circ$	$\theta_\oplus^\circ$	θ_{13}°
30	30	∼31	43	.06–.4
30	60	31–32	∼47	.6–2.5

Right single flops with $\mathcal{V}_{23}$ and $\mathcal{V}_{12}$ produce:

$\eta_\odot^\circ$	$\eta_\oplus^\circ$	$\theta_\odot^\circ$	$\theta_\oplus^\circ$	θ_{13}°
30	60	30–31	48–50	3–10
15	45	∼30.3	44–45	2–4
45	45	∼32	44–46	1–3

A right double flop with $\mathcal{V}_{\text{double}}$:

$\eta_\odot^\circ$	$\eta_\oplus^\circ$	$\theta_\odot^\circ$	$\theta_\oplus^\circ$	θ_{13}°
15	60	∼30.3	45–51	1–6
45	60	32	48–53	6–12

Finally a left double flop with $\mathcal{V}_{\text{double}}$:

$\eta_\odot^\circ$	$\eta_\oplus^\circ$	$\theta_\odot^\circ$	$\theta_\oplus^\circ$	θ_{13}°
45	30	∼34	40–46	5–8

We see that double flops can produce a larger CHOOZ angle. Also a left flop from a family-symmetric Yukawa,

$$\mathcal{U}_{MNS} = \begin{pmatrix} 1 & \lambda & \lambda^3 \\ \lambda & 1 & \lambda^2 \\ \lambda^3 & \lambda^2 & 1 \end{pmatrix}$$

$$\begin{pmatrix} \cos\eta_\odot & \sin\eta_\odot & 0 \\ -\cos\eta_\oplus\sin\eta_\odot & \cos\eta_\oplus\cos\eta_\odot & \sin\eta_\oplus \\ \sin\eta_\oplus\sin\eta_\odot & -\sin\eta_\oplus\cos\eta_\odot & \cos\eta_\oplus \end{pmatrix},$$

yields $\eta_\odot \sim 40°$; $\eta_\oplus \sim 45°$; $\theta_{13} \sim 0.7\lambda \sim 9°$, which we have already seen. Finally we note that CP-violation effects can be much larger than in the quark sector. This is because the CP-violating lepton invariant [29, 30] is

$$J \sim (\lambda - \lambda^3)\sin\delta,$$

to be compared with that in the quark sector which is of order λ^6. If the limit of zero Cabibbo mixing is meaningful for theory, analyses of the type we have just presented will assume some importance. One important conclusion emerges: precision measurements of the MNS matrix is quite important for theory.

8. Conclusions

The recent harvest of neutrino mass and mixing data has been, if anything, spectacular. Alas, such brilliance has not been matched by theorists, this author included. Much work needs to be done before a credible theory of flavor is devised. Through the Seesaw mechanism, the new data must be translated in terms of the physics of right-handed neutrinos. With no electroweak quantum numbers, they could hold the key to the flavor puzzles and could well prove to be (like proton decay, if ever observed) a window to Planck scale physics. Already, the large solar angle suggest that flavor hierarchy is independent of electroweak breaking, and naturally occurs at those scales.

Acknowledgments

I wish to thank Professor P. O. Hulth and the organizing committee for giving me the opportunity to discuss physics in the intellectually rich company of the participants and also in the plush surroundings of Haga Slott.

References

1. Glashow, S. L., Nucl. Phys. **22**, 579 (1961); Weinberg, S., Phys. Rev. Lett. **19**, 1264 (1967); Salam, A., "Svartholm: Elementary Particle Theory, Proceedings of The Nobel Symposium" (1968), Lerum, Sweden.
2. Pati, J. C. and Salam, A., Phys. Rev. D**10**, 275 (1974).
3. Froggatt, C. and Nielsen, H. B., Nucl. Phys. B **147**, 277 (1979).
4. King, S. F. and Ross, G. G., Phys. Lett. B**574**, 239 (2003).
5. Gell-Mann, M., Ramond, P. and Slansky, R., in Sanibel Talk, CALT-68-709, Feb. 1979, hep-ph/9809459 (retroprint), and in "Supergravity" (North Holland, Amsterdam 1979); Yanagida, T., in Proc. Workshop on Unified Theory and Baryon Number of the Universe, KEK, Japan, Feb. (1979).
6. Minkowski, P., Phys. Lett. B**67**, 421 (1977), in which the seesaw matrix is first proposed without the link to Planck scale physics.
7. Datta, A., Everett, L. and Ramond, P., in preparation.
8. Davis Jr., R., Harmer, D. S. and Hoffman, K. C., Phys. Rev. Lett. **20**, 1205 (1968).
9. Hampel, W. *et al.*, GALLEX collaboration, Phys. Lett. B**447**, 127 (1999).
10. Abdurashitov, J. N. *et al.*, SAGE collaboration, Phys. Rev. C**60**, 055801 (1999).
11. The Super-Kamiokande Collaboration, Phys. Lett. B**539**, 179 (2002).
12. The SNO Collaboration, Phys. Rev. Lett. **89**, 011301 (2002).
13. The KamLAND Collaboration, Phys. Rev. Lett. **90**, 021802 (2003).
14. The Super-Kamiokande Collaboration, Phys. Rev. Lett. **85**, 3999 (2000).
15. The K2K Collaboration, Phys. Rev. Lett. **90**, 041801 (2003).
16. Bennett, C. L. *et al.*, Astrophys. J. Suppl. **148**, 1 (2003); Spergel, D. N. *et al.*, Astrophys. J. Suppl. **148**, 175 (2003).
17. Apollonio, M. *et al.*, Phys. Lett. B **338**, 383 (1998); Phys. Lett. B **420**, 397 (1998).

18. Fukugita, M. and Yanagida, T., Phys. Lett. B **174**, 45 (1986).
19. Datta, A., Ling, F-S. and Ramond, P., Nucl. Phys. B **671**, 383 (2003).
20. Georgi, H. and Glashow, S. L., Phys. Rev. Lett. **32**, 438 (1974).
21. Fritzsch, H. and Minkowski, P., Ann. Phys. **93**, 193 (1975); Georgi, H., Invited Talk at Williamsburg Conference, (1975).
22. Gürsey, F., Ramond, P. and Sikivie, P., Phys. Lett. **B60**, 177 (1976).
23. Wolfenstein, L., Phys. Rev. Lett. **51**, 1945 (1983).
24. Harvey, J., Ramond, P. and Reiss, D., Phys. Lett. B **92**, 309 (1980); Nucl. Phys. B**199**, 223 (1982).
25. Irges, N., Lavignac, S. and Ramond, P., Phys. Rev. D **58**, 035003 (1998).
26. Ramond, P., "Windows to Planck Physics", [HEP-PH 0401001], in "Quantum Theory and Symmetries", September (2003), Cincinnati, (P. C. Argyres, T. J. Hodges, F. Mansouri, J. J. Scanio, P. Suranyi and L. C. R. Wijewardhana, editors), (World Scientific).
27. Ramond, P., "Neutrino Mass and Seesaw Mechanism", Fujihara Seminar at KEK, Japan, February (2004)." Also "Seesaw and the Riddle of Mass", Institut Henri Poincaré, Paris June (2004).
28. Minakata, H. and Smirnov, A. Yu., hep-ph/0405088; Petcov, S. and Smirnov, A., Phys. Lett. B **322**, 109 (1994).
29. Jarlskog, C., Phys. Rev. Lett. **55**, 1039 (1985).
30. Dunietz, I., Greenberg, O. W. and Dan-di Wu., Phys. Rev. Lett. **55**, 2935 (1985).

Physica Scripta. Vol. T121, 178–184, 2005

Neutrino Mass Models and Leptogenesis

S. F. King[1,*]

[1] School of Physics and Astronomy, University of Southampton, Southampton, SO17 1BJ, United Kingdom

Received November 26, 2004; accepted December 27, 2004

PACS numbers: 14.60.Pq, 11.30.Pb, 98.80.Cq

Abstract

In this talk we show how a natural neutrino mass hierarchy can follow from the type I see-saw mechanism, and a natural neutrino mass degeneracy from the type II see-saw mechanism, where the bi-large mixing angles can arise from either the neutrino or charged lepton sector. We summarize the phenomenological implications of such natural models, and discuss the model building applications of the approach, focussing on the $SU(3) \times SO(10)$ model. We also show that in such type II models the leptogenesis asymmetry parameter becomes proportional to the neutrino mass scale, in sharp contrast to the type I case, which leads to an upper bound on the neutrino mass scale, allowing lighter right-handed neutrinos and hence making leptogenesis more consistent with the gravitino constraints in supersymmetric models.

1. Introduction

The discovery of neutrino mass and mixing at the end of the last century implies that the Standard Model is incomplete and needs to be extended, but how [1]? In attempting to answer this question, it is useful to being by classifying models in terms of the mechanisms responsible for small neutrino mass, and large lepton mixing, as a first step towards finding the Next Standard Model. Amongst the most elegant mechanisms for small neutrino mass is the see-saw mechanism [2]. However the see-saw mechanism by itself does not provide an explanation for bi-large lepton mixing for either hierarchical or denegerate neutrinos.

In this talk we discuss model independent approaches to accounting for bi-large mixing in a natural way, based on the see-saw mechanism, which are valid for both hierarchical or denegenerate neutrino mass spectra. For the case of hierarchical neutrino masses arising from the type I see-saw mechanism, it is shown how the neutrino mass hierarchy and bi-large mixing angles could originate from the sequential dominance of right-handed neutrinos [3]. It is then shown how to obtain partially degenerate neutrinos in a natural way by including a type II contribution proportional to the unit mass matrix, with the neutrino mass splittings and mixing angles controlled by type I contributions and sequential dominance [4]. The bi-large mixing angles could originate either from the neutrino or the charged lepton sector [5]. For a review see [6]. We summarize the phenomenological implications of such natural models, and discuss the model building applications of the approach, focussing on the $SU(3) \times SO(10)$ model. We also discuss leptogenesis in such type II models. The leptogenesis asymmetry parameter becomes proportional to the neutrino mass scale, in sharp contrast to the type I case, which leads to an upper bound on the neutrino mass scale, allowing lighter right-handed neutrinos and hence making leptogenesis more consistent with the gravitino constraints in supersymmetric models [7].

*Email address: sfk@hep.phys.soton.ac.uk

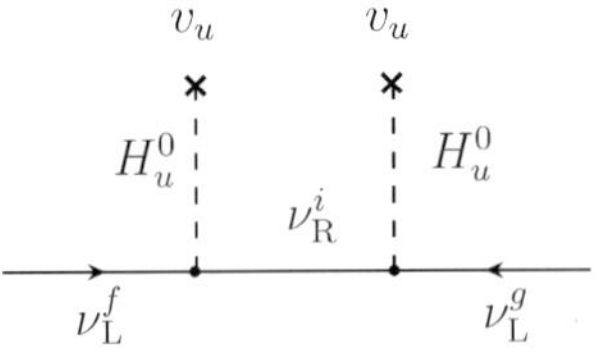

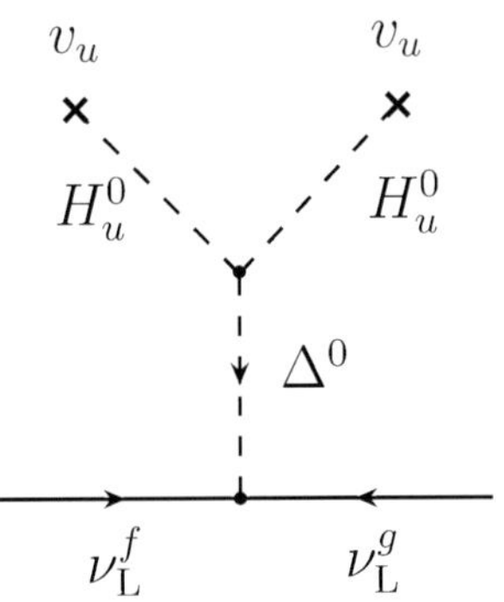

Fig. 1. Diagram (a) shows the contribution from the exchange of a heavy right-handed neutrino as in the type I see-saw mechanism. Diagram (b) illustrates the contribution from an induced vev of the triplet Δ. At low energy, they can be viewed as contributions to the effective neutrino mass operator from integrating out the heavy fields ν_R^i and Δ^0, respectively.

2. See-saw mechanism

The most commonly discussed version of the see-saw mechanism is sometimes called the type I see-saw mechanism [2]. The type I see-saw mechanism is illustrated diagramatically in Fig. 1(a).

In models with a left-right symmetric particle content like minimal left-right symmetric models, Pati-Salam models or grand unified theories (GUTs) based on SO(10), the type I see-saw mechanism is often generalized to a type II see-saw (see e.g. [8]), where an additional direct mass term m_{LL}^{II} for the light neutrinos is present.

With such an additional direct mass term, the general neutrino mass matrix is given by

$$\begin{pmatrix} \overline{\nu_L} & \overline{\nu_R^C} \end{pmatrix} \begin{pmatrix} m_{LL}^{II} & m_{LR}^\nu \\ m_{LR}^{\nu T} & M_{RR} \end{pmatrix} \begin{pmatrix} \nu_L^C \\ \nu_R \end{pmatrix}. \tag{1}$$

Under the assumption that the mass eigenvalues M_{Ri} of M_{RR} are very large compared to the components of m_{LL}^{II} and m_{LR}, the mass matrix can approximately be diagonalized yielding effective Majorana masses

$$m_{LL}^\nu \approx m_{LL}^{II} + m_{LL}^{I} \tag{2}$$

with

$$m_{LL}^{I} \approx -m_{LR}^\nu M_{RR}^{-1} m_{LR}^{\nu T} \tag{3}$$

for the light neutrinos. The direct mass term m_{LL}^{II} can also provide a naturally small contribution to the light neutrino masses if it stems e.g. from a see-saw suppressed induced vev. The type II contribution may be induced via the exchange of heavy Higgs triplets of $SU(2)_L$ as illustrated diagramatically in Fig. 1(b).

3. A natural neutrino mass hierarchy

In this section we discuss an elegant and natural way of accounting for a neutrino mass hierarchy and two large mixing angles in the type I see-saw mechanism. The starting point is to assume that one of the right-handed neutrinos contributes dominantly to the see-saw mechanism and determines the atmospheric neutrino mass and mixing. A second right-handed neutrino contributes sub-dominantly and determines the solar neutrino mass and mixing. The third right-handed neutrino is effectively decoupled from the see-saw mechanism.

The above Sequential Dominance mechanism [3] is most simply described assuming three right-handed neutrinos in the basis where the right-handed neutrino mass matrix is diagonal although it can also be developed in other bases. In this basis we write the input see-saw matrices as

$$
M_{RR} = \begin{pmatrix} X & 0 & 0 \\ 0 & Y & 0 \\ 0 & 0 & Z \end{pmatrix}, \qquad m_{LR}^{v} = \begin{pmatrix} a & d & p \\ b & e & q \\ c & f & r \end{pmatrix}.
\tag{4}
$$

Each right-handed neutrino in the basis of Eq. (4) couples to a particular column of m_{LR}^{v}. There is no mass ordering of X, Y, Z implied in Eq. (4). The dominant right-handed neutrino may be taken to be the one with mass Y without loss of generality. Sequential dominance occurs when the right-handed neutrinos dominate sequentially [3]:

$$
\frac{|e^2|, |f^2|, |ef|}{Y} \gg \frac{|xy|}{X} \gg \frac{|x'y'|}{Z},
\tag{5}
$$

where $x, y \in a, b, c$ and $x', y' \in p, q, r$. This leads to a full neutrino mass hierarchy $m_3^2 \gg m_2^2 \gg m_1^2$. Ignoring phases, in the case that $d = 0$, corresponding to a Yukawa 11 texture zero in Eq. (4), we have:

$$
m_1 \sim O\left(\frac{x'y'}{Z}\right), \qquad m_2 \approx \frac{|a|^2}{Xs_{12}^2}, \qquad m_3 \approx \frac{|e|^2 + |f|^2}{Y}
\tag{6}
$$

where $s_{12} = \sin\theta_{12}$ is given below. Note that each neutrino mass is generated by a separate right-handed neutrino, and the sequential dominance condition naturally results in a neutrino mass hierarchy $m_1 \ll m_2 \ll m_3$. The neutrino mixing angles are given to leading order in m_2/m_3 by [3]:

$$
\tan\theta_{23} \approx \frac{e}{f}, \qquad \tan\theta_{12} \approx \frac{a}{c_{23}b - s_{23}c},
$$
$$
\theta_{13} \approx s_{12}c_{12}\frac{(s_{23}b + c_{23}c)\,m_2}{(c_{23}b - s_{23}c)\,m_3}.
\tag{7}
$$

Physically these results show that in sequential dominance the atmospheric neutrino mass m_3 and mixing θ_{23} is determined by the couplings of the dominant right-handed neutrino of mass Y. The angle θ_{13} is generically of order $\theta_{13} \sim O(m_2/m_3) \sim 0.2$. However the coefficient in Eq. (7) can be arbitrarily small, since to leading order as $b \to -c$, $\theta_{13} \to 0$, but θ_{12} remains large. The solar neutrino mass m_2 and mixing θ_{12} is determined by the couplings of the sub-dominant right-handed neutrino of mass X. From Eq. (7), the solar angle only depends on the

sub-dominant couplings and the simple requirement for large solar angle is $a \sim b - c$. The third right-handed neutrino of mass Z is effectively decoupled from the see-saw mechanism and leads to the vanishingly small mass $m_1 \approx 0$.

4. A natural neutrino mass degeneracy

We now show that it is possible to obtain a (partially) degenerate neutrino mass spectrum by essentially adding a type II direct neutrino mass contribution proportional to the unit matrix: Thus we shall consider a type II extension [4], where the mass matrix of the light neutrinos has the form:

$$
m_{LL}^{v} \approx m^{II} \begin{pmatrix} 1 & 0 & 0 \\ 0 & 1 & 0 \\ 0 & 0 & 1 \end{pmatrix} + m_{LL}^{I}.
\tag{8}
$$

Assuming here that the type I mass matrix m_{LL}^{I} is real, the full neutrino mass matrix is diagonalised by the same matrix that diagonalises the type I matrix:

$$
(m_{LL}^{v})_{\text{diag}} = m^{II} VV^T + V m_{LL}^{I} V^T = m^{II} \mathbb{1} + diag(m_1^{I}, m_2^{I}, m_3^{I}).
\tag{9}
$$

In this case the neutrino mass scale is controlled by the type II mass scale m^{II}, while the neutrino mass splittings are determined by the type I mass eigenvalues:

$$
m_1 \approx |m^{II} - m_2^{I}|, \qquad m_2 \approx |m^{II} - m_2^{I}|, \qquad m_3 \approx |m^{II} - m_3^{I}|.
\tag{10}
$$

Sequential dominance in the type I sector naturally predicts $m_1^{I} \ll m_2^{I} \ll m_3^{I}$. Hence the very small mass splittings required for degenerate neutrinos can be achieved naturally by sequential dominance. The predictions for the mixings are determined from the type I mass matrix (the type II unit matrix is irrelevant). In particular the atmospheric and solar angles are given by the sequential dominance estimates in Eq. (7) and are independent of the type II neutrino mass scale m^{II}. However the angle θ_{13} in Eq. (7) is now of order $\theta_{13} \sim O(m_2^{I}/m_3^{I})$. This is now much smaller than the type I result $\theta_{13} \sim O(m_2/m_3)$ since the neutrino mass splittings, controlled by the type I masses, are very small for partially degenerate neutrinos.

5. Mixing angles from the charged leptons?

In this section we show how bi-large mixing could originate from the charged lepton sector using a generalization of sequential right-handed neutrino dominance to all right-handed leptons [5]. We write the mass matrices for the charged leptons m_E as

$$
m_E = \begin{pmatrix} p' & d' & a' \\ q' & e' & b' \\ r' & f' & c' \end{pmatrix}.
\tag{11}
$$

In our notation, each right-handed charged lepton couples to a column in m_E. For the charged leptons, the sequential dominance conditions are [5]:

$$
|a'|, |b'|, |c'| \gg |d'|, |e'|, |f'| \gg |p'|, |q'|, |r'|.
\tag{12}
$$

They imply the desired hierarchy for the charged lepton masses $m_\tau \gg m_\mu \gg m_e$ and small right-handed mixing of U_{e_R}. We assume zero mixing from the neutrino sector. A natural possibility for obtaining a small θ_{13} is [5]

$$
|d'|, |e'| \ll |f'|.
\tag{13}
$$

In leading order in $|d'|/|f'|$ and $|e'|/|f'|$, for the mixing angles θ_{12}, θ_{23} and θ_{13}, we obtain (again ignoring phases here)

$$\tan\theta_{12} \approx \frac{a'}{b'}, \qquad \tan\theta_{23} \approx \frac{s_{12}\,a' + c_{12}\,b'}{c'},$$

$$\tan\theta_{13} \approx \frac{s_{12}\,e' - c_{12}d'}{f'}. \tag{14}$$

θ_{13} only depends on d'/f' and e'/f' from the Yukawa couplings to the sub-dominant right-handed muon and on θ_{12}. We find that in the limit $|d'|, |e'| \ll |f'|$, the two large mixing angles θ_{12} and θ_{23} approximately depend only on a'/c' and b'/c' from the right-handed tau Yukawa couplings. Both mixing angles are large if a', b' and c' are of the same order.

6. Phenomenological implications

We now summarize the phenomenological consequences of type I see-saw models with a natural mass hierarchy and their type II extensions with a natural mass degeneracy, for the low energy neutrino parameters and high-energy mechanisms as leptogenesis and minimal lepton flavour violation (LFV). In order to compare the predictions of such natural see-saw models (based on sequential dominance) with the experimental data obtained at low energy, the renormalization group (RG) running of the effective neutrino mass matrix has to be taken into account.

6.1. *Renormalization group corrections*

For type I models with sequential dominance, the running of the mixing angles is generically small [14] since the mass scheme is strongly hierarchical. When the neutrino mass scale is lifted, e.g. via a type II upgrade, a careful treatment of the RG running of the neutrino parameters, including the energy ranges between and above the see-saw scale [14], is required. Dependent on $\tan\beta$ in the MSSM, on the size of the neutrino Yukawa couplings and on the neutrino mass scale, the RG effects can be sizable or cause only small corrections.

6.2. *Dirac and Majorana CP phases and neutrinoless double beta decay*

At present, the CP phases in the lepton sector are unconstrained by experiment. In type I see-saw models based on sequential dominance, there is no restriction on them from a theoretical point of view. The type-II-extension however predicts that all observable CP phases, i.e. the Dirac CP phase δ relevant for neutrino oscillations and the Majorana CP phases β_2 and β_3, become small as the neutrino mass scale increases.

The key process for measuring the neutrino mass scale could be neutrinoless double beta decay. The decay rates depend on an effective Majorana mass defined by $\langle m_v \rangle = \left| \sum_i (U_{\mathrm{MNS}})_{1i}^2\, m_i \right|$. Future experiments which are under consideration at present might increase the sensitivity to $\langle m_v \rangle$ by more than an order of magnitude. For type I models with sequential dominance, which have a hierarchical mass scheme, $\langle m_v \rangle$ can be very small, below the accessible sensitivity.

For models where the neutrino mass scale is lifted via a type II extension [4], there is a close relation between the neutrino mass scale, i.e. the mass of the lightest neutrino and $\langle m_v \rangle$. Since the CP phases are small, there can be no significant cancellations in $\langle m_v \rangle$. This implies that the effective mass for neutrinoless double beta decay is approximately equal to the neutrino mass scale $\langle m_v \rangle \approx m^{\mathrm{II}}$ and therefore neutrinoless double beta decay will be observable in the next round of experiments if the neutrino mass spectrum is partially degenerate.

6.3. *Theoretical expectations for the mixing angles*

In order to discriminate between models, precision measurements of the neutrino mixing angles have the potential to play an important role.

One important parameter is the value of the mixing angle θ_{13}, which is at present only bounded from above to be smaller than approximately $13°$. In the type I sequential dominance case, the mixing angle θ_{13} is typically of the order $\mathcal{O}(m_2^{\mathrm{I}}/m_3^{\mathrm{I}})$. In the type-II-upgrade scenario this ratio decreases with increasing neutrino mass scale and is smaller than $\approx 5°$ for partially degenerate neutrinos even if it was quite large in the type I limit. Sizable RG corrections, which are usually expected for partially degenerate neutrinos, are suppressed in the type-II-upgrade scenario due to small CP phases β_2, β_3 and δ.

Another important parameter is θ_{23}. Its present best-fit value is close to $45°$, however comparably large deviations are experimentally allowed as well. With sequential dominance, we expect minimal deviations of θ_{23} from $45°$ of the order $\mathcal{O}(m_2^{\mathrm{I}}/m_3^{\mathrm{I}})$, which could be observed by future long-baseline experiments in the type I see-saw case. In the type II upgraded version, the corrections can be significantly smaller since the ratio $m_2^{\mathrm{I}}/m_3^{\mathrm{I}}$ decreases with increasing neutrino mass scale [4]. For large $\tan\beta$ in the MSSM, the major source for the corrections can be RG effects, which are un-suppressed for small CP phases.

6.4. *Minimal Lepton Flavour Violation*

At leading order in a mass insertion approximation the branching fractions of LFV processes are given by[†]

$$\mathrm{BR}(l_i \to l_j\gamma) \approx \frac{\alpha^3}{G_F^2} f(M_2, \mu, m_{\tilde{\nu}}) |m_{\tilde{L}_{ij}}^2|^2 \tan^2\beta, \tag{15}$$

where $l_1 = e, l_2 = \mu, l_3 = \tau$, and where the off-diagonal slepton doublet mass squared is given in the leading log approximation (LLA) by

$$m_{\tilde{L}_{ij}}^{2(LLA)} \approx -\frac{(3m_0^2 + A_0^2)}{8\pi^2} C_{ij}. \tag{16}$$

With sequential dominance, using the notation of Eqs. (4), the leading log coefficients relevant for $\mu \to e\gamma$ and $\tau \to \mu\gamma$ are given approximately as

$$C_{21} = ab\ln\frac{M_U}{X} + de\ln\frac{M_U}{Y},$$

$$C_{32} = bc\ln\frac{M_U}{X} + ef\ln\frac{M_U}{Y}. \tag{17}$$

Large rates of $\tau \to \mu\gamma$ which is the characteristic expectation of lop-sided models in general [9]. Such models occur if the dominant right-handed neutrino is the heaviest one. A global analysis of LFV has been performed in the constrained minimal supersymmetric standard model (CMSSM) for the case of sequential dominance, focussing on the two cases where the dominant right-handed neutrino is either the heaviest one or the lightest one [9]. If the dominant right-handed neutrino is the

---

[†]The mass insertion approximation given in Eq. (15) is for illustrative purposes only. The conclusions quoted below from [9] do not rely on this approximation.

lightest one then the rate for $\tau \to \mu\gamma$ is well below observable values. Therefore $\tau \to \mu\gamma$ provides a good discriminator between the two cases of dominance. In [9] it is shown that the rate for $\mu \to e\gamma$ may determine the order of the sub-dominant neutrino Yukawa couplings in the flavour basis.

7. Model building applications

7.1. Effective two right-handed neutrino models

In sequential dominance we have seen that one of the right-handed neutrinos effectively decouples from the see-saw mechanism. If the decoupled right-handed neutrino is also the heaviest then it would be expected to play no part in phenomenology. In this case sequential dominance reduces to effectively two right-handed neutrino models [3]. Recently there have been several studies based on the "minimal see-saw" involving two right-handed neutrinos [10], and it is worth bearing in mind that such models could naturally arise as the limiting case of sequential dominance.

7.2. Sneutrino inflation models

Sequential dominance has recently also been applied to sneutrino inflation [15], [16]. Requiring a low reheat temperature after inflation, in order to solve the gravitino problem, forces the sneutrino inflaton to couple very weakly to ordinary matter and its superpartner almost to decouple from the see-saw mechanism. This decoupling of a right-handed neutrino from the see-saw mechanism is a characteristic of sequential dominance.

7.3. GUT and family symmetry models

There are many models in the literature based on sequential dominance. A Pati-Salam model with $U(1)$ family symmetry was considered in [11]. Single right-handed neutrino dominance has also been applied to SO(10) GUT models involving a $U(2)$ family symmetry [12]. Sequential dominance with $SU(3)$ family symmetry and $SO(10)$ GUTs has been considered in [13]. Type II up-gradable models based on sequential dominance of the ISD type with $SO(3)$ family symmetry have been considered in [5, 4]. For GUT models the renormalisation group corrections need to be taken into account, although for a natural hierarchy such corrections are only a few per cent [14].

As an example of a model based on a non-Abelian family symmetry, we briefly review the model proposed in [17]. The model uses the largest family symmetry $SU(3)$ consistent with $SO(10)$ GUTs. An important further motivation for $SU(3)$ family symmetry is, in the framework of sequential dominance, to relate the second and third entries of the Yukawa matrix, as required to obtain an almost maximal 23 mixing in the atmospheric neutrino sector [13]. The theoretical requirements that the neutrino Yukawa matrix resembles the quark Yukawa matrices, and therefore has a large 33 element with no large off-diagonal elements and a texture zero in the 11 position leads uniquely to the dominant right-handed neutrino being the first (lightest) one. Assuming this then the atmospheric neutrino mixing angle is given by $\tan \theta_{23}^{\nu} \approx Y_{21}^{\nu}/Y_{31}^{\nu} \approx 1$. The sequential dominance conditions which were assumed earlier will here be derived from the symmetries of the model. Thus this model provides an example of the application of sequential dominance to realistic models of flavour, and shows how the conditions of sequential dominance which were simply assumed earlier can motivate models based on GUTs and family symmetry which are capable of explaining these conditions. In other words, the conditions for sequential

dominance can provide clues to the nature of the underlying flavour theory.

The starting point of the model is the observation that an excellent fit to all quark data is given by the approximately symmetric form of quark Yukawa matrices [18]

$$Y^u \propto \begin{pmatrix} 0 & \epsilon^3 & O(\epsilon^3) \\ . & \epsilon^2 & O(\epsilon^2) \\ . & . & 1 \end{pmatrix}, \qquad Y^d \propto \begin{pmatrix} 0 & 1.5\bar{\epsilon}^3 & 0.4\bar{\epsilon}^3 \\ . & \bar{\epsilon}^2 & 1.3\bar{\epsilon}^2 \\ . & . & 1 \end{pmatrix} \qquad (18)$$

where the expansion parameters ϵ and $\bar{\epsilon}$ are given by

$$\epsilon \approx 0.05, \qquad \bar{\epsilon} \approx 0.15. \qquad (19)$$

This motivates a particular model in which the three families are unified as triplets under an $SU(3)$ family symmetry, and $16's$ under an $SO(10)$ GUT [13, 17],

$$\psi_i = (3, 16), \qquad (20)$$

where the $SO(10)$ is broken via the Pati-Salam group resulting in:

$$\psi_i = (3, 4, 2, 1), \qquad \bar{\psi}_i = (3, \bar{4}, 1, \bar{2}). \qquad (21)$$

Further symmetries $R \times Z_2 \times U(1)$ are assumed to ensure that the vacuum alignment leads to a universal form of Dirac mass matrices for the neutrinos, charged leptons and quarks [17]. To build a viable model we also need spontaneous breaking of the family symmetry

$$SU(3) \longrightarrow SU(2) \longrightarrow \text{Nothing} \qquad (22)$$

To achieve this symmetry breaking additional Higgs fields ϕ_3, $\bar{\phi}_3$, ϕ_{23} and $\bar{\phi}_{23}$ are required. The largeness of the third family fermion masses implies that $SU(3)$ must be strongly broken by new Higgs antitriplet fields ϕ_3 which develop a vev in the third $SU(3)$ component $\langle\phi_3\rangle^T = (0, 0, a_3)$ as in [13]. ϕ_3^i transforms under $SU(2)_R$ as $3 \oplus 1$ rather than being $SU(2)_R$ singlets as assumed in [13], and develops vevs in the $SU(3) \times SU(2)_R$ directions

$$\langle\phi_3\rangle = \langle\bar{\phi}_3\rangle = \begin{pmatrix} 0 \\ 0 \\ 1 \end{pmatrix} \otimes \begin{pmatrix} a_3^u & 0 \\ 0 & a_3^d \end{pmatrix}. \qquad (23)$$

The symmetry breaking also involves the $SU(3)$ antitriplets ϕ_{23} which develop vevs [13]

$$\langle\phi_{23}\rangle = \begin{pmatrix} 0 \\ 1 \\ 1 \end{pmatrix} b, \qquad (24)$$

where, as in [13], vacuum alignment ensures that the vevs are aligned in the 23 direction. Due to D-flatness there must also be accompanying Higgs triplets such as $\bar{\phi}_{23}$ which develop vevs [13]

$$\langle\bar{\phi}_{23}\rangle = \begin{pmatrix} 0 \\ 1 \\ 1 \end{pmatrix} b. \qquad (25)$$

We also introduce an adjoint Σ field which develops vevs in the $SU(4)_{PS} \times SU(2)_R$ direction which preserves the hypercharge generator $Y = T_{3R} + (B - L)/2$, and implies that any coupling of the Σ to a fermion and a messenger such as $\Sigma_{b\beta}^{a\alpha}\psi_{a\alpha}^c\chi^{b\beta}$, where the $SU(2)_R$ and $SU(4)_{PS}$ indices have been displayed explicitly, is proportional to the hypercharge Y of the particular fermion component of ψ^c times the vev σ. In addition a θ field is required for the construction of Majorana neutrino masses.

The leading operators allowed by the symmetries are

$$P_{\text{Yuk}} \sim \frac{1}{M^2} \psi_i \phi_3^i \bar{\psi}_j \phi_3^j h \tag{26}$$

$$+ \frac{\Sigma}{M^3} \psi_i \phi_{23}^i \bar{\psi}_j \phi_{23}^j h, \tag{27}$$

$$P_{\text{Maj}} \sim \frac{1}{M} \bar{\psi}_i \theta^i \theta^j \bar{\psi}_j \tag{28}$$

where the operator mass scales, generically denoted by M may differ and we have suppressed couplings of $O(1)$.

The final form of the Yukawa matrices and heavy Majorana matrix after inserting a particular choice of order unity coefficients is [17]

$$Y^u \approx \begin{pmatrix} 0 & 1.2\epsilon^3 & 0.9\epsilon^3 \\ -1.2\epsilon^3 & -\frac{2}{3}\epsilon^2 & -\frac{2}{3}\epsilon^2 \\ -0.9\epsilon^3 & -\frac{2}{3}\epsilon^2 & 1 \end{pmatrix} \bar{\epsilon}, \tag{29}$$

$$Y^d \approx \begin{pmatrix} 0 & 1.6\bar{\epsilon}^3 & 0.7\bar{\epsilon}^3 \\ -1.6\bar{\epsilon}^3 & \bar{\epsilon}^2 & \bar{\epsilon}^2 + \bar{\epsilon}^{\frac{5}{2}} \\ -0.7\bar{\epsilon}^3 & \bar{\epsilon}^2 & 1 \end{pmatrix} \bar{\epsilon}, \tag{30}$$

$$Y^e \approx \begin{pmatrix} 0 & 1.6\bar{\epsilon}^3 & 0.7\bar{\epsilon}^3 \\ -1.6\bar{\epsilon}^3 & 3\bar{\epsilon}^2 & 3\bar{\epsilon}^2 \\ -0.7\bar{\epsilon}^3 & 3\bar{\epsilon}^2 & 1 \end{pmatrix} \bar{\epsilon}, \tag{31}$$

$$Y^\nu \approx \begin{pmatrix} 0 & 1.2\epsilon^3 & 0.9\epsilon^3 \\ -1.2\epsilon^3 & -\alpha\epsilon^2 & -\alpha\epsilon^2 \\ -0.9\epsilon^3 & -\alpha\epsilon^2 - \epsilon^3 & 1 \end{pmatrix} \bar{\epsilon}. \tag{32}$$

$$M_{RR} \approx \begin{pmatrix} \epsilon^6\bar{\epsilon}^3 & 0 & 0 \\ 0 & \epsilon^6\bar{\epsilon}^2 & 0 \\ 0 & 0 & 1 \end{pmatrix} M_3. \tag{33}$$

The model gives excellent agreement with the quark and lepton masses and mixing angles. For the up and down quarks the form of Y^u and Y^d given in Eq. (29, 30) is consistent with the phenomenological fit in Eq. (18). The charged lepton mass matrix is of the Georgi-Jarslkog [19] form which, after including radiative corrections, gives an excellent description of the charged lepton masses. In the neutrino sector the parameters satisfy the conditions of sequential dominance in Eq. (5), with the lightest right-handed neutrino giving the dominant contribution to the heaviest physical neutrino mass, and the second right-handed neutrino giving the leading subdominant contribution, providing that $\alpha \sim \epsilon$.

Analytic estimates of neutrino masses and mixing angles for sequential dominance were derived in Section 3 [3]. With the dominant right-handed neutrino of mass Y being the lightest one, the matrices in Eq. (4) should be re-ordered as follows before comparing to the neutrino matrices in Eqs. (32) and (33):

$$M_{RR} = \begin{pmatrix} Y & 0 & 0 \\ 0 & X & 0 \\ 0 & 0 & Z \end{pmatrix}, \quad m_{\text{LR}}^\nu = Y^\nu v_2 = \begin{pmatrix} 0 & a & p \\ e & b & q \\ f & c & r \end{pmatrix}. \tag{34}$$

The above re-ordering of course leaves the results in Eqs. (6) and (7) unchanged. The analytic estimates for the neutrino masses are obtained by comparing the matrices in Eqs. (34) to those in

Eqs. (32) and (33) then using the results in Eqs. (6) and (7):

$$m_1 \sim \bar{\epsilon}^2 \frac{v_2^2}{M_3}, \tag{35}$$

$$m_2 \approx 5.8 \frac{v_2^2}{M_3}, \tag{36}$$

$$m_3 \approx 15 \frac{v_2^2}{M_3} \tag{37}$$

and neutrino mixing angles:

$$\tan \theta_{23}^\nu \approx 1.3, \tag{38}$$

$$\tan \theta_{12}^\nu \approx 0.66, \tag{39}$$

$$\theta_{13}^\nu \approx 1.6\bar{\epsilon}. \tag{40}$$

The physical lepton mixing angle θ_{13} receives a large contribution from the neutrino sector $\theta_{13}^\nu \sim 0.3$ at the high energy scale, for this choice of parameters, compared to the current CHOOZ limit $\theta_{13} \leq 0.2$. However the physical mixing angles will receive charged lepton contributions [3] and all the parameters are subject to radiative corrections in running from the high energy scale to low energies, although in sequential dominance models these corrections are only a few per cent [14]. Thus the model predicts that θ_{13} is close to the current CHOOZ limit, and could be observed by the next generation of long baseline experiments such as MINOS or OPERA.

8. Leptogenesis

Neutrino mass allows the possibility that the baryon asymmetry of the universe is generated by out-of-equilibrium decay of lepton-number violating Majorana right-handed (s)neutrinos, whose decays result in a net lepton number which is subsequently converted to a net baryon number by sphaleron transitions. This mechanism is known as leptogenesis [20]. In models which give a natural neutrino mass hierarchy, if the dominant right-handed neutrino is the lightest one then the washout parameter $\tilde{m}_1 \sim O(m_3)$, which is rather too large compared to the optimal value of around 10^{-3} eV, while if the dominant right-handed neutrino is either the intermediate one or the heaviest one then one finds $\tilde{m}_1 \sim O(m_2)$ or arbitrary $\tilde{m}_1$, which can be closer to the desired value [21].

It is interesting to note that if the dominant right-handed neutrino is the lightest one, and there is a 11 texture zero, as is the case in the $SU(3) \times SO(10)$ model discussed in the last section, then there is a link between the CP violation required for leptogenesis, ϕ_{COSMO}, and the phase δ measurable in accurate neutrino oscillation experiments [22]. δ turns out to be a function of the same two see-saw phases that also determine ϕ_{COSMO}. If both the see-saw phases are zero, then both ϕ_{COSMO} and δ are necessarily zero. This feature is absolutely crucial. It means that, barring cancellations, measurement of a non-zero value for the phase δ at a neutrino factory will be a signal of a non-zero value of the leptogenesis phase ϕ_{COSMO}. We also find the remarkable result

$$|\phi_{\text{COSMO}}| = |\phi_{\beta\beta0\nu}|. \tag{41}$$

where $\phi_{\beta\beta0\nu}$ is the phase which enters the rate for neutrinoless double beta decay [22].

We now discuss the consequences of the neutrino mass scale for leptogenesis via the out-of-equilibrium decay of the lightest right-handed (s)neutrinos in type II see-saw models [7]. In [7] we calculated the type II contributions to the decay asymmetries

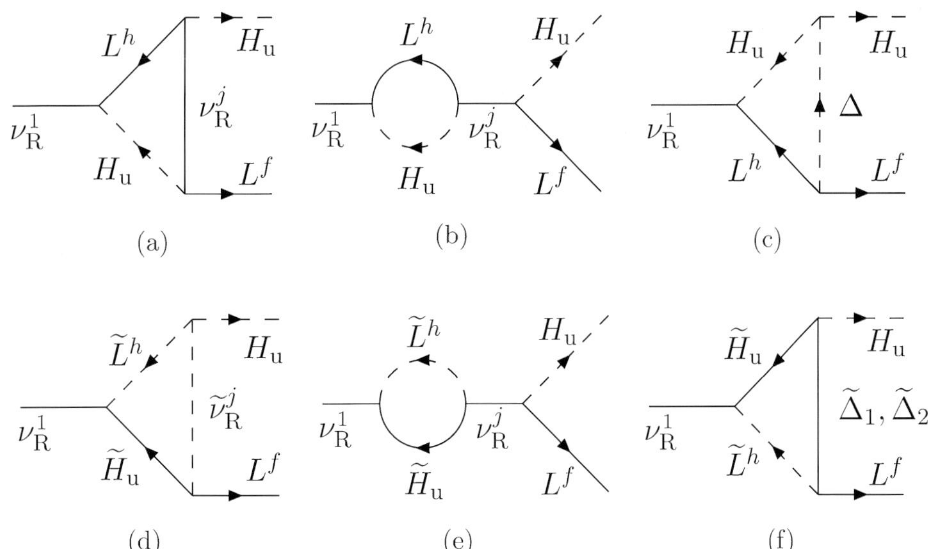

Fig. 2. Loop diagrams in the MSSM which contribute to the decay $v_{\mathrm{R}}^1 \to L_a^f H_{\mathrm{u}b}$ for the case of a type II see-saw mechanism where the direct mass term for the neutrinos stems from the induced vev of a Higgs triplet. In diagram (f), $\widetilde{\Delta}_1$ and $\widetilde{\Delta}_2$ are the mass eigenstates corresponding to the superpartners of the $SU(2)_{\mathrm{L}}$-triplet scalar fields Δ and $\bar{\Delta}$. The SM diagrams are the ones where no superpartners (marked by a tilde) are involved and where H_{u} is renamed to the SM Higgs.

for minimal scenarios based on the Standard Model (SM) and on the Minimal Supersymmetric Standard Model (MSSM), where the additional direct mass term for the neutrinos stems from the induced vev of a triplet Higgs. The diagrams are shown in Figure 2. The result we obtained for the supersymmetric case is new and we corrected the previous result in the scenario based on the Standard Model [23].

We subsequently derived a general upper bound on the decay asymmetry and found that it increases with the neutrino mass scale:

$$|\varepsilon_1^{\mathrm{SM}}| \leq \frac{3}{16\pi} \frac{M_{\mathrm{R}1}}{v_{\mathrm{u}}^2} m_{\max}^{\nu}, \tag{42a}$$

$$|\varepsilon_1^{\mathrm{MSSM}}| < \frac{3}{8\pi} \frac{M_{\mathrm{R}1}}{v_{\mathrm{u}}^2} m_{\max}^{\nu}. \tag{42b}$$

It leads to a lower bound on the mass of the lightest right-handed neutrino, which is significantly below the type I bound for partially degenerate neutrinos. It is worth emphasizing that these results are in sharp contrast to the type I see-saw mechanism where an upper bound on the neutrino mass scale is predicted. Here we find no upper limit on the neutrino mass scale which may be increased arbitrarily. Indeed we find that the lower bound on the mass of the lightest right-handed neutrino decreases as the physical neutrino mass scale increases. This allows a lower reheat temperature, making thermal leptogenesis more consistent with the gravitino constraints in supersymmetric models.

9. Conclusion

In this talk we have discussed how a natural neutrino mass hierarchy can follow from the type I see-saw mechanism, and a natural neutrino mass degeneracy from the type II see-saw mechanism, where the bi-large mixing angles can arise from either the neutrino or charged lepton sector. The key to achieving naturalness is the idea of sequential dominance of right-handed neutrinos, namely that in the see-saw mechanism one of the right-handed neutrinos dominates and couples with approximately equal strength to the τ and μ families, leading to an approximately maximal atmospheric mixing angle. A second right-handed neutrino then plays the leading sub-dominant role, and couples with approximately equal strength to all three families, leading to a large solar mixing angle. We have shown that this can lead either to a hierchical neutrino mass spectrum or, if a type II contribution proportional to the unit matrix is considered, to approximately degenerate neutrino masses. We have summarised the phenomenological implications of such a natural approach to neutrino model building, and have discussed some of the model building applications, focussing on the $SU(3) \times SO(10)$ model. We then turned to leptogenesis and mentioned the possible link between the leptogenesis phase and the phase measurable in neutrino oscillation experiments. We then pointed out that in natural type II models the leptogenesis asymmetry parameter becomes proportional to the neutrino mass scale, in sharp contrast to the type I case, which leads to an upper bound on the neutrino mass scale, allowing lighter right-handed neutrinos and hence making leptogenesis more consistent with the gravitino constraints in supersymmetric models.

Acknowledgements

I would like to thank Stefan Antusch for his collaboration in some of the work presented here. I would also like to thank the organisers of the Nobel Symposium 129 for a very enjoyable and fruitful meeting.

References

1. For a review see for example: King, S. F., Rept. Prog. Phys. **67**, 107 (2004) [arXiv:hep-ph/0310204], and references therein.
2. Minkowski, P., Phys. Lett. B **67**, 421 (1977); Yanagida, T., in Proc. of the Workshop on Unified Theory and Baryon Number of the Universe, KEK, Japan, (1979); Glashow, S. L., Cargese Lectures (1979); Gell-Mann, M., Ramond, P. and Slansky, R., in Sanibel Talk, CALT-68-709, Feb (1979), and in "Supergravity", (North Holland, Amsterdam 1979); Mohapatra, R. N. and Senjanovic, G., Phys. Rev. Lett. **44**, 912 (1980); Schechter, J. and Valle, J. W. F., Phys. Rev. D **22**, 2227 (1980).
3. King, S. F., Phys. Lett. B **439**, 350 (1998) [arXiv:hep-ph/9806440]; King, S. F., Nucl. Phys. B **562**, 57 (1999) [arXiv:hep-ph/9904210]; King, S. F., Nucl. Phys. B **576**, 85 (2000) [arXiv:hep-ph/9912492]; King, S. F., J. Exp. Theor. Phys. **0209**, 011 (2002) [arXiv:hep-ph/0204360].
4. Antusch, S. and King, S. F., arXiv:hep-ph/0402121.
5. Antusch, S. and King, S. F., Phys. Lett. B **591**, 104 (2004) [arXiv:hep-ph/0403053].

6. Antusch, S. and King, S. F., arXiv:hep-ph/0405272.
7. Antusch, S. and King, S. F., Phys. Lett. B **597**, 199 (2004) [arXiv:hep-ph/0405093].
8. Lazarides, G., Shafi, Q. and Wetterich, C., Nucl. Phys. B **181**, 287 (1981); Mohapatra, R. N. and Senjanović, G., Phys. Rev. **D23**, 165 (1981); Wetterich, C., Nucl. Phys. **B187**, 343 (1981).
9. Blazek, T. and King, S. F., Phys. Lett. B **518**, 109 (2001) [arXiv:hep-ph/0105005]; Blazek, T. and King, S. F., Nucl. Phys. B **662**, 359 (2003) [arXiv:hep-ph/0211368].
10. Frampton, P. H., Glashow, S. L. and Yanagida, T., Phys. Lett. B **548**, 119 (2002) [arXiv:hep-ph/0208157]; Raidal, M. and Strumia, A., Phys. Lett. B **553**, 72 (2003) [arXiv:hep-ph/0210021]; King, S. F., Phys. Rev. D **67**, 113010 (2003) [arXiv:hep-ph/0211228]; Ibarra, A. and Ross, G. G., [arXiv:hep-ph/0312138].
11. King, S. F. and Oliveira, M., Phys. Rev. D **63**, 095004 (2001) [arXiv:hep-ph/0009287]; Blazek, T., King, S. F. and Parry, J. K., J. High. En. Phys. **0305**, 016 (2003) [arXiv:hep-ph/0303192].
12. Barbieri, R., Creminelli, P. and Romanino, A., Nucl. Phys. B **559**, 17 (1999) [arXiv:hep-ph/9903460]; Raby, S., Phys. Lett. B **561**, 119 (2003) [arXiv:hep-ph/0302027].
13. King, S. F. and Ross, G. G., Phys. Lett. B **520**, 243 (2001) [arXiv:hep-ph/0108112].
14. King, S. F. and Singh, N. N., Nucl. Phys. B **591**, 3 (2000) [arXiv:hep-ph/0006229]; Antusch, S., these proceedings, arXiv:hep-ph/0409229.
15. Ellis, J. R., Raidal, M. and Yanagida, T., Phys. Lett. B **581**, 9 (2004) [arXiv:hep-ph/0303242]; Chankowski, P. H., Ellis, J. R., Pokorski, S., Raidal, M. and Turzynski, K., arXiv:hep-ph/0403180.
16. Antusch, S., Bastero-Gil, M., King, S. F. and Shafi, Q., arXiv:hep-ph/0411298.
17. King, S. F. and Ross, G. G., "Fermion Masses and Mixing Angles from SU(3) Family Symmetry and Unification," arXiv:hep-ph/0307190.
18. Roberts, R. G., Romanino, A., Ross, G. G. and Velasco-Sevilla, L., Nucl. Phys. B **615**, 358 (2001) [arXiv:hep-ph/0104088].
19. Georgi, H. and Jarlskog, C., Phys. Lett. B **86**, 297 (1979).
20. Fukugita, M. and Yanagida, T., Phys. Lett. **B174**, 45 (1986).
21. Hirsch, M. and King, S. F., Phys. Rev. D **64**, 113005 (2001) [arXiv:hep-ph/0107014].
22. King, S. F., Phys. Rev. D **67**, 113010 (2003) [arXiv:hep-ph/0211228].
23. Hambye, T. and Senjanovic, G., Phys. Lett. B **582**, 73 (2004) [arXiv:hep-ph/0307237].

Physica Scripta. Vol. T121, 185–191, 2005

Neutrino Mass and Grand Unification

R. N. Mohapatra

Department of Physics, University of Maryland, College Park, MD-20742, USA

Received November 23, 2004; accepted December 29, 2004

PACS number: 14.60.Pq

Abstract

Seesaw mechanism appears to be the simplest and most appealing way to understand small neutrino masses observed in recent experiments. It introduces three right handed neutrinos with heavy masses to the standard model, with at least one mass required by data to be close to the scale of conventional grand unified theories. This may be a hint that the new physics scale implied by neutrino masses and grand unification of forces are one and the same. Taking this point of view seriously, I explore different ways to resolve the puzzle of large neutrino mixings in grand unified theories such as SO(10) and models based on its subgroup $SU(2)_L \times SU(2)_R \times SU(4)_c$.

1. Introduction

The discovery of neutrino masses and mixings has been an important milestone in the history of particle physics and rightly qualifies as the first evidence for new physics beyond the standard model. The amount of new information on neutrinos already established from various oscillation experiments has provided very strong clues to new symmetries of particles and forces and new directions for unification. Enough puzzles have emerged making this field a hotbed for research with implications ranging from ideas such as supersymmetry and grand unification to cosmology and astrophysics.

A major cornerstone for the theory research in this field has been the seesaw mechanism introduced in the late seventies [1] to understand why neutrino masses are so much smaller than the masses of other fermions of the standard model. Even though there was no evidence for neutrino masses then, there were very well motivated extensions of the standard models that led to nonzero masses for neutrinos. It was therefore incumbent on those models that they have a mechanism for understanding why upper limits on neutrino masses known at that time were so small. Seesaw mechanism introduces three right handed neutrinos into the standard model with very large Majorana masses and predicts that observed neutrinos are their own anti-particles. A very appealing aspect of this mechanism is not only the beauty and elegance it brings to the standard model by restoring quark-lepton symmetry but also the new insight it provides into such questions as the origin of parity violation and Dirac vrs Majorana nature of the neutrino.

The first conclusive evidence for nonzero neutrino masses appeared in 1998. During the past six years, we have learnt that neutrinos not only have mass but they also mix among themselves with a pattern which is very different from that among quarks. The equation below summarizes our present knowledge about neutrino masses and mixings [2] in the notation $|v_\alpha\rangle = \sum U_{\alpha i}|v_i\rangle$ (where $\alpha = e, \mu.\tau$ is the flavor index and $i = 1, 2, 3$ denotes the mass eigenstate index). For the CP conserving case $U_{\alpha i}$ are functions of three angles, θ_{ij} and for these angles we have:

$$\begin{aligned}
&\sin^2 2\theta_A \equiv \sin^2 2\theta_{23} \geq 0.89, \\
&\Delta m_A^2 \simeq 1.4 \times 10^{-3}\text{eV}^2 - 3.3 \times 10^{-3}\text{eV}^2, \\
&\sin^2 \theta_\odot \equiv \sin^2 \theta_{12} \simeq 0.23 - 0.37, \\
&\Delta m_\odot^2 \simeq 7.3 \times 10^{-5}\text{eV}^2 - 9.1 \times 10^{-5}\text{eV}^2, \\
&\sin^2 \theta_{13} \leq 0.047.
\end{aligned} \tag{1}$$

For the sake of comparision, note the corresponding quark mixing angles i.e. $\theta_{12}^q \simeq 0.22$; $\theta_{23}^q \simeq 0.04$ and $\theta_{13}^q \simeq 0.004$. Clearly, the mixing pattern in the lepton sector is very different from that among quarks.

It is also important to point that while the mass differences among neutrinos are fairly well determined, the situation with respect to absolute values of masses is far from clear. This is another major gap in our understanding of neutrinos compared to quarks. At present, there are three equally viable mass arrangements among the neutrinos:

(*i*) Normal hierarchy i.e. $m_1 \ll m_2 \ll m_3$. In this case, we can deduce the value of $m_3 \simeq \sqrt{\Delta m_{23}^2} \equiv \sqrt{\Delta m_A^2} \simeq 0.03$–0.07 eV. In this case $\Delta m_{23}^2 \equiv m_3^2 - m_2^2 > 0$. The solar neutrino oscillation involves the two lighter levels. The mass of the lightest neutrino is unconstrained. If $m_1 \ll m_2$, then we get the value of $m_2 \simeq 0.008$ eV.

(*ii*) Inverted hierarchy i.e. $m_1 \simeq m_2 \gg m_3$ with $m_{1,2} \simeq \sqrt{\Delta m_{23}^2} \simeq 0.03 - 0.07$ eV. In this case, solar neutrino oscillation takes place between the heavier levels and we have $\Delta m_{23}^2 \equiv m_3^2 - m_2^2 < 0$.

(*iii*) Degenerate neutrinos i.e. $m_1 \simeq m_2 \simeq m_3$.

There are a large number of experiments in the planning stage to improve our knowledge of mixings, to determine the mass ordering and also to find out whether neutrinos are Majorana (i.e. their own antiparticles) or Dirac fermions. These are not only crucial pieces of information about the neutrinos that we need to know to elevate our knowledge of them to the same level as the quarks but it is becoming increasingly clear that they will also point very clearly to the direction of new physics beyond standard model. For instance if neutrinos are established to be Dirac fermions, seesaw mechanism in its simplest form will not be able to describe their masses and a major theoretical idea will be disproved.

If we accept the seesaw mechanism as the explanation for the smallness of neutrino masses, the next major challenge for theory is to understand the unusual mixing pattern among them. The hope is that in the process of understanding the mixings we will find out which of the mass patterns is realized in Nature and more importantly, will get a definite clue to the nature of new physics.

In this talk I will give some promising possibilities for this new physics and discuss their experimental tests. In particular, I will argue that the seesaw mechanism for small neutrino masses requires a scale of new physics close to the traditional scale of grand unification where all forces and matter are supposed to become unified and a new symmetry B-L which naturally arises if the gauge group is assumed to be SO(10) [5]. I will then show that a minimal version of supersymmetric SO(10) provides a very natural way to understand the large solar as well as atmospheric neutrino mixing angles while predicting a value for the mixing angle $\theta_{13} \sim 0.1$–0.18 depending on details. This prediction can be tested by the various planned reactor [3] and long baseline experiments [4]. I will also discuss two other related ideas which are outside the SO(10) framework but are based on one of the maximal subgroups of SO(10) i.e. $SU(2)_L \times SU(2)_R \times SU(4)_c$ [6] that also unifies quarks and leptons and then argue that measurement of the parameter θ_{13} may provide crucial insight into the question of whether there is quark-lepton unification at high scale.

While in this talk I will assume that there are only three neurtinos, we do not know for sure how many neutrinos there are. In particular if the LSND results are confirmed by the Mini Boone experiment [7], we will have evidence that there are more neutrinos and the discussions presented here will have to be extended.

2. Seesaw mechanism, B-L and left-right symmetry

In order to introduce the seesaw mechanism, which will form the anchor for the main body of the talk, let us start with a discussion of neutrino mass in the standard model. It is based on the gauge group $SU(3)_c \times SU(2)_L \times U(1)_Y$ group under which the quarks and leptons transform as follows: Quarks: $Q_L(3, 2, \frac{1}{3})$; $u_R(3, 1, \frac{4}{3})$; $d_R(3, 1, -\frac{2}{3})$; Leptons $L(1, 2 - 1)$; $e_R(1, 1, -2)$; Higgs Boson $\mathbf{H}(1, 2, +1)$; Gluons $G_a(8, 1, 0)$ and Weak Gauge Fields $W^\pm, Z, \gamma$. The electroweak symmetry $SU(2)_L \times U(1)_Y$ is broken by the vacuum expectation of the Higgs doublet $\langle H^0 \rangle = v_{wk} \simeq 180\,\text{GeV}$, which gives mass to the gauge bosons and all fermions except the neutrino. The model had been a complete success in describing all known low energy phenomena, until the evidence for neutrino masses appeared.

Note that there is no right handed neutrino in the standard model and this directly leads to massless neutrinos at the tree level. The situation remains the same not only to all orders in perturbation theory but also when nonperturbative effects are taken into account. This is due to existence of an exact B-L symmetry in the theory and the absence of the right handed neutrino, N_R. The absence of the right handed neutrino from the standard model of course destroys the symmetry between quarks and leptons that is so obvious in weak interactions.

Once the right handed neutrinos (N_R) are included in the standard model, new Yukawa couplings of the form $h_\nu \bar{L} H N_R$ are allowed which after electroweak symmetry breaking lead to a neutrino mass, $M_D \equiv h_\nu v_{wk}$. Since h_ν is expected to be of same order as the charged fermion Yukawa couplings in the model, these masses are much too large to describe neutrino oscillations. Luckily, since the N_R's are singlets under the standard model gauge group, they are allowed to have Majorana masses unlike the charged fermions. We denote them by $M_R N_R^T C^{-1} N_R$ (where C is the Dirac charge conjugation matrix). The masses M_R are not constrained by the gauge symmetry and can therefore

be arbitrarily large (i.e. $M_R \gg h_\nu v_{wk}$). This together with mass induced by Yukawa couplings (called the Dirac mass) leads to a the mass matrix for the neutrinos (left and right handed neutrinos together) which has the form

$$\mathcal{M}_\nu = \begin{pmatrix} 0 & M_D \\ M_D^T & M_R \end{pmatrix} \tag{2}$$

where M_D and M_R are 3×3 matrices. Diagonalizing this mass matrix, one gets the mass matrix for the light neutrinos (the seesaw formula) as:

$$\mathcal{M}_\nu = -M_D^T M_R^{-1} M_D. \tag{3}$$

Since as already noted M_R can be much larger than M_D, one finds that $m_\nu \ll m_{e,u,d}$ very naturally.

Seesaw mechanism of course raises its own questions:

- Is there a natural reason for the existence of the right handed neutrinos other than quark-lepton symmetry?
- What determines the scale of M_R?
- Is the seesaw mechanism by itself enough to explain all aspects of neutrino masses and mixings?

Below, we try to answer some of these questions. Restoration of quark-lepton symmetry and unification of quarks and leptons within a single gauge theory framework provided the first inspiration to bring the right handed neutrino into particle physics [6]. It is easy to see that in the presence of the N_R's, the minimal anomaly free gauge group of weak interactions expands beyond the standard model and becomes the left-right symmetric group $SU(2)_L \times SU(2)_R \times U(1)_{B-L}$ [8] which is a subgroup of the $SU(2)_L \times SU(2)_R \times SU(4)_c$ group. This makes the weak interactions parity conserving at short distances [8], providing another appealing feature of adding the right handed neutrino. To see this explicitly, we give in Table I the assignment of fermions and Higgs fields to the left-right gauge group.

It is clear that this theory leads to a weak interaction Lagrangian of the form

$$\mathcal{L}_{wk} = \frac{g}{2}(\vec{j}_L^\mu \cdot \vec{W}_{L,\mu} + \vec{j}_R^\mu \cdot \vec{W}_{R,\mu}) \tag{4}$$

which is parity conserving prior to symmetry breaking. Furthermore, the electric charge formula is given by [9]:

$$Q = I_{3L} + I_{3R} + \frac{B - L}{2}. \tag{5}$$

where all the terms have physical meaning unlike the case of the standard model.

Table I. *Assignment of fermions and Higgs field to the left-right gauge group.*

Fields	$SU(2)_L \times SU(2)_R \times U(1)_{B-L}$ representation
$Q_L \equiv \begin{pmatrix} u_L \\ d_L \end{pmatrix}$	$(2, 1, +\frac{1}{3})$
$Q_R \equiv \begin{pmatrix} u_R \\ d_R \end{pmatrix}$	$(1, 2, \frac{1}{3})$
$L_L \equiv \begin{pmatrix} \nu_L \\ e_L \end{pmatrix}$	$(2, 1, -1)$
$L_R \equiv \begin{pmatrix} \nu_R \\ e_R \end{pmatrix}$	$(1, 2, -1)$
ϕ	$(2, 2, 0)$
Δ_L	$(3, 1, +2)$
Δ_R	$(1, 3, +2)$

The left-right symmetric theories face two challenges: (*i*) how does the predominantly V-A nature of weak interactions emerge in such a theory and (*ii*) how does one understand the small neutrino masses since $SU(2)_R$ makes both the electron and the neutrino much more similar than they were in the standard model. We will see that both these challenges are met in one stroke i.e. breakdown of $SU(2)_R \times U(1)_{B-L}$ symmetry to $U(1)_Y$ not only explains the V-A nature of weak interactions but it also explain why $m_\nu \ll m_e$ via the seesaw mechanism. The seesaw scale then becomes the scale of parity violation. Furthermore, when the gauge symmetry $SU(2)_R \times U(1)_{B-L}$ is broken down while keeping the standard model symmetry unbroken, one finds from Eq. (5) the relation $\Delta I_{3R} = -\Delta\frac{B-L}{2}$. This connects $B-L$ breaking to the breakdown of parity symmetry i.e. $\Delta I_{3R} \neq 0$ and clearly implies that neutrinos must be Majorana particles.

To see this explicitly, we break the gauge symmetry of the left-right model in two stages : in stage I, vacuum expectation values (vev) of the Higgs multiplets $\Delta_R(1, 3, 2)$ breaks the left-right gauge symmetry to the standard model gauge group and in stage II by the bidoublet $\phi(2, 2, 0)$ vev breaks the standard model group to $SU(3)_c \times U(1)_{em}$. In the first stage of symmetry breaking, the right handed neutrino picks up a mass of order $f\langle\Delta_R^0\rangle \equiv fv_R$. Denoting the left and right handed neutrino by (ν, N) (in a two component notation), the mass matrix for neutrinos at this stage looks like

$$\mathcal{M}_\nu^0 = \begin{pmatrix} 0 & 0 \\ 0 & fv_R \end{pmatrix}. \tag{6}$$

At this stage, familiar standard model particles remain massless. As soon as the standard model symmetry is broken by the bidoublet ϕ i.e. $\langle\phi\rangle \equiv \begin{pmatrix} \kappa & 0 \\ 0 & \kappa' \end{pmatrix}$, the W and Z boson as well as the fermions pick up mass. I will generically denote κ, κ' by a common symbol v_{wk}. The contribution to neutrino mass at this stage look like

$$\mathcal{M}_\nu^0 = \begin{pmatrix} fv_L & hv_{wk} \\ hv_{wk} & fv_R \end{pmatrix}. \tag{7}$$

Note the appearance of a new term in the neutrino mass matrix i.e. $v_L = \frac{v_{wk}^2}{v_R}$ compared to the seesaw matrix given in Eq. (1). This is a reflection of parity invariance of the model. Diagonalizing this matrix, we get a modified seesaw formula for the light neutrino mass matrix

$$\mathcal{M}_\nu = fv_L - h_\nu^T f_R^{-1} h_\nu \left(\frac{v_{wk}^2}{v_R}\right). \tag{8}$$

The important point to note is that v_L is suppressed by the same factor as the second term so that despite the new contribution to neutrino masses, seesaw suppression remains [10]. This is called the type II seesaw in contrast with the formula in Eq. (2) which is called type I seesaw formula.

An important physical meaning of the seesaw formula is brought out when it is viewed in the context of left-right models. Note that $m_\nu \to 0$ when v_R goes to infinity. In the same limit the weak interactions become pure V-A type. Therefore, left-right model derivation of the seesaw formula smoothly connects smallness of neutrino mass with suppression of V + A part of the weak interactions providing an important clarification of a major puzzle of the standard model i.e. why are weak interactions are near maximally parity violating ? The answer is that they are near

maximally parity violating because the neutrino mass happens to be small.

In a subsequent section, we will discuss the connection of the seesaw mass scale with the scale of grand unification. SO(10) is the simplest gauge group that contains the right handed neutrino needed to implement the seesaw mechanism and also it is important to note that the left-right symmetric gauge group is a subgroup of the SO(10) group, which therefore provides an attractive over all grand unified framework for the discussion of neutrino masses. The extra bonus one may expect is that since bigger symmetries tend to relate different parameters of a theory, one may be able to predict neutrino masses and mixings. We will present a model where indeed this happens.

Before proceeding further, it is important to point out that type I and type II seesaw can be tested by the nature of neutrino spectrum in a model independent way. Since type I seesaw involves the Dirac mass of the neutrino, a general expectation is that it scales with generation the same way as the charged fermions of the standard model. In this case, unless there is extreme hierarchy among the right handed neutrinos, one would expect the spectrum to be hierarchical. On the other hand, it has been realized for a long time [12] that if neutrino masses are quasi-degenerate, it is a tell-tale sign of type II seesaw with the triplet vev term being the dominant one. However, a normal hierarchy can also arise with type II seesaw as we discuss in the example below. Therefore, whereas a normal hierarchy cannot distinguish between type I and type II seesaw, a quasi-degenerate spectrum is a definite sign of type II kind.

3. Seesaw and large neutrino mixings

While seesaw mechanism provides a simple framework for understanding the smallness of neutrino masses, it does not throw any light on the question of why neutrino mixings are large. The point is that mixings are a consequence of the structure of the light neutrino mass matrix and the seesaw mechanism is only statement about the scale of new physics. This can also be understood by doing a simple parameter counting. If we work in a basis where the right handed neutrino masses are diagonal, there are 18 parameters describing the seesaw formula for neutrino masses – three RH neutrino masses and 15 parameters in the Dirac mass matrix. On the other hand, there are only nine observables (three masses, three mixing angle and three phases) describing low energy neutrino sector. Thus there are twice as many parameters as observables. As a result, understanding neutrino mixings needs inputs beyond the simple seesaw mechanism to fix the neutrino mass matrix. Nonetheless, since the large mixings could arise from the physics involving the seesaw formula e.g. flavor structure of M_R, the large mixings are not in obvious contradiction with quark lepton unification. This becomes clear in the examples given below.

Many seesaw models for large mixings have been considered in the literature [11]. In the following section, I will focus on a recently discussed minimal SO(10) model, where without any assumption other than SO(10) grand unification, one can indeed predict all but one neutrino parameters. I will then consider a case where assumption of quasi-degeneracy in the neutrino spectrum at high scale leads in a natural way via radiative corrections to large mixings at low energies as well as briefly describe a model of quark-lepton complementarity.

To understand the fundamental physics behind neutrino mixings, we first write down the neutrino mass matrix that leads to maximal solar and atmospheric mixing for the case of normal hierarchy:

$$\mathcal{M}_\nu = \frac{\sqrt{\Delta m_A^2}}{2} \begin{pmatrix} c\epsilon & b\epsilon & d\epsilon \\ b\epsilon & 1+a\epsilon & -1 \\ d\epsilon & -1 & 1+\epsilon \end{pmatrix} \qquad (9)$$

where $\epsilon \simeq \sqrt{\frac{\Delta m_\odot^2}{\Delta m_A^2}}$ and parameters a, b, c, d are of order one. Any theory of neutrino which attempts to explain the observed mixing pattern for the case of normal hierarchy must strive to get a mass matrix of this form.

It is important to point out that the above mass matrix when $a = 1$ and $b = d$, becomes symmetric under the interchange of μ and τ and yields $\theta_{13} = 0$. It was shown in two recent papers that [13], if $\mu - \tau$ symmetry is broken via $a \neq 1$ with $b = d$, then typically $\theta_{13} \sim \frac{\Delta m_\odot^2}{\Delta m_A^2}$ whereas if we have $b \neq d$, one gets $\theta_{13} \sim \sqrt{\frac{\Delta m_\odot^2}{\Delta m_A^2}}$. It turns out that most grand unified (or quark-lepton unified) theories lead to $\theta_{13} \sim \sqrt{\frac{\Delta m_\odot^2}{\Delta m_A^2}}$ (see examples below). Therefore, measurement of the mixing parameter θ_{13} may provide a way to test for possible quark-lepton unification at high scales.

4. A predictive minimal SO(10) theory for neutrinos

The main reason for considering SO(10) for neutrino masses is that its **16** dimensional spinor representation consists of all fifteen standard model fermions plus the right handed neutrino arranged according to the it $SU(2)_L \times SU(2)_R \times SU(4)_c$ [6] subgroup as follows:

$$\Psi = \begin{pmatrix} u_1 & u_2 & u_3 & \nu \\ d_1 & d_2 & d_3 & e \end{pmatrix}. \qquad (10)$$

There are three such spinors for three fermion families.

In order to implement the seesaw mechanism in the SO(10) model, one must break the B-L symmetry. In supersymmetric SO(10) models, how B-L breaks has profound consequences for low energy physics. For instance, if B-L is broken by a Higgs field belonging to the **16** dimensional Higgs field (to be denoted by Ψ_H), then the field that acquires a nonzero vev has the quantum numbers of the ν_R field i.e. B-L breaks by one unit. In this case higher dimensional operators of the form $\Psi\Psi\Psi\Psi_H$ will lead to R-parity violating operators in the effective low energy MSSM theory such as $QLd^c, u^c d^c d^c$ etc which can lead to large breaking of lepton and baryon number symmetry and hence unacceptable rates for proton decay. This theory also has no dark matter candidate.

On the other hand, if one breaks B-L by a **126** dimensional Higgs field, the member of this multiplet that acquires vev has $B-L = 2$. R-parity is therefore left as an automatic symmetry of the low energy Lagrangian. There is a naturally stable dark matter in this case. It has recently been shown that this class of models lead to a very predictive scenario for neutrino mixings [14, 15, 16, 17]. We summarize this model below.

As already noted earlier, any theory with asymptotic parity symmetry leads to type II seesaw formula and if B-L is broken by a **126** field, then the first term in the type II seesaw formula can in principle dominate in the seesaw formula. We will discuss a model of this type below.

The basic ingredients of this model are that one considers only two Higgs multiplets that contribute to fermion masses i.e. one **10** and one **126**. A unique property of the **126** multiplet is that it not only breaks the B-L symmetry and therefore contributes to right handed neutrino masses, but it also contributes to charged fermion masses by virtue of the fact that it contains MSSM doublets which mix with those from the **10** dimensional multiplets and survive down to the MSSM scale. This leads to a tremendous reduction of the number of arbitrary parameters, as we will see below.

There are only two Yukawa coupling matrices in this model: (*i*) h for the **10** Higgs and (*ii*) f for the **126** Higgs. SO(10) has the property that the Yukawa couplings involving the **10** and **126** Higgs representations are symmetric. Therefore if we assume that CP violation arises from other sectors of the theory (e.g. squark masses) and work in a basis where one of these two sets of Yukawa coupling matrices is diagonal, then it will have only nine coupling parameters. Noting the fact that the (2, 2, 15) submultiplet of **126** has a pair of standard model doublets that contributes to charged fermion masses, one can write the quark and lepton mass matrices as follows [14]:

$$\begin{aligned} M_u &= h\kappa_u + f v_u, \\ M_d &= h\kappa_d + f v_d, \\ M_\ell &= h\kappa_d - 3 f v_d, \\ M_{\nu_D} &= h\kappa_u - 3 f v_u \end{aligned} \qquad (11)$$

where $\kappa_{u,d}$ are the vev's of the up and down standard model type Higgs fields in the **10** multiplet and $v_{u,d}$ are the corresponding vevs for the same doublets in **126**. The vevs added to the Yukawa couplings give a total of 13 parameters in the theory. They are determined by 13 inputs (six quark masses, three lepton masses and three quark mixing angles and weak scale). There is therefore no free parameter in the neutrino sector except for an overall seesaw scale.

To determine the light neutrino masses, we use the seesaw formula in Eq. (7), where the **f** is nothing but the **126** Yukawa coupling. These models were extensively discussed in the last decade [15] using type I seesaw formula. It was pointed out in Ref. [16] that if the direct triplet term in type II seesaw dominates, then it provides a very natural understanding of the large atmospheric mixing angle for the case of two generations without invoking any symmetries. Subsequently it was shown in Ref. [17] that the same $b - \tau$ mass convergence also provides an explanation of large solar mixing as well as small θ_{13} making the model realistic and experimentally interesting.

A simple way to see how large mixings arise in this model is to note that when the triplet term dominates the seesaw formula, we have the neutrino mass matrix $M_\nu \propto f$, where f matrix is the **126** coupling to fermions discussed earlier. Using the above equations, one can derive the following sumrule :

$$M_\nu = c(M_d - M_\ell). \qquad (12)$$

To see how this leads to large atmospheric and solar mixing, let us work in the basis where the down quark mass matrix is diagonal. All the quark mixing effects are then in the up quark mass matrix i.e. $M_u = U_{CKM}^T M_u^d U_{CKM}$. Note further that the minimality of the Higgs content leads to the following sumrule among the mass matrices:

$$k\tilde{M}_\ell = r\tilde{M}_d + \tilde{M}_u \qquad (13)$$

where the tilde denotes the fact that we have made the mass matrices dimensionless by dividing them by the heaviest mass of the species i.e. up quark mass matrix by m_t, down quark mass matrix by m_b etc. k, r are functions of the symmetry breaking parameters of the model. Using the hierarchical pattern of quark mixings, we can conclude that that we have

$$M_{d,\ell} \approx m_{b,\tau} \begin{pmatrix} \lambda^3 & \lambda^3 & \lambda^3 \\ \lambda^3 & \lambda^2 & \lambda^2 \\ \lambda^3 & \lambda^2 & 1 \end{pmatrix} \tag{14}$$

where $\lambda \sim 0.22$ and the matrix elements are supposed to give only the approximate order of magnitude. An important consequence of the relation between the charged lepton and the quark mass matrices in Eq. (12) is that the charged lepton contribution to the neutrino mixing matrix i.e. $U_\ell \simeq \mathbf{1} + O(\lambda)$ or close to identity matrix. As a result the neutrino mixing matrix is given by $U_{PMNS} = U_\ell^\dagger U_\nu \simeq U_\nu$. Thus the dominant contribution to large mixings will come from U_ν, which in turn will be dictated by the sum rule in Eq. (11).

To show that U_ν has two large mixings, we extrapolate the quark masses to the GUT scale and use the well known fact that $m_b - m_\tau \approx m_\tau \lambda^2$ for a wide range of values of $\tan\beta$. Using this the neutrino mass matrix $M_\nu = c(M_d - M_\ell)$ roughly takes the form

$$M_\nu = c(M_d - M_\ell) \approx m_0 \begin{pmatrix} \lambda^3 & \lambda^3 & \lambda^3 \\ \lambda^3 & \lambda^2 & \lambda^2 \\ \lambda^3 & \lambda^2 & \lambda^2 \end{pmatrix}. \tag{15}$$

This mass matrix is in the form discussed in Eq. (8) and it is easy to see that both the θ_{12} (solar angle) and θ_{23} (the atmospheric angle) are now large. The detailed magnitudes of these angles of course depend on the details of the quark masses at the GUT scale. Using the extrapolated values of the quark masses and mixing angles to the GUT scale, the predictions of this model for various oscillation parameters are given in Fig. 1, 2 and 3 in a self expalanatory notation. The predictions for the solar and atmospheric mixing angles fall within 3σ range of the present central values. Note specifically the prediction in Fig. 3 for $U_{e3} \simeq 0.18$ which can be tested in MINOS as well as other planned Long Base Line neutrino experiments such as Numi-Off-Axis, JPARC etc. This model has been the subject of many investigations, which we do not discuss here [18].

4.1. *CP violation in the minimal SO(10) model*

In the discussion given above, it was assumed that CP violation is non-CKM type and resides in the soft SUSY breaking terms of the Lagrangian. The overwhelming evidence from experiments seem to be that CP violation is perhaps is of CKM type. It has recently been pointed out that with slight modification, one can include CKM CP violation in the model [19]. The basic idea is to include

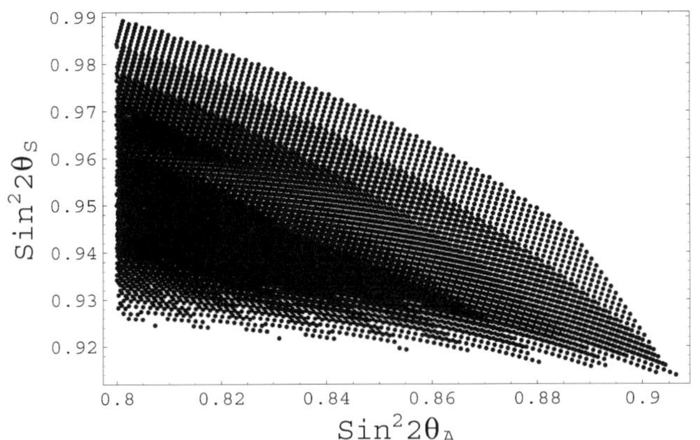

Fig. 1. The figure shows the predictions of the minimal SO(10) model for $\sin^2 2\theta_\odot$ and $\sin^2 2\theta_A$ for the presently range of quark masses. Note that $\sin^2 2\theta_\odot \geq 0.9$ and $\sin^2 2\theta_A \leq 0.9$.

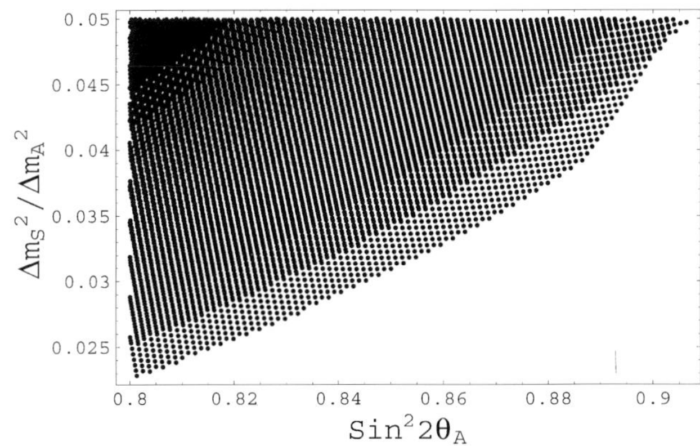

Fig. 2. The figure shows the predictions of the minimal SO(10) model for $\sin^2 2\theta_A$ and $\Delta m_\odot^2 / \Delta m_A^2$ for the range of quark masses and mixings that fit charged lepton masses.

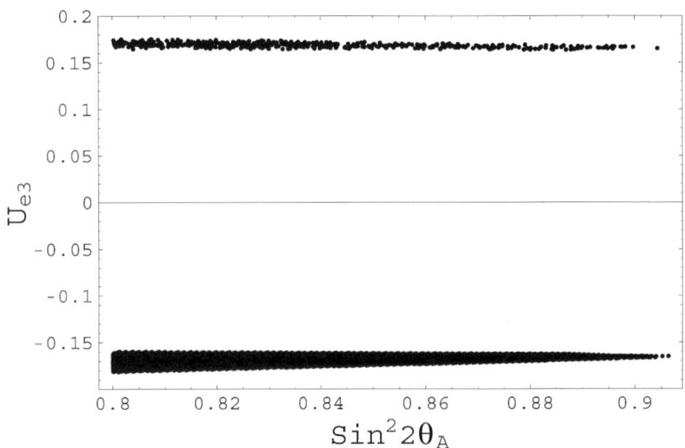

Fig. 3. The figure shows the predictions of the minimal SO(10) model for $\sin^2 2\theta_A$ and U_{e3} for the allowed range of parameters in the model. Note that U_{e3} is very close to the upper limit allowed by the existing reactor experiments.

all higher dimensional operators of type $h'\Psi\Psi\bar{\Delta}\Sigma/M$ where $\bar{\Delta}$ and Σ denote respectively the **126** and the **210** dimensional representation. It is then clear that those operators transforming as **10** and **126** representations will simply redefine the h, f coupling matrices and add no new physics. On the other hand the higher dimensional operator that transforms like an effective **120** representation will add a new piece to all fermion masses. Now suppose we introduce a parity symmetry into the theory which transforms Ψ to Ψ^{c*}, then it turns out that the couplings h and f become real and symmetric matrices whereas the **120** coupling (denoted by h') becomes imaginary and antisymmetric. This process introduces three new parameters into the theory and the charged fermion masses are related to the fundamental couplings in the theory as follows:

$$M_u = h\kappa_u + fv_u + h'v_u,$$

$$M_d = h\kappa_d + fv_d + h'v_d, \tag{16}$$

$$M_\ell = h\kappa_d - 3fv_d - 3h'v_d,$$

$$M_{\nu_D} = h\kappa_u - 3fv_u - 3h'v_u. \tag{17}$$

Note that the extra contribution compared to Eq. (10) is antisymmetric which therefore does not interfere with the mechanism that lead to $\mathcal{M}_{\nu,33}$ becoming small as a result of $b - \tau$ convergence. Hence the natural way that θ_A became large in the CP conserving case remains.

Let us discuss if the new model is still predictive in the neutrino sector. Of the three new parameters, one is determined by the CP

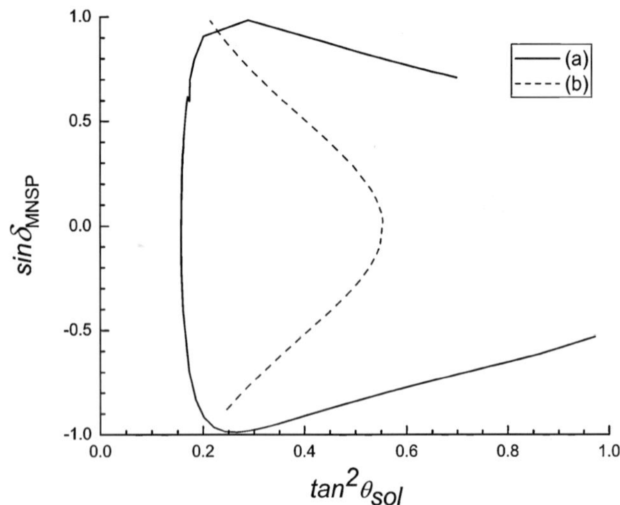

Fig. 4. The prediction of MNSP phase is plotted as a function of the solar mixing angle. The two lines (a) and (b) correspond to two choices of signs of the fermion masses. The phases are plotted for $\Delta m^2_{sol}/\Delta m^2_A = 0.02$.

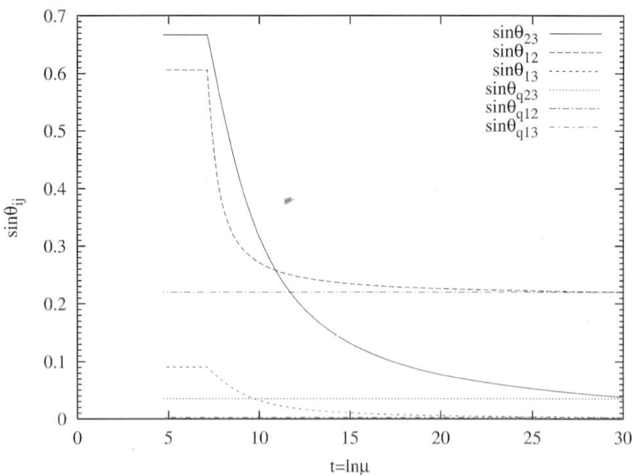

Fig. 5. Radiative magnification of small quark-like neutrino mixings at the seesaw scale to bilarge values at low energies. The solid, dashed and dotted lines represent $\sin\theta_{23}$, $\sin\theta_{13}$, and $\sin\theta_{12}$, respectively.

violating quark phase. the two others are determined by the solar mixing angle and the solar mass difference squared. Therefore we lose the prediction for these parameters. However, we can predict in addition to θ_A which is close to maximal, $\theta_{13} \geq 0.1$ and the Dirac phase for the neutrinos. We show the predictions for Dirac phase in Fig. 4. This is a unique property of the model that it can predict the leptonic CP phase.

5. Radiative generation of large mixings: another application of type II seesaw

As alluded before, type II seesaw liberates the neutrinos from obeying normal generational hierarchy and instead could easily be quasi-degenerate in mass. This raises a new way to understand the large mixings instead of having to generate them in the original seesaw theory as is normally done. The basic idea is that at the seesaw scale, all mixings angles are small. Since the observed neutrino mixings are the weak scale observables, one must extrapolate [20] the seesaw scale mass matrices to the weak scale and recalculate the mixing angles. The extrapolation formula is

$$M_\nu(M_Z) = \mathbf{I} M_\nu(v_R)\mathbf{I},\tag{18}$$

where

$$\mathbf{I}_{\alpha\alpha} = \left(1 - \frac{h_\alpha^2}{16\pi^2}\right).\tag{19}$$

Note that since $h_\alpha = \sqrt{2}m_\alpha/v_{wk}$ (α being the charged lepton index), in the extrapolation only the τ-lepton makes a difference. In the MSSM, this increases the $M_{\tau\tau}$ entry of the neutrino mass matrix and essentially leaves the others unchanged. It was shown in ref. [21] that if the muon and the tau neutrinos are nearly degenerate but not degenerate enough in mass at the seesaw scale, the radiative corrections can become large enough so that at the weak scale the two diagonal elements of M_ν become much more degenerate. This leads to an enhancement of the mixing angle to become almost maximal value. This can also be seen from the renormalization group equations when they are written in the mass basis [22]. Denoting the mixing angles as θ_{ij} where i, j stand for

generations, the equations are:

$$\frac{ds_{23}}{dt} = -F_\tau c_{23}{}^2\left(-s_{12}U_{\tau 1}D_{31} + c_{12}U_{\tau 2}D_{32}\right),\tag{20}$$

$$\frac{ds_{13}}{dt} = -F_\tau c_{23}c_{13}{}^2\left(c_{12}U_{\tau 1}D_{31} + s_{12}U_{\tau 2}D_{32}\right),\tag{21}$$

$$\frac{ds_{12}}{dt} = -F_\tau c_{12}\left(c_{23}s_{13}s_{12}U_{\tau 1}D_{31} - c_{23}s_{13}c_{12}U_{\tau 2}D_{32}\right.$$
$$\left. + U_{\tau 1}U_{\tau 2}D_{21}\right),\tag{22}$$

where $D_{ij} = (m_i + m_j)/(m_i - m_j)$ and $U_{\tau 1,2,3}$ are functions of the neutrino mixings angles. The presence of $(m_i - m_j)$ in the denominator makes it clear that as $m_i \simeq m_j$, that particular coefficient becomes large and as we extrapolate from the GUT scale to the weak scale, small mixing angles at GUT scale become large at the weak scale. It has been shown recently that indeed such a mechanism for understanding large mixings can work for three generations [23]. It was shown that if we identify the seesaw scale neutrino mixing angles with the corresponding quark mixings and assume quasi-degenerate neutrinos, the weak scale solar and atmospheric angles get magnified to the desired level while due to the extreme smallness of V_{ub}, the magnified value of U_{e3} remains within its present upper limit. In figure 5, we show the evolution of the mixing angles to the weak scale. A requirement for this scenario to work is that the common mass of neutrinos must be larger than 0.1 eV, a result that can be tested in neutrinoless double beta experiments.

6. Quark-lepton complementarity and large solar mixing

There has been a recent suggestion [24] that perhaps the large but not maximal solar mixing angle is related to physics of the quark sector. According to this, the deviation from maximality of the solar mixing may be related to the quark mixing angle $\theta_C \equiv \theta^q_{12}$ and is based on the observation that the mixing angle responsible for solar neutrino oscillations, $\theta_\odot \equiv \theta^\nu_{12}$ satisfies an interesting complementarity relation with the corresponding angle in the quark sector $\theta_{Cabibbo} \equiv \theta^q_{12}$ i.e. $\theta^\nu_{12} + \theta^q_{12} \simeq \pi/4$. While it is quite possible that this relation is purely accidental or due to some other dynamical effects, it is interesting to pursue the possibility that there is a deep meaning behind it and see where it leads. It has been shown in a recent paper that if Nature is quark lepton unified at high scale, then a relation between θ_{12}^{ν} and θ_{12}^{q} can be obtained in a natural manner provided the neutrinos obey the inverse hierarchy [25]. It predicts $sin^2\theta_{\odot} \simeq 0.34$ which agrees with present data at the 2σ level. It also predicts a large $\theta_{13} \sim 0.18$, both of which are predictions that can be tested experimentally in the near future.

7. Conclusion

In summary, the seesaw mechanism is by far the simplest and most appealing way to understand neutrino masses. It not only improves the aesthetic appeal of the standard model by restoring quark-lepton symmetry but it also makes weak interactions asymptotically parity conserving. Further more it connects neutrino masses with the hypothesis of grand unification. In this talk I have discussed three ways to understand the large solar and atmospheric neutrino mixings within the frameworks that unify quarks and lepton and in one case into a grand unified model based on SO(10). All three models predict large values for θ_{13} and can therefore be tested in forthcoming experiments. The SO(10) model appears to be most promising since it not only resolves the difficulties of the minimal SUSY SU(5) GUT but is also a minimal predictive model for neutrinos.

From these examples, one is also tempted to conclude that a large θ_{13} could be a generic feature of models that unify quarks and leptons, which if true will be a unique window to a very important question in beyond the standard model physics.

Acknowledgments

This work is partially supported by the National Science Foundation Grant No. PHY-0354401. I would like to thank the organizers of the Nobel symposium 129 at Haga Slott for creating a very pleasant environment for physics.

References

1. Minkowski, P., Phys. Lett. **B67**, 421 (1977); Gell-Mann, M., Ramond, P. and Slansky, R., "Supergravity" (P. van Nieuwenhuizen *et al.*, eds.), (North Holland, Amsterdam, 1980), p. 315; Yanagida, T., in Proceedings of the Workshop on the Unified Theory and the Baryon Number in the Universe (O. Sawada and A. Sugamoto, eds.), (KEK, Tsukuba, Japan, 1979), p. 95; Glashow, S. L., "The future of elementary particle physics," in Proc. 1979 Cargèse Summer Institute on Quarks and Leptons (M. Lévy *et al.*, eds.), (Plenum Press, New York, 1980), pp. 687–713; Mohapatra, R. N. and Senjanović, G., Phys. Rev. Lett. **44**, 912 (1980).

2. For recent reviews, see Bahcall, J. N., Gonzalez-Garcia, M. C. and Pena-Garay, C., J. High En. Phys. **0408**, 016 (2004) [arXiv:hep-ph/0406294]; Gonzalez-Garcia and Maltoni, M., hep-ph/0406056; Maltoni, M., Schwetz, T., Tortola, M. and Valle, J. W. F., hep-ph/0405172.

3. Anderson, K. *et al.*, arXiv:hep-ex/0402041; Apollonio, M. *et al.*, Eur. Phys. J. C **27**, 331 (2003) arXiv:hep-ex/0301017.

4. Diwan, M. V. *et al.*, Phys. Rev. D **68**, 012002 (2003) arXiv:hep-ph/0303081; Ayrea, D. *et al.*, hep-ex/0210005; Itow, Y. *et al.*, (T2K collaboration) hep-ex/0106019; Ambats, I. *et al.*, (NOvA Collaboration), FERMILAB-PROPOSAL-0929.

5. Georgi, H., in "Particles and Fields," (ed. C. E. Carlson), A. I. P. (1975); Fritzsch, H. and Minkowski, P., Ann. Phys. **93**, 193 (1975).

6. Pati, J. C. and Salam, A., Phys. Rev. **D10**, 275 (1974).

7. Conrad, J., these proceedings.

8. Pati, J. C. and Salam, A., [6]; Mohapatra, R. N. and Pati, J. C., Phys. Rev. **D 11**, 566, 2558 (1975); Senjanović, G. and Mohapatra, R. N., Phys. Rev. **D 12**, 1502 (1975).

9. Mohapatra, R. N. and Marshak, R. E., Phys. Lett. **B 91**, 222 (1980); Davidson, A., Phys. Rev. **D20**, 776 (1979).

10. Lazarides, G., Shafi, Q. and Wetterich, C., Nucl. Phys. **B181**, 287 (1981); Mohapatra, R. N. and Senjanović, G., Phys. Rev. **D 23**, 165 (1981).

11. Smirnov, A., hep-ph/0311259; King, S. F., Rept. Prog. Phys. **67**, 107 (2004); Altarelli, G. and Feruglio, F., hep-ph/0405048; Mohapatra, R. N., hep-ph/0211252; New J. Phys. **6**, 82 (2004).

12. See for instance Caldwell, D. O. and Mohapatra, R. N., Phys. Rev. D **48**, 3259 (1993); Lee, D. G. and Mohapatra, R. N., Phys. Lett. B **329**, 463 (1994); Antusch, S. and King, S. F., hep-ph/0402121.

13. Mohapatra, R. N., Slac Summer Inst. lecture; http://www-conf.slac.stanford.edu/ssi/2004; hep-ph/0408187; JHEP, **10**, 027 (2004); Grimus, W. *et al.*, hep-ph/0408123.

14. Babu, K. S. and Mohapatra, R. N., Phys. Rev. Lett. **70**, 2845 (1993).

15. Lee, D. G. and Mohapatra, R. N., Phys. Rev. **D 51**, 1353 (1995); Lavoura, L., Phys. Rev. **D 48**, 5440 (1993); Brahmachari, B. and Mohapatra, R. N., Phys. Rev. **D 58**, 015001 (1998); Oda, K., Takasugi, E., Tanaka, M. and Yoshimura, M., Phys. Rev. **D 58**, 055001 (1999); Matsuda, K., Koide, Y., Fukuyama, T. and Okada, N., Phys. Rev. **D 65**, 033008 (2002); Fukuyama, T. and Okada, N., hep-ph/0205066; Oshimo, N., hep-ph/0305166.

16. Bajc, B., Senjanovic, G. and Vissani, F., Phys. Rev. Lett. **90**, 051802 (2003) [arXiv:hep-ph/0210207]

17. Goh, H. S., Mohapatra, R. N. and Ng, S. P., Phys. Lett. B **570**, 215 (2003) [arXiv:hep-ph/0303055].

18. Aulakh, C. S., Bajc, B., Melfo, A., Senjanovic, G. and Vissani, F., Phys. Lett. B **588**, 196 (2004) [hep-ph/0306242]; Aulakh, C. S. and Giridhar, A., hep-ph/0204097; Fukuyama, T., Ilakovac, A., Kikuchi, T., Meljanac, S. and Okada, N., arXiv:hep-ph/0401213; Bertolini, S., Frigerio, M. and Malinsky, M., hep-ph/0406117; Dutta, B., Mimura, Y. and Mohapatra, R. N., hep-ph/0406262; Yang, W. M. and Wang, Z. G., hep-ph/0406221; Goh, H. S., Mohapatra, R. N., Nasri, S. and Ng, S. P., Phys. Lett. B **587**, 105 (2004); Goh, H. S., Mohapatra, R. N. and Nasri, S., arXiv:hep-ph/0408139; Phys. Rev. **D 70**, 075002 (2004).

19. Dutta, B., Mimura, Y. and Mohapatra, R. N., hep-ph/0406262; Phys. Lett. **B** (2004).

20. Babu, K. S., Leung, C. N. and Pantaleone, J., Phys. Lett. **B319**, 191 (1993); Chankowski, P. and Pluciennik, Z., Phys. Lett. **B316**, 312 (1993).

21. Balaji, K. S., Dighe, A., Mohapatra, R. N. and Parida, M. K., Phys. Rev. Lett. **84**, 5034 (2000); Phys. Lett. **B481** , 33 (2000).

22. Casas, J., Espinoza, J., Ibara, J. and Navaro, S., hep-ph/9905381; Nucl. Phys. **B 573**, 652 (2000).

23. Mohapatra, R. N., Parida, M. K. and Rajasekaran, G., hep-ph/0301234; Phys.Rev. **D69**, 053007 (2004).

24. Raidal, M., hep-ph/0404046; Minakata, H. and Smirnov., A. Y., hep-ph/0405088.

25. Frampton, P. and Mohapatra, R. N., hep-ph/0407139.